IET ELECTRICAL MEASUREMENT SERIES 12

Microwave Measurements
3rd Edition

Other volumes in this series:

Volume 4 **The current comparator** W.J.M. Moore and P.N. Miljanic
Volume 5 **Principles of microwave measurements** G.H. Bryant
Volume 7 **Radio frequency and microwave power measurement** A.E. Fantom
Volume 8 **A handbook for EMC testing and measurement** D. Morgan
Volume 9 **Microwave circuit theory and foundations of microwave metrology** G. Engen
Volume 11 **Digital and analogue instrumentation: testing and measurement** N. Kularatna

Microwave Measurements
3rd Edition

Edited by R.J. Collier and A.D. Skinner

The Institution of Engineering and Technology

Published by The Institution of Engineering and Technology, London, United Kingdom

© 1985, 1989 Peter Peregrinus Ltd
© 2007 The Institution of Engineering and Technology

First published 1985 (0 86341 048 0)
Second edition 1989 (0 86341 184 3)
Third edition 2007 (978 0 86341 735 1)
Reprinted 2012

This publication is copyright under the Berne Convention and the Universal Copyright Convention. All rights reserved. Apart from any fair dealing for the purposes of research or private study, or criticism or review, as permitted under the Copyright, Designs and Patents Act, 1988, this publication may be reproduced, stored or transmitted, in any form or by any means, only with the prior permission in writing of the publishers, or in the case of reprographic reproduction in accordance with the terms of licences issued by the Copyright Licensing Agency. Inquiries concerning reproduction outside those terms should be sent to the publishers at the undermentioned address:

The Institution of Engineering and Technology
Michael Faraday House
Six Hills Way, Stevenage
Herts, SG1 2AY, United Kingdom

www.theiet.org

While the authors and the publishers believe that the information and guidance given in this work are correct, all parties must rely upon their own skill and judgement when making use of them. Neither the authors nor the publishers assume any liability to anyone for any loss or damage caused by any error or omission in the work, whether such error or omission is the result of negligence or any other cause. Any and all such liability is disclaimed.

The moral rights of the author to be identified as author of this work have been asserted by him in accordance with the Copyright, Designs and Patents Act 1988.

British Library Cataloguing in Publication Data
Microwave measurements. – 3rd ed.
   1. Microwave measurements
   I. Collier, Richard II. Skinner, Douglas III. Institution of Engineering and Technology
621.3′813

ISBN 978-0-86341-735-1

Typeset in India by Newgen Imaging Systems (P) Ltd, Chennai
First printed in the UK by Athenaeum Press Ltd, Gateshead, Tyne & Wear
Reprinted in India by Replika Press Pvt. Ltd.

Contents

List of contributors	xvii
Preface	xix

1 Transmission lines – basic principles   1
R. J. Collier

1.1	Introduction		1
1.2	Lossless two-conductor transmission lines – equivalent circuit and velocity of propagation		1
	1.2.1	Characteristic impedance	4
	1.2.2	Reflection coefficient	5
	1.2.3	Phase velocity and phase constant for sinusoidal waves	5
	1.2.4	Power flow for sinusoidal waves	6
	1.2.5	Standing waves resulting from sinusoidal waves	7
1.3	Two-conductor transmission lines with losses – equivalent circuit and low-loss approximation		8
	1.3.1	Pulses on transmission lines with losses	9
	1.3.2	Sinusoidal waves on transmission lines with losses	10
1.4	Lossless waveguides		10
	1.4.1	Plane (or transverse) electromagnetic waves	10
	1.4.2	Rectangular metallic waveguides	12
	1.4.3	The cut-off condition	14
	1.4.4	The phase velocity	15
	1.4.5	The wave impedance	15
	1.4.6	The group velocity	16
	1.4.7	General solution	16
	Further reading		17

2 Scattering parameters and circuit analysis   19
P. R. Young

2.1	Introduction	19
2.2	One-port devices	19

vi Contents

2.3		Generalised scattering parameters	22
2.4		Impedance and admittance parameters	24
	2.4.1	Examples of S-parameter matrices	27
2.5		Cascade parameters	27
2.6		Renormalisation of S-parameters	28
2.7		De-embedding of S-parameters	29
2.8		Characteristic impedance	30
	2.8.1	Characteristic impedance in real transmission lines	30
	2.8.2	Characteristic impedance in non-TEM waveguides	33
	2.8.3	Measurement of Z_0	35
2.9		Signal flow graphs	36
		Appendices	37
	2.A	Reciprocity	37
	2.B	Losslessness	39
	2.C	Two-port transforms	40
		References	41
		Further reading	41

3 Uncertainty and confidence in measurements   43
John Hurll

3.1		Introduction	43
3.2		Sources of uncertainty in RF and microwave measurements	52
	3.2.1	RF mismatch errors and uncertainty	52
	3.2.2	Directivity	54
	3.2.3	Test port match	54
	3.2.4	RF connector repeatability	54
	3.2.5	Example – calibration of a coaxial power sensor at a frequency of 18 GHz	54
		References	56

4 Using coaxial connectors in measurement   59
Doug Skinner

4.1		Introduction	59
	4.1.1	Coaxial line sizes	60
4.2		Connector repeatability	61
	4.2.1	Handling of airlines	61
	4.2.2	Assessment of connector repeatability	61
4.3		Coaxial connector specifications	62
4.4		Interface dimensions and gauging	62
	4.4.1	Gauging connectors	62
4.5		Connector cleaning	63
	4.5.1	Cleaning procedure	64
	4.5.2	Cleaning connectors on static sensitive devices	64
4.6		Connector life	65

Contents vii

	4.7	Adaptors	65
	4.8	Connector recession	65
	4.9	Conclusions	66
	4.A	Appendix A	66
	4.B	Appendix B	66
	4.C	Appendix C	85
	4.D	Appendix D	86
	4.E	Appendix E	87
		Further reading	88

5 Attenuation measurement 91
Alan Coster

	5.1	Introduction		91
	5.2	Basic principles		91
	5.3	Measurement systems		93
		5.3.1	Power ratio method	94
		5.3.2	Voltage ratio method	97
		5.3.3	The inductive voltage divider	98
		5.3.4	AF substitution method	104
		5.3.5	IF substitution method	105
		5.3.6	RF substitution method	107
		5.3.7	The automatic network analyser	108
	5.4	Important considerations when making attenuation measurements		110
		5.4.1	Mismatch uncertainty	110
		5.4.2	RF leakage	112
		5.4.3	Detector linearity	112
		5.4.4	Detector linearity measurement uncertainty budget	114
		5.4.5	System resolution	115
		5.4.6	System noise	115
		5.4.7	Stability and drift	115
		5.4.8	Repeatability	115
		5.4.9	Calibration standard	116
	5.5	A worked example of a 30 dB attenuation measurement		116
		5.5.1	Contributions to measurement uncertainty	117
		References		119
		Further reading		120

6 RF voltage measurement 121
Paul C. A. Roberts

	6.1	Introduction		121
	6.2	RF voltage measuring instruments		122
		6.2.1	Wideband AC voltmeters	122
		6.2.2	Fast sampling and digitising DMMs	124

viii Contents

		6.2.3	RF millivoltmeters	125
		6.2.4	Sampling RF voltmeters	126
		6.2.5	Oscilloscopes	127
		6.2.6	Switched input impedance oscilloscopes	129
		6.2.7	Instrument input impedance effects	130
		6.2.8	Source loading and bandwidth	132
	6.3	AC and RF/microwave traceability		133
		6.3.1	Thermal converters and micropotentiometers	133
	6.4	Impedance matching and mismatch errors		135
		6.4.1	Uncertainty analysis considerations	136
		6.4.2	Example: Oscilloscope bandwidth test	137
		6.4.3	Harmonic content errors	137
		6.4.4	Example: Oscilloscope calibrator calibration	138
		6.4.5	RF millivoltmeter calibration	140
		Further reading		143

7 Structures and properties of transmission lines   **147**
R. J. Collier

	7.1	Introduction	147
	7.2	Coaxial lines	148
	7.3	Rectangular waveguides	150
	7.4	Ridged waveguide	150
	7.5	Microstrip	151
	7.6	Slot guide	152
	7.7	Coplanar waveguide	153
	7.8	Finline	154
	7.9	Dielectric waveguide	154
		References	155
		Further reading	156

8 Noise measurements   **157**
David Adamson

	8.1	Introduction		157
	8.2	Types of noise		158
		8.2.1	Thermal noise	158
		8.2.2	Shot noise	159
		8.2.3	Flicker noise	159
	8.3	Definitions		160
	8.4	Types of noise source		162
		8.4.1	Thermal noise sources	162
		8.4.2	The temperature-limited diode	163
		8.4.3	Gas discharge tubes	163
		8.4.4	Avalanche diode noise sources	163

			Contents	ix

	8.5	Measuring noise	164
		8.5.1 The total power radiometer	164
		8.5.2 Radiometer sensitivity	166
	8.6	Measurement accuracy	166
		8.6.1 Cascaded receivers	169
		8.6.2 Noise from passive two-ports	169
	8.7	Mismatch effects	171
		8.7.1 Measurement of receivers and amplifiers	172
	8.8	Automated noise measurements	174
		8.8.1 Noise figure meters or analysers	175
		8.8.2 On-wafer measurements	175
	8.9	Conclusion	176
		Acknowledgements	176
		References	176

9 Connectors, air lines and RF impedance 179
N. M. Ridler

	9.1	Introduction	179
	9.2	Historical perspective	180
		9.2.1 Coaxial connectors	180
		9.2.2 Coaxial air lines	181
		9.2.3 RF impedance	181
	9.3	Connectors	182
		9.3.1 Types of coaxial connector	182
		9.3.2 Mechanical characteristics	185
		9.3.3 Electrical characteristics	187
	9.4	Air lines	188
		9.4.1 Types of precision air line	189
		9.4.2 Air line standards	190
		9.4.3 Conductor imperfections	192
	9.5	RF impedance	193
		9.5.1 Air lines	194
		9.5.2 Terminations	198
	9.6	Future developments	200
		Appendix: 7/16 connectors	201
		References	203

10 Microwave network analysers 207
Roger D. Pollard

	10.1	Introduction	207
	10.2	Reference plane	208
		10.2.1 Elements of a microwave network analyser	208
	10.3	Network analyser block diagram	214
		Further reading	216

x   Contents

11 RFIC and MMIC measurement techniques   217
Stepan Lucyszyn

- 11.1 Introduction   217
- 11.2 Test fixture measurements   218
  - 11.2.1 Two-tier calibration   220
  - 11.2.2 One-tier calibration   229
  - 11.2.3 Test fixture design considerations   230
- 11.3 Probe station measurements   230
  - 11.3.1 Passive microwave probe design   231
  - 11.3.2 Probe calibration   236
  - 11.3.3 Measurement errors   240
  - 11.3.4 DC biasing   240
  - 11.3.5 MMIC layout considerations   241
  - 11.3.6 Low-cost multiple DC biasing technique   243
  - 11.3.7 Upper-millimetre-wave measurements   243
- 11.4 Thermal and cryogenic measurements   246
  - 11.4.1 Thermal measurements   246
  - 11.4.2 Cryogenic measurements   247
- 11.5 Experimental field probing techniques   249
  - 11.5.1 Electromagnetic-field probing   249
  - 11.5.2 Magnetic-field probing   250
  - 11.5.3 Electric-field probing   251
- 11.6 Summary   254
- References   255

12 Calibration of automatic network analysers   263
Ian Instone

- 12.1 Introduction   263
- 12.2 Definition of calibration   263
- 12.3 Scalar network analysers   263
- 12.4 Vector network analyser   266
- 12.5 Calibration of a scalar network analyser   267
  - 12.5.1 Transmission measurements   267
  - 12.5.2 Reflection measurements   267
- 12.6 Problems associated with scalar network analyser measurements   269
- 12.7 Calibration of a vector network analyser   269
- 12.8 Accuracy enhancement   270
  - 12.8.1 What causes measurement errors?   270
  - 12.8.2 Directivity   270
  - 12.8.3 Source match   271
  - 12.8.4 Load match   272
  - 12.8.5 Isolation (crosstalk)   273
  - 12.8.6 Frequency response (tracking)   273

12.9	Characterising microwave systematic errors		273
	12.9.1	One-port error model	273
12.10	One-port device measurement		276
12.11	Two-port error model		279
12.12	TRL calibration		284
	12.12.1	TRL terminology	284
	12.12.2	True TRL/LRL	286
	12.12.3	The TRL calibration procedure	287
12.13	Data-based calibrations		289
	References		289

13 Verification of automatic network analysers — 291
Ian Instone

13.1	Introduction		291
13.2	Definition of verification		291
13.3	Types of verification		292
	13.3.1	Verification of error terms	292
	13.3.2	Verification of measurements	292
13.4	Calibration scheme		293
13.5	Error term verification		293
	13.5.1	Effective directivity	293
	13.5.2	Effective source match	296
	13.5.3	Effective load match	299
	13.5.4	Effective isolation	299
	13.5.5	Transmission and reflection tracking	299
	13.5.6	Effective linearity	300
13.6	Verification of measurements		301
	13.6.1	Customised verification example	301
	13.6.2	Manufacturer supplied verification example	302
	References		304

14 Balanced device characterisation — 305
Bernd A. Schincke

14.1	Introduction		305
	14.1.1	Physical background of differential structures	306
14.2	Characterisation of balanced structures		309
	14.2.1	Balanced device characterisation using network analysis	310
	14.2.2	Characterisation using physical transformers	310
	14.2.3	Modal decomposition method	312
	14.2.4	Mixed-mode-S-parameter-matrix	318
	14.2.5	Characterisation of single-ended to balanced devices	319
	14.2.6	Typical measurements	320

	14.3	Measurement examples	321
		14.3.1 Example 1: Differential through connection	321
		14.3.2 Example 2: SAW-filter measurement	326
	14.4	(De)Embedding for balanced device characterisation	326
		Further reading	328
15	**RF power measurement**		**329**
	James Miall		
	15.1	Introduction	329
	15.2	Theory	329
		15.2.1 Basic theory	329
		15.2.2 Mismatch uncertainty	332
	15.3	Power sensors	333
		15.3.1 Thermocouples and other thermoelectric sensors	333
		15.3.2 Diode sensors	333
		15.3.3 Thermistors and other bolometers	333
		15.3.4 Calorimeters	334
		15.3.5 Force and field based sensors	336
		15.3.6 Acoustic meter	336
	15.4	Power measurements and calibration	337
		15.4.1 Direct power measurement	337
		15.4.2 Uncertainty budgets	337
	15.5	Calibration and transfer standards	338
		15.5.1 Ratio measurements	338
	15.6	Power splitters	339
		15.6.1 Typical power splitter properties	340
		15.6.2 Measurement of splitter output match	340
		15.6.3 The direct method of measuring splitter output	341
	15.7	Couplers and reflectometers	343
		15.7.1 Reflectometers	343
	15.8	Pulsed power	344
	15.9	Conclusion	346
	15.10	Acknowledgements	346
		References	347
16	**Spectrum analyser measurements and applications**		**349**
	Doug Skinner		
	16.1	Part 1: Introduction	349
		16.1.1 Signal analysis using a spectrum analyser	349
		16.1.2 Measurement domains	350
		16.1.3 The oscilloscope display	350
		16.1.4 The spectrum analyser display	351
		16.1.5 Analysing an amplitude-modulated signal	351

16.2	Part 2: How the spectrum analyser works		354
	16.2.1	Basic spectrum analyser block diagram	354
	16.2.2	Microwave spectrum analyser with harmonic mixer	354
	16.2.3	The problem of multiple responses	355
	16.2.4	Microwave spectrum analyser with a tracking preselector	356
	16.2.5	Effect of the preselector	356
	16.2.6	Microwave spectrum analyser block diagram	356
	16.2.7	Spectrum analyser with tracking generator	358
16.3	Part 3: Spectrum analyser important specification points		359
	16.3.1	The input attenuator and IF gain controls	360
	16.3.2	Sweep speed control	360
	16.3.3	Resolution bandwidth	361
	16.3.4	Shape factor of the resolution filter	362
	16.3.5	Video bandwidth controls	365
	16.3.6	Measuring low-level signals – noise	366
	16.3.7	Dynamic range	366
	16.3.8	Amplitude accuracy	372
	16.3.9	Effect of input VSWR	372
	16.3.10	Sideband noise characteristics	373
	16.3.11	Residual responses	373
	16.3.12	Residual FM	374
	16.3.13	Uncertainty contributions	375
	16.3.14	Display detection mode	376
16.4	Spectrum analyser applications		376
	16.4.1	Measurement of harmonic distortion	378
	16.4.2	Example of a tracking generator measurement	378
	16.4.3	Zero span	378
	16.4.4	The use of zero span	379
	16.4.5	Meter Mode	379
	16.4.6	Intermodulation measurement	380
	16.4.7	Intermodulation analysis	381
	16.4.8	Intermodulation intercept point	382
	16.4.9	Nomograph to determine intermodulation products using intercept point method	383
	16.4.10	Amplitude modulation	383
	16.4.11	AM spectrum with modulation distortion	384
	16.4.12	Frequency modulation	384
	16.4.13	FM measurement using the Bessel zero method	385
	16.4.14	FM demodulation	386
	16.4.15	FM demodulation display	387
	16.4.16	Modulation asymmetry – combined AM and FM	387
	16.4.17	Spectrum of a square wave	388
	16.4.18	Pulse modulation	389
	16.4.19	Varying the pulse modulation conditions	389

xiv Contents

		16.4.20	'Line' and 'Pulse' modes	391
		16.4.21	Extending the range of microwave spectrum analysers	392
		16.4.22	EMC measurements	392
		16.4.23	Overloading a spectrum analyser	392
	16.5	Conclusion		393
		Further reading		394

17 Measurement of frequency stability and phase noise — 395
David Owen

17.1	Measuring phase noise	396
17.2	Spectrum analysers	397
17.3	Use of preselecting filter with spectrum analysers	399
17.4	Delay line discriminator	400
17.5	Quadrature technique	401
17.6	FM discriminator method	404
17.7	Measurement uncertainty issues	405
17.8	Future method of measurements	406
17.9	Summary	406

18 Measurement of the dielectric properties of materials at RF and microwave frequencies — 409
Bob Clarke

18.1	Introduction		409
18.2	Dielectrics – basic parameters		410
18.3	Basic dielectric measurement theory		413
	18.3.1	Lumped-impedance methods	414
	18.3.2	Wave methods	414
	18.3.3	Resonators, cavities and standing-wave methods	416
	18.3.4	The frequency coverage of measurement techniques	417
18.4	Loss processes: conduction, dielectric relaxation, resonances		418
18.5	International standard measurement methods for dielectrics		422
18.6	Preliminary considerations for practical dielectric measurements		422
	18.6.1	Do we need to measure our dielectric materials at all?	422
	18.6.2	Matching the measurement method to the dielectric material	423
18.7	Some common themes in dielectric measurement		425
	18.7.1	Electronic instrumentation: sources and detectors	425
	18.7.2	Measurement cells	426
	18.7.3	Q-factor and its measurement	427
18.8	Good practices in RF and MW dielectric measurements		429

18.9	A survey of measurement methods		430
	18.9.1	Admittance methods in general and two- and three-terminal admittance cells	430
	18.9.2	Resonant admittance cells and their derivatives	432
	18.9.3	TE_{01}-mode cavities	434
	18.9.4	Split-post dielectric resonators	436
	18.9.5	Substrate methods, including ring resonators	437
	18.9.6	Coaxial and waveguide transmission lines	437
	18.9.7	Coaxial probes, waveguide and other dielectric probes	439
	18.9.8	Dielectric resonators	442
	18.9.9	Free-field methods	444
	18.9.10	The resonator perturbation technique	446
	18.9.11	Open-resonators	446
	18.9.12	Time domain techniques	448
18.10	How should one choose the best measurement technique?		449
18.11	Further information		449
	References		450

19 Calibration of ELF to UHF wire antennas, primarily for EMC testing   459
M. J. Alexander

19.1	Introduction		459
19.2	Traceability of E-field strength		460
	19.2.1	High feed impedance half wave dipole	460
	19.2.2	Three-antenna method	461
	19.2.3	Calculability of coupling between two resonant dipole antennas	462
	19.2.4	Calculable field in a transverse electromagnetic (TEM) cell	462
	19.2.5	Uncertainty budget for EMC-radiated E-field emission	462
19.3	Antenna factors		464
	19.3.1	Measurement of free-space AFs	466
	19.3.2	The calculable dipole antenna	466
	19.3.3	Calibration of biconical antennas in the frequency range 20–300 MHz	466
	19.3.4	Calibration of LPDA antennas in the frequency range 200 MHz to 5 GHz	467
	19.3.5	Calibration of hybrid antennas	467
	19.3.6	Calibration of rod antennas	467
	19.3.7	Calibration of loop antennas	468
	19.3.8	Other antenna characteristics	468

19.4 Electro-optic sensors and traceability of fields
in TEM cells 469
Acknowledgements 470
References 470

Index **473**

Contributors

David Adamson
National Physical Laboratory
Teddington, Middlesex, TW11 0LW
david.adamson@npl.co.uk

Martin J. Alexander
National Physical Laboratory
Teddington, Middlesex, TW11 0LW
martin.alexander@npl.co.uk

Bob Clarke
National Physical Laboratory
Teddington, Middlesex, TW11 0LW
bob.clarke@npl.co.uk

Richard J. Collier
Corpus Christi College
University of Cambridge
Trumpington Street
Cambridge, CB2 1RH
rjc48@cam.ac.uk

Alan Coster
Consultant
5 Fieldfare Avenue, Yateley
Hampshire, GU46 6PD
ajcoster@theiet.org

John Hurll
United Kingdom Accreditation Service
21–47 High Street, Feltham
Middlesex, TW13 4UN
john.hurll@ukas.com

Ian Instone
Agilent Technologies UK Ltd
Scotstoun Avenue, South Queensferry
West Lothian, EH30 9TG
ian_instone@agilent.com

Stepan Lucyszyn
Dept Electrical and Electronic
   Engineering
Imperial College London
Exhibition Road, London, SW7 2AZ
s.lucyszyn@imperial.ac.uk

James Miall
National Physical Laboratory
Teddington, Middlesex, TW11 0LW
james.miall@npl.co.uk

David Owen
Business Development Manager
Pickering Interfaces
183A Poynters Road, Dunstable
Bedfordshire, LU5 4SH
david.owen@pickeringtest.com

Roger D. Pollard
School of Electronic and
  Electrical Engineering
University of Leeds
Leeds, LS2 9JT
r.d.pollard@leeds.ac.uk

Nick M. Ridler
National Physical Laboratory
Teddington, Middlesex, TW11 0LW
nick.ridler@npl.co.uk

Paul C.A. Roberts
Fluke Precision Measurement Ltd
Hurricane Way
Norwich, NR6 6JB
paul.roberts@fluke.com

Bernd Schincke
Rohde and Schwarz Training Centre
Germany
bernd.schincke@rohde-schwarz.com

Doug Skinner
Metrology Consultant
doug.skinner@theiet.org

Paul R. Young
Department of Electronics
University of Kent
Canterbury, Kent, CT2 7NT
P.R.Young@kent.ac.uk

Preface

This book contains most of the lecture notes used during the 14th IET Training Course on Microwave Measurements in May 2005 and is intended for use at the next course in 2007. These courses began in 1970 at the University of Kent with the title 'RF Electrical Measurements' and were held subsequently at the University of Surrey (1973) and the University of Lancaster (1976 and 1979). In 1983 the course returned to the University of Kent with the new title 'Microwave Measurements'. In recent years it has been held appropriately at the National Physical Laboratory in Teddington, where the National Microwave Measurement Facilities and Standards are housed. In 1985 and again in 1989, the late A.E. Bailey was the editor of a publication of these notes in book form. This third edition has been jointly edited by myself and A.D. Skinner and includes a large number of new authors since the last edition. Although the book is primarily intended for the course, it has proved popular over the years to anyone starting to measure microwaves.

The book begins with some revision chapters on transmission lines and scattering parameters, and these topics are used in the later chapters. Although these topics were included in most Physics and Electronics degree programmes in 1970, this is not the case now. As a result the reader who is totally unfamiliar with electromagnetic waves is advised to consult one of the many first-rate introductory books beforehand. The uncertainty of measurement is introduced next and this is followed by the techniques for the measurement of attenuation, voltage and noise. Many of these measurements are made using a microwave network analyser, and this remarkable instrument is discussed in the next few chapters. After this, the measurement of power, the use of spectrum analysers and aspects of digital modulation and phase noise measurements are covered in separate chapters. Finally, with the measurement of material properties, and both antenna and free field measurements, the wide range of topics is completed.

A feature of the training course is that in addition to the lectures there are workshops and demonstrations using the excellent facilities available in the National Physical Laboratory. The course is organised by a small committee of the IET, many of whom are involved in both the lectures and the workshops. Without their hard work, the course would not have survived the endless changes that have occurred since 1970. Although the book is a good reference source for those starting in the field of microwave measurements, the course is strongly recommended, as not only will the lectures and workshops enliven the subject, but meeting others on the course will also help in forming useful links for the future.

Finally, today microwaves are being used more extensively than ever before and yet there is a serious shortage of microwave metrologists. This book, and the course linked with it, are intended to help redress this imbalance.

R.J. Collier
Chairman – Organising Committee of the IET Training Course on Microwave Measurements

Chapter 1

Transmission lines – basic principles

R. J. Collier

1.1 Introduction

The aim of this chapter is to revise the basic principles of transmission lines in preparation for many of the chapters that follow in this book. Obviously one chapter cannot cover such a wide topic in any depth so at the end some textbooks are listed that may prove useful for those wishing to go further into the subject.

Microwave measurements involve transmission lines because many of the circuits used are larger than the wavelength of the signals being measured. In such circuits, the propagation time for the signals is not negligible as it is at lower frequencies. Therefore, some knowledge of transmission lines is essential before sensible measurements can be made at microwave frequencies.

For many of the transmission lines, such as coaxial cable and twisted pair lines, there are two separate conductors separated by an insulating dielectric. These lines can be described by using voltages and currents in an equivalent circuit. However, another group of transmission lines, often called waveguides, such as metallic waveguide and optical fibre have no equivalent circuit and these are described in terms of their electric and magnetic fields. This chapter will describe the two-conductor transmission lines first, followed by a description of waveguides and will end with some general comments about attenuation, dispersion and power. A subsequent chapter will describe the properties of some transmission lines that are in common use today.

1.2 Lossless two-conductor transmission lines – equivalent circuit and velocity of propagation

All two-conductor transmission lines can be described by using a distributed equivalent circuit. To simplify the treatment, the lines with no losses will be considered

Microwave measurements

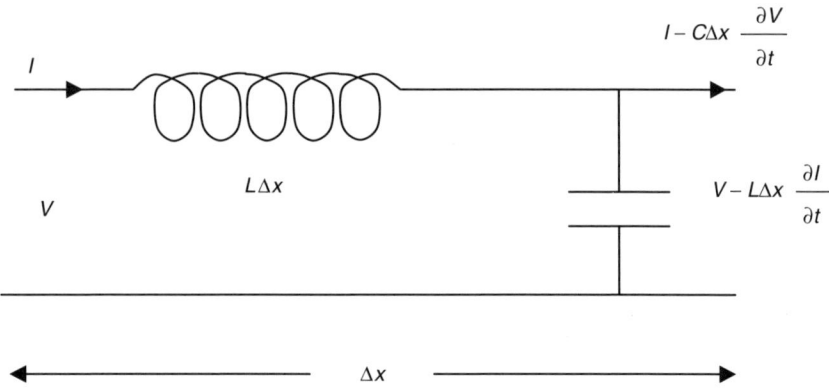

Figure 1.1 The equivalent circuit of a short length of transmission line with no losses

first. The lines have an inductance per metre, L, because the current going along one conductor and returning along the other produces a magnetic flux linking the conductors. Normally at high frequencies, the skin effect reduces the self-inductance of the conductors to zero so that only this 'loop' inductance remains. The wires will also have a capacitance per metre, C, because any charges on one conductor will induce equal and opposite charges on the other. This capacitance between the conductors is the dominant term and is much larger than any self-capacitance. The equivalent circuit of a short length of transmission line is shown in Figure 1.1.

If a voltage, V, is applied to the left-hand side of the equivalent circuit, the voltage at the right-hand side will be reduced by the voltage drop across the inductance. In mathematical terms

$$V \text{ becomes } V - L\Delta x \frac{\partial I}{\partial t} \text{ in a distance } \Delta x$$

Therefore, the change ΔV in that distance is given by

$$\Delta V = -L\Delta x \frac{\partial I}{\partial t}$$

Hence

$$\frac{\Delta V}{\Delta x} = -L \frac{\partial I}{\partial t}$$

and

$$\operatorname*{Lim}_{\Delta x \to 0} \frac{\Delta V}{\Delta x} = \frac{\partial V}{\partial x} = -L \frac{\partial I}{\partial t}$$

In a similar manner the current, I, entering the circuit on the left-hand side is reduced by the small current going through the capacitor. Again, in mathematical terms

$$I \text{ becomes } I - C\Delta x \frac{\partial V}{\partial t} \text{ in a distance } \Delta x$$

Therefore, the change ΔI in that distance is given by

$$\Delta I = -C\Delta x \frac{\partial V}{\partial t}$$

Hence

$$\frac{\Delta I}{\Delta x} = -C\frac{\partial V}{\partial t}$$

and

$$\operatorname*{Lim}_{\Delta x \to 0} \frac{\Delta I}{\Delta x} = \frac{\partial I}{\partial x} = -C\frac{\partial V}{\partial t}$$

The following equations are called Telegraphists' equations.

$$\frac{\partial V}{\partial x} = -L\frac{\partial I}{\partial t}$$
$$\frac{\partial I}{\partial x} = -C\frac{\partial V}{\partial t}$$
(1.1)

Differentiating these equations with respect to both x and t gives

$$\frac{\partial^2 V}{\partial x^2} = -L\frac{\partial^2 I}{\partial x \partial t}; \quad \frac{\partial^2 I}{\partial x^2} = -C\frac{\partial^2 V}{\partial x \partial t}$$
$$\frac{\partial^2 V}{\partial t \partial x} = -L\frac{\partial^2 I}{\partial t^2}; \quad \frac{\partial^2 I}{\partial t \partial x} = -C\frac{\partial^2 V}{\partial t^2}$$

Given that x and t are independent variables, the order of the differentiation is not important; thus, the equations can be reformed into wave equations.

$$\frac{\partial^2 V}{\partial x^2} = LC\frac{\partial^2 V}{\partial t^2}; \quad \frac{\partial^2 I}{\partial x^2} = LC\frac{\partial^2 I}{\partial t^2}$$
(1.2)

The equations have general solutions of the form of any function of the variable $(t \mp x/v)$. So if any signal, which is a function of time, is introduced at one end of a lossless transmission line then at a distance x down the line, this function will be delayed by x/v. If the signal is travelling in the opposite direction then the delay will be the same except that x will be negative and the positive sign in the variable will be necessary.

If the function $f(t - x/v)$ is substituted into either of the wave equations this gives

$$\frac{1}{v^2} f''\left(t - \frac{x}{v}\right) = LC f''\left(t - \frac{x}{v}\right)$$

This shows that for all the types of signal – pulse, triangular and sinusoidal – there is a unique velocity on lossless lines, v, given by

$$v = \frac{1}{\sqrt{LC}} \qquad (1.3)$$

This is the velocity for both the current and voltage waveforms, as the same wave equation governs both parameters.

1.2.1 Characteristic impedance

The relationship between the voltage waveform and the current waveform is derived from the Telegraphists' equations.

If the voltage waveform is

$$V = V_0 f\left(t - \frac{x}{v}\right)$$

then

$$\frac{\partial V}{\partial x} = -\frac{V_0}{v} f'\left(t - \frac{x}{v}\right)$$

Using the first Telegraphists' equation

$$\frac{\partial I}{\partial t} = \frac{V_0}{Lv} f'\left(t - \frac{x}{v}\right)$$

Integrating with respect to time gives

$$I = \frac{V_0}{Lv} f\left(t - \frac{x}{v}\right) = \frac{V}{Lv}$$

Therefore

$$\frac{V}{I} = Lv = \sqrt{\frac{L}{C}} \qquad (1.4)$$

This ratio is called the characteristic impedance, Z_0 and for lossless lines

$$\frac{V}{I} = Z_0 \qquad (1.5)$$

This is for waves travelling in a positive x direction. If the wave was travelling in a negative x direction, i.e. a reverse or backward wave, then the ratio of V to I would be equal to $-Z_0$.

1.2.2 Reflection coefficient

A transmission line may have at its end an impedance, Z_L, which is not equal to the characteristic impedance of the line Z_0. Thus, a wave on the line faces the dilemma of obeying two different Ohm's laws. To achieve this, a reflected wave is formed. Giving positive suffices to the incident waves and negative suffices to the reflected waves, the Ohm's law relationships become

$$\frac{V_+}{I_+} = Z_0; \quad \frac{V_-}{I_-} = -Z_0$$

$$\frac{V_L}{I_L} = \frac{V_+ + V_-}{I_+ + I_-} = Z_L$$

where V_L and I_L are the voltage and current in the terminating impedance Z_L. A reflection coefficient, ρ or Γ, is defined as the ratio of the reflected wave to the incident wave. Thus

$$\Gamma = \frac{V_-}{V_+} = -\frac{I_-}{I_+} \tag{1.6}$$

So

$$Z_L = \frac{V_+(1+\Gamma)}{I_+(1-\Gamma)} = \frac{Z_0(1+\Gamma)}{(1-\Gamma)}$$

and

$$\Gamma = \frac{Z_L - Z_0}{Z_L + Z_0} \tag{1.7}$$

As Z_0 for lossless lines is real and Z_L may be complex, Γ, in general, will also be complex. One of the main parts of microwave impedance measurement is to measure the value of Γ and hence Z_L.

1.2.3 Phase velocity and phase constant for sinusoidal waves

So far, the treatment has been perfectly general for any shape of wave. In this section, just the sine waves will be considered. In Figure 1.2, a sine wave is shown at one instant in time. As the waves move down the line with a velocity, v, the phase of the waves further down the line will be delayed compared with the phase of the oscillator on the left-hand side of Figure 1.2.

The phase delay for a whole wavelength is equal to 2π. The phase delay per metre is called β and is given by

$$\beta = \frac{2\pi}{\lambda} \tag{1.8}$$

6 Microwave measurements

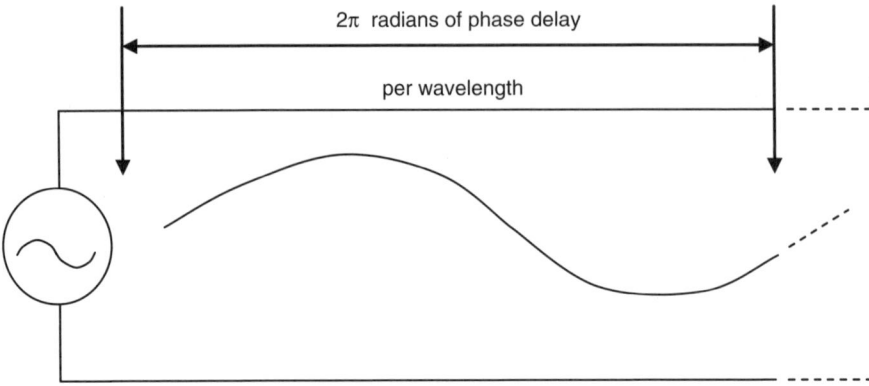

Figure 1.2 Sine waves on transmission lines

Multiplying numerator and denominator of the right-hand side by frequency gives

$$\beta = \frac{2\pi f}{\lambda f} = \frac{\omega}{v} = \omega\sqrt{LC} \qquad (1.9)$$

where v is now the phase velocity, i.e. the velocity of a point of constant phase and is the same velocity as that given in Section 1.2.

1.2.4 Power flow for sinusoidal waves

If a transmission line is terminated in impedance equal to Z_0 then all the power in the wave will be dissipated in the matching terminating impedance. For lossless lines, a sinusoidal wave with amplitude V_1 the power in the termination would be

$$\frac{V_1^2}{2Z_0}$$

This is also the power in the wave arriving at the matched termination.

If a transmission line is not matched then part of the incident power is reflected (see Section 1.2.2) and if the amplitude of the reflected wave is V_2 then the reflected power is

$$\frac{V_2^2}{2Z_0}$$

Since

$$|\Gamma| = \frac{V_2}{V_1}$$

then

$$|\Gamma|^2 = \frac{\text{Power reflected}}{\text{Incident power}} \qquad (1.10)$$

Clearly, for a good match the value of $|\Gamma|$ should be near to zero. The return loss is often used to express the match

$$\text{Return loss} = 10 \log_{10} \frac{1}{|\Gamma|^2} \qquad (1.11)$$

In microwave circuits a return loss of greater than 20 dB means that less than 1% of the incident power is reflected.

Finally, the power transmitted into the load is equal to the incident power minus the reflected power. A transmission coefficient, τ, is used as follows:

$$\frac{\text{Transmitted power}}{\text{Incident power}} = |\tau|^2 = 1 - |\Gamma|^2 \qquad (1.12)$$

1.2.5 Standing waves resulting from sinusoidal waves

When a sinusoidal wave is reflected by terminating impedance, which is not equal to Z_0, the incident and reflected waves form together a standing wave.

If the incident wave is

$$V_+ = V_1 \sin(\omega t - \beta x)$$

and the reflected wave is

$$V_- = V_2 \sin(\omega t + \beta x)$$

where $x = 0$ at the termination, then at some points on the line the two waves will be in phase and the voltage will be

$$V_{\text{MAX}} = (V_1 + V_2)$$

where V_{MAX} is the maximum of the standing wave pattern. At other points on the line the two waves will be out of phase and the voltage will be

$$V_{\text{MIN}} = (V_1 - V_2)$$

where V_{MIN} is the minimum of the standing wave pattern. The Voltage Standing Wave Ratio (VSWR) or S is defined as

$$S = \frac{V_{\text{MAX}}}{V_{\text{MIN}}} \qquad (1.13)$$

Now

$$S = \frac{V_1 + V_2}{V_1 - V_2} = \frac{V_1(1 + |\Gamma|)}{V_1(1 - |\Gamma|)}$$

8 Microwave measurements

So

$$S = \frac{1+|\Gamma|}{1-|\Gamma|} \qquad (1.14)$$

or

$$|\Gamma| = \frac{S-1}{S+1} \qquad (1.15)$$

Measuring S is relatively easy and, therefore, a value for $|\Gamma|$ can be obtained. From the position of the maxima and minima the argument or phase of Γ can be found. For instance, if a minimum of the standing wave pattern occurs at a distance D from a termination then the phase difference between the incident and the reflected waves at that point must be $n\pi$ ($n = 1, 3, 5, \ldots$). Now the phase delay as the incident wave goes from that point to the termination is βD. The phase change on reflection is the argument of Γ. Finally, the further phase delay as the reflected wave travels back to D is also βD. Hence

$$n\pi = 2\beta D + \arg(\Gamma) \qquad (1.16)$$

Therefore by a measurement of D and from knowledge of β, the phase of Γ can also be measured.

1.3 Two-conductor transmission lines with losses – equivalent circuit and low-loss approximation

In many two-conductor transmission lines there are two sources of loss, which cause the waves to be attenuated as they travel along the line. One source of loss is the ohmic resistance of the conductors. This can be added to the equivalent circuit by using a distributed resistance, R, whose units are Ohms per metre. Another source of loss is the ohmic resistance of the dielectric between the lines. Since, this is in parallel with the capacitance it is usually added to the equivalent circuit using a distributed conductance, G, whose units are Siemens per metre. The full equivalent circuit is shown in Figure 1.3.

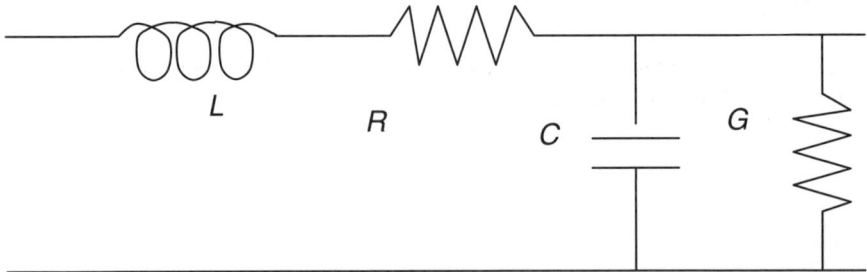

Figure 1.3   *The equivalent circuit of a line with losses*

The Telegraphists' equations become

$$\frac{\partial V}{\partial x} = -L\frac{\partial I}{\partial t} - RI$$
$$\frac{\partial I}{\partial x} = -C\frac{\partial V}{\partial t} - GV \quad (1.17)$$

Again, wave equations for lines with losses can be found by differentiating with respect to both x and t.

$$\frac{\partial^2 V}{\partial x^2} = LC\frac{\partial^2 V}{\partial t^2} + (LG + RC)\frac{\partial V}{\partial t} + RGV$$
$$\frac{\partial^2 I}{\partial x^2} = LC\frac{\partial^2 I}{\partial t^2} + (LG + RC)\frac{\partial I}{\partial t} + RGI \quad (1.18)$$

These equations are not easy to solve in the general case. However, for sinusoidal waves on lines with small losses, i.e. $\omega L \gg R$; $\omega C \gg G$, there is a solution of the form:

$$V = V_0 \exp(-\alpha x) \sin(\omega t - \beta x)$$

where

$$v = \frac{1}{\sqrt{LC}} \text{ m s}^{-1} \text{ (same as for lossless lines)} \quad (1.19)$$

$$\alpha = \frac{R}{2Z_0} + \frac{GZ_0}{2} \text{ nepers m}^{-1} \quad (1.20)$$

or

$$\alpha = 8.686 \left[\frac{R}{2Z_0} + \frac{GZ_0}{2}\right] \text{ dB m}^{-1} \quad (1.21)$$

$$\beta = \omega\sqrt{LC} \text{ radians m}^{-1} \text{ (same as for lossless lines)} \quad (1.22)$$

$$Z_0 = \sqrt{\frac{L}{C}} \text{ (same as for lossless lines)} \quad (1.23)$$

1.3.1 Pulses on transmission lines with losses

As well as attenuation, a pulse on a transmission line with losses will also change its shape. This is mainly caused by the fact that a full solution of the wave equations for lines with losses gives a pulse shape which is time dependent. In addition, the components of the transmission lines L, C, G and R are often different functions of frequency. Therefore if the sinusoidal components of the pulse are considered separately, they all travel at different velocities and with different attenuation. This frequency dependence is called dispersion. For a limited range of frequencies, it is sometimes possible to describe a group velocity, which is the velocity of the pulse rather than the velocity of the individual sine waves that make up the pulse. One effect

1.3.2 Sinusoidal waves on transmission lines with losses

For sinusoidal waves, there is a general solution of the wave equation for lines with losses and it is

$$\alpha + j\beta = \sqrt{(R+j\omega L)(G+j\omega C)} \quad \text{and} \quad Z_0 = \sqrt{\frac{R+j\omega L}{G+j\omega C}} \quad (1.24)$$

In general, α, β and Z_0 are all functions of frequency. In particular, Z_0 at low frequencies can be complex and deviate considerably from its high-frequency value. As R, G, L and C also vary with frequency, a careful measurement of these properties at each frequency is required to characterise completely the frequency variation of Z_0.

1.4 Lossless waveguides

These transmission lines cannot be easily described in terms of voltage and current as they sometimes have only one conductor (e.g. metallic waveguide) or no conductor (e.g. optical fibre). The only way to describe their electrical properties is in terms of the electromagnetic fields that exist in and, in some cases, around their structure. This section will begin with a revision of the properties of a plane or transverse electromagnetic (TEM) wave. The characteristics of metallic waveguides will then be described using these waves. The properties of other wave guiding structures will be given in Chapter 7.

1.4.1 Plane (or transverse) electromagnetic waves

A plane or TEM wave has two fields that are perpendicular or transverse to the direction of propagation. One of the fields is the electric field and the direction of this field is usually called the direction of polarisation (e.g. vertical or horizontal). The other field which is at right angles to both the electric field and the direction of propagation is the magnetic field. These two fields together form the electromagnetic wave. The electromagnetic wave equations for waves propagating in the z direction are as follows:

$$\frac{\partial^2 E}{\partial z^2} = \mu\varepsilon \frac{\partial^2 E}{\partial t^2}$$
$$\frac{\partial^2 H}{\partial z^2} = \mu\varepsilon \frac{\partial^2 H}{\partial t^2} \quad (1.25)$$

where μ is the permeability of the medium. If μ_R is the relative permeability then $\mu = \mu_R \mu_0$ and μ_0 is the free space permeability and has a value of $4\pi \times 10^{-7}\,\text{H m}^{-1}$.

Similarly, ε is the permittivity of the medium and if ε_R is the relative permittivity then $\varepsilon = \varepsilon_R \varepsilon_0$ and ε_0 is the free space permittivity and has a value of 8.854×10^{-12} F m^{-1}. These wave equations are analogous to those in Section 1.2. The variables V, I, L and C are replaced with the new variables E, H, μ and ε and the same results follow. For a plane wave the velocity of the wave, v, in the z direction is given by

$$v = \frac{1}{\sqrt{\mu\varepsilon}} \left(\text{see Section 2 where } v = \frac{1}{\sqrt{LC}} \right) \tag{1.26}$$

If $\mu = \mu_0$ and $\varepsilon = \varepsilon_0$, then $v_0 = 2.99792458 \times 10^8$ m s^{-1}.

The ratio of the amplitude of the electric field to the magnetic field is called the intrinsic impedance and has the symbol η.

$$\eta = \sqrt{\frac{\mu}{\varepsilon}} = \frac{E}{H} \left(\text{see Section 1.2 where } Z_0 = \sqrt{\frac{L}{C}} \right) \tag{1.27}$$

If $\mu = \mu_0$ and $\varepsilon = \varepsilon_0$, then $\eta_0 = 376.61$ Ω or $120\,\pi\,\Omega$.

As in Section 1.2, fields propagating in the negative z direction are related by using $-\eta$. The only difference is the orthogonality of the two fields in space, which comes from Maxwell's equations. For an electric field polarised in the x direction

$$\frac{\partial E_x}{\partial z} = -\mu \frac{\partial H_y}{\partial t}$$

If E_x is a function of $(t - \frac{z}{v})$ as before then

$$\frac{\partial E_x}{\partial z} = -\frac{1}{v} E_0 f'\left(t - \frac{z}{v}\right) = -\mu \frac{\partial H_y}{\partial t}$$

$$H_y = \frac{1}{\mu v} E_0 f\left(t - \frac{z}{v}\right) = \frac{1}{\mu v} E_x$$

Therefore

$$\frac{E_x}{H_y} = \mu v = \sqrt{\frac{\mu}{\varepsilon}} = \eta \tag{1.28}$$

If an E_y field was chosen, the magnetic field would be in the negative x direction. For sinusoidal waves, the phase constant is called the wave number and a symbol k is assigned.

$$k = \omega\sqrt{\mu\varepsilon} \left(\text{see Section 1.2 where } \beta = \omega\sqrt{LC} \right) \tag{1.29}$$

Inside all two-conductor transmission lines are various shapes of plane waves and it is possible to describe them completely in terms of fields rather than voltages and currents. The electromagnetic wave description is more fundamental but the equivalent circuit description is often easier to use. At high frequencies, two-conductor transmission lines also have higher order modes and the equivalent circuit model for

1.4.2 Rectangular metallic waveguides

Figure 1.4 shows a rectangular metallic waveguide. If a plane wave enters the waveguide such that its electric field is in the y (or vertical) direction and its direction of propagation is not in the z direction then it will be reflected back and forth by the metal walls in the y direction. Each time the wave is reflected it will have its phase reversed so that the sum of the electric fields on the surfaces of the walls in the y direction is zero. This is consistent with the walls being metallic and therefore good conductors capable of short circuiting any electric fields. The walls in the x direction are also good conductors but are able to sustain these electric fields perpendicular to their surfaces. Now, if the wave after two reflections has its peaks and troughs in the same positions as the original wave then the waves will add together and form a mode. If there is a slight difference in the phase, then the vector addition after many reflections will be zero and so no mode is formed. The condition for forming a mode is thus a phase condition and it can be found as follows.

Figure 1.5 shows a plane wave in a rectangular metallic waveguide with its electric field in the y direction and the direction of propagation at an angle θ to the z direction. The rate of change of phase in the direction of propagation is k_0, the free space wave number. As this wave is incident on the right wall of the waveguide in Figure 1.5, it will be reflected according to the usual laws of reflection as shown in Figure 1.6.

On further reflection, this wave must 'rejoin' the original wave to form a mode. Therefore, Figure 1.6 also shows the sum of all the reflections forming two waves – one incident on the right wall and the other on the left wall. The waves form a mode if they are linked together in phase.

Consider the line AB. This is a line of constant phase for the wave moving towards the right. Part of that wave at B reflects and moves along BA to A where it reflects again and rejoins the wave with the same phase. At the first reflection there is a phase shift of π. Then along BA there is a phase delay followed by another phase shift of π at the second reflection. The phase condition is

$$2\pi + \text{phase delay along BA for wave moving to the left} = 2m\pi$$

where $m = 0, 1, 2, 3, \ldots$.

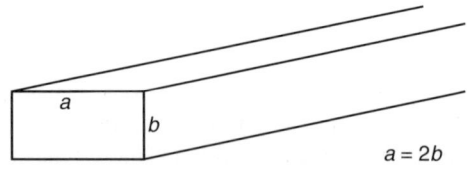

Figure 1.4   A rectangular metallic waveguide

Transmission lines – basic principles   13

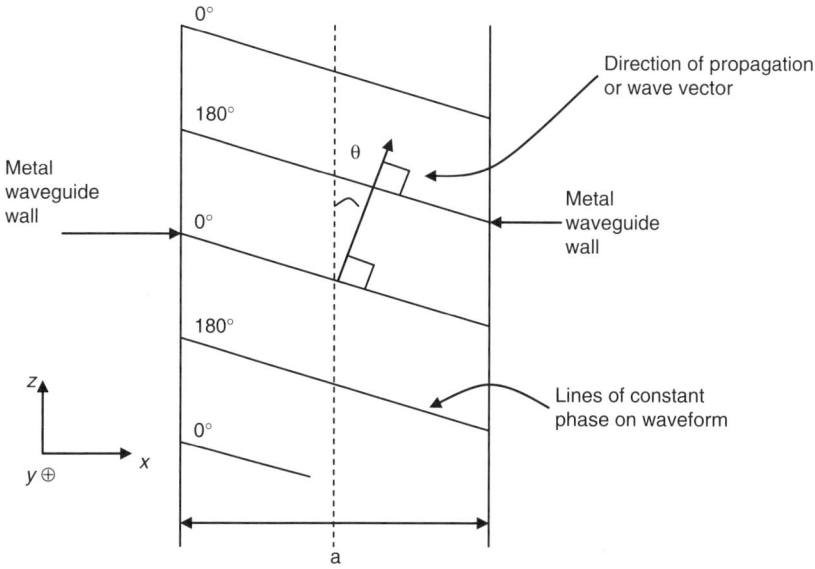

Figure 1.5   *A plane wave in a rectangular metallic waveguide*

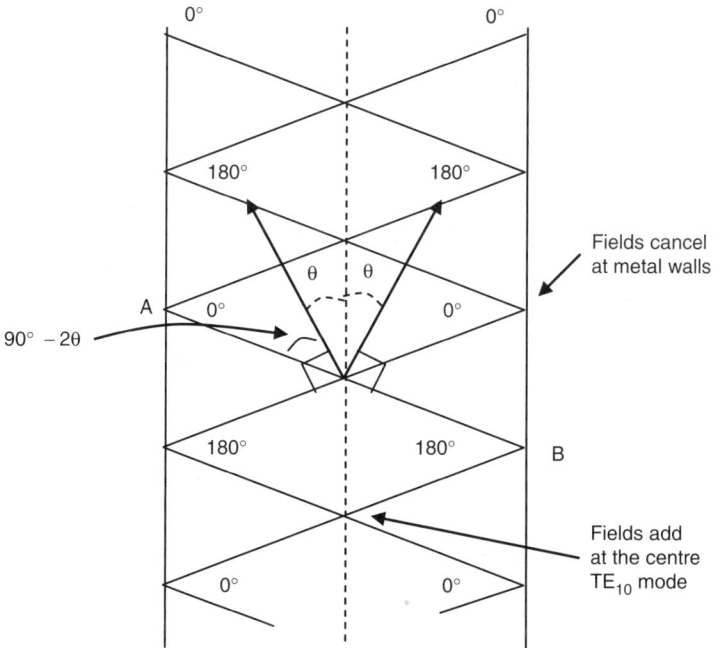

Figure 1.6   *Two plane waves in a rectangular metallic waveguide. The phases $0°$ and $180°$ refer to the lines below each figure*

14 Microwave measurements

The phase delay can be found by resolving the wave number along BA and is

$k_0 \sin 2\theta$ radians m^{-1}

If the walls in the y direction are separated by a distance a, then

$$AB = \frac{a}{\cos \theta}$$

Therefore the phase delay is $k_0 \sin 2\theta (a/\cos \theta)$ or $2k_0 \sin \theta$. Hence, the phase condition is

$$k_0 a \sin \theta = m\pi \quad \text{where } m = 0, 1, 2, 3, \ldots \tag{1.30}$$

The solutions to this phase condition give the various waveguide modes for waves with fields only in the y direction, i.e. the TE$_{mo}$ modes, 'm' is the number of half sine variations in the x direction.

1.4.3 The cut-off condition

Using the above phase condition, if $k_0 = \omega/v_0$ then

$$\omega \sin \theta = \frac{v_0 m \pi}{a}$$

The terms on the right-hand side are constant. For very high frequencies the value of θ tends to zero and the two waves just propagate in the z direction. However, if ω reduces in value the largest value of $\sin \theta$ is 1 and at this point the mode is cut-off and can no longer propagate. The cut-off frequency is ω_c and is given by

$$\omega_c = \frac{v_0 m \pi}{a}$$

or

$$f_c = \frac{v_0 m}{2a} \tag{1.31}$$

where f_c is the cut-off frequency.
If $v_0 = \lambda_c f_c$

then $$\lambda_c = \frac{2a}{m} \tag{1.32}$$

where λ_c is the cut-off wavelength.

Transmission lines – basic principles 15

A simple rule for TE_{mo} modes is that at cut-off, the wave just fits in 'sideways'. Indeed, since $\theta = 90°$ at cut-off, the two plane waves are propagating from side to side with a perfect standing wave between the walls.

1.4.4 The phase velocity

All waveguide modes can be considered in terms of plane waves. As the simpler modes just considered consist of only two plane waves they form a standing wave pattern in the x direction and form a travelling wave in the z direction. As the phase velocity in the z direction is related to the rate of change of phase, i.e. the wave number, then

$$\text{Velocity in the } z \text{ direction} = \frac{\omega}{\text{Wave number in the } z \text{ direction}}$$

Using Figure 1.5 or 1.6 the wave number in the z direction is

$$k_0 \cos \theta$$

Now from the phase condition $\sin \theta = m\pi / k_0 a$.

So $k_0 \cos \theta = k_0 \left(1 - \left(\frac{m\pi}{k_0 a}\right)^2\right)^{1/2}$

Also $k_0 = \frac{2\pi}{\lambda_0} = \frac{2\pi f}{\lambda_0 f} = \frac{\omega}{v_0}$

where v_0 is the free space velocity $= 1/\sqrt{\mu_0 \varepsilon_0}$.

Hence the velocity in the z direction

$$v_z = \frac{\omega}{k_0 \cos \theta}$$

$$v_z = \frac{v_0}{\left[1 - \left(\frac{m\lambda_0}{2a}\right)^2\right]^{1/2}} = \frac{v_0}{\left[1 - \left(\frac{\lambda_0}{\lambda_c}\right)^2\right]^{1/2}} \quad (1.33)$$

As can be seen from this condition, when λ_0 is equal to the cut-off wavelength (see Section 1.4.3) then v_z is infinite. As λ_0 gets smaller than λ_c then the velocity approaches v_0. Thus the phase velocity is always greater than v_0. Waveguides are not normally operated near cut-off as the high rate of change of velocity means both impossible design criteria and high dispersion.

1.4.5 The wave impedance

The ratio of the electric field to the magnetic field for a plane wave has already been discussed in Section 1.4.1. Although the waveguide has two plane waves in it the wave impedance is defined as the ratio of the transverse electric and magnetic fields. For TE modes this is represented by using the symbol Z_{TE}.

$$Z_{TE_{mo}} = \frac{E_y}{H_x} = \frac{E_0}{H_0 \cos\theta} = \frac{\eta}{\cos\theta}$$

where E_0 and H_0 refer to the plane waves. The electric fields of the plane waves are in the y direction but the magnetic fields are at an angle θ to the x direction. Therefore

$$Z_{TE_{mo}} = \frac{\eta}{\left[1 - \left(\frac{m\lambda_0}{2a}\right)^2\right]^{1/2}} = \frac{\eta}{\left[1 - \left(\frac{\lambda_0}{\lambda_c}\right)^2\right]^{1/2}} \tag{1.34}$$

Thus, for the $\lambda_0 \leqslant \lambda_c$ the value of $Z_{TE_{mo}}$ is always greater than η_0. A typical value might be 500 Ω.

1.4.6 The group velocity

Since a plane wave in air has no frequency-dependent parameters similar to those of the two-conductor transmission lines, i.e. μ_0 and ε_0 are constant, then there is no dispersion and so the phase velocity is equal to the group velocity. A pulse in a waveguide therefore would travel at v_0 at an angle of θ to the z-axis. The group velocity, v_g, along the z-axis is given by

$$v_g = v_0 \cos\theta$$

$$= v_0 \left(1 - \left(\frac{m\lambda_0}{2a}\right)^2\right)^{1/2} \tag{1.35}$$

The group velocity is always less than v_0 and is a function of frequency. For rectangular metallic waveguides

$$\text{Phase velocity} \times \text{Group velocity} = v_0^2 \tag{1.36}$$

1.4.7 General solution

To obtain all the possible modes in a rectangular metallic waveguide the plane wave must also have an angle ψ to the z-axis in the y–z plane. This will involve the wave reflecting from all four walls. If the two walls in the x direction are separated by a distance b then the following are valid for all modes. If

$$A = \left[1 - \left(\frac{m\lambda_0}{2a}\right)^2 - \left(\frac{n\lambda_0}{2b}\right)^2\right]^{1/2} \tag{1.37}$$

$m = 0, 1, 2, \ldots$ and $n = 0, 1, 2, \ldots$

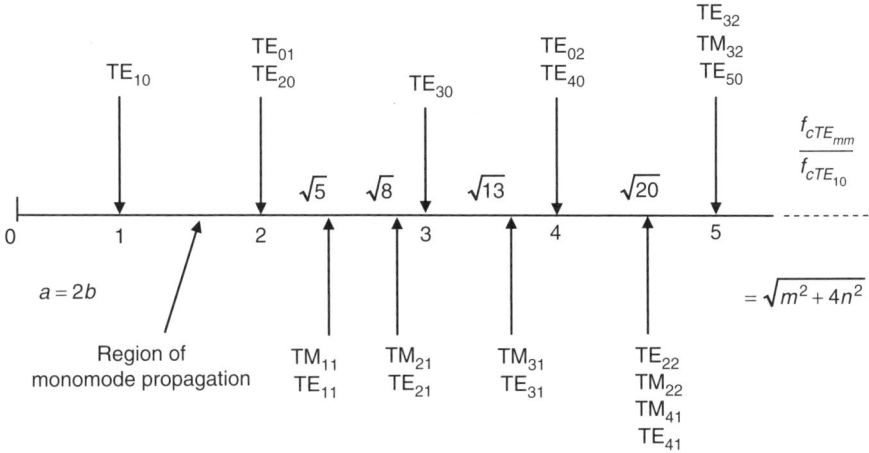

Figure 1.7 Relative cut-off frequencies for rectangular metallic waveguides

then the velocity

$$v = \frac{v_0}{A}$$

$$Z_{TE} = \frac{\eta}{A}$$

$$v_g = Av_0 \qquad (1.38)$$

The modes with the magnetic field in the y direction – the dual of TE modes – are called transverse magnetic modes or TM modes. They have a constraint that neither m nor n can be zero as the electric field for these modes has to be zero at all four walls.

$$Z_{TM} = \eta A \qquad (1.39)$$

The relative cut-off frequencies are shown in Figure 1.7, which also shows that monomode propagation using the TE_{10} mode is possible at up to twice the cut-off frequency. However, the full octave bandwidth is not used because propagation near cut-off is difficult and just below the next mode it can be hampered by energy coupling into that mode as well.

Further reading

1 Ramo, S., Whinnery, J. R., and VanDuzer, T.: *Fields and Waves in Communication Electronics*, 3rd edn (Wiley, New York, 1994)
2 Marcowitz, N.: *Waveguide Handbook* (Peter Peregrinus, London, 1986)
3 Cheng, D. K.: *Field and Wave Electromagnetics*, 2nd edn (Addison-Wesley, New York, 1989)

4 Magid, L. M.: *Electromagnetic Fields and Waves* (Wiley, New York, 1972)
5 Kraus, J. D., and Fleisch, D. A.: *Electromagnetics with Applications* (McGraw-Hill, Singapore, 1999)
6 Jordan, E. C., and Balmain, K. G.: *Electromagnetic Waves and Radiating Systems*, 2nd edn (Prentice-Hall, New Jersey, 1968)
7 Chipman, R. A.: *Transmission Lines*, Schaum's Outline Series (McGraw-Hill, New York, 1968)

Chapter 2
Scattering parameters and circuit analysis
P. R. Young

2.1 Introduction

Scattering parameters or scattering coefficients are fundamental to the design, analysis and measurement of all microwave and millimetre-wave circuits and systems. Scattering parameters define the forward and reverse wave amplitudes at the inputs and outputs of a network. Microwave networks take on various forms and can be as simple as a shunt capacitor or as complicated as a complete system. Common microwave networks are one-, two-, three- or four-port devices.

The definition of a scattering parameter is intrinsically linked to the form of transmission medium used at the ports of the network. Transmission lines and waveguides come in four distinct classes: (1) transverse electromagnetic (TEM), which includes coaxial lines and parallel pairs; (2) quasi-TEM lines, such as microstrip and coplanar waveguide (CPW); (3) transverse electric (TE) and transverse magnetic (TM) waveguides, such as rectangular waveguide; and (4) hybrid waveguides, which include dielectric guides and most lossy transmission lines and waveguides.

For simplicity, the one-port scattering parameter or reflection coefficient will be defined first for TEM lines before the more complicated multi-port and waveguide networks are analysed.

2.2 One-port devices

Consider a simple two-conductor transmission line, such as a coaxial cable or parallel pair. These types of transmission lines support TEM waves allowing the wave transmission to be expressed purely in terms of voltage between the conductors and the current flowing through the conductors. If the line is terminated by a load Z (Figure 2.1), which is not perfectly matched with the transmission line, then some

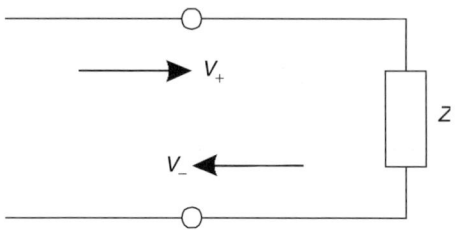

Figure 2.1 Transmission line terminated in a mismatched load

of the incident waves will be reflected back from the load. In terms of voltage and current along the line we have, at a point z,

$$V(z) = V_+ e^{-j\beta z} + V_- e^{+j\beta z} \qquad (2.1)$$

and

$$I(z) = I_+ e^{-j\beta z} - I_- e^{+j\beta z} \quad \text{or} \quad I(z) = \frac{1}{Z_0}\left(V_+ e^{-j\beta z} - V_- e^{+j\beta z}\right) \qquad (2.2)$$

where

$$Z_0 = \frac{V_+}{I_+} = \frac{V_-}{I_-}$$

β is the phase constant (rad m^{-1}), V_+ and I_+ are, respectively, amplitude of the voltage and current of the forward propagating wave; V_- and I_- are that of the reverse. The $e^{-j\beta z}$ terms denote forward propagation (towards the load), whereas the $e^{+j\beta z}$ terms denote reverse propagation (away from the load). Z_0 is the characteristic impedance of the transmission line and is dependent on the geometry and material of the structure. For simplicity, the time dependence has been omitted from Equations 2.1 and 2.2. The actual voltage and current are given by $\mathrm{Re}\{V(z)e^{j\omega t}\}$ and $\mathrm{Re}\{I(z)e^{j\omega t}\}$, respectively.

Suppose we choose a reference plane at the termination where we set $z = 0$. We define the reflection coefficient at this point by

$$\Gamma = \frac{V_-}{V_+} \qquad (2.3)$$

Since Ohm's law must apply at the termination

$$Z = \frac{V}{I} \qquad (2.4)$$

Note that we are free to set the reference plane anywhere along the line; hence, this might be at the connector interface, at the load element or some distance along the line.

Substituting (2.1) and (2.2) into (2.4) gives the well-known relationship between the reflection coefficient of the termination and its impedance at the reference plane

$$Z = Z_0 \frac{V_+ + V_-}{V_+ - V_-} = Z_0 \frac{1 + \Gamma}{1 - \Gamma} \tag{2.5}$$

Similarly, the relationship between the admittance of the termination and its reflection coefficient is given by

$$Y = Z^{-1} = Y_0 \frac{V_+ - V_-}{V_+ + V_-} = Y_0 \frac{1 - \Gamma}{1 + \Gamma} \tag{2.6}$$

where Y_0 is the admittance of the transmission line, $Y_0 = Z_0^{-1}$. Another useful expression is given by solving for Γ in (2.5)

$$\Gamma = \frac{Z - Z_0}{Z + Z_0} = \frac{Y_0 - Y}{Y_0 + Y} \tag{2.7}$$

Clearly, Γ is dependent on the impedance of the termination but we note that it is also dependent on the characteristic impedance of the line. A knowledge of Z_0 is therefore required to define Γ.

Relationships for the power flow can also be defined. It is well known that, using phasor notation, the root mean square power is given by

$$P = \frac{1}{2} \text{Re}\{VI^*\}$$

where '*' denotes the complex conjugate. Substituting for V and I from (2.1) and (2.2) yields, for $z = 0$,

$$P = \frac{1}{2Z_0} \text{Re}\left\{(V_+ + V_-)(V_+ - V_-)^*\right\}$$

which gives

$$P = \frac{1}{2Z_0} \left(|V_+|^2 - |V_-|^2\right)$$

where we have made use of the fact that $V_+^* V_- - V_+ V_-^*$ is purely imaginary and $V_+ V_+^* = |V_+|^2$. We also assume that Z_0 is purely real. Note that the power is dependent on the characteristic impedance. We can, however, define the network in terms of another set of amplitude constants such that the impedance is not required in power calculations. Let

$$V = \sqrt{Z_0} \left(ae^{-j\beta z} + be^{+j\beta z}\right) \tag{2.8}$$

and

$$I = \frac{1}{\sqrt{Z_0}} \left(ae^{-j\beta z} - be^{+j\beta z} \right) \quad (2.9)$$

where a and b are defined as the wave amplitudes of the forward and backward propagating waves. With reference to (2.1) it is easy to show that $a = V_+/\sqrt{Z_0}$ and $b = V_-/\sqrt{Z_0}$. We find now, if Z_0 is purely real, that the power is simply given by

$$P = \frac{1}{2} \left(|a|^2 - |b|^2 \right) \quad (2.10)$$

$|a|^2/2$ is the power in the forward propagating wave and $|b|^2/2$ is the power in the backward propagating wave. Equation (2.10) is a very satisfying result since it allows propagation to be defined in terms of wave amplitudes that are directly related to the power in the wave. This is particularly useful for measurement purposes since power is more easy to measure than voltage or current. In fact we shall see that for many microwave networks, voltage and current cannot be measured or even defined.

The analysis so far has dealt with one-port devices. These are completely specified by their impedance Z or reflection coefficient Γ (with respect to Z_0). The more important case of the two-port, or multi-port, device requires a more complicated model.

2.3 Generalised scattering parameters

Consider the two-port network shown in Figure 2.2. There will, in general, be waves propagating into and out of each of the ports. If the device is linear, the output waves can be defined in terms of the input waves. Thus,

$$b_1 = S_{11}a_1 + S_{12}a_2 \quad (2.11)$$
$$b_2 = S_{21}a_1 + S_{22}a_2 \quad (2.12)$$

where b_1 and b_2 are the wave amplitudes of the waves flowing out of ports 1 and 2, respectively. Similarly, a_1 and a_2 are the wave amplitudes of the waves flowing into ports 1 and 2, respectively. S_{11}, S_{21}, S_{12} and S_{22} are the scattering coefficients or

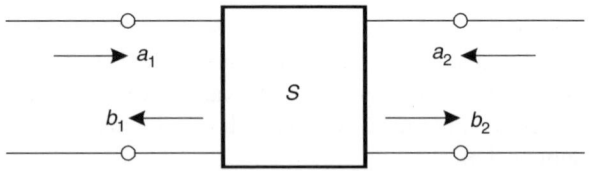

Figure 2.2 Two-port device represented by S-parameter matrix

Scattering parameters and circuit analysis 23

scattering parameters. Using the definition of wave amplitude the voltages at port 1 and port 2 are given by

$$V_1 = \sqrt{Z_{01}}(a_1 + b_1) \quad \text{and} \quad V_2 = \sqrt{Z_{02}}(a_2 + b_2)$$

respectively, where it is assumed that the characteristic impedance is different at each port: Z_{01} at port 1 and Z_{02} at port 2.

Similarly, the currents entering port 1 and port 2 are

$$I_1 = \frac{1}{\sqrt{Z_{01}}}(a_1 - b_1) \quad \text{and} \quad I_2 = \frac{1}{\sqrt{Z_{02}}}(a_2 - b_2)$$

respectively. Equations 2.11 and 2.12 can be more neatly written in matrix notation

$$\begin{bmatrix} b_1 \\ b_2 \end{bmatrix} = \begin{bmatrix} S_{11} & S_{12} \\ S_{21} & S_{22} \end{bmatrix} \begin{bmatrix} a_1 \\ a_2 \end{bmatrix} \tag{2.13}$$

or $\mathbf{b} = \mathbf{Sa}$, where

$$\mathbf{a} = \begin{bmatrix} a_1 \\ a_2 \end{bmatrix}, \mathbf{b} = \begin{bmatrix} b_1 \\ b_2 \end{bmatrix} \quad \text{and} \quad \mathbf{S} = \begin{bmatrix} S_{11} & S_{12} \\ S_{21} & S_{22} \end{bmatrix}$$

where $\mathbf{S}$ is the scattering matrix or S-parameter matrix of the two-port network.

If port 2 is terminated by a perfect match of impedance Z_{02}, that is, all of the incident power is absorbed in the termination, then we have the following properties:

$$S_{11} = \left. \frac{b_1}{a_1} \right|_{a_2=0} \quad \text{and} \quad S_{21} = \left. \frac{b_2}{a_1} \right|_{a_2=0}$$

Similarly, if port 1 is terminated by a perfect match of impedance Z_{01} then

$$S_{22} = \left. \frac{b_2}{a_2} \right|_{a_1=0} \quad \text{and} \quad S_{12} = \left. \frac{b_1}{a_2} \right|_{a_1=0}$$

By using the above definitions we can obtain some insight into the meaning of the individual S-parameters. S_{11} is the reflection coefficient at port 1 with port 2 terminated in a matched load. It, therefore, gives a measure of the mismatch due to the network and not any other devices that may be connected to port 2. S_{21} is the transmission coefficient from port 1 to port 2 with port 2 terminated in a perfect match. It gives a measure of the amount of signal that is transmitted from port 1 to port 2. S_{22} and S_{12} are similarly defined with S_{22} giving the reflection from port 2 and S_{12} the transmission from port 2 to port 1.

24   Microwave measurements

The S-parameter representation equally applies to multi-port devices. For an n-port device, the S-parameter matrix is given by

$$\begin{bmatrix} b_1 \\ b_2 \\ \vdots \\ b_n \end{bmatrix} = \begin{bmatrix} S_{11} & S_{12} & \cdots & S_{1n} \\ S_{21} & S_{22} & \cdots & S_{2n} \\ \vdots & \vdots & \ddots & \vdots \\ S_{n1} & S_{n2} & \cdots & S_{nn} \end{bmatrix} \begin{bmatrix} a_1 \\ a_2 \\ \vdots \\ a_n \end{bmatrix}$$

where b_k is the amplitude of the wave travelling away from the junction at port k. Similarly, a_k is the wave amplitude travelling into the junction at port k. The S-parameters are defined as

$$S_{ij} = \left. \frac{b_i}{a_j} \right|_{a_k=0 \text{ for } k \neq j}$$

2.4   Impedance and admittance parameters

Expressions similar to (2.5) and (2.7) can be obtained for n-port devices. If we have an n-port device then the voltage and current at the reference plane of port k are given by

$$V_k = \sqrt{Z_{0k}}(a_k + b_k) \tag{2.14}$$

and

$$I_k = \frac{1}{\sqrt{Z_{0k}}}(a_k - b_k) \tag{2.15}$$

where Z_{0k} is the characteristic impedance of the transmission line connected to port k. Equations 2.14 and 2.15 can be written in matrix notation as

$$\mathbf{V} = Z_0^{1/2}(\mathbf{a} + \mathbf{b}) \tag{2.16}$$

and

$$\mathbf{I} = Z_0^{-1/2}(\mathbf{a} - \mathbf{b}) \tag{2.17}$$

respectively. Where $\mathbf{a}$ and $\mathbf{b}$ are column vectors containing the wave amplitudes and $\mathbf{V}$ and $\mathbf{I}$ are column vectors containing the port voltages and currents

$$\mathbf{a} = \begin{bmatrix} a_1 \\ a_2 \\ \vdots \\ a_n \end{bmatrix}, \mathbf{b} = \begin{bmatrix} b_1 \\ b_2 \\ \vdots \\ b_n \end{bmatrix}, \mathbf{V} = \begin{bmatrix} V_1 \\ V_2 \\ \vdots \\ V_n \end{bmatrix}, \text{ and } \mathbf{I} = \begin{bmatrix} I_1 \\ I_2 \\ \vdots \\ I_n \end{bmatrix}$$

Z_0 is a diagonal matrix with Z_{0k} as its diagonal elements

$$Z_0 = \begin{bmatrix} Z_{01} & 0 & \cdots & 0 \\ 0 & Z_{02} & \cdots & 0 \\ \vdots & \vdots & \ddots & \vdots \\ 0 & 0 & \cdots & Z_{0n} \end{bmatrix} \tag{2.18}$$

$Z_0^{1/2}$ denotes a diagonal matrix with $\sqrt{Z_{0k}}$ as its diagonal elements. Often, the characteristic impedance of each of the ports is identical, in which case each of the diagonal elements is equal. From (2.8) and (2.9), with $z = 0$, we have

$$\mathbf{a} = \frac{1}{2} Z_0^{-1/2} (\mathbf{V} + Z_0 \mathbf{I}) \tag{2.19}$$

and

$$\mathbf{b} = \frac{1}{2} Z_0^{-1/2} (\mathbf{V} - Z_0 \mathbf{I}) \tag{2.20}$$

Let $\mathbf{V} = Z\mathbf{I}$, where Z is the impedance matrix, extensively used in electrical circuit theory

$$Z = \begin{bmatrix} Z_{11} & Z_{12} & \cdots & Z_{1n} \\ Z_{21} & Z_{22} & \cdots & Z_{2n} \\ \vdots & \vdots & \ddots & \vdots \\ Z_{n1} & Z_{n2} & \cdots & Z_{nn} \end{bmatrix}$$

Also let $\mathbf{I} = Y\mathbf{V}$, where Y is the admittance matrix

$$Y = \begin{bmatrix} Y_{11} & Y_{12} & \cdots & Y_{1n} \\ Y_{21} & Y_{22} & \cdots & Y_{2n} \\ \vdots & \vdots & \ddots & \vdots \\ Y_{n1} & Y_{n2} & \cdots & Y_{nn} \end{bmatrix}.$$

We note that $Z = Y^{-1}$. The individual elements of the impedance and admittance matrices are defined as follows:

$$Z_{ij} = \left. \frac{V_i}{I_j} \right|_{I_k = 0 \text{ for } k \neq j} \quad \text{and} \quad Y_{ij} = \left. \frac{I_i}{V_j} \right|_{V_k = 0 \text{ for } k \neq j}$$

That is, Z_{ij} is the ratio of voltage at port i to the current at port j with all other port currents set to zero, that is, short circuit. Y_{ij} is defined as the ratio of current at port i to the voltage at port j with all other port voltages set to zero, that is, open circuit. Substituting $\mathbf{V} = Z\mathbf{I}$ and $\mathbf{b} = S\mathbf{a}$ into (2.19) and (2.20) yields

$$Z = Z_0^{1/2} (U - S)^{-1} (U + S) Z_0^{1/2} = Y^{-1} \tag{2.21}$$

Table 2.1 Network parameters for common microwave networks

Circuit	Network parameters
Lossless transmission line of length L, phase constant β and characteristic impedance Z_0	$S = \begin{bmatrix} 0 & e^{-j\beta L} \\ e^{-j\beta L} & 0 \end{bmatrix}$
Shunt admittance Y	$S = \dfrac{1}{Y + 2Y_0} \begin{bmatrix} -Y & 2Y_0 \\ 2Y_0 & -Y \end{bmatrix}$ $Z = \begin{bmatrix} Z & Z \\ Z & Z \end{bmatrix}$ where $Z = Y^{-1}$
Series impedance Z	$S = \dfrac{1}{2Z_0 + Z} \begin{bmatrix} Z & 2Z_0 \\ 2Z_0 & Z \end{bmatrix}$ $Y = \begin{bmatrix} Y & -Y \\ -Y & Y \end{bmatrix}$ where $Y = Z^{-1}$
π network	$Y = \begin{bmatrix} Y_A + Y_C & -Y_C \\ -Y_C & Y_B + Y_C \end{bmatrix}$
T network	$Z = \begin{bmatrix} Z_A + Z_C & Z_C \\ Z_C & Z_B + Z_C \end{bmatrix}$

or solving for S

$$S = Z_0^{-1/2}(Z - Z_0)(Z + Z_0)^{-1} Z_0^{1/2} \quad (2.22)$$

where U is the unit matrix:

$$U = \begin{bmatrix} 1 & 0 & \cdots & 0 \\ 0 & 1 & \cdots & 0 \\ \vdots & \vdots & \ddots & \vdots \\ 0 & 0 & \cdots & 1 \end{bmatrix}$$

Examples of (2.21) and (2.22) for two-port networks are given in Appendix 2.C.

Z and Y parameters can be very useful in the analysis of microwave networks since they can be related directly to simple π or T networks (refer to Table 2.1). These circuits are fundamental in lumped element circuits, such as attenuators, and are also important in equivalent circuits for waveguide junctions and discontinuities.

2.4.1 Examples of S-parameter matrices

Table 2.1 shows some common examples of microwave networks and their network parameters. Parameters are only shown for the simplest form. The associated S, Z or Y parameters can be determined using (2.21) and (2.22). In each case it is assumed that the characteristic impedance is identical at each port and equal to $Z_0 = Y_0^{-1}$.

We notice from the table that the S-parameter matrices are symmetrical, that is, $S_{mn} = S_{nm}$. This is a demonstration of reciprocity in microwave networks and applies to most networks (see Appendix 2.A). A property of lossless scattering matrices is also seen for the line section. Here, $S^T S^* = U$, which applies to all lossless networks; refer to Appendix 2.B.

2.5 Cascade parameters

Another useful transformation of the S-parameter matrix is the cascade matrix. The two-port cascade matrix is given by

$$\begin{bmatrix} a_1 \\ b_1 \end{bmatrix} = \begin{bmatrix} T_{11} & T_{12} \\ T_{21} & T_{22} \end{bmatrix} \begin{bmatrix} b_2 \\ a_2 \end{bmatrix} \quad (2.23)$$

where we notice that the wave amplitudes on port 1 are given in terms of the wave amplitudes on port 2. Note that some textbooks interchange a_1 with b_1 and b_2 with a_2. Comparing (2.23) with (2.13) gives the following relationships between the cascade matrix elements and the scattering coefficients:

$$T = \begin{bmatrix} T_{11} & T_{12} \\ T_{21} & T_{22} \end{bmatrix} = \frac{1}{S_{21}} \begin{bmatrix} 1 & -S_{22} \\ S_{11} & S_{12}S_{21} - S_{11}S_{22} \end{bmatrix} \quad (2.24)$$

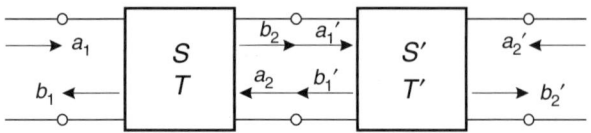

Figure 2.3 Two two-port networks cascaded together

Similarly, the reverse transform is given by

$$S = \frac{1}{T_{11}} \begin{bmatrix} T_{21} & T_{22}T_{11} - T_{12}T_{21} \\ 1 & -T_{12} \end{bmatrix} \qquad (2.25)$$

Suppose we have two two-port devices cascaded together (refer to Figure 2.3). The first network is given by

$$\begin{bmatrix} a_1 \\ b_1 \end{bmatrix} = \begin{bmatrix} T_{11} & T_{12} \\ T_{21} & T_{22} \end{bmatrix} \begin{bmatrix} b_2 \\ a_2 \end{bmatrix}$$

and the second network

$$\begin{bmatrix} a_1' \\ b_1' \end{bmatrix} = \begin{bmatrix} T_{11}' & T_{12}' \\ T_{21}' & T_{22}' \end{bmatrix} \begin{bmatrix} b_2' \\ a_2' \end{bmatrix}$$

where, by inspecting Figure 2.3 we see that

$$\begin{bmatrix} b_2 \\ a_2 \end{bmatrix} = \begin{bmatrix} a_1' \\ b_1' \end{bmatrix}$$

Therefore,

$$\begin{bmatrix} a_1 \\ b_1 \end{bmatrix} = \begin{bmatrix} T_{11} & T_{12} \\ T_{21} & T_{22} \end{bmatrix} \begin{bmatrix} T_{11}' & T_{12}' \\ T_{21}' & T_{22}' \end{bmatrix} \begin{bmatrix} b_2' \\ a_2' \end{bmatrix}$$

We see that in order to calculate the input wave amplitudes in terms of the output amplitudes we simply multiply the cascade matrices together. Often the cascaded two port is converted back to an S-parameter matrix using (2.25). Any number of cascaded two-port networks can then be replaced by a single equivalent two-port network.

2.6 Renormalisation of S-parameters

We have already seen that S-parameters are defined with respect to a reference characteristic impedance at each of the network's ports. Often we require that the S-parameters are renormalised to another set of port characteristic impedances. This is important as measured S-parameters are usually with respect to the transmission line Z_0 or matched load impedance of the calibration items used in the measurement system. These often differ from the nominal 50 Ω. To convert an S-parameter matrix S, that is, with respect to the port impedance matrix

$$Z_0 = \begin{bmatrix} Z_{01} & 0 & \cdots & 0 \\ 0 & Z_{02} & \cdots & 0 \\ \vdots & \vdots & \ddots & \vdots \\ 0 & 0 & \cdots & Z_{0n} \end{bmatrix}$$

we, first, transform S to an impedance matrix using (2.21)

$$Z = Z_0^{1/2} (U - S)^{-1} (U + S) Z_0^{1/2}$$

Next the impedance matrix is transformed into the S-parameter matrix S'

$$S' = Z_0'^{-1/2} (Z - Z_0') (Z + Z_0')^{-1} Z_0'^{1/2}$$

where now a reference impedance matrix of Z_0' is used

$$Z_0' = \begin{bmatrix} Z_{01}' & 0 & \cdots & 0 \\ 0 & Z_{02}' & \cdots & 0 \\ \vdots & \vdots & \ddots & \vdots \\ 0 & 0 & \cdots & Z_{0n}' \end{bmatrix}$$

S' is then with respect to Z'$_0$. Often the renormalised S-parameters are with respect to 50 Ω in which case all the diagonal elements of Z'$_0$ are equal to 50 Ω.

2.7 De-embedding of S-parameters

Another very important operation on an S-parameter matrix is the de-embedding of a length of transmission line from each of the ports. This is extremely important in measurement since often the device under test is connected to the measurement instrument by a length of transmission line and, therefore, the actual measured value includes the phase and attenuation of the line. It can be shown that the measured n-port S-parameters S' are related to the network's actual S-parameters S by

$$S' = \Theta S \Theta$$

where

$$\Theta = \begin{bmatrix} e^{-\gamma_1 L_1} & 0 & \cdots & 0 \\ 0 & e^{-\gamma_2 L_2} & \cdots & 0 \\ \vdots & \vdots & \ddots & \vdots \\ 0 & 0 & \cdots & e^{-\gamma_N L_N} \end{bmatrix}$$

It is assumed that all of the lines are matched to their respective ports. If we know the length of line L_k at each port and the complex propagation constant $\gamma_k = \alpha_k + j\beta_k$ then we can de-embed the effect of the lines. Thus,

$$S = \Theta^{-1} S' \Theta^{-1}$$

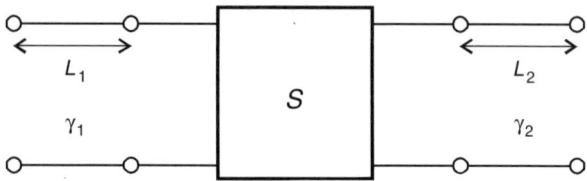

Figure 2.4 Two-port network with feeding transmission lines at each port

Due to the diagonal nature of Θ, the inverse operation Θ^{-1} simply changes the $-\gamma_k L_k$ terms to $+\gamma_k L_k$.

Figure 2.4 shows a typical two-port network with lines connected to both ports. In this case the actual network parameters S are related to the S-parameters S' by

$$S = \begin{bmatrix} S'_{11}e^{+2\gamma_1 L_1} & S'_{12}e^{+\gamma_1 L_1+\gamma_2 L_2} \\ S'_{21}e^{+\gamma_1 L_1+\gamma_2 L_2} & S'_{22}e^{+2\gamma_2 L_2} \end{bmatrix}$$

In the lossless case γ_k would degenerate to $j\beta_k$ and only a phase shift would be introduced by the lines.

2.8 Characteristic impedance

We have seen that a microwave network can be characterised in terms of its S-parameters and that the S-parameters are defined with respect to the characteristic impedance at the ports of the network. A fundamental understanding of the nature of Z_0 is therefore essential in microwave circuit analysis and measurement. Unfortunately, the true nature of characteristic impedance is often overlooked by microwave engineers and Z_0 is usually considered to be a real-valued constant, such as 50 Ω. In many cases this is a very good assumption. However, the careful metrologist does not make assumptions and the true nature of the characteristic impedance is imperative in precision microwave measurements. In fact without the knowledge of characteristic impedance, S-parameter measurements have little meaning and this lack of knowledge is so often the cause of poor measurements. This is particularly important in measured S-parameters from network analysers. S-parameters measured on a network analyser are with respect to either the Z_0 of the calibration items or impedance of the matched element used to calibrate the analyser. If this value is ill-defined then so are the measured S-parameters.

2.8.1 Characteristic impedance in real transmission lines

If a TEM or quasi-TEM line contains dielectric and conductive losses then (2.1) and (2.2) become

$$V(z) = V_+ e^{-\gamma z} + V_- e^{+\gamma z} \tag{2.26}$$

and

$$I(z) = \frac{1}{Z_0}\left(V_+ e^{-\gamma z} - V_- e^{+\gamma z}\right) \qquad (2.27)$$

where the complex propagation constant is defined as $\gamma = \alpha + j\beta$. Equations 2.26 and 2.27 are very similar to (2.1) and (2.2); however, the attenuation constant α adds an exponential decay to the wave's amplitude as it propagates along the line. It is easy to show that in terms of the transmission line's per length series impedance Z and shunt admittance Y the propagation constant γ and characteristic impedance Z_0 are given by [1]

$$\gamma = \sqrt{ZY} \quad \text{and} \quad Z_0 = \sqrt{\frac{Z}{Y}}$$

In the lossless case $Z = j\omega L$ and $Y = j\omega C$, representing the series inductance of the conductors and the shunt capacitance between them. The complex propagation constant then reduces to the familiar phase constant $j\beta$ and Z_0 degenerates to a real-valued constant dependent only on L and C

$$\gamma = j\beta = j\omega\sqrt{LC} \quad \text{and} \quad Z_0 = \sqrt{\frac{L}{C}}$$

In lossy lines there is a series resistive component due to conduction losses and a shunt conductance due to dielectric losses. Hence, $Z = R + j\omega L$ and $Y = G + j\omega C$ and therefore,

$$\gamma = \sqrt{(R + j\omega L)(G + j\omega C)} = Z_0(G + j\omega C) \qquad (2.28)$$

and

$$Z_0 = \sqrt{\frac{R + j\omega L}{G + j\omega C}} = \frac{R + j\omega L}{\gamma} \qquad (2.29)$$

where R, L, G and C are often functions of ω. Two important facts about Z_0 are immediately evident from (2.29): Z_0 is complex and a function of frequency. Therefore, the assumption that Z_0 is a real-valued constant which is independent of frequency is only an approximation. Fortunately, for many transmission lines the loss is small. In this case $R \ll \omega L$ and $G \ll \omega C$ and an approximate expression for the propagation constant is obtained by using a first-order binomial expansion. Thus,

$$\alpha \approx \frac{1}{2}\sqrt{LC}\left(\frac{R}{L} + \frac{G}{C}\right), \quad \beta \approx \omega\sqrt{LC} \quad \text{and} \quad Z_0 \approx \sqrt{\frac{L}{C}}$$

We see that first-order Z_0 is identical to the lossless expression and, hence, the assumption that Z_0 is a real-valued constant is often used. However, there are many cases when this approximation is far from valid. For example, transmission lines and waveguides operating at millimetre-wave frequencies often have very large losses due to the increase in conduction and dielectric loss with frequency. In these cases, precision measurements must consider the complex nature of the transmission line. Furthermore at low frequencies where ω is small, we find that $R \gg \omega L$ and $G \gg \omega C$. The complex nature of both γ and Z_0 then plays a very important role.

By way of example, Figures 2.5 and 2.6 show how the real and imaginary parts of Z_0 vary with frequency for a CPW. The parameters of the line are typical for a microwave monolithic integrated circuit (MMIC) with a 400 μm thick gallium arsenide (GaAs) substrate and gold conductors of 1.2 μm thickness. We see that above a few GHz the real component of Z_0 approaches the nominal 50 Ω of the

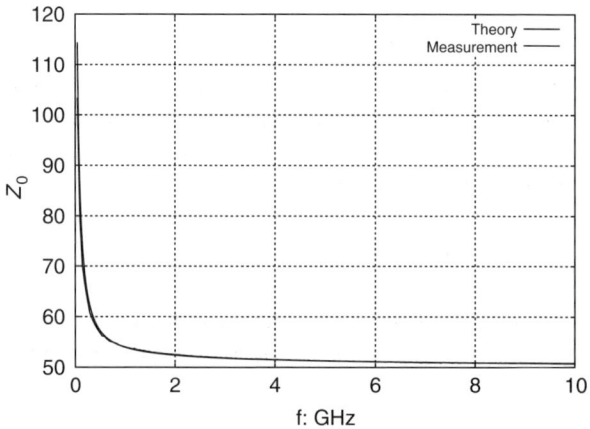

Figure 2.5   Real part of characteristic impedance of CPW transmission line on GaAs

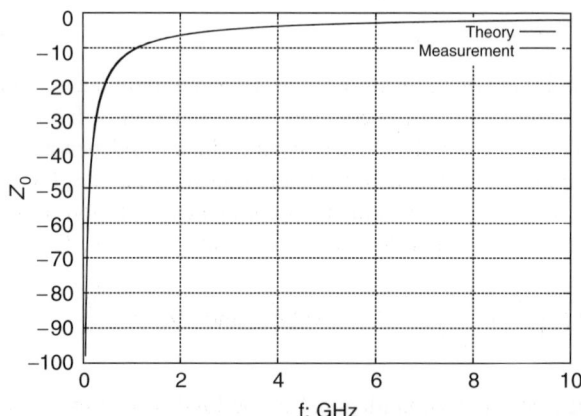

Figure 2.6   Imaginary part of characteristic impedance of CPW transmission line on GaAs

design but with a small imaginary part of a few ohms. At low frequencies the picture is very different. We see that as the frequency decreases there is a rapid increase in the magnitude of both the real and imaginary component of Z_0. Although the results are shown for CPW, similar results would be seen for microstrip, stripline and even coaxial cable.

If Z_0 has an appreciable imaginary part associated with it then a more complicated network analysis is required. The normal definitions of a and b are

$$a = V_+/\sqrt{Z_0} \quad \text{and} \quad b = V_-/\sqrt{Z_0}$$

which have units of $W^{1/2}$. If the mode travelling on the transmission line carries power p_0 then a and b can be written as

$$a = C_+\sqrt{p_0} \quad \text{and} \quad b = C_-\sqrt{p_0}$$

where C_+ and C_- are constants with $\Gamma = C_-/C_+$. If Z_0 is complex then it follows that the power p_0 will also be complex having a real, travelling component and an imaginary, stored component. As a and b are travelling wave amplitudes they must be defined in terms of the real component of the power. Hence,

$$a = C_+\text{Re}\left(\sqrt{p_0}\right) \quad \text{and} \quad b = C_-\text{Re}\left(\sqrt{p_0}\right)$$

Equation 2.10 then becomes

$$P = \frac{1}{2}\left(|a|^2 - |b|^2\right) + \text{Im}(ab^*)\frac{\text{Im}(Z_0)}{\text{Re}(Z_0)}$$

That is, the power is not simply the difference in the forward and reverse waves unless $\text{Im}(Z_0) = 0$. This results in a more complicated network theory where the definitions of impedance and admittance matrices have to be modified, resulting in different expressions for the conversion and renormalisation equations. The interested reader should consult Reference 2 for a thorough study of network theory with complex characteristic impedance.

2.8.2 Characteristic impedance in non-TEM waveguides

The usual definition of characteristic impedance is the ratio of the forward voltage to forward current. These are easily determined for simple TEM transmission lines, such as coaxial cable, where the voltage between the two conductors and the current flowing through them is uniquely defined. However, it can be more difficult to define voltages and currents in quasi-TEM transmission lines, such as microstrip and CPW, due to their hybrid nature. In fact, many waveguides used in microwave systems may only have a single conductor, such as rectangular waveguide, or no conductors at all as in a dielectric waveguide. In these cases, it becomes impossible to define a unique voltage or current and guides of this type are better explained in terms of their electric and magnetic fields

$$\mathbf{E}(x,y,z) = C_+\mathbf{e_t}(x,y)e^{-\gamma z} + C_-\mathbf{e_t}(x,y)e^{+\gamma z} \tag{2.30}$$

$$\mathbf{H}(x,y,z) = C_+\mathbf{h_t}(x,y)e^{-\gamma z} - C_-\mathbf{h_t}(x,y)e^{+\gamma z} \tag{2.31}$$

where e_t and $\mathbf{h}_t$ are the electric and magnetic fields in the transverse plane, respectively. C_- and C_+ are complex-valued constants. In general, all transmission lines are described by (2.30) and (2.31) and not by (2.1) and (2.2). Equations 2.30 and 2.31 can be expressed as [2]

$$\mathbf{E} = \frac{V(z)}{v_0} \mathbf{e}_t \tag{2.32}$$

and

$$\mathbf{H} = \frac{I(z)}{i_0} \mathbf{h}_t \tag{2.33}$$

where the equivalent waveguide voltage and current are given by

$$V(z) = v_0 \left(C_+ e^{-\gamma z} + C_- e^{+\gamma z} \right) \tag{2.34}$$

and

$$I(z) = i_0 \left(C_+ e^{-\gamma z} - C_- e^{+\gamma z} \right) \tag{2.35}$$

respectively. v_0 and i_0 are normalisation constants such that

$$Z_0 = \frac{v_0}{i_0} \tag{2.36}$$

Both $V(z)$ and v_0 have units of voltage and $I(z)$ and i_0 have units of current.

In order to extend the concept of voltage and current to the general waveguide structure, (2.34) and (2.35) must satisfy the same power relationships as (2.8) and (2.9). It can be shown that the power flow in a waveguide across a transverse surface S is given by [1]

$$P = \frac{1}{2} \operatorname{Re} \left\{ \int_S \mathbf{E} \times \mathbf{H}^* \cdot d\mathbf{S} \right\} = \frac{1}{2} \operatorname{Re} \left\{ \frac{V(z) I(z)^*}{v_0 i_0^*} p_0 \right\} \tag{2.37}$$

with modal power

$$p_0 = \int_S \mathbf{e}_t \times \mathbf{h}_t^* \cdot d\mathbf{S} \tag{2.38}$$

Therefore, in order to retain the analogy with (2.8) and (2.9) we require

$$P = \frac{1}{2} \operatorname{Re}\{V(z) I(z)^*\} \tag{2.39}$$

and thus

$$v_0 i_0^* = p_0 \tag{2.40}$$

We see that the magnitude of Z_0 is not uniquely defined since we are free to choose any value of v_0 and i_0 as long as (2.40) is satisfied. For example, $|Z_0|$ is often set to the wave impedance of the propagating mode. Another popular choice, used in network analysers, is $|Z_0| = 1$. Note, however, that the phase of Z_0 is set by (2.40) and is an inherent characteristic of the propagating mode.

To ensure that the characteristic impedance satisfies causality, Z_0 should be equal to, within a constant multiplier, the TE or TM wave impedance of the waveguide [3]. This is essential in time-domain analysis and synthesis where responses cannot precede inputs. For hybrid structures, such as dielectric waveguide, causality is ensured if the following condition is satisfied:

$$|Z_0(\omega)| = \lambda e^{-H(\arg[p_0(\omega)])}$$

where λ is an arbitrary constant and H is the Hilbert transform [4].

Since we cannot define a unique value of Z_0, we cannot define S-parameter measurements with respect to a nominal characteristic impedance. This is not a problem for standard rectangular waveguide and coaxial cable which have set dimensions, since we can specify measurements with respect to WG-22 or APC7, etc. However, if we are using non-standard waveguides, such as image or dielectric waveguide then all we can say is our S-parameters are with respect to the propagating mode on the structure.

Another important difference in waveguide networks is that in general a waveguide will support more than one mode. Multimode structures can be analysed using multimodal S-parameters [5]. Fortunately, under usual operating conditions, only the fundamental mode propagates. However, at a discontinuity, evanescent modes will always be present. These exponentially decaying fields will exist in the vicinity of the discontinuity and are required to completely explain the waveguide fields and network parameters.

It is important to remember that (2.34) and (2.35) are only equivalent waveguide voltages and currents, which do not have all the properties of (2.1) and (2.2). For example, Z_0 is dependent on normalisation and therefore we could define two different values of Z_0 for the same waveguide. Furthermore, even if we do use the same normalisation scheme it is possible for two different waveguides to have the same Z_0. Clearly, a transition from one of these guides to the other will not result in a reflectionless transition, as conventional transmission theory would suggest. We also have to be very careful when converting to Z-parameters using (2.21), since Z is related to Z_0. Since Z_0 is not uniquely defined the absolute value of Z-parameters cannot be determined. This is not surprising since impedance is intrinsically linked to current and voltage. However, even though absolute values cannot be defined, they can be useful in the development of equivalent circuit models for waveguide devices and junctions.

2.8.3 Measurement of Z_0

For precision air-filled coaxial lines, the characteristic impedance can be approximately obtained from measurements of the geometry of the line. For quasi-TEM lines several techniques can be used to measure Z_0. These include the constant capacitance technique [6,7] and the calibration comparison technique [8]. In the constant capacitance technique the characteristic impedance is derived from measured values of the propagation constant using the thru-reflect-line (TRL) technique [9] and a DC resistance measurement of the line. In the calibration comparison technique a two-tier calibration is performed, first in a known transmission line and then in the unknown line. This allows an 'error-box' to be determined, which acts as an impedance transformer from the known transmission line impedance to that of the lines under test.

2.9 Signal flow graphs

The analysis so far has relied on matrix algebra. However, another important technique can also be used to analyse microwave circuits, or indeed their low-frequency counterparts. This technique is known as the signal flow graph. Signal flow graphs express the network pictorially (Figure 2.7). The wave amplitudes are denoted by nodes, with the S-parameters being the gain achieved by the paths between nodes. To analyse signal flow graphs the following rules can be applied [10].

Rule 1. Two series branches, joined by a common node, can be replaced by one branch with gain equal to the product of the individual branches.

Rule 2. Two parallel branches joining two common nodes can be replaced with a single branch with gain equal to the sum of the two individual branches.

Rule 3. A branch that begins and ends on a single node can be eliminated by dividing the gains of all branches entering the node by one minus the gain of the loop.

Rule 4. Any node can be duplicated as long as all paths are retained.

These four rules are illustrated in Figure 2.8.

Figure 2.9 gives an example of applying the above rules to analyse a microwave circuit. The network is a simple two-port network terminated by an impedance with

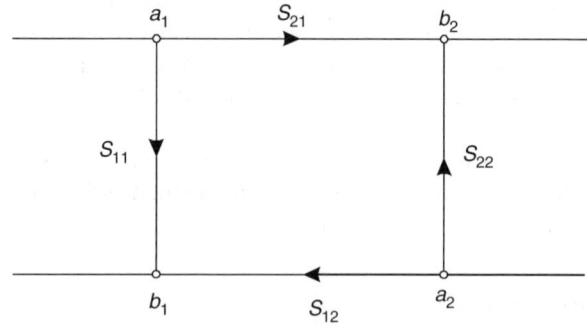

Figure 2.7 Signal flow graph for two-port network

Scattering parameters and circuit analysis 37

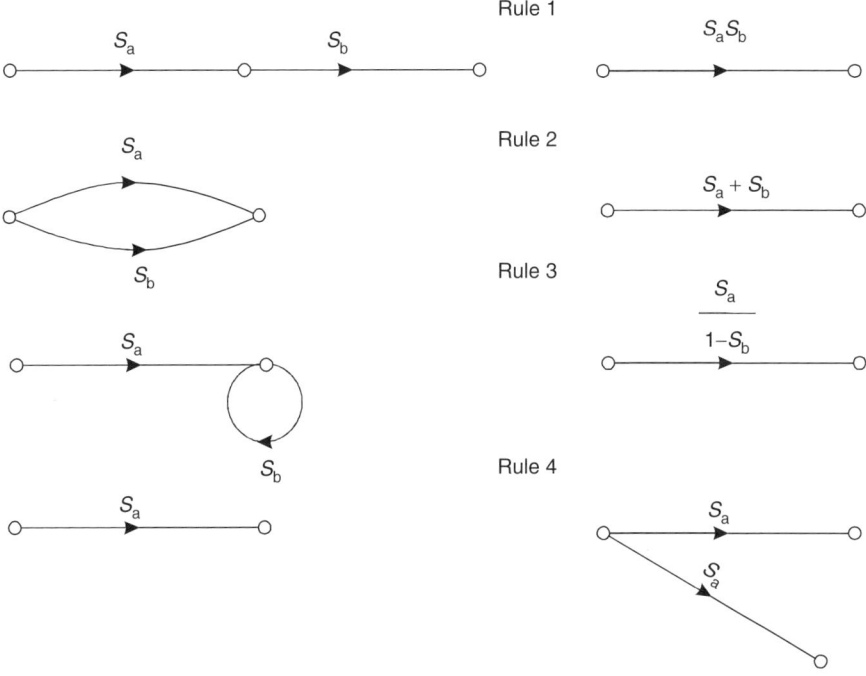

Figure 2.8 Kuhn's rules for signal flow graph analysis

reflection coefficient Γ_L. We wish to calculate the reflection coefficient at the input terminal, i.e. b_1/a_1. First, we apply Rule 4 to duplicate the a_2 node. Then, using Rule 1, we eliminate both the a_2 nodes. The closed loop, $S_{22}\Gamma_L$ is eliminated using Rule 3. Next, Rule 1 is applied to eliminate node b_2. Finally, applying Rule 2, we obtain a value for b_1/a_1.

Clearly, for larger networks, the signal flow graph technique can be very difficult to apply. However, it can often be useful for analysing simple networks – giving a more intuitive approach to the problem.

2.10 Appendix

The following relationships are true only for purely real port characteristic impedances. For complex characteristic impedance the reader is referred to Reference [2].

2.A *Reciprocity*

Using the Lorentz reciprocity relation [11] it can be shown that, in general, $Z_{mn} = Z_{nm}$. This is only true for networks that do not contain anisotropic media, such as ferrites.

38   *Microwave measurements*

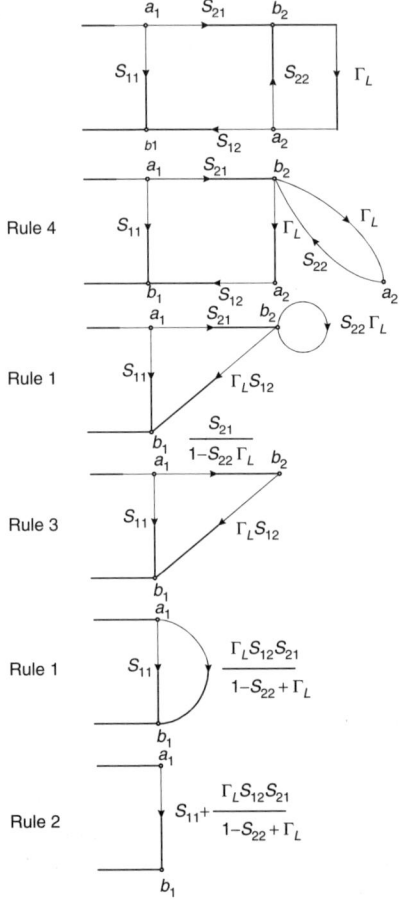

Figure 2.9   *Example of the use of signal flow graphs to analyse a microwave network*

Therefore, in matrix notation we have

$$Z = Z^T$$

where Z^T is the transpose of Z. From (2.21) it becomes apparent that if the characteristic impedance is identical at every port then

$$(U - S)^{-1}(U + S) = \left(U + S^T\right)\left(U - S^T\right)^{-1}$$

Therefore, $S = S^T$ and provided the impedance matrices are symmetrical, $S_{mn} = S_{nm}$.

2.B Losslessness

An n-port network can be described by an $n \times n$ S-parameter matrix:

$$\begin{bmatrix} b_1 \\ b_2 \\ \vdots \\ b_n \end{bmatrix} = \begin{bmatrix} S_{11} & S_{12} & \cdots & S_{1n} \\ S_{21} & S_{22} & \cdots & S_{2n} \\ \vdots & \vdots & \ddots & \vdots \\ S_{n1} & S_{n2} & \cdots & S_{nn} \end{bmatrix} \begin{bmatrix} a_1 \\ a_2 \\ \vdots \\ a_n \end{bmatrix} \tag{2B.1}$$

$$\mathbf{b} = \mathbf{S}\mathbf{a}$$

If the network is lossless, then the power entering the network must be equal to the power flowing out of the network. Therefore, from (2.10) we have

$$\sum_{i=1}^{n} |b_i|^2 = \sum_{i=1}^{n} |a_i|^2 \tag{2B.2}$$

But $|b_i|^2 = b_i b_i^*$, therefore,

$$\sum_{i=1}^{n} |b_i|^2 = (b_1, b_2, \ldots, b_n) \begin{bmatrix} b_1^* \\ b_2^* \\ \vdots \\ b_n^* \end{bmatrix}$$

We see that the column matrix in the above equation is given by the conjugate of the right-hand side of (2B.1), that is, $\mathbf{S}^*\mathbf{a}^*$. Similarly, the row matrix is given by the transpose of the right-hand side of (2B.1), that is, $(\mathbf{S}\mathbf{a})^T$. Therefore, we have

$$\sum_{i=1}^{n} |b_i|^2 = (\mathbf{S}\mathbf{a})^T \mathbf{S}^* \mathbf{a}^* = \mathbf{a}^T \mathbf{S}^T \mathbf{S}^* \mathbf{a}^* \tag{2B.3}$$

where we have used the fact that the transpose of the product of two matrices is equal to the product of the transposes in reverse. Substituting (2B.3) into (2B.2) yields

$$\mathbf{a}^T \mathbf{S}^T \mathbf{S}^* \mathbf{a}^* = \mathbf{a}^T \mathbf{a}^* \tag{2B.4}$$

where we have used the fact that

$$\sum_{i=1}^{n} |a_i|^2 = \mathbf{a}^T \mathbf{a}^*$$

Equation (2B.4) can be written as

$$\mathbf{a}^T \left(\mathbf{S}^T \mathbf{S}^* - \mathbf{U} \right) \mathbf{a}^* = 0$$

where U is the unit matrix. Therefore,

$$\mathbf{S}^T \mathbf{S}^* = \mathbf{U}$$

In other words, the product of the transpose of the S-parameter matrix S with its complex conjugate is equal to the unit matrix U. In the case of a two port we have

$$S_{11}S_{11}^* + S_{21}S_{21}^* = 1$$
$$S_{12}S_{12}^* + S_{22}S_{22}^* = 1$$
$$S_{11}S_{12}^* + S_{21}S_{22}^* = 0$$

and

$$S_{12}S_{11}^* + S_{22}S_{21}^* = 0$$

2.C Two-port transforms

Using (2.21) for a two-port network gives

$$S = \frac{1}{\Delta Z} \begin{bmatrix} \Delta Z_1 & 2\sqrt{Z_{01}Z_{02}}Z_{12} \\ 2\sqrt{Z_{01}Z_{02}}Z_{21} & \Delta Z_2 \end{bmatrix}$$

where

$$\Delta Z = (Z_{11} + Z_{01})(Z_{22} + Z_{02}) - Z_{12}Z_{21}$$
$$\Delta Z_1 = (Z_{11} - Z_{01})(Z_{22} + Z_{02}) - Z_{12}Z_{21}$$

and

$$\Delta Z_2 = (Z_{11} + Z_{01})(Z_{22} - Z_{02}) - Z_{12}Z_{21}$$

Similarly, using (2.22)

$$Z = \frac{1}{\Delta S} \begin{bmatrix} Z_{01}\Delta S_1 & 2Z_{01}S_{21} \\ 2Z_{02}S_{12} & Z_{02}\Delta S_2 \end{bmatrix}$$

where

$$\Delta S = (1 - S_{11})(1 - S_{22}) - S_{12}S_{21}$$
$$\Delta S_1 = (1 + S_{11})(1 - S_{22}) + S_{12}S_{21}$$

and

$$\Delta S_2 = (1 - S_{11})(1 + S_{22}) + S_{12}S_{21}$$

References

1 Ramo, S., Whinnery, J. R., and Van Duzer, T.: *Fields and Waves in Communication Electronics*, 2nd edn (John Wiley & Sons Inc., 1984)
2 Marks, R. B., and Williams, D. F.: 'A general waveguide circuit theory', *Journal of Research of the National Institute of Standards and Technology*, 1992;**97**: 533–62

3 Williams, D. F., and Alpert, B. K.: 'Characteristic impedance, power and causality', *IEEE Microwave Guided Wave Letters*, 1999;**9** (5):181–3
4 Williams, D. F., and Alpert, B. K.: 'Causality and waveguide circuit theory', *IEEE Transactions on Microwave Theory and Techniques*, 2001;**49** (4):615–23
5 Shibata, T., and Itoh, T.: 'Generalized-scattering-matrix modelling of waveguide circuits using FDTD field simulations' *IEEE Transactions on Microwave Theory and Techniques*, 1998;**46**(11):1742–51
6 Marks, R. B., and Williams, D. F.: 'Characteristic impedance determination using propagation constant measurement', *IEEE Microwave Guided Wave Letters*, 1991;**1** (6):141–3
7 Williams, D. F., and Marks, R. B.: 'Transmission line capacitance measurement', *IEEE Microwave Guided Wave Letters*, 1991;**1** (9):243–5
8 Williams, D. F., Arz, U., and Grabinski, H.: 'Characteristic-impedance measurement error on lossy substrates', *IEEE Microwave Wireless Components Letters*, 2001;**1** (7):299–301
9 Engen, G. F., and Hoer, C. A.: 'Thru-reflect-line: an improved technique for calibrating the dual six-port automatic network analyser', *IEEE Transactions on Microwave Theory and Techniques*, 1979;**MTT-27**:987–93
10 Kuhn, N.: 'Simplified signal flow graph analysis', *Microwave Journal*, 1963;**6**:59–66
11 Collin, R. E.: *Foundations for Microwave Engineering*, 2nd edn (McGraw-Hill, New York, 1966)

Further reading

1 Bryant, G. H.: *Principles of Microwave Measurements*, IEE Electrical measurement series 5 (Peter Peregrinus, London, 1997)
2 Engen, G. F.: *Microwave Circuit Theory and Foundations of Microwave Metrology*, IEE Electrical Measurement Series 9 (Peter Peregrinus, London, 1992)
3 Kerns, D. M., and Beatty, R. W.: *Basic Theory of Waveguide Junctions and Introductory Microwave Network Analysis*, International series of monographs in electromagnetic waves volume 13 (Pergamon Press, Oxford, 1969)
4 Somlo, P. I., and Hunter, J. D.: *Microwave Impedance Measurement*, IEE Electrical Measurement Series 2 (Peter Peregrinus, London, 1985)

Chapter 3

Uncertainty and confidence in measurements

John Hurll

3.1 Introduction

The objective of a measurement is to determine the value of the measurand, that is, the specific quantity subject to measurement. A measurement begins with an appropriate specification of the measurand, the generic method of measurement and the specific detailed measurement procedure. Knowledge of the influence quantities involved for a given procedure is important so that the sources of uncertainty can be identified. Each of these sources of uncertainty will contribute to the uncertainty associated with the value assigned to the measurand.

The guidance in this chapter is based on information in the *Guide to the Expression of Uncertainty in Measurement* [1], hereinafter referred to as the GUM. The reader is also referred to the UKAS document M3003 Edition 2 [2], which uses terminology and methodology that are compatible with the GUM. M3003 is available as a free download at www.ukas.com.

A quantity (Q) is a property of a phenomenon, body or substance to which a magnitude can be assigned. The purpose of a measurement is to assign a magnitude to the measurand: the quantity intended to be measured. The assigned magnitude is considered to be the best estimate of the value of the measurand.

The uncertainty evaluation process will encompass a number of influence quantities that affect the result obtained for the measurand. These influence, or input, quantities are referred to as X and the output quantity, that is, the measurand, is referred to as Y.

As there will usually be several influence quantities, they are differentiated from each other by the subscript i, so there will be several input quantities called X_i, where i represents integer values from 1 to N (N being the number of such quantities). In other words, there will be input quantities of $X_1, X_2, \ldots, X_N$. One of the first steps is to establish the mathematical relationship between the values of the input quantities,

x_i, and that of the measurand, y. Details about the derivation of the mathematical model can be found in Appendix D of M3003 [2].

Each of these input quantities will have a corresponding value. For example, one quantity might be the temperature of the environment – this will have a value, say 23 °C. A lower-case 'x' represents the values of the quantities. Hence the value of X_1 will be x_1, that of X_2 will be x_2, and so on.

The purpose of the measurement is to determine the value of the measurand, Y. As with the input uncertainties, the value of the measurand is represented by the lower-case letter, that is, y. The uncertainty associated with y will comprise a combination of the input, or x_i, uncertainties.

The values x_i of the input quantities X_i will all have an associated uncertainty. This is referred to as $u(x_i)$, that is, 'the uncertainty of x_i'. These values of $u(x_i)$ are, in fact, something known as the *standard uncertainty*.

Some uncertainties, particularly those associated with evaluation of repeatability, have to be evaluated by statistical methods. Others have been evaluated by examining other information, such as data in calibration certificates, evaluation of long-term drift, consideration of the effects of environment, etc.

The GUM [1] differentiates between statistical evaluations and those using other methods. It categorises them into two types – *Type A* and *Type B*.

Type A evaluation of uncertainty is carried out using statistical analysis of a series of observations.

Type B evaluation of uncertainty is carried out using methods other than statistical analysis of a series of observations.

In paragraph 3.3.4 of the GUM [1] it is stated that the purpose of the Type A and Type B classification is to indicate the two different ways of evaluating uncertainty components, and is for convenience in discussion only. Whether components of uncertainty are classified as 'random' or 'systematic' in relation to a specific measurement process, or described as *Type A* or *Type B* depending on the method of evaluation, all components regardless of classification are modelled by probability distributions quantified by variances or standard deviations. Therefore any convention as to how they are classified does not affect the estimation of the total uncertainty. But it should always be remembered that when the terms 'random' and 'systematic' are used they refer to the effects of uncertainty on a specific measurement process. It is the usual case that random components require Type A evaluations and systematic components require Type B evaluations, but there are exceptions.

For example, a random effect can produce a fluctuation in an instrument's indication, which is both noise-like in character and significant in terms of uncertainty. It may then only be possible to estimate limits to the range of indicated values. This is not a common situation but when it occurs, a Type B evaluation of the uncertainty component will be required. This is done by assigning limit values and an associated probability distribution, as in the case of other Type B uncertainties.

The input uncertainties, associated with the values of the influence quantities X_i, arise in a number of forms. Some may be characterised as limit values within which

Uncertainty and confidence in measurements 45

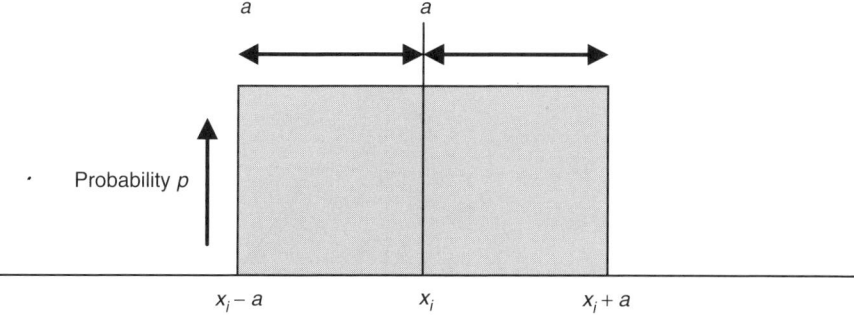

Figure 3.1   The expectation value x_i lies in the centre of a distribution of possible values with a half-width, or semi-range, of a

little is known about the most likely place within the limits where the 'true' value may lie. It therefore has to be assumed that the 'true' value is equally likely to lie anywhere within the assigned limits.

This concept is illustrated in Figure 3.1, from which it can be seen that there is equal probability of the value of x_i being anywhere within the range $x_i - a$ to $x_i + a$ and zero probability of it being outside these limits.

Thus, a contribution of uncertainty from the influence quantity can be characterised as a *probability distribution*, that is, a range of possible values with information about the most likely value of the input quantity x_i. In this example, it is not possible to say that any particular position of x_i within the range is more or less likely than any other. This is because there is no information available upon which to make such a judgement.

The probability distributions associated with the input uncertainties are therefore a reflection of the *available knowledge* about that particular quantity. In many cases, there will be insufficient information available to make a reasonable judgement and therefore a uniform, or rectangular, probability distribution has to be assumed. Figure 3.1 is an example of such a distribution.

If more information is available, it may be possible to assign a different probability distribution to the value of a particular input quantity. For example, a measurement may be taken as the *difference* in readings on a digital scale – typically, the zero reading will be subtracted from a reading taken further up the scale. If the scale is linear, both of these readings will have an associated rectangular distribution of identical size. If two identical rectangular distributions, each of magnitude $\pm a$, are combined then the resulting distribution will be triangular with a semi-range of $\pm 2a$ (Figure 3.2).

There are other possible distributions that may be encountered. For example, when making measurements of radio-frequency power, an uncertainty arises due to imperfect matching between the source and the termination. The imperfect match usually involves an unknown phase angle. This means that a cosine function characterises the probability distribution for the uncertainty. Harris and Warner [3] have

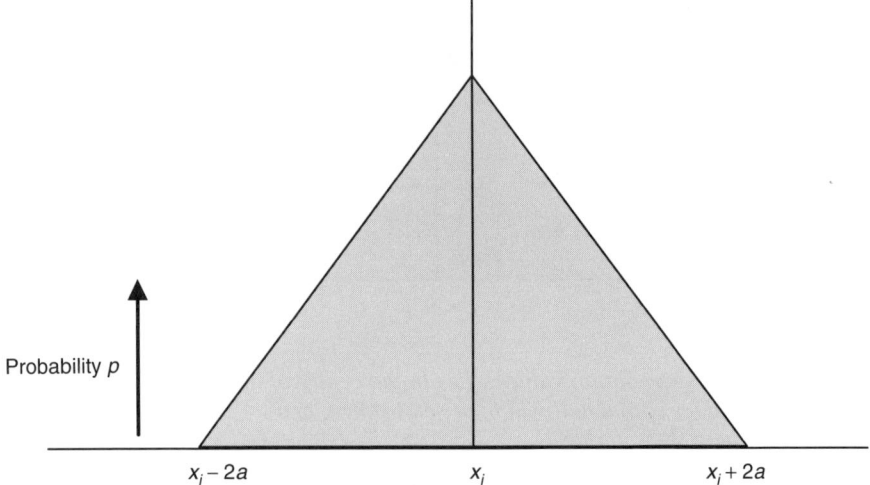

Figure 3.2 Combination of two identical rectangular distributions, each with semi-range limits of ±a, yields a triangular distribution with a semi-range of ±2a

shown that a symmetrical U-shaped probability distribution arises from this effect. In this example, the distribution has been evaluated from a theoretical analysis of the principles involved (Figure 3.3).

An evaluation of the effects of non-repeatability, performed by statistical methods, will usually yield a Gaussian or normal distribution.

When a number of distributions of whatever form are combined it can be shown that, apart from in one exceptional case, the resulting probability distribution tends to the normal form in accordance with the Central Limit Theorem. The importance of this is that it makes it possible to assign a level of confidence in terms of probability to the combined uncertainty. The exceptional case arises when one contribution to the total uncertainty dominates; in this circumstance, the resulting distribution departs little from that of the dominant contribution.

Note: If the dominant contribution is itself normal in form, then clearly the resulting distribution will also be normal.

When the input uncertainties are combined, a normal distribution will usually be obtained. The normal distribution is described in terms of a standard deviation. It will therefore be necessary to express the input uncertainties in terms that, when combined, will cause the resulting normal distribution to be expressed at the one standard deviation level, like the example in Figure 3.4.

As some of the input uncertainties are expressed as limit values (e.g. the rectangular distribution), some processing is needed to convert them into this form, which is known as a standard uncertainty and is referred to as $u(x_i)$.

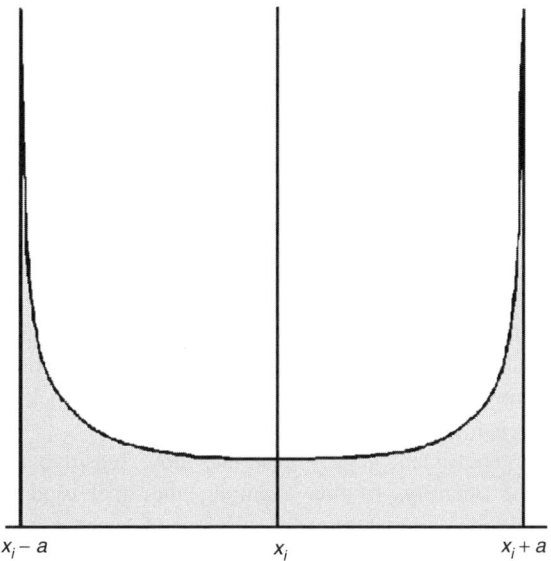

Figure 3.3 U-shaped distribution, associated with RF mismatch uncertainty. For this situation, x_i is likely to be close to one or other of the edges of the distribution

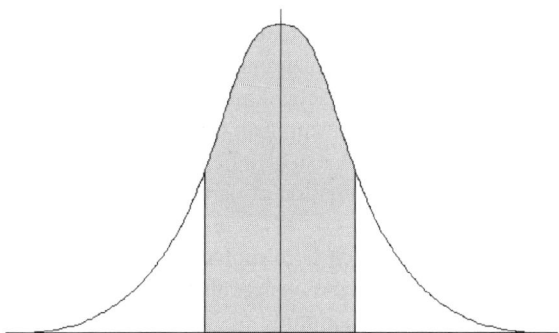

Figure 3.4 The normal, or Gaussian, probability distribution. This is obtained when a number of distributions, of any form, are combined and the conditions of the Central Limit Theorem are met. In practice, if three or more distributions of similar magnitude are present, they will combine to form a reasonable approximation to the normal distribution. The size of the distribution is described in terms of a standard deviation. The shaded area represents ±1 standard deviation from the centre of the distribution. This corresponds to approximately 68 per cent of the area under the curve

When it is possible to assess only the upper and lower bounds of an error, a rectangular probability distribution should be assumed for the uncertainty associated with this error. Then, if a_i is the semi-range limit, the standard uncertainty is given by $u(x_i) = a_i/\sqrt{3}$. Table 3.1 gives the expressions for this and for other situations.

The quantities X_i that affect the measurand Y may not have a direct, one to one, relationship with it. Indeed, they may be entirely different units altogether. For example, a dimensional laboratory may use steel gauge blocks for calibration of measuring tools. A significant influence quantity is temperature. Because the gauge blocks have a significant temperature coefficient of expansion, there is an uncertainty that arises in their length due to an uncertainty in temperature units.

In order to translate the temperature uncertainty into an uncertainty in length units, it is necessary to know how *sensitive* the length of the gauge block is to temperature. In other words, a *sensitivity coefficient* is required.

The sensitivity coefficient simply describes how sensitive the result is to a particular influence quantity. In this example, the steel used in the manufacture of gauge blocks has a temperature coefficient of expansion of approximately $+11.5 \times 10^{-6}$ per °C. So, in this case, this figure can be used as the sensitivity coefficient.

The sensitivity coefficient associated with each input quantity X_i is referred to as c_i. It is the partial derivative $\partial f/\partial x_i$, where f is the functional relationship between the input quantities and the measurand. In other words, it describes how the output estimate y varies with a corresponding small change in an input estimate x_i.

The calculations required to obtain sensitivity coefficients by partial differentiation can be a lengthy process, particularly when there are many input contributions and uncertainty estimates are needed for a range of values. If the functional relationship is not known for a particular measurement system the sensitivity coefficients can sometimes be obtained by the practical approach of changing one of the input variables by a known amount, while keeping all other inputs constant, with no change in the output estimate. This approach can also be used if f is known, but if f is not a straightforward function the determination of partial derivatives required is likely to be error-prone.

A more straightforward approach is to replace the partial derivative $\partial f/\partial x_i$ by the quotient $\Delta f/\Delta x_i$, where Δf is the change in f resulting from a change Δx_i in x_i. It is important to choose the magnitude of the change Δx_i carefully. It should be balanced between being sufficiently large to obtain adequate numerical accuracy in Δf and sufficiently small to provide a mathematically sound approximation to the partial derivative. The example in Figure 3.5 illustrates this, and why it is necessary to know the functional relationship between the influence quantities and the measurand. If the uncertainty in d is, say, ± 0.1 m then the estimate of h could be anywhere between $(7.0 - 0.1) \tan(37)$ and $(7.0 + 0.1) \tan(37)$, that is, between 5.200 and 5.350 m. A change of ± 0.1 m in the input quantity x_i has resulted in a change of ± 0.075 m in the output estimate y. The sensitivity coefficient is therefore $(0.075/0.1) = 0.75$.

Table 3.1   Probability distributions and standard uncertainties

Assumed probability distribution	Expression used to obtain the standard uncertainty	Comments or examples
Rectangular	$u(x_i) = \dfrac{a_i}{\sqrt{3}}$	A digital thermometer has a least significant digit of $0.1\,°C$. The numeric rounding caused by finite resolution will have semi-range limits of $0.05\,°C$. Thus the corresponding standard uncertainty will be $u(x_i) = \dfrac{a_i}{\sqrt{3}} = \dfrac{0.05}{1.732} = 0.029\,°C.$
U-shaped	$u(x_i) = \dfrac{a_i}{\sqrt{2}}$	A mismatch uncertainty associated with the calibration of an RF power sensor has been evaluated as having semi-range limits of 1.3%. Thus the corresponding standard uncertainty will be $u(x_i) = \dfrac{a_i}{\sqrt{2}} = \dfrac{1.3}{1.414} = 0.92\%.$
Triangular	$u(x_i) = \dfrac{a_i}{\sqrt{6}}$	A tensile testing machine is used in a testing laboratory where the air temperature can vary randomly but does not depart from the nominal value by more than $3\,°C$. The machine has a large thermal mass and is therefore most likely to be at the mean air temperature, with no probability of being outside the $3\,°C$ limits. It is reasonable to assume a triangular distribution, therefore the standard uncertainty for its temperature is $u(x_i) = \dfrac{a_i}{\sqrt{6}} = \dfrac{3}{2.449} = 1.2\,°C.$
Normal (from repeatability evaluation)	$u(x_i) = s(\bar{q})$	A statistical evaluation of repeatability gives the result in terms of one standard deviation; therefore no further processing is required.
Normal (from a calibration certificate)	$u(x_i) = \dfrac{U}{k}$	A calibration certificate normally quotes an expanded uncertainty U at a specified, high level of confidence. A coverage factor, k, will have been used to obtain this expanded uncertainty from the combination of standard uncertainties. It is therefore necessary to divide the expanded uncertainty by the same coverage factor to obtain the standard uncertainty.
Normal (from a manufacturer's specification)	$u(x_i) = \dfrac{\text{Tolerance limit}}{k}$	Some manufacturers' specifications are quoted at a given confidence level, for example, 95% or 99%. In such cases, a normal distribution can be assumed and the tolerance limit is divided by the coverage factor k for the stated confidence level. For a confidence level of 95%, $k = 2$ and for a confidence level of 99%, $k = 2.58.$
		If a confidence level is not stated then a rectangular distribution should be assumed.

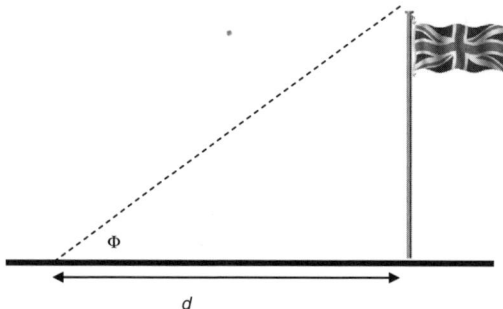

Figure 3.5   *The height h of a flagpole is determined by measuring the angle obtained when observing the top of the pole at a specified distance d. Thus $h = d \tan \Phi$. Both h and d are in units of length but the related by $\tan \Phi$. In other words, $h = f(d) = d \tan \Phi$. If the measured distance is 7.0 m and the measured angle is 37°, the estimated height is $7.0 \times \tan(37) = 5.275$ m.*

Similar reasoning can be applied to the uncertainty in the angle Φ. If the uncertainty in Φ is $\pm 0.5°$, then the estimate of h could be anywhere between 7.0 tan (36.5) and 7.0 tan (37.5), that is, between 5.179 and 5.371 m. A change of $\pm 0.5°$ in the input quantity x_i has resulted in a change of ± 0.096 m in the output estimate y. The sensitivity coefficient is therefore $(0.096/0.5) = 0.192$ m per degree.

Once the standard uncertainties x_i and the sensitivity coefficients c_i have been evaluated, the uncertainties have to be combined in order to give a single value of uncertainty to be associated with the estimate y of the measurand Y. This is known as the *combined standard uncertainty* and is given the symbol $u_c(y)$.

The combined standard uncertainty is calculated as follows:

$$u_c(y) = \sqrt{\sum_{i=1}^{N} c_i^2 u^2(x_i)} \equiv \sqrt{\sum_{i=1}^{N} u_i^2(y)}. \tag{3.1}$$

In other words, the individual standard uncertainties, expressed in terms of the measurand, are squared; these squared values are added and the square root is taken.

An example of this process is presented in Table 3.2, using the data from the measurement of the flagpole height described previously. For the purposes of the example, it is assumed that the repeatability of the process has been evaluated by making repeat measurements of the flagpole height, giving an estimated standard deviation of the mean of 0.05 m.

In accordance with the Central Limit Theorem, this combined standard uncertainty takes the form of a normal distribution. As the input uncertainties had been expressed in terms of a standard uncertainty, the resulting normal distribution is expressed as one standard deviation, as illustrated in Figure 3.6.

Table 3.2 Calculation of combined standard uncertainty $u_i(y)$ from standard uncertainties $y_i(y)$

Source of uncertainty	Value	Probability distribution	Divisor	Sensitivity coefficient	Standard uncertainty $u_i(y)$
Distance from flagpole	0.1 m	Rectangular	$\sqrt{3}$	0.75	$\dfrac{0.1}{\sqrt{3}} \times 0.75 = 0.0433\,\text{m}$
Angle measurement	0.5 m per degree	Rectangular	$\sqrt{3}$	0.192 m per degree	$\dfrac{0.5}{\sqrt{3}} \times 0.192 = 0.0554\,\text{m}$
Repeatability	0.05 m	Normal	1	1	$\dfrac{0.05}{1} \times 1 = 0.05\,\text{m}$
Combined standard uncertainty $u_c(y) = \sqrt{0.0433^2 + 0.0554^2 + 0.05^2} = 0.0863\,\text{m}$					

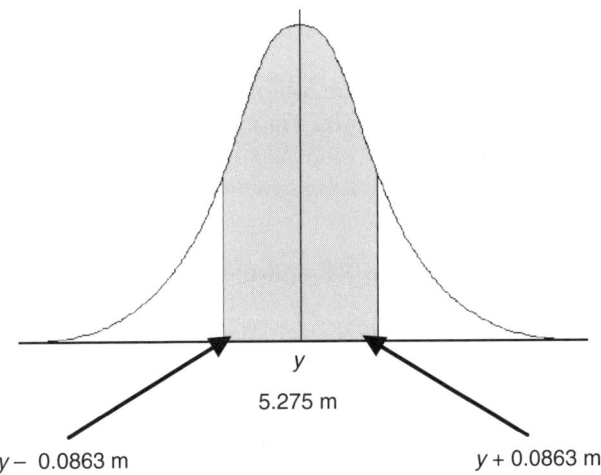

Figure 3.6 The measured value y is at the centre of a normal distribution with a standard deviation equal to $u_c(y)$. The figures shown relate to the example discussed in the text

For a normal distribution, the 1 standard deviation limits encompass 68.27 per cent of the area under the curve. This means that there is about 68 per cent confidence that the measured value y lies within the stated limits.

The GUM [1] recognises the need for providing a high level of confidence associated with an uncertainty and uses the term *expanded uncertainty*, U, which is obtained by multiplying the combined standard uncertainty by a *coverage factor*. The coverage factor is given the symbol k, thus the expanded uncertainty is given by

$$U = ku_c(y)$$

In accordance with generally accepted international practice, it is recommended that a coverage factor of $k = 2$ is used to calculate the expanded uncertainty. This value of k will give a level of confidence, or *coverage probability*, of approximately 95 per cent, assuming a normal distribution.

Note: A coverage factor of $k = 2$ actually provides a coverage probability of 95.45 per cent for a normal distribution. For convenience this is approximated to 95 per cent which relates to a coverage factor of $k = 1.96$. However, the difference is not generally significant since, in practice, the level of confidence is based on conservative assumptions and approximations to the *true* probability distributions.

Example: The measurement of the height of the flagpole had a combined standard uncertainty $u_c(y)$ of 0.0863 m. Hence the expanded uncertainty $U = ku_c(y) = 2 \times 0.0863 = 0.173$ m.

There may be situations where a normal distribution cannot be assumed and a different coverage factor may be needed in order to obtain a confidence level of approximately 95 per cent. This is done by obtaining a new coverage factor based on the effective degrees of freedom of $u_c(y)$. Details of this process can be found in Appendix B of M3003 [2].

There may also be situations where a normal distribution can be assumed, but a different level of confidence is required. For example, in safety-critical situations a higher coverage probability may be more appropriate. Table 3.3 gives the coverage factor necessary to obtain various levels of confidence for a normal distribution.

3.2 Sources of uncertainty in RF and microwave measurements

3.2.1 RF mismatch errors and uncertainty

At RF and microwave frequencies the mismatch of components to the characteristic impedance of the measurement system transmission line can be one of the

Table 3.3 Coverage factor necessary to obtain various levels of confidence for a normal distribution

Coverage probability p (%)	Coverage factor k
90	1.65
95	1.96
95.45	2.00
99	2.58
99.73	3.00

most important sources of error and of the systematic component of uncertainty in power and attenuation measurements. This is because the phases of voltage reflection coefficients (VRCs) are not usually known and hence corrections cannot be applied.

In a power measurement system, the power, P_0, that would be absorbed in a load equal to the characteristic impedance of the transmission line has been shown [3] to be related to the actual power, P_L, absorbed in a wattmeter terminating the line by the equation

$$P_0 = \frac{P_L}{1 - |\Gamma_L|^2}(1 - 2|\Gamma_G||\Gamma_L|\cos\varphi + |\Gamma_G|^2|\Gamma_L|^2) \qquad (3.2)$$

where φ is the relative phase of the generator and load VRCs Γ_G and Γ_L. When Γ_G and Γ_L are small, this becomes

$$P_0 = \frac{P_L}{1 - |\Gamma_L|^2}(1 - 2|\Gamma_G||\Gamma_L|\cos\varphi) \qquad (3.3)$$

When φ is unknown, this expression for absorbed power can have limits.

$$P_0(\text{limits}) = \frac{P_L}{1 - |\Gamma_L|^2}(1 \pm 2|\Gamma_G||\Gamma_L|) \qquad (3.4)$$

The calculable mismatch error is $1 - |\Gamma_L|^2$ and is accounted for in the calibration factor, while the limits of mismatch uncertainty are $\pm 2|\Gamma_L||\Gamma_G|$. Because a cosine function characterises the probability distribution for the uncertainty, Harris and Warner [3] showed that the distribution is U-shaped with a standard deviation given by

$$u(\text{mismatch}) = \frac{2|\Gamma_G||\Gamma_L|}{\sqrt{2}} = 1.414\Gamma_G\Gamma_L \qquad (3.5)$$

When a measurement is made of the attenuation of a two-port component inserted between a generator and load that are not perfectly matched to the transmission line, Harris and Warner [3] have shown that the standard deviation of mismatch, M, expressed in dB is approximated by

$$M = \frac{8.686}{\sqrt{2}}[|\Gamma_G|^2(|s_{11a}|^2 + |s_{11b}|^2) + |\Gamma_L|^2(|s_{22a}|^2 + |s_{22b}|^2)$$
$$+ |\Gamma_G|^2 \cdot |\Gamma_L|^2(|s_{21a}|^4 + |s_{21b}|^4)]^{0.5} \qquad (3.6)$$

where Γ_G and Γ_L are the source and load VRCs, respectively, and s_{11}, s_{22}, s_{21} are the scattering coefficients of the two-port component with the suffix a referring to the starting value of the attenuator and b referring to the finishing value of the attenuator. Harris and Warner [3] concluded that the distribution for M would approximate to that of a normal distribution due to the combination of its component distributions.

The values of Γ_G and Γ_L used in (3.4) and (3.5) and the scattering coefficients used in (3.6) will themselves be subject to uncertainty because they are derived from measurements. This uncertainty has to be considered when calculating the mismatch uncertainty and it is recommended that this is done by adding it in quadrature with the measured or derived value of the reflection coefficient; for example, if the measured value of Γ_L is 0.03 ± 0.02 then the value of Γ_L that should be used to calculate the mismatch uncertainty is $\sqrt{0.03^2 + 0.02^2}$, that is, 0.036.

3.2.2 Directivity

When making VRC measurements at RF and microwave frequencies, the finite directivity of the bridge or reflectometer gives rise to an uncertainty in the measured value of the VRC, if only the magnitude and not the phase of the directivity component is known. The uncertainty will be equal to the directivity, expressed in linear terms; for example, a directivity of 30 dB is equivalent to an uncertainty of ± 0.0316 VRC.

As above, it is recommended that the uncertainty in the measurement of directivity is taken into account by adding the measured value in quadrature with the uncertainty, in linear quantities; for example, if the measured directivity of a bridge is 36 dB (0.016) and has an uncertainty of $+8$ dB -4 dB (± 0.01) then the directivity to be used is $\sqrt{0.016^2 + 0.01^2} = 0.019$ (34.4 dB).

3.2.3 Test port match

The test port match of a bridge or reflectometer used for reflection coefficient measurements will give rise to an error in the measured VRC due to re-reflection. The uncertainty, $u(\text{TP})$, is calculated from $u(\text{TP}) = \text{TP} \cdot \Gamma_X^2$ where TP is the test port match, expressed as a VRC, and Γ_X is the measured reflection coefficient. When a directional coupler is used to monitor incident power in the calibration of a power meter, it is the effective source match of the coupler that defines the value of Γ_G. The measured value of test port match will have an uncertainty that should be taken into account by using quadrature summation.

3.2.4 RF connector repeatability

The lack of repeatability of coaxial pair insertion loss and, to a lesser extent, VRC is a problem when calibrating devices in a coaxial line measurement system and subsequently using them in some other system. Although connecting and disconnecting the device can evaluate the repeatability of particular connector pairs in use, these connector pairs are only samples from a whole population. To obtain representative data for all the various types of connector in use is beyond the resources of most measurement laboratories; however, some useful guidance can be obtained from the ANAMET Connector Guide [4].

3.2.5 Example – calibration of a coaxial power sensor at a frequency of 18 GHz

The measurement involves the calibration of an unknown power sensor against a standard power sensor by substitution on a stable, monitored source of known source impedance. The measurement is made in terms of Calibration factor, defined as

$$\frac{\text{Incident power at reference frequency}}{\text{Incident power at calibration frequency}}$$

for the same power sensor response and is determined from the following:

$$\text{Calibration factor, } K_X = (K_s + D_s) \times \delta DC \times \delta M \times \delta REF \tag{3.7}$$

where K_S is the calibration factor of the standard sensor, D_S is the drift in standard sensor since the previous calibration, δDC is the ratio of DC voltage outputs, δM is the ratio of mismatch losses and δREF is the ratio of reference power source (short-term stability of 50 MHz reference).

Four separate measurements were made which involved disconnection and reconnection of both the unknown sensor and the standard sensor on a power transfer system. All measurements were made in terms of voltage ratios that are proportional to calibration factor.

There will be mismatch uncertainties associated with the source/standard sensor combination and with the source/unknown sensor combination. These will be $200\Gamma_G\Gamma_S$ per cent and $200\Gamma_G\Gamma_X$ per cent, respectively, where

$\Gamma_G = 0.02$ at 50 MHz and 0.07 at 18 GHz

$\Gamma_S = 0.02$ at 50 MHz and 0.10 at 18 GHz

$\Gamma_X = 0.02$ at 50 MHz and 0.12 at 18 GHz.

These values are assumed to include the uncertainty in the measurement of Γ.

The standard power sensor was calibrated by an accredited laboratory 6 months before use; the expanded uncertainty of ±1.1 per cent was quoted for a coverage factor $k = 2$.

The long-term stability of the standard sensor was estimated from the results of five annual calibrations to have rectangular limits not greater than ±0.4 per cent per year. A value of ±0.2 per cent is assumed as the previous calibration was within 6 months.

The instrumentation linearity uncertainty was estimated from measurements against a reference attenuation standard. The expanded uncertainty for $k = 2$ of ±0.1 per cent applies to ratios up to 2:1.

Type A evaluation

The four measurements resulted in the following values of Calibration factor: 93.45, 92.20, 93.95 and 93.02 per cent.

The mean value $\bar{K}_X = 93.16$ per cent.

The standard deviation of the mean, $u(K_R) = s(\bar{K}_X) = 0.7415/\sqrt{4} = 0.3707$ per cent.

Table 3.4 Uncertainty budget

Symbol	Source of uncertainty	Value ±%	Probability distribution	Divisor	c_i	$u_i(K_x)$%	v_i or v_{eff}
K_S	Calibration factor of standard	1.1	Normal	2.0	1.0	0.55	∞
D_S	Drift since last calibration	0.2	Rectangular	$\sqrt{3}$	1.0	0.116	∞
δDC	Instrumentation linearity	0.1	Normal	2.0	1.0	0.05	∞
δM	Stability of 50 MHz reference	0.2	Rectangular	$\sqrt{3}$	1.0	0.116	∞
	Mismatch:						
M_1	Standard sensor at 50 MHz	0.08	U-shaped	$\sqrt{2}$	1.0	0.06	∞
M_2	Unknown sensor at 50 MHz	0.08	U-shaped	$\sqrt{2}$	1.0	0.06	∞
M_3	Standard sensor at 18 GHz	1.40	U-shaped	$\sqrt{2}$	1.0	0.99	∞
M_4	Unknown sensor at 18 GHz	1.68	U-shaped	$\sqrt{2}$	1.0	1.19	∞
K_R	Repeatability of indication	0.37	Normal	1.0	1.0	0.37	3
$u(K_X)$	Combined standard uncertainty		Normal			1.69	>500
U	Expanded uncertainty		Normal ($k=2$)			3.39	>500

3.2.5.1 Reported result

The measured calibration factor at 18 GHz is 93.2% ± 3.4%.

The reported expanded uncertainty (Table 3.4) is based on a standard uncertainty multiplied by a coverage factor $k = 2$, providing a coverage probability of approximately
95 per cent.

3.2.5.2 Notes

(1) For the measurement of calibration factor, the uncertainty in the absolute value of the 50 MHz reference source need not be included if the standard and unknown sensors are calibrated using the same source, within the timescale allowed for its short-term stability.

(2) This example illustrates the significance of mismatch uncertainty in measurements at relatively high frequencies.
(3) In a subsequent use of a sensor further random components of uncertainty may arise due to the use of different connector pairs.

References

1 BIPM, IEC, IFCC, ISO, IUPAC, IUPAP, OIML. *Guide to the Expression of Uncertainty in Measurement*. International Organisation for Standardization, Geneva, Switzerland. ISBN 92-67-10188-9, First Edition 1993. BSI Equivalent: BSI PD 6461: 1995, *Vocabulary of Metrology, Part 3. Guide to the Expression of Uncertainty in Measurement*. BSI ISBN 0 580 23482 7
2 United Kingdom Accreditation Service, *The Expression of Uncertainty and Confidence in Measurement*, M3003, 2nd edn, January 2007
3 Harris, I. A., and Warner, F. L.: 'Re-examination of mismatch uncertainty when measuring microwave power and attenuation', *IEE Proc. H, Microw. Opt. Antennas*, 1981;**128** (1):35–41
4 National Physical Laboratory, *ANAMET Connector Guide*, 3rd edn, 2007

Chapter 4
Using coaxial connectors in measurement
Doug Skinner

Disclaimer

Every effort has been made to ensure that this chapter contains accurate information obtained from many sources that are acknowledged where possible. However, the NPL, ANAMET and the Compiler cannot accept liability for any errors, omissions or misleading statements in the information. The compiler would also like to thank all those members of ANAMET who have supplied information, commented on and given advice on the preparation of this chapter.

4.1 Introduction

The importance of the correct use of coaxial connectors not only applies at radio and microwave frequencies but also at DC and low frequency. The requirement for 'Traceability to National Standards' for measurements throughout the industry may depend on several different calibration systems 'seeing' the same values for the parameters presented by a device at its coaxial terminals. It is not possible to include all the many different types of connector in this guide and the selection has been made on those connectors used on measuring instruments and for metrology use.

It is of vital importance to note that mechanical damage can be inflicted on a connector when a connection or disconnection is made at any time during its use.

The common types of precision and general-purpose coaxial connector that are in volume use worldwide are the Type N, GPC 3.5 mm, Type K, 7/16, TNC, BNC and SMA connectors. These connectors are employed for interconnection of components and cables in military, space, industrial and domestic applications.

The simple concept of a coaxial connector comprises an outer conductor contact, an inner conductor contact, and means for mechanical coupling to a cable and/or to another connector. Most connectors, in particular the general-purpose types, are composed of a pin and socket construction.

There are basically two types of coaxial connector in use and they are known as laboratory precision connectors (LPC), using only air dielectric, and general precision connectors (GPC), having a self contained, low reflection dielectric support. There are also a number of general-purpose connectors in use within the GPC group but they are not recommended for metrology use.

Some connectors are hermaphroditic (non-sexed), particularly some of the laboratory precision types, and any two connectors may be joined together. They have planar butt contacts and are principally employed for use on measurement standards and on equipment and calibration systems where the best possible uncertainty of measurement is essential. Most of the non-sexed connectors have a reference plane that is common to both the outer and inner conductors. The mechanical and electrical reference planes coincide and, in the case of the precision connectors, a physically realised reference plane is clearly defined. Hermaphroditic connectors are used where the electrical length and the characteristic impedance are required at the highest accuracy. The 14 mm and 7 mm connectors are examples of this type of construction and they are expensive.

GPCs of the plug and socket construction look similar but the materials used in the construction can vary. The best quality connectors are more robust, which use stainless steel, and the mechanical tolerances are more precise. It is important to be clear on the quality of the connector being used. The GPC types in common use are the Type N 7 mm, GPC 3.5 mm, Type K 2.92 mm, GPC 2.4 mm and Type V 1.85 mm.

The choice of connectors, from the range of established designs, must be appropriate to the proposed function and specification of the device or measurement system. Often a user requirement is for a long-life quality connector with minimum effect on the performance of the device it is used on and the repeatability of the connection is generally one of the most important parameters.

4.1.1 Coaxial line sizes

Some coaxial line sizes for establishing a characteristic impedance of 50 Ω are shown in Table 4.1. They are chosen to achieve the desired performance over their operating frequency range up to 110 GHz.

Table 4.1 Coaxial line sizes for 50 Ω characteristic impedance

Inside diameter of the outer conductor in mm (nominal)	14.29	7.00	3.50	2.92	2.40	1.85	1.00
Rated minimum upper frequency limit in GHz	8.5	18.0	33.0	40.0	50.0	65.0	110.0
Theoretical limit in GHz for the onset of the TE_{11} (H_{11}) mode	9.5	19.4	38.8	46.5	56.5	73.3	135.7

4.2 Connector repeatability

Connectors in use on test apparatus and measuring instruments at all levels need to be maintained in pristine condition in order to retain the performance of the test apparatus. The connector repeatability is a key contribution to the performance of a measurement system.

Connector repeatability can be greatly impaired because of careless assembly, misalignment, over-tightening, inappropriate handling, poor storage and unclean working conditions. In extreme cases, permanent damage can be caused to the connectors concerned and possibly to other originally sound connectors to which they are coupled.

Connectors should never be rotated relative to one another when being connected and disconnected. Special care should be taken to avoid rotating the mating plane surfaces against one another.

4.2.1 Handling of airlines

When handling or using airlines and similar devices used in automatic network analyser, calibration and verification kits, it is extremely important to avoid contamination of the component parts due to moisture and finger marks on the lines. Protective lint-free cotton gloves should always be worn. The failure to follow this advice may significantly reduce performance and useful life of the airlines.

4.2.2 Assessment of connector repeatability

In a particular calibration or measurement system, repeatability of the coaxial interconnections can be assessed from measurements made after repeatedly disconnecting and reconnecting the device. It is clearly necessary to ensure that all the other conditions likely to influence the alignment are maintained as constant as possible.

In some measurement situations, it is important that the number of repeat connections made uses the same positional alignment of the connectors.

In other situations, it is best to rotate one connector relative to the other between connections and reconnections. For example, when calibrating or using devices fitted with Type N connectors (e.g. power sensors or attenuator pads) three rotations of 120° or five rotations of 72° are made. However, it is important to remember to make the rotation before making the contact.

The repeatability determination will normally be carried out when trying to achieve the best measurement capability on a particular device, or when initially calibrating a measuring system. The number of reconnections and rotations can then be recommended in the measurement procedure.

Repeatability of the insertion loss of coaxial connectors introduces a major contribution to the Type A component of uncertainty in a measurement process. If a measurement involving connectors is repeated several times, the Type A uncertainty contribution deduced from the results will include that arising from the connector repeatability provided that the connection concerned is broken and remade at each repetition.

It should be remembered that a connection has to be made at least once when connecting an item under test to the test equipment and this gives rise to a contribution to the Type A uncertainty contribution associated with the connector repeatability.

Experience has shown that there is little difference in performance between precision and ordinary connectors (when new) so far as the repeatability of connection is concerned, but with many connections and disconnections the ordinary connector performance will become progressively inferior when compared with the precision connector.

4.3 Coaxial connector specifications

The following specifications are some of those that provide information and define the parameters of established designs of coaxial connectors and they should be consulted for full information on electrical performance, mechanical dimensions and mechanical tolerances.

>IEEE Standard 287-2007
>
>IEC Publication 457
>
>MIL-STD-348A incorporates MIL-STD-39012C
>
>IEC Publication 169
>
>CECC 22000
>
>British Standard 9210
>
>DIN Standards

Users of coaxial connectors should also take into account any manufacturer's performance specifications relating to a particular connector type.

4.4 Interface dimensions and gauging

It is of utmost importance that connectors do not damage the test equipment interfaces to which they are offered for calibration. Poor performance of many coaxial devices and cable assemblies can often be traced to poor construction and non-compliance with the mechanical specifications. The mechanical gauging of connectors is essential to ensure correct fit and to achieve the best performance. This means that all coaxial connectors fitted on all equipments, cables and terminations should be gauged on a routine basis in order to detect any out of tolerance conditions that may impair the electrical performance.

4.4.1 Gauging connectors

A connector should be gauged before it is used for the very first time or if someone else has used the device on which it is fitted.

If the connector is to be used on another item of equipment, the connector on the equipment to be tested should also be gauged.

Connectors should never be forced together when making a connection since forcing often indicates incorrectness and incompatibility. Many connector screw coupling mechanisms, for instance, rarely need to be more than finger-tight for electrical calibration purposes; most coaxial connectors usually function satisfactorily, giving adequately repeatable results, unless damaged. There are some dimensions that are critical for the mechanical integrity, non-destructive mating and electrical performance of the connector.

Connector gauge kits are available for many connector types but it is also easy to manufacture simple low-cost test pieces for use with a micrometer depth gauge or other device to ensure that the important dimensions can be measured or verified.

The mechanical gauging of coaxial connectors will detect and prevent the following problems.

Inner conductor protrusion. This may result in buckling of the socket contacts or damage the internal structure of a device due to the axial forces generated.

Inner conductor recession. This will result in poor voltage reflection coefficient, possibly unreliable contact and could even cause breakdown under peak power conditions.

Appendix 4.A shows a list of the most common types of coaxial connector in use. Appendix 4.B gives information on the various connector types including the critical mechanical dimensions that need to be measured for the selected connector types.

4.5 Connector cleaning

To ensure a long and reliable connector life, careful and regular inspection of connectors is necessary and cleaning of connectors is essential to maintaining good performance.

Connectors should be inspected initially for dents, raised edges, and scratches on the mating surfaces. Connectors that have dents on the mating surfaces will usually also have raised edges around them and will make less than perfect contact; further to this, raised edges on mating interfaces will make dents in other connectors to which they are mated. Connectors should be replaced unless the damage is very slight.

Awareness of the advantage of ensuring good connector repeatability and its effect on the overall uncertainty of a measurement procedure should encourage careful inspection, interface gauging and handling of coaxial connectors.

Prior to use, a visual examination should be made of a connector or adaptor, particularly for concentricity of the centre contacts and for dirt on the dielectric. It is essential that the axial position of the centre contact of all items offered for calibration should be gauged because the butting surfaces of mated centre contacts must not touch. If the centre contacts do touch, there could be damage to the connector or possibly to other parts of the device to which the connector is fitted. For precision hermaphroditic connectors the two centre conductor petals do butt up and the dimensions are critical for safe connections.

Small particles, usually of metal, are often found on the inside connector mating planes, threads, and on the dielectric. They should be removed to prevent damage to the connector surfaces. The items required for cleaning connectors and the procedure to be followed is described in the next section.

4.5.1 Cleaning procedure

Items required:

(1) Low-pressure compressed air (solvent free);
(2) cotton swabs (special swabs can be obtained for this purpose);
(3) lint-free cleaning cloth;
(4) isopropanol and
(5) illuminated magnifier or a jeweller's eye glass.

Note: Isopropanol that contains additives should not be used for cleaning connectors as it may cause damage to plastic dielectric support beads in coaxial and microwave connectors. It is important to take any necessary safety precautions when using chemicals or solvents.

4.5.1.1 First step

Remove loose particles on the mating surfaces and threads using low-pressure compressed air. A wooden cocktail stick can be used to carefully remove any small particles that the compressed air does not remove.

4.5.1.2 Second step

Clean surfaces using isopropanol on cotton swabs or lint-free cloth. Use only sufficient solvent to clean the surface. When using swabs or lint-free cloth, use the least possible pressure to avoid damaging connector surfaces. Do not spray solvents directly on to connector surfaces or use contaminated solvents.

4.5.1.3 Third step

Use the low-pressure compressed air once again to remove any remaining small particles and to dry the surfaces of the connector to complete the cleaning process before using the connector.

4.5.2 Cleaning connectors on static sensitive devices

Special care is required when cleaning connectors on test equipment containing static sensitive devices. When cleaning such connectors always wear a grounded wrist strap and observe correct procedures. The cleaning should be carried out in a special handling area. These precautions will prevent electrostatic discharge (ESD) and possible damage to circuits.

4.6 Connector life

The number of times that a connector can be used is very difficult to predict and it is quite clear that the number of connections and disconnections that can be achieved is dependent on the use, environmental conditions and the care taken when making a connection. Some connector bodies such as those used on the Type N connector are made using stainless steel and are generally more rugged, have a superior mechanical performance and a longer useable life. The inner connections are often gold plated to give improved electrical performance. For many connector types the manufacturer's specification will quote the number of connections and disconnections that can be made. The figure quoted may be as high or greater than 5000 times but this figure assumes that the connectors are maintained in pristine condition and correctly used. For example, the Type SMA connector was developed for making interconnections within equipment and its connector life is therefore relatively short in repetitive use situations.

However, by following the guidance given in this document it should be possible to maximise the lifetime of a connector used in the laboratory.

4.7 Adaptors

Buffer adaptors or 'connector savers' can be used in order to reduce possible damage to output connectors on signal sources and other similar devices. It should be remembered that the use of buffer adaptors and connector savers may have an adverse effect on the performance of a measurement system and may result in significant contributions to uncertainty budgets. Adaptors are often used for the following reasons:

(1) To reduce wear on expensive or difficult to replace connectors on measuring instruments where the reduction in performance can be tolerated.
(2) When measuring a coaxial device that is fitted with an SMA connector.

4.8 Connector recession

The ideal connector pair would be constructed in such a way in order to eliminate any discontinuities in the transmission line system into which the connector pair is connected. In practice, due to mechanical tolerances there will almost always be a small gap between the mated plug and socket connectors. This small gap is often referred to as 'recession'. It may be that both the connectors will have some recession because of the mechanical tolerances and the combined effect of the recession is to produce a very small section of transmission line that will have different characteristic impedance than the remainder of the line causing a discontinuity.

The effect of the recession could be calculated but there are a number of other effects present in the mechanical construction of connectors that could make the result unreliable. Some practical experimental work has been carried out at Agilent Technologies on the effect of recession. For more information a reference is given in Further reading to an ANAMET paper where the results of some practical measurements have been published.

The connector specifications give limit values for the recession of the plug and socket connectors when joined (e.g. see Appendix 4.B for the Type N connector). The effect on electrical performance caused by recession in connectors is a subject of special interest to users of network analysers and more experimental work needs to be carried out.

4.9 Conclusions

The importance of the interconnections in measurement work should never be underestimated and the replacement of a connector may enable the Type A uncertainty contribution in a measurement process to be reduced significantly. Careful consideration must be given, when choosing a connector, to select the correct connector for the measurement task. In modern measuring instruments, such as power meters, spectrum analysers and signal generators, the coaxial connector socket on the front panel is often an integral part of a complex sub-assembly and any damage to this connector may result in a very expensive repair.

It is particularly important when using coaxial cables, with connectors that are locally fitted or repaired that they are tested before use to ensure that the connector complies with the relevant mechanical specification limits. All cables, even those obtained from specialist manufacturers, should be tested before use. Any connector that does not pass the relevant mechanical tests should be rejected and replaced.

Further information on coaxial connectors can be obtained direct from manufacturers. Many connector manufacturers have a website and there are other manufacturers, documents and specifications that can be found by using World Wide Web and searching by connector type.

4.A Appendix A

4.A.1 Frequency range of some common coaxial connectors

Table 4A.1 lists common types of coaxial connector used on measurement systems showing the frequency range over which they are often used and the approximate upper frequency limit for the various line sizes.

4.B Appendix B

4.B.1 The 14 mm precision connector

The 14 mm precision connector was developed in the early 1960s by the General Radio Company and is known as the GR900 connector. It has limited usage and is mainly used in primary standards laboratories and in military metrology. It is probably the best coaxial connector ever built in terms of its performance and it has

Table 4A.1 Common types of coaxial connectors used on measurement systems

Title	Line size	Impedance	Upper frequency range (for normal use)	Upper frequency limit (approximate value)
Precision non-sexed connectors				
GPC 14	14.2875 mm	50 Ω	8.5 GHz	9 GHz
GPC 14	14.2875 mm	75 Ω	3.0 GHz*	8.5 GHz
GPC 7	7.0 mm	50 Ω	18 GHz	18 GHz
Precision sexed connectors				
Type N	7.0 mm	50 Ω	18 GHz	22 GHz
GPC 3.5	3.5 mm	50 Ω	26.5 GHz	34 GHz
Type K	2.92 mm	50 Ω	40 GHz	46 GHz
Type Q	2.4 mm	50 Ω	50 GHz	60 GHz
Type V	1.85 mm	50 Ω	65 GHz	75 GHz
Type W	1.0 mm	50 Ω	110 GHz	110 GHz
1.0 mm	1.0 mm	50 Ω	110 GHz	110 GHz
Generel purpose conncetors				
Type N	7.0 mm	50 Ω	18 GHz	22 GHz
Type N	7.0 mm	75 Ω	3 GHz*	22 GHz
7/16	16.0 mm	50 Ω	7.5 GHz	9 GHz
SMA	3.5 mm	50 Ω	26.5 GHz	34 GHz

Note: *Measurements made in 75 Ω impedance are normally restricted to an upper frequency limit of 3 GHz.

low insertion loss, low reflection and extremely good repeatability. However, it is bulky and expensive.

The interface dimensions for the GPC 14 mm connector are given in IEEE Standard 287 and IEC Publication 457.

Before use, a visual examination, particularly of the centre contacts, should be made. Contact in the centre is made through sprung inserts and these should be examined carefully. A flat smooth disc pressed against the interface can be used to verify correct functioning of the centre contact. The disc must fit inside the castellated coupling ring that protects the end surface of the outer connector and ensures correct alignment of the two connectors when mated. The inner connector should be gauged with the collet removed. There is a special tool kit available for use with GR900 connectors.

There are 50 Ω and 75 Ω versions of the GR900 connector available. The GR900 14 mm connector is made in two types.

    LPC   Laboratory precision connector   Air dielectric
    GPC General precision connector   Dielectric support

The LPC version is usually fitted to devices such as precision airlines for use in calibration and verification kits for automatic network analysers and reflectometers (Figure 4B.1).

There is also a lower performance version of the GR900 connector designated the GR890. The GR890 connector can be identified by the marking on the locking ring and it has a much reduced frequency range of operation, for example, approx. 3 GHz.

4.B.2 The 7 mm precision connector

This connector series was developed to meet the need for precision connectors for use in laboratory measurements over the frequency range DC to 18 GHz. This connector is often known as the Type GPC7 connector and it is designed as a hermaphroditic connector with an elaborate coupling mechanism. The connector interface features a butt coplanar contact for the inner and outer contacts, with both the mechanical and electrical interfaces at the same location. A feature of the GPC7 connector is its ruggedness and good repeatability over multiple connections in a laboratory environment. The connector is made in two types.

GPC General precision connector   Dielectric support
LPC Laboratory precision connector   Air dielectric

The LPC version is usually fitted to devices such as precision airlines for use in calibration and verification kits for automatic network analysers and reflectometers.

The interface dimensions for the GPC7 connector are given in IEEE Standard 287 and IEC 457. The most common connector of this type in the UK has a centre contact comprising a slotted resilient insert within a fixed centre conductor. The solid part of the centre conductor must not protrude beyond the planar connector reference plane, although the resilient inserts must protrude beyond the reference plane. However, the inserts must be capable of taking up coplanar position under pressure. A flat, smooth plate or disc, pressed against the interfaces can verify correct functioning of the centre contact.

There are two versions of the collet for this connector: one has four slots and the other has six slots. For best performance it is good practice to replace the four-slot version with the six-slot type (see Figure 4B.2).

Figure 4B.2a shows the construction of the GPC7 connector and the location of the outer conductor mating plane. The use of GPC7 connector is normally restricted to making precision measurements in calibration laboratories.

4.B.2.1 Connection and disconnection of GPC7 connectors

It is important to use the correct procedure when connecting or disconnecting GPC7 connectors to prevent damage and to ensure a long working life and consistent electrical performance. The following procedure is recommended for use with GPC7 connectors.

Using coaxial connectors in measurement 69

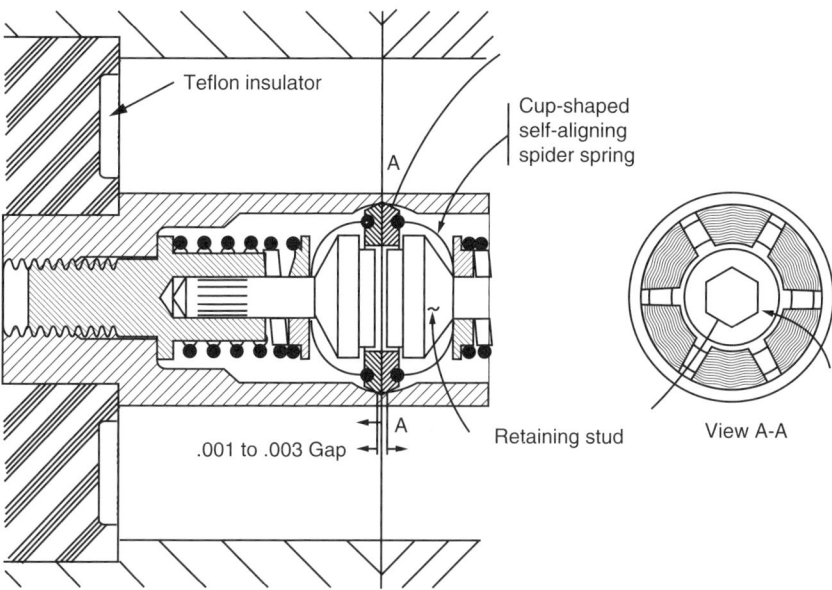

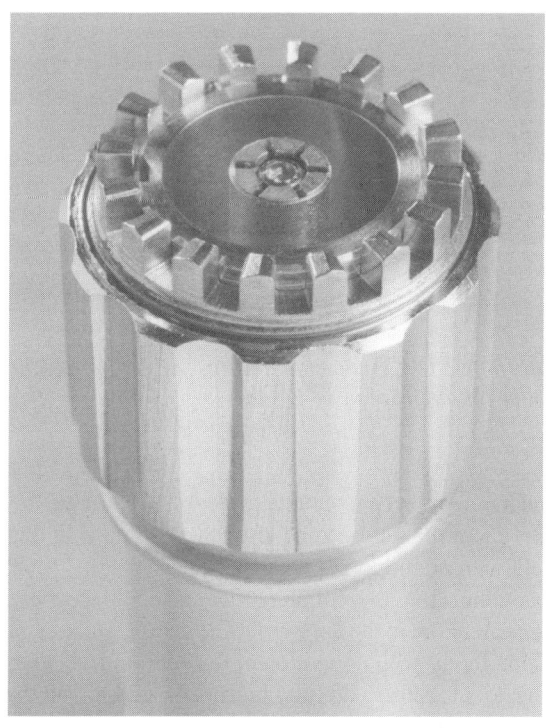

Figure 4B.1   The GR 900 14 mm connector [photograph NPL]

70 *Microwave measurements*

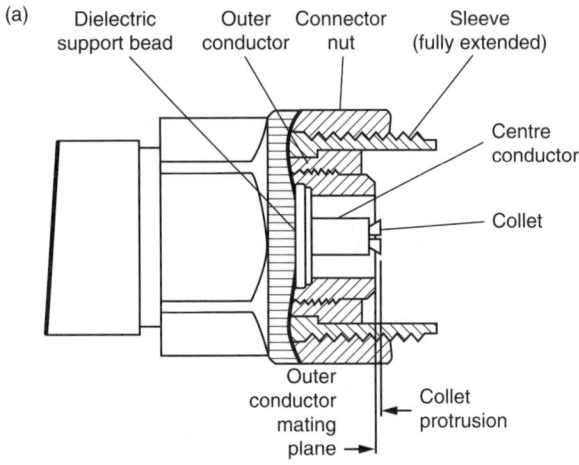

Figure 4B.2 *(a) The GPC 7 connector and (b) GPC 7 connector with the 6 slot collet [photograph A.D. Skinner]*

Connection
(1) On one connector, retract the coupling sleeve by turning the coupling nut until the sleeve and the nut become disengaged. The coupling nut can then be spun freely with no motion of the coupling sleeve.
(2) On the other connector, the coupling sleeve should be fully extended by turning the coupling nut in the appropriate direction. Once again the coupling nut can be spun freely with no motion of the coupling sleeve.
(3) Put the connectors together carefully but firmly, and thread the coupling nut of the connector with the retracted sleeve over the extended sleeve. Finally tighten using a torque spanner set to the correct torque (see Appendix D).

Disconnection

(1) Loosen the fixed coupling nut of the connector showing the wide gold band behind the coupling nut. This is the one that had the coupling sleeve fully retracted when connected.
(2) Part the connectors carefully to prevent damage to the inner conductor collet.

It is a common but bad practice with hermaphroditic connectors, to screw the second coupling ring against the first in the belief that there should be no loose parts in the coupled pair. This reduces the pressure between the two outer contacts of the connectors, leading to higher contact resistance and less reliable contact.

When connecting terminations or mismatches do not allow the body of the termination to rotate. To avoid damage, connectors with retractable sleeves (e.g. GPC7) should not be placed face down on their reference plane on work surfaces. When not in use withdraw the threaded sleeve from under the coupling nut and fit the plastic protective caps.

4.B.3 The Type N 7 mm connector

The Type N connector (Figure 4B.3a) is a rugged connector that is often used on portable equipment and military systems because of its large size and robust nature. The design of the connector makes it relatively immune to accidental damage due to misalignment during mating (subject to it being made and aligned correctly). The Type N connector is made in both 50 and 75 Ω versions and both types are in common use (Figure 4B.3).

Two different types of inner socket are at present available for Type N socket connectors. They are referred to as 'slotted' or 'slotless' sockets. The slotted Type N (Figure 4B.3c) normally has either four or six slots cut along the inner conductor axis to form the socket. This means that the diameter and, therefore, the characteristic impedance are determined by the diameter of the mating pin and they are easy to damage or distort. The development of the slotless socket (Figure 4B.3b) by Agilent has resulted in a solid inner conductor with internal contacts and is independent of the

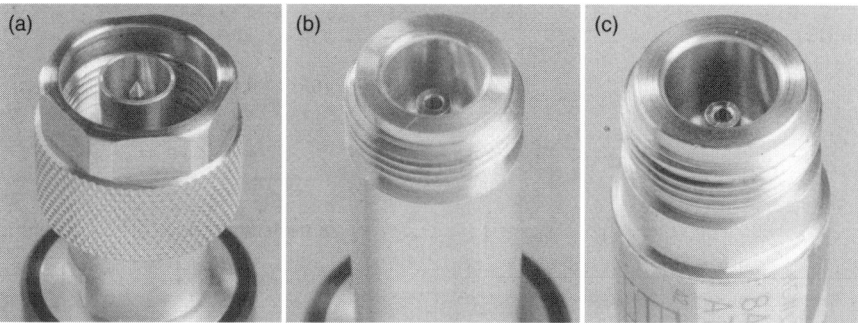

Figure 4B.3 *(a) Type N plug, (b) Type N socket slotless and (c) Type N socket slotted [photographs NPL]*

mating pin providing improved and more consistent performance. The slotted inner is normally only fitted to general-purpose versions of the Type N connector.

The reference plane for the Type N connector is the junction surface of the outer conductors. Unlike some other pin and socket connectors the junction surface of the inner connector is offset from the reference plane by 5.258 mm (0.207 inches). The offset is designed this way in order to reduce the possibility of mechanical damage due to misalignment during the connection process.

The construction and mechanical gauging requirements for the Type N connector are shown in Figure 4B.4. The offset specifications can vary and the different values shown in the table for A and B in the diagram show various values depending on the specification used. The electrical performance of a beadless airline is particularly dependent on the size of the gap at the inner connector junction due to manufacturing tolerances in both the airline and the test port connectors. Therefore for the best performance it is important to minimise the gap to significantly improve the electrical performance. It is important to note the manufacturer's specification for any particular Type N connector being used. To meet the present MIL-STD-348A requirement the minimum recession on the pin centre contact is 5.283 mm (0.208 inches).

For some applications Type N connectors need only be connected finger-tight but torque settings are given in Appendix D that should be used in metrology applications.

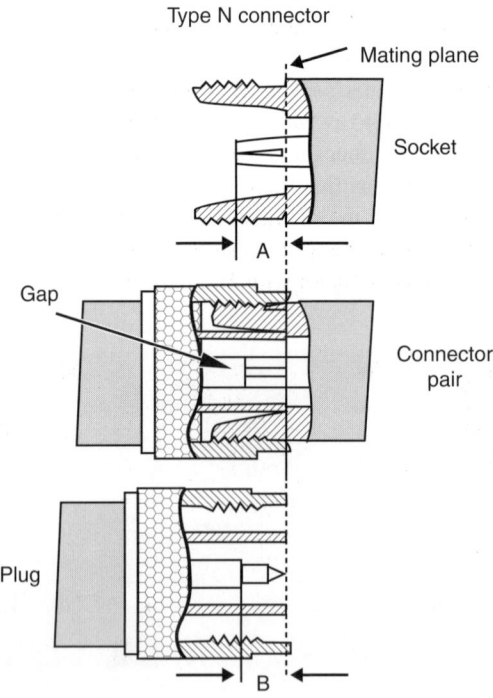

Figure 4B.4   The Type N connector

Using coaxial connectors in measurement 73

The gauging limits are listed in Table 4B.1 and apply to both 50 and 75 Ω connector types. For the convenience of users the dimensions are given in Imperial and Metric Units. The metric values are shown in brackets (Table 4B.1).

The Type N connector is designed to operate up to 18 GHz but special versions are available that can operate up to 22 GHz and also to 26.5 GHz. (Traceability of measurement is not at present available for devices fitted with 7 mm connectors above 18 GHz.)

4.B.3.1 Gauging a plug Type N connector

When gauging a plug Type N connector a clockwise deflection of the gauge pointer (a 'plus') indicates that the shoulder of the plug contact pin is recessed less than the minimum recession of 0.207 inches behind the outer conductor mating plane. This will cause damage to other connectors to which it is mated.

4.B.3.2 Gauging a socket Type N connector

When gauging a socket Type N connector a clockwise deflection of the gauge pointer (a 'plus') indicates that the tip of the socket mating fingers are protruding more than the maximum of 0.207 inches in front of the outer conductor mating plane. This will cause damage to other connectors to which it is mated.

> *Warning*
>
> **75 Ω Type N connectors.** On the 75 Ω connector the centre contact of the socket can be physically destroyed by a 50 Ω pin centre contact so that cross coupling of 50 and 75 Ω connectors is not admissible. Special adaptors can be purchased, which are commonly known as 'short transitions', to enable the connection to be made if necessary, but these transitions should be used with caution. If possible it is best to use a minimum loss attenuation pad to change the impedance to another value.

4.B.4 The 7/16 connector

This connector was developed in Germany during the 1960s for high-performance military systems and was later developed for commercial applications in analogue cellular systems and GSM base station installations.

This connector is now being widely used in the telecommunications industry and it has a frequency range covering from DC to 7.5 GHz. The '7/16' represents a nominal value of 16 mm at the interface for the internal diameter of the external conductor, and a nominal value of 7 mm for the external diameter of the internal conductor to achieve 50 Ω.

High-quality 7/16 connectors are available to be used as standards for the calibration of automatic network analysers, reflection analysers and other similar devices.

A range of push on adaptors is available to eliminate the time-consuming need for tightening, and disconnecting using a torque spanner.

Table 4B.1 Type N connector

Type N specification	Dimensions in inches (mm)		Gap between mated centre contacts		
	Socket (A)	Plug (B)	Min.	Nom.	Max.
MMC precision also HP precision	**0.000** *(0.000)* **0.207** *(5.2578)* **−0.003** *(−0.0762)*	**+0.003** *(+0.0762)* **0.207** *(5.2578)* **0.000** *(0.0000)*	**0.000** *(0.000)*	**0.000** *(0.000)*	**0.006** *(0.1524)*
MIL-STD-348A standard test	**0.000** *(0.0000)* **0.207** *(5.2578)* **−0.003** *(−0.0762)*	**+0.003** *(+0.0762)* **0.208** *(5.2832)* **0.000** *(0.0000)*	**0.001** *(0.0254)*	**0.001** *(0.0254)*	**0.007** *(0.1778)*
MIL-STD-348A Class 2 present Type N	**0.207 max** *(5.2578)*	**0.208 min** *(5.2832)*	–	**0.001** *(0.0254)*	–
MMC Type N equivalent to MIL-C-71B	**+0.005** *(+0.1270)* **0.197** *(5.0038)* **−0.005** *(−0.1270)*	**+0.005** *(+0.1270)* **0.223** *(5.6642)* **−0.005** *(−0.1270)*	**0.016** *(0.4064)*	**0.026** *(0.6604)*	**0.036** *(0.9144)*
MIL-C-71B old Type N	**+0.010** *(+0.254)* **0.197** *(5.0038)* **−0.010** *(−0.2540)*	**+0.010** *(+0.2540)* **0.223** *(5.6642)* **−0.010** *(−0.2540)*	**0.006** *(0.1524)*	**0.026** *(0.6604)*	**0.046** *(1.1684)*

It is a repeatable long-life connector with a low return loss. It also has a good specification for inter-modulation performance and a high-power handling capability.

Terminations, mismatches, open and short circuits are also made and back-to-back adaptors, such as plug-to-plug, socket-to-plug and socket-to-socket are also available.

They are designed to a DIN specification number 47223 (Figure 4B.5). For further information on the 7/16 connector the article by Paynter and Smith is recommended. This article describes the 7/16 connector and discusses whether to use Type N connector technology or to replace it with the 7/16 DIN interface for use in mobile radio GSM base stations.

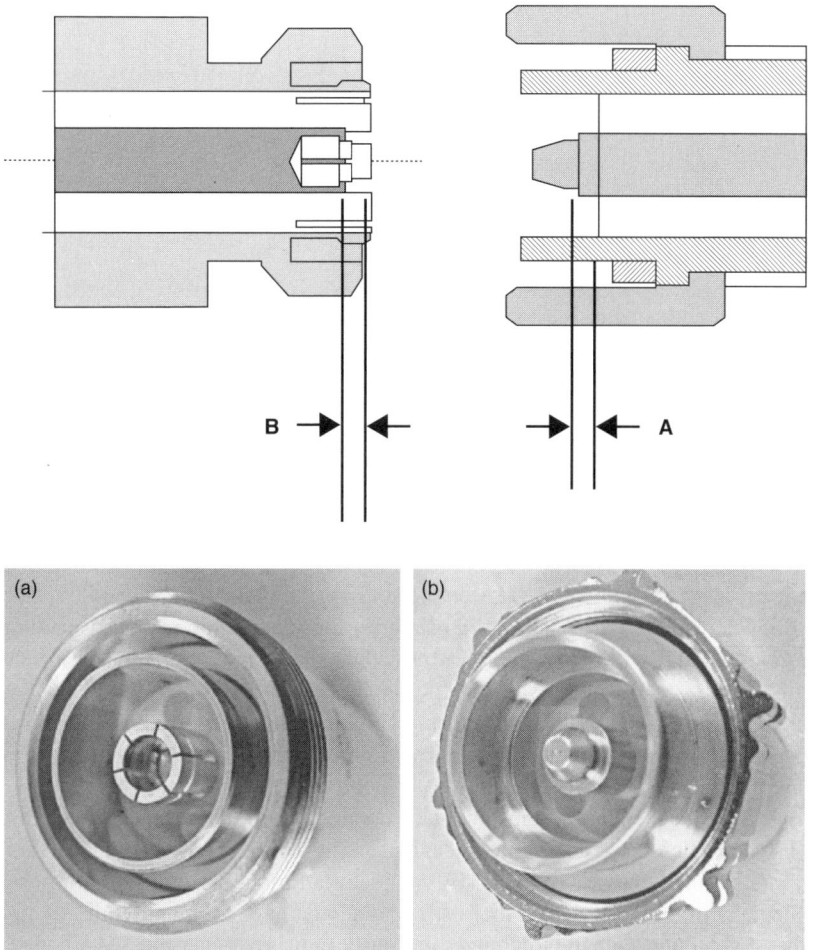

Figure 4B.5 (a) The 7/16 socket connector and (b) the 7/16 plug connector [photographs A.D. Skinner]

Table 4B.2 7/16 connector

7-16	Dimensions							
	A				B			
	Plug (Inches)		Plug (mm)		Socket (Inches)		Socket (mm)	
Specification	min.	max.	min.	max.	min.	max.	min.	max.
General purpose	0.0579	0.0697	1.47	1.77	0.697	0.815	1.77	2.07
Reference/test	0.0681	0.0689	1.73	1.75	0.0705	0.0713	1.79	1.81

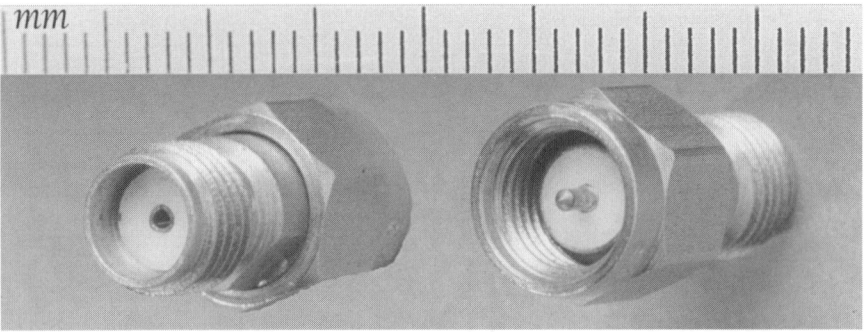

Figure 4B.6 The SMA connector [photograph NPL]

The mechanical gauging requirements for the 7/16 connector are shown in Table 4B.2.

4.B.5 The SMA connector

The interface dimensions for SMA connectors are listed in MIL-STD 348A.

BS 9210 N0006 Part 2 published primarily for manufacturers and inspectorates, also gives details for the SMA interface but some of the requirements and specification details differ. For example, wall thickness may be a little thinner and hence a little weaker. However, MIL-STD-348A does not preclude thin walls in connectors meeting this specification although the physical requirements and arrangements will probably ensure that thicker walls are used for both specifications (Figure 4B.6).

The SMA connector is a semi-precision connector and should be carefully gauged and inspected before use as the tolerances and quality can vary between manufacturers. The user should be aware of the SMA connector's limitations and look for possible problems with the solid plastic dielectric and any damage to the plug pin. In a good quality SMA connector the tolerances are fairly tight. However the SMA connector is not designed for repeated connections and they can wear out quickly, be out of specification, and potentially destructive to other connectors. The SMA connector is widely used in many applications as it is a very cost-effective connector and suitable for many purposes; however, precision metrology is not normally possible using SMA connectors.

Connector users are advised that manufacturers' specifications vary in the value of coupling torque needed to make a good connection. Unsatisfactory performance with hand tightening can indicate damage or dirty connector interfaces. It is common, but bad practice to use ordinary spanners to tighten SMA connectors. However, excessive tightening (>15 lb in) can easily cause collapse of the tubular portion of the pin connector.

Destructive interference may result if the contacts protrude beyond the outer conductor mating planes; this may cause buckling of the socket contact fingers or damage to associated equipment during mating.

The dielectric interface is also critical since protrusion beyond the outer conductor mating plane may prevent proper electrical contact, whereas an excessively recessed condition can introduce unwanted reflections in a mated pair.

The critical axial interface of SMA type connectors is shown in Figures 4B.7a and b and Table 4B.3 where the dimensions are given in inches, with the equivalent in millimetres shown in parentheses.

The specification allows dielectric to protrude past the outer conductor mating plane to 0.002 inches (0.0508 mm) maximum. However, there is some doubt if the SMA standards permit the dielectric to protrude beyond the reference plane. There is a high voltage version that does allow the dielectric to protrude beyond the reference plane, but it does not claim to be compatible with the SMA standard.

4.B.6 The 3.5 mm connector

This connector is physically compatible with the SMA connector and is often known as the GPC 3.5 mm connector. It has an air dielectric interface and closely controlled centre conductor support bead providing mechanical interface tolerances similar to hermaphroditic connectors. However, although in some ways planar, it is not an IEEE 287 precision connector. There is a discontinuity capacitance when coupled with SMA connectors.

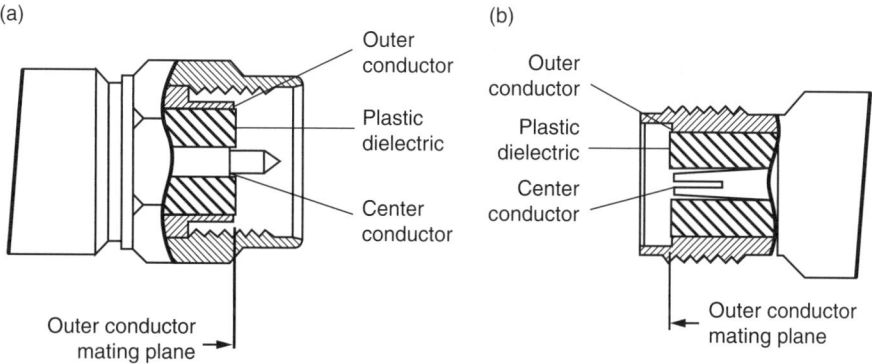

Figure 4B.7   *(a) SMA plug connector and (b) SMA socket connector*

Table 4B.3 SMA connector

SMA specification	Pin depth in inches (mm)			
	Socket pin	Socket dielectric	Plug pin	Plug dielectric
MIL-STD-348A Class 2	+0.030 *(0.7620)* 0.000 *(0.000)* 0.000 *(0.000)*	−0.002 max	0.000 *(0.000)* min	−0.002 max
MMC Standard	+0.005 *(0.1270)* 0.000 *(0.000)* 0.000 *(0.000)*	+0.002 *(0.0508)* 0.000 *(0.000)* −0.002 *(−0.0508)*	+0.005 *(0.1270)* 0.000 *(0.000)* 0.000 *(0.000)*	+0.002 *(0.0508)* 0.000 *(0.000)* −0.002 *(−0.0508)*
MMC precision	+0.005 *(0.1270)* 0.000 *(0.000)* 0.000 *(0.000)*	+0.002 *(−0.0508)* 0.000 *(0.000)* 0.000 *(0.000)*	+0.005 *(0.1270)* 0.000 *(0.000)* 0.000 *(0.000)*	+0.002 *(0.0508)* 0.000 *(0.000)* 0.000 *(0.000)*
MIL-STD-348A standard test	+0.003 *(0.0762)* 0.000 *(0.000)* 0.000 *(0.000)*	+0.002 *(0.0508)* 0.000 *(0.000)* 0.000 *(0.000)*	+0.003 *(0.0762)* 0.000 *(0.000)* 0.000 *(0.000)*	+0.002 *(0.0508)* 0.000 *(0.000)* 0.000 *(0.000)*

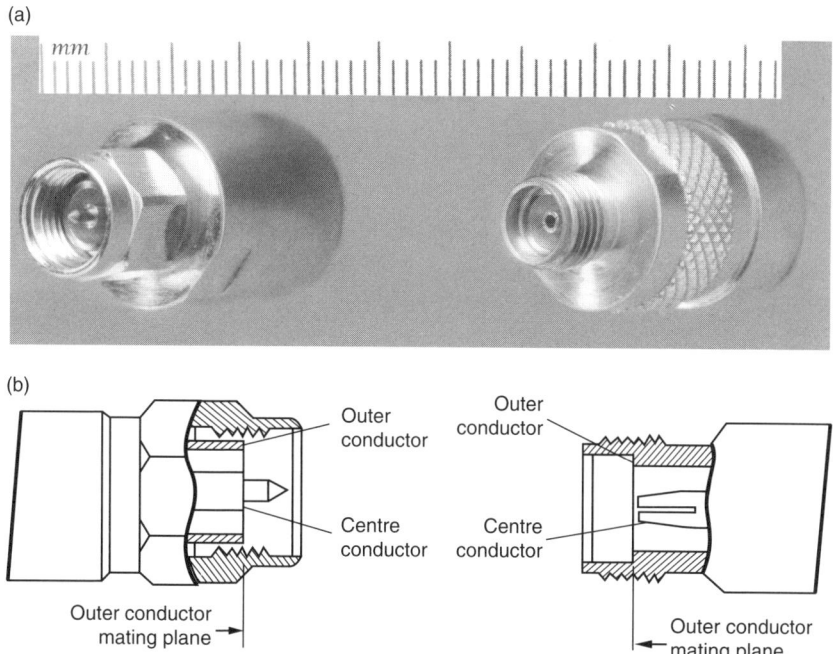

Figure 4B.8 (a) GPC 3.5 mm plug connector and (b) GPC 3.5 mm socket connector [photograph NPL]

Table 4B.4 The 3.5 mm connector

3.5 mm Specification	Pin depth in inches (mm)	
	Socket	Plug
LPC	0 to +0.0005 (+0.0127)	0 to +0.0005 (+0.0127)
GPC	0 to +0.002 (+0.0508)	0 to +0.002 (+0.0508)

Note: A plus (+) tolerance indicates a recessed condition below the outer mating plane.

A special version of the GPC 3.5 mm connector has been designed. The design incorporates a shortened plug pin and allows the centre conductors to be pre-aligned before contact thus considerably reducing the likelihood of damage when connecting or disconnecting the 3.5 mm connector. Figure 4B.8 shows the plug and socket types of GPC 3.5 mm connector and Table 4B.4 shows the gauging dimensions.

4.B.7 The 2.92 mm connector

The 2.92 mm connector is a reliable connector that operates up to 46 GHz and it is used in measurement systems and on high-performance components, calibration and

80 *Microwave measurements*

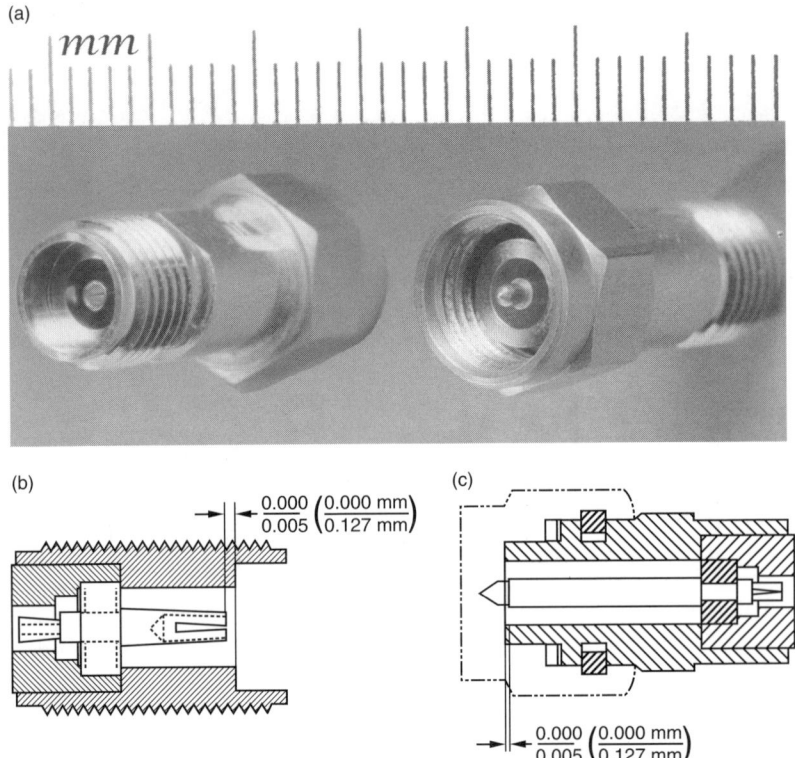

Figure 4B.9 *(a) Type K plug and socket connector [photograph NPL], (b) Type K socket connector and (c) Type K plug connector*

verification standards. It is also known as the Type K™ connector. The K connector interfaces mechanically with 3.5 mm and SMA connectors. However, when mated with the 3.5 mm or SMA connector the junction creates a discontinuity that must be accounted for in use.

Compared with the 3.5 mm and the SMA connector the 2.92 mm connector has a shorter pin that allows the outer conductor alignment before the pin encounters the socket contact when mating a connector pair. The type K connector is therefore less prone to damage in industrial use.

Figure 4B.9 shows the diagram of the Type K connector and Table 4B.5 gives the important gauging dimensions.

4.B.8 The 2.4 mm connector

The 2.4 mm connector was designed by the Hewlett Packard Company (now Agilent Technologies) and the connector assures mode free operation up to 60 GHz. It is also known as the Type Q connector. The 2.4 mm connector is a pin and socket type

Table 4B.5   Type K connector

2.92 mm Specification	Pin depth in inches (mm)	
	Socket	Plug
LPC	0 to +0.0005 (+0.0127)	0 to +0.0005 (+0.0127)
GPC	0 to +0.002 (+0.0508)	0 to +0.002 (+0.0508)

Table 4B.6   Type 2.4 mm connector

2.4 mm Specification	Pin depth in inches (mm)	
	Socket	Plug
LPC	0 to +0.0005 (+0.0127)	0 to +0.0005 (+0.0127)
GPC	0 to +0.002 (+0.0508)	0 to +0.002 (+0.0508)

connector that utilises an air dielectric filled interface. The 2.4 mm interface is also mechanically compatible with the 1.85 mm connector.

Note: The manufacturers of small coaxial connectors have agreed to the mechanical dimensions so that they can be mated non-destructively. This has led to the use of the term 'mechanically compatible' and because both lines are nominally 50 Ω it has been assumed that 'mechanically compatible' equates to electrical compatibility. The effect of the electrical compatibility of mechanically mateable coaxial lines is discussed in ANAlyse Note No. 3 January 1994 included in the list of further reading at the end of this chapter.

As for other connectors of this type, the coupling engagement of the outer conductors is designed to ensure that the outer conductors are coupled together before the inner conductors can engage to prevent damage to the inner conductor. Figure 4B.10 shows the diagram of the 2.4 mm connector and Table 4B.6 gives the important gauging dimensions.

4.B.9   The 1.85 mm connector

The 1.85 mm connector was designed by Anritsu and the connector assures mode-free operation up to 75 GHz. It is also known as the Type V™ connector. The 1.85 mm connector is a pin and socket type connector that uses an air dielectric filled interface.

The coupling engagement of the outer conductors is designed to ensure that the outer conductors are coupled before the inner conductors can engage to ensure a damage-free fit. Figure 4B.11 shows the diagram of the 1.85 mm connector and Table 4B.7 shows the important gauging dimensions.

82 Microwave measurements

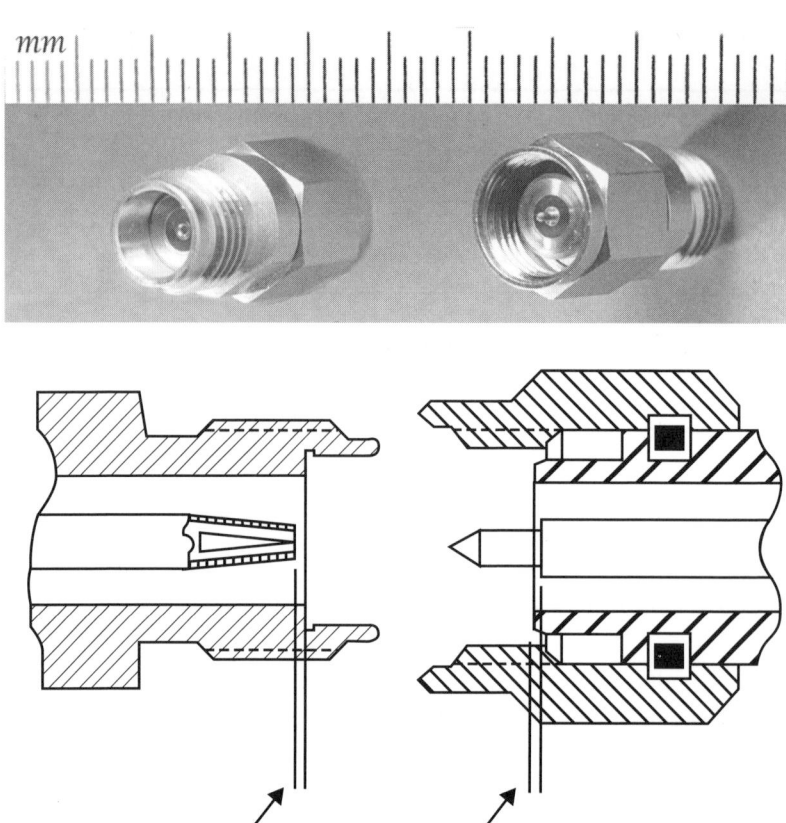

0.000 (0.000) to + 0.002 (+ 0.0508)

Figure 4B.10 The 2.4 mm socket and plug connector [photograph NPL]

Table 4B.7 Type 1.85 mm connector

1.85 mm Specification	Dimensions in inches (mm)	
	Socket	Plug
LPC	0 to +0.0005 (+0.0127)	0 to +0.0005 (+0.0127)
GPC	0 to +0.002 (+0.0508)	0 to +0.002 (+0.0508)

Note: A plus (+) tolerance indicates a recessed condition below the outer mating plane.

4.B.10 The 1.0 mm connector

The 1.0 mm connector was designed by Hewlett Packard (now Agilent Technologies) and is described in IEEE 287 Standard. No patent applications were filed to protect the design of the 1.0 mm connector as it is intended by Agilent to allow free use of

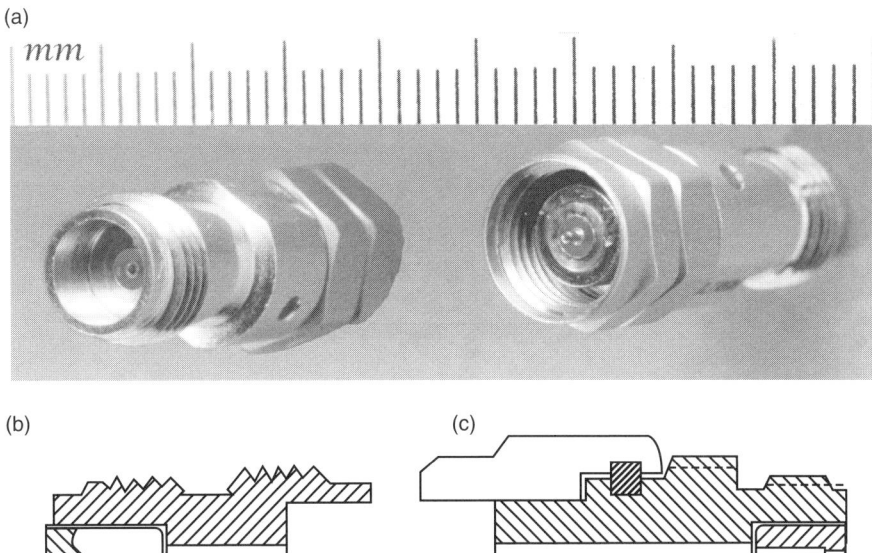

Figure 4B.11 *(a) The 1.85 mm socket and plug connector [photograph NPL], (b) Type V socket connector and (c) Type V plug connector*

the interface by everyone. Any manufacturer of connectors is free to manufacture its own version of the 1.0 mm connector.

The 1.0 mm connector is also known as the Type W Connector. It is a pin and socket type connector that utilises an air dielectric filled interface and assures mode-free operation up to 110 GHz.

The coupling diameter and thread size are chosen to maximise strength and increase durability. The coupling engagement of the outer conductors is designed to ensure that they are coupled together before the inner conductors can engage, toe ensure a damage-free fit.

Figures 4B.12 and 4B.13 show two versions of the 1.0 mm connector available from Agilent Technologies and the Anritsu Company. They are based on the dimensions shown in IEEE 287.

Figure 4B.14 is a diagram of Anritsu's W connector; a 1.0 mm connector based on the dimensions in the IEEE 287 Standard.

Table 4B.8 shows the important gauging dimensions for a 1.0 mm connector.

84 *Microwave measurements*

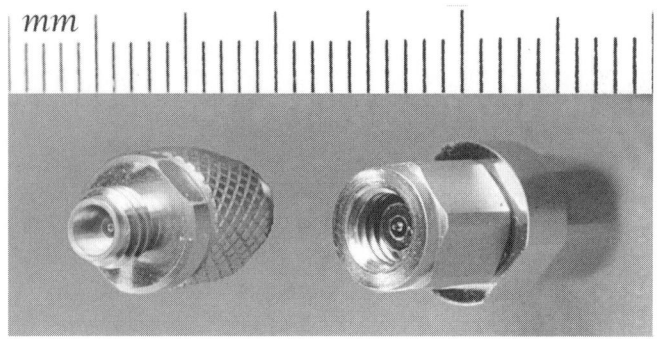

Figure 4B.12   The Agilent 1.0 mm socket and plug connector [photograph NPL]

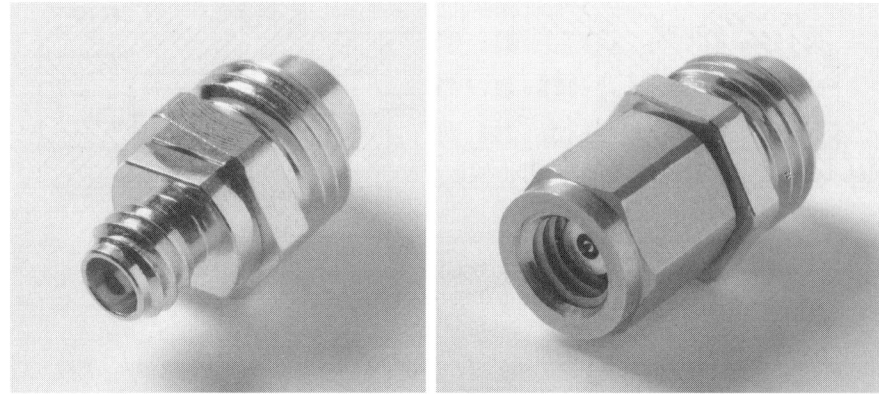

Figure 4B.13   The Agilent 1.0 mm socket and plug connector [photograph Anritsu]

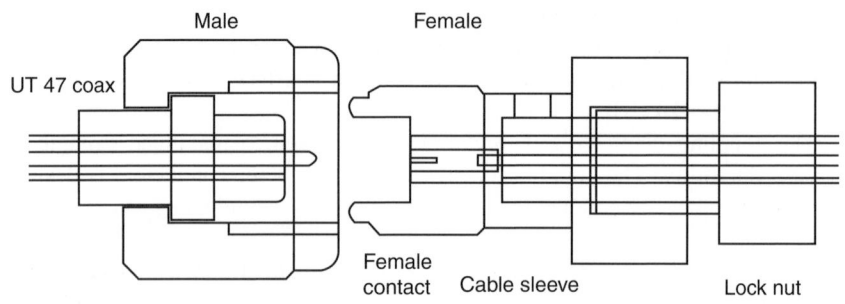

Figure 4B.14   Diagram of the plug and socket arrangement

Table 4B.8   Maximum pin depth for a 1.0 mm connector

1.0 mm Specification	Pin depth in inches (*mm*)	
	Socket	Plug
LPC	0 to +0.0005 (+*0.0127*)	0 to +0.0005 (+*0.0127*)
GPC	0 to +0.002 (+*0.0508*)	0 to +0.002 (+*0.0508*)

Note: A plus (+) tolerance indicates a recessed condition below the outer conductor mating plate.

4.C Appendix C

4.C.1 Repeatability of connector pair insertion loss

The values shown in Table 4C.1 show some insertion loss repeatability (dB) figures provided that the connector-pairs are in good mechanical condition and clean; further, that in use, they are not subjected to stress and strain due to misalignment or transverse loads. For any particular measurement process the connector repeatability in the uncertainty budget is calculated in the same units of the final measurements. For example, when measuring the calibration factor of a power sensor the repeatability is measured in per cent.

These guidance figures will serve two purposes:

(1) They show limits for connector repeatability for normal use in uncertainty estimates where unknown connectors may be involved.
(2) Provide a measure against which a 'real' repeatability assessment can be judged.

The figures in Table 4C.1 are based on practical measurement experience at NPL, SESC, and in UKAS Calibration Laboratories. In practice, connector repeatability is an important contribution to measurement uncertainties and should be carefully determined when verifying measurement systems or calculated for each set of measurements made. In some cases values better than those shown in Table 4C.1 can be obtained.

Table 4C.1   Typical connector insertion loss repeatability

Connector	Connector Insertion loss repeatability dB		
GR900 – 14 mm	0.001 (DC to 0.5 GHz)	0.002 (0.5–8.5 GHz)	
GPC7 – 7 mm	0.001 (DC to 2 GHz)	0.004 (2–8 GHz)	0.006 (8–18 GHz)
Type N – 7 mm	0.001 (DC to 1 GHz)	0.004 (1–12 GHz)	0.008 (12–18 GHz)
GPC3.5 – 3.5 mm	0.002 (DC to 1 GHz)	0.006 (1–12 GHz)	
SMA 3.5 mm	0.002 (DC to 1 GHz)	0.006 (1–12 GHz)	

4.D Appendix D

4.D.1 Torque wrench setting values for coaxial connectors

Table 4D.1 gives a list of recommended connector tightening torque values to be used for metrology purposes for each connector type.

This list is based on the best available information from various sources and should be used with care. Some manufacturers recommend slightly different values for the torque settings in their published performance data. Where this is the case the manufacturers' data should be used. With all torque spanners, it is possible to get substantially the wrong torque by twisting the handle axially and by a variety of other incorrect methods of using the torque spanner.

There are also some differences on the torque settings used when making a permanent connection (within an instrument) rather than for metrology purposes. Many manufacturers quote a maximum coupling torque which if exceeded will result in permanent mechanical damage to the connector.

For combinations of GPC3.5/SMA connectors the torque should be set to the lower value, for example, 5 in-lb. Torque spanners used should be regularly calibrated, and set to the correct torque settings for the connector in use and clearly marked.

On some torque spanners, the handles are colour coded to represent the torque value set for ease of identification, for example, 12 in-lb (1.36 N-m) blue and 8 in-lb (0.90 N-m) red.

However, for safety, always check the torque setting before use especially if it is a spanner not owned by or normally used every day in the laboratory (Table 4D.1).

Table 4D.1   Torque spanner setting values

Connector		Torque	
Type	Size (mm)	in-lb	N-m
GR900	14	12	1.36
GPC 7	7	12	1.36
N	7	12	1.36
7/16	16.5	20	2.26
GPC 3.5	3.5	8	0.90
SMA	3.5	5	0.56
K	2.92	5–8	0.56–0.90
Q	2.4	8	0.90
V	1.85	8	0.90
W1	1.1	4	0.45
W	1.0	3	0.34

4.E Appendix E

4.E.1 Calibrating dial gauges and test pieces

There are a number of different types of dial gauge and gauge calibration block used for gauging connectors. They require regular calibration to ensure that they are performing correctly. There is a British Standard BS 907: 'Specification for dial gauges for linear measurement' dated 1965 that covers the procedure for the calibration of dial gauges and this should be used. However, the calibration of the gauge calibration blocks is not covered by a British Standard, but they can be measured in a mechanical metrology laboratory. It is important to use the correct gauge for each connector type to avoid damage to the connector under test. Some gauges have very strong gauge plunger springs that, if used on the wrong connector, can push the centre block through the connector resulting in damage. Also if gauges are used incorrectly they can compress the centre conductor collet in precision GPC 7 mm connectors, during a measurement, resulting in inaccurate readings when measuring the collet protrusion.

4.E.2 Types of dial gauge

Dial gauges used for the testing of connectors for correct mechanical compliance are basically of two types:

4.E.2.1 Push on type

The push on type is used for measuring the general-purpose type of connector. For plug and socket connectors two gauges are normally used (one plug and one socket) or a single gauge with plug and socket adaptor bushings.

4.E.2.2 Screw on type

The screw on type is mainly used (except GR 900) in calibration kits for network analysers and reflectometers. They are used for the GPC7 and sexed connectors and for the latter they are made in both plug and socket versions. The screw on type is made in the form of a connector of the opposite sex to the one being measured.

When a gauge block is used to initially calibrate the dial gauge, a torque spanner should be used to tighten up the connection to the correct torque.

4.E.3 Connector gauge measurement resolution

Because of connector gauge measurement resolution uncertainties (one small division on the dial) and variations in measurement technique from user to user connector dimensions may be difficult to measure. Dirt and contamination can cause differences of 0.0001 inch (0.00254 mm) and in addition the way that the gauge is used can result

88 Microwave measurements

in larger variations. When using a gauge system for mechanical compliance testing of connectors carry out the following procedures each time:

(1) carefully inspect the connector to be tested and clean if necessary;
(2) clean and inspect the dial gauge, and the gauge calibration block;
(3) carefully zero the dial gauge with the gauge calibration block in place;
(4) remove the gauge calibration block;
(5) measure the connector using the dial gauge and note the reading and
(6) repeat the process at least once or more times as necessary.

4.E.4 Gauge calibration blocks

Every connector gauge requires a gauge calibration block that is used to zero the gauge to a pre-set value before use.

The diagram in Figure 4E.1 shows a set of dial gauges and gauge calibration blocks for a Type N connector screw on type gauge.

The diagram in Figure 4E.2 shows an SMA dial gauge of the push on type with its gauge calibration block.

There are a number of different types and manufacturers of connector gauge kits in general use and the manufacturer's specification and calibration instructions should be used.

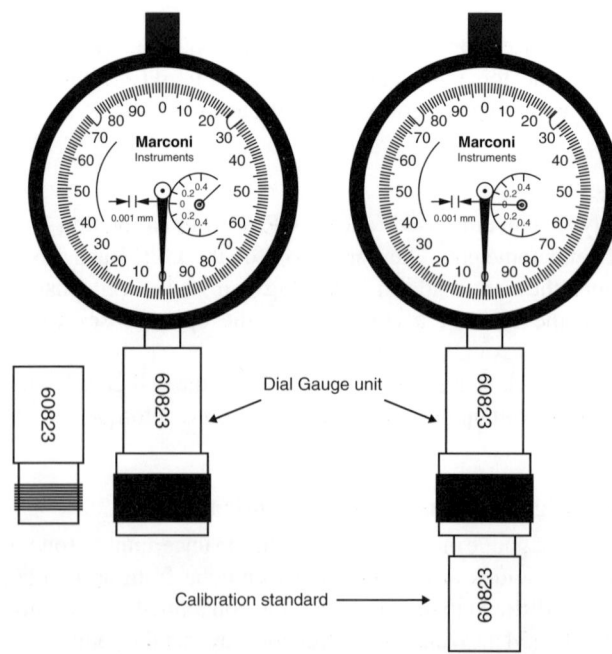

Figure 4E.1 Type N screw on dial gauge and calibration block

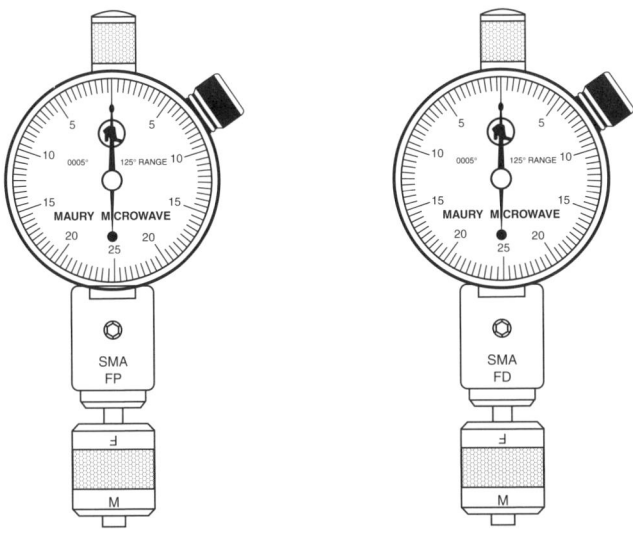

Figure 4E.2   Type SMA push – on type dial gauge for socket pin depth (FP) and dielectric FD with calibration block

Further reading

Uncertainties of measurement

UKAS: *The expression of uncertainty and confidence in measurement*, M3003, 2nd edn (HMSO, London)

ANAMET Reports[1]

Ridler, N. M., and Medley, J. C.: *ANAMET-962: dial gauge comparison exercise*, ANAMET Report, no. 001, Jul 1996

Ridler, N. M., and Medley, J. C.: ANAMET-963 *live dial gauge comparison exercise*: ANAMET Report, no. 007, May 1997

Ridler, N. M., and Graham, C.: *An investigation into the variation of torque values obtained using coaxial connector torque spanners*, ANAMET Report, no. 018, Sep 1998.

French, G. J.: *ANAMET-982: live torque comparison exercise*, ANAMET Report, no. 022, Feb 1999

Ridler, N. M., and Morgan, A. G.: *ANAMET-032: 'live' dial gauge measurement investigation using Type-N connectors*, ANAMET Report, no. 041, Nov 2003

[1] **Other reports which are fore-runners of the 2nd edition of the ANAMET Connector Guide**

Skinner, A. D.: *Guidance on using coaxial connectors in measurement – draft for comment*, ANAMET Report, no. 015, Feb 1998

Skinner, A. D.: *ANAMET connector guide*, ANAMET Report, no. 032, Jan 2001, Revised Mar 2006

ANAlyse notes

Ide, J. P. L.: *A study of the electrical compatibility of mechanically mateable coaxial lines*, ANAlyse Note, no. 3, Jan 1994

Ridler, N. M.: *How much variation should we expect from coaxial connector dial gauge measurements?*, ANAlyse Note, no. 14, Feb 1996

ANA tips notes

Smith, A. J. A., and Ridler, N. M.: *Gauge compatibility for the smaller coaxial line sizes*, ANA-tips Note, no. 1, Oct 1999

Woolliams, P. D. and Ridler, N. M.: *Tips on using coaxial connector torque spanners*, ANA-tips Note, no. 2, Jan 2000

ANAMET news articles

[The first three items in this list are short, amusing, articles (albeit containing important information).]

Ide, J. P.: 'Are two collets better than one?', *ANAMET News*, Issue 2, Spring 1994, p. 3

Ide, J. P.: 'Masters of the microverse', *ANAMET News*, Issue 2, Spring 1994, p. 2

Ide, J. P.: 'More from the gotcha! files: out of my depth', *ANAMET News*, Issue 9, Autumn 1997, p. 9

Instone, I.: 'The effects of port recession on ANA accuracy', *ANAMET News*, Issue 11, Autumn 1998, pp. 4–6

For further publications on connectors

Ridler, N. M.: 'Connectors, air lines and RF impedance', notes to accompany the IEE training course on Microwave Measurements, Milton Keynes, UK, 13–17 May 2002

Connector repeatability

Bergfield, D., and Fischer, H.: 'Insertion loss repeatability versus life of some coaxial connectors'. *IEEE Transactions on Instrumentation and Measurement* Nov 1970, vol. Im-19, no. 4, pp. 349–53

Type 7/16 coaxial connector

Paynter, J. D., and Smith, R.: 'Coaxial connectors: 7/16 DIN and Type N', *Mobile Radio Technology*, April 1995 (Intertec Publishing Corp.)

Ridler, N. M.: 'Traceability to National Standards for S-parameter measurements of devices fitted with precision 1.85 mm coaxial connectors', presented at 68th ARFTG Conference, Broomfield, Colorado, Dec 2006

Chapter 5
Attenuation measurement
Alan Coster

5.1 Introduction

Accurate attenuation measurement is an important part of characterising radio frequency (RF) or microwave circuits and devices. For example, attenuation measurement of the component parts of a radar system will enable a designer to calculate the power delivered to the antenna from the transmitter, the noise figure of the receiver and hence the fidelity or bit error rate of the system. A precision power measurement system, such as the calorimeter described by Oldfield [1], requires the transmission line preceding the measurement element to be characterised to determine the effective efficiency of the system. The thermal electrical noise standard described by Sinclair [2] requires accurate attenuation measurement of the transition or thermal block between the hot termination and the ambient temperature output connector to determine its excess noise ratio.

5.2 Basic principles

With reference to Figure 5.1, when a generator with a reflection coefficient Γ_G is connected directly to a load of reflection coefficient Γ_L, let the power dissipated in the load be denoted by P_1. Now if a two-port network is connected between the same generator and load, let the power dissipated in the load be reduced to P_2.

Insertion loss in decibels of this two-port network is defined as follows:

$$L(\text{dB}) = 10 \log_{10} \frac{P_1}{P_2} \tag{5.1}$$

Attenuation is defined as the insertion loss where the reflection coefficients Γ_G and $\Gamma_L = 0$.

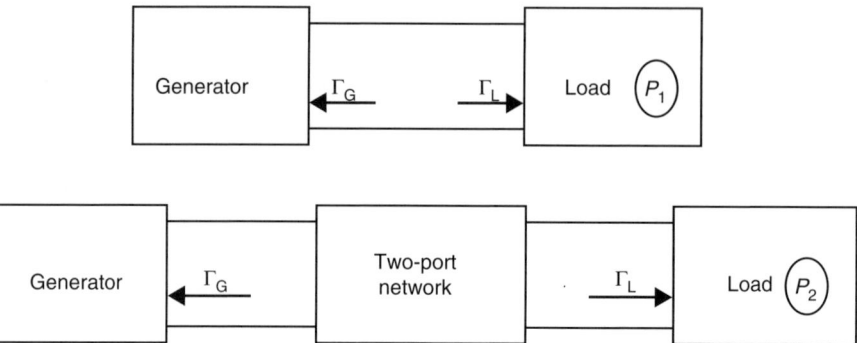

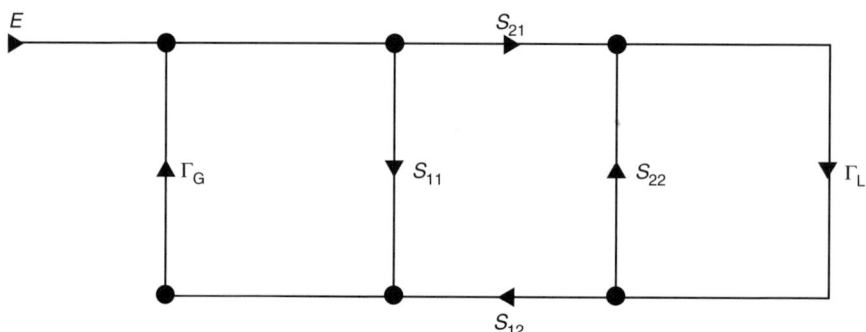

Figure 5.1 Insertion loss

Figure 5.2 Signal flow

Note that insertion loss depends on the value of Γ_G and Γ_L, whereas attenuation depends only on the two-port network. If the source and load are not perfectly matched in an attenuation measurement, there will be an error associated with the result. This error is called the 'mismatch error' and it is defined as the difference between the insertion loss and attenuation. Hence

$$\text{Mismatch error } (M) = L - A \tag{5.2}$$

Figure 5.2 shows a signal flow diagram of a two-port network between a generator and load where S_{11} is the voltage reflection coefficient looking into the input port when the output port is perfectly matched. S_{22} is the voltage reflection coefficient looking into the output port when the input port is perfectly matched. S_{21} is the ratio of the complex wave amplitude emerging from the output port to that incident upon the input port when the output port is perfectly matched. S_{12} is the ratio of the complex wave amplitude emerging from the input port to that incident upon the output port when the input port is perfectly matched.

From the above definitions, the equation from Warner [3] for insertion loss is given as follows:

$$L = 20 \log_{10} \frac{|(1 - \Gamma_G S_{11})(1 - \Gamma_L S_{22}) - \Gamma_L \Gamma_G S_{12} S_{21}|}{|S_{21}| \cdot |1 - \Gamma_G \Gamma_L|} \quad (5.3)$$

It can now be seen that the insertion loss is dependent upon Γ_G and Γ_L as well as the S-parameters of the two-port network. When Γ_G and Γ_L are matched (Γ_G and $\Gamma_L = 0$), then (5.3) is simplified to

$$A = 20 \log_{10} \frac{1}{|S_{21}|} \quad (5.4)$$

where A represents attenuation (dB).

From (5.2), the mismatch error M is the difference between (5.3) and (5.4).

$$M = 20 \log_{10} \frac{|(1 - \Gamma_G S_{11})(1 - \Gamma_L S_{22}) - \Gamma_G \Gamma_L S_{12} S_{21}|}{|1 - \Gamma_G \Gamma_L|} \quad (5.5)$$

Note that all the independent variables of (5.5) are complex. In practice, it may be difficult to measure the phase relationships where the magnitudes are small. Where this is the case, and only magnitudes are known, the mismatch uncertainty is given as a maximum and minimum limit. Thus

$$M(\text{limit}) = 20 \log_{10} \frac{1 \pm (|\Gamma_G S_{11}| + |\Gamma_L S_{22}| + |\Gamma_G \Gamma_L S_{11} S_{22}| + |\Gamma_G \Gamma_L S_{12} S_{21}|)}{1 \mp |\Gamma_G \Gamma_L|} \quad (5.6)$$

If the two-port network is a variable attenuator, then the mismatch limit is expanded from (5.6) to give

$$M(\text{limit}) = 20 \log_{10}$$
$$\times \frac{1 \pm (|\Gamma_G S_{11e}| + |\Gamma_L S_{22e}| + |\Gamma_G \Gamma_L S_{11e} S_{22e}| + |\Gamma_G \Gamma_L S_{12e} S_{21e}|)}{1 \mp (|\Gamma_G S_{11b}| + |\Gamma_L S_{22b}| + |\Gamma_G \Gamma_L S_{11b} S_{22b}| + |\Gamma_G \Gamma_L S_{12b} S_{21b}|)} \quad (5.7)$$

where suffix b denotes the attenuator at zero or datum position (residual attenuation) and suffix e denotes the attenuator incremented to another setting (incremental attenuation).

5.3 Measurement systems

Many different and ingenious ways of measuring attenuation have been developed over the years, and most methods in use today embody the following principles:

(1) Power ratio
(2) Voltage ratio
(3) AF substitution
(4) IF substitution
(5) RF substitution

94   Microwave measurements

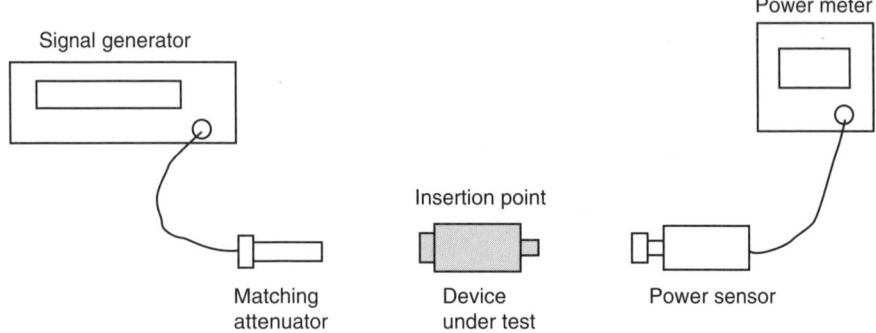

Figure 5.3   Power ratio

5.3.1 Power ratio method

The power ratio method of measuring attenuation is perhaps one of the easiest to configure. Figure 5.3 represents a simple power ratio configuration. First, the power sensor is connected directly to the matching attenuator and the power meter indication noted P_1. Next, the device under test is inserted between the matching pad and power sensor and the power meter indication again noted P_2. Insertion loss is then calculated using

$$L(\text{dB}) = 10 \log_{10} \frac{P_1}{P_2} \tag{5.8}$$

Note that unless the reflection coefficient of the generator and load at the insertion point is known to be zero, or that the mismatch factor has been calculated and taken into consideration, measured insertion loss and not attenuation is quoted.

This simple method has some limitations:

(1) Amplitude stability and drift of the signal generator
(2) Power linearity of the power sensor
(3) Zero carry over
(4) Range switching and resolution

Amplitude drift of the signal generator.   Measurement accuracy is directly proportional to the signal generator output amplitude drift.

Power linearity of the power sensor.   The modern semiconductor thermocouple power sensor embodies a tantalum nitride film resistor, shaped so that it is thin in the centre and thick at the outside, such that when RF power is absorbed, there is a temperature gradient giving rise to a thermoelectric emf. The RF match due to the deposited resistor is extremely good and the sensor will operate over a 50 dB range (+20 dBm to −30 dBm) but there is considerable departure from linearity (10 per cent at 100 mW), and it is necessary to compensate for this and the temperature dependence of the sensitivity by electronics.

Attenuation measurement 95

Diode power sensors, described by Cherry *et al.* [4] may be modelled by the following equation

$$P_{in} = kV_{dc} \exp(yV_{dc}) \tag{5.9}$$

where P_{in} is the incident power, V_{dc} is the rectified dc voltage, and k and y are constants which are functions of parameters such as temperature, ideality factor and video impedance.

At levels below 1 μW (-30 dBm), the exponent tends to zero, leading to a linear relationship between diode output voltage and input power (dc output voltage proportional to the square of the rms RF input voltage). For power levels above 10 μW (-20 dBm), correction for linearity must be made. Modern (smart) diode power sensors embody an RF attenuator preceding the sensing element. The attenuator is electronically switched in or out to maintain best sensor linearity over a wide input level range. These sensors have a claimed dynamic range of 90 dB ($+20$ dBm to -70 dBm) and may be corrected for power linearity, frequency response and temperature coefficient by using the manufacturer's calibration data stored within the sensor e2prom.

Experiments by Orford and Abbot [5] show that power meters based on the thermistor mount and self compensating bridge are extremely linear. Here the thermistor forms one arm of a Wheatstone bridge which is powered by dc current, heating the thermistor until its resistance is such that the bridge balances. The RF power changes the thermistor resistance but the bridge is automatically rebalanced by reducing the applied dc current. This reduction in dc power to the bridge is called retracted power and is directly proportional to the RF power absorbed in the thermistor. The advantage of this system is that the thermistor impedance is maintained constant as the RF power changes. Thermistors, however, have a slower response time than thermocouple and diode power sensors and have a useful dynamic range of only 30 dB.

Zero carry over. Due to the electronic circuits, small errors may occur when a power meter is zeroed on one range and then used on another.

Range switching and resolution. Most power meters operate over several ranges, each of approximately 5 dB. With a near full scale reading, the resolution and noise of the power meter indication will be good. However, the resolution and noise contribution at a low scale reading may be an order worse (this should be borne in mind when making measurements and all effort made to ensure that the power meter is at near full scale for at least one of the two power measurements P_1, P_2).

Figure 5.4 shows a dual channel power ratio attenuation measurement system, which uses a two-resistor power splitter to improve the source match and monitor the source output level. First, power sensor A is connected directly to the two-resistor splitter and the power meter indication is noted as PA_1, PB_1.

Next the device under test is inserted between power sensor A and the two-resistor splitter and again the power meter indication is noted as PA_2, PB_2. Insertion loss may

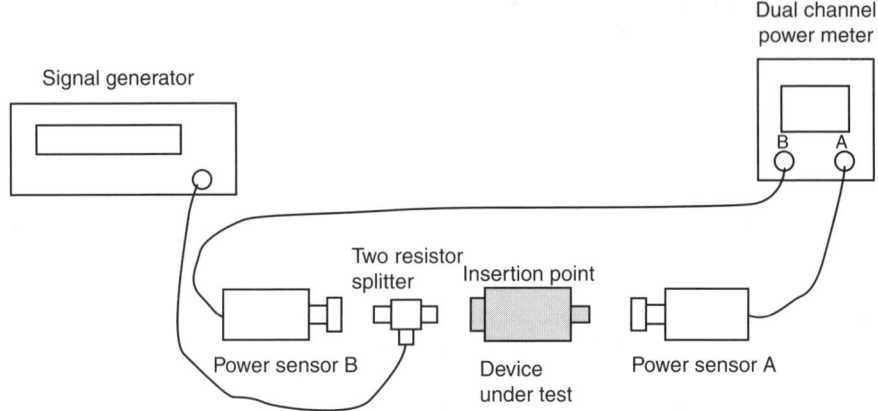

Figure 5.4   Dual channel power ratio

now be calculated using

$$L(\text{dB}) = 10\log_{10}\frac{\text{PA}_1}{\text{PB}_1} \cdot \frac{\text{PB}_2}{\text{PA}_2} \tag{5.10}$$

The dual channel power ratio method has two advantages over the simple system of Figure 5.3.

(1) The signal level is constantly monitored by power sensor B, reducing the error due to signal generator RF output level drift.
(2) Using a two-resistor power splitter or high directivity coupler will improve the source match [6,7].

The fixed frequency performance of this system may be improved by using tuners or isolators to reduce the generator and detector mismatch.

A system has been described by Stelzried and co-workers [8,9] that is capable of measuring attenuation in waveguide, WG22 (26.5–40 GHz), with an uncertainty of ±0.005 dB/10 dB to 30 dB.

The author has assembled a measurement system similar to that shown in Figure 5.4, with the equipment computer controlled through the general-purpose interface bus (GPIB). The system operates from 10 MHz to 18 GHz and has a measurement uncertainty of ±0.03 dB up to 30 dB, and ±0.06 to ±0.3 dB from 30 dB up to 70 dB. Return loss of the device under test is measured using an RF bridge at the insertion point. Insertion loss is then measured as described above and the mismatch error is calculated so that the attenuation may be quoted.

Figure 5.5 shows a scalar network analyser attenuation and voltage standing wave ratio (VSWR) measurement system. The analyser comprises a levelled swept frequency generator, three detector channels and a display. The DC output from the detectors is digitised and a microprocessor is used to make temperature, frequency response and linearity corrections for the detectors. Mathematical functions such

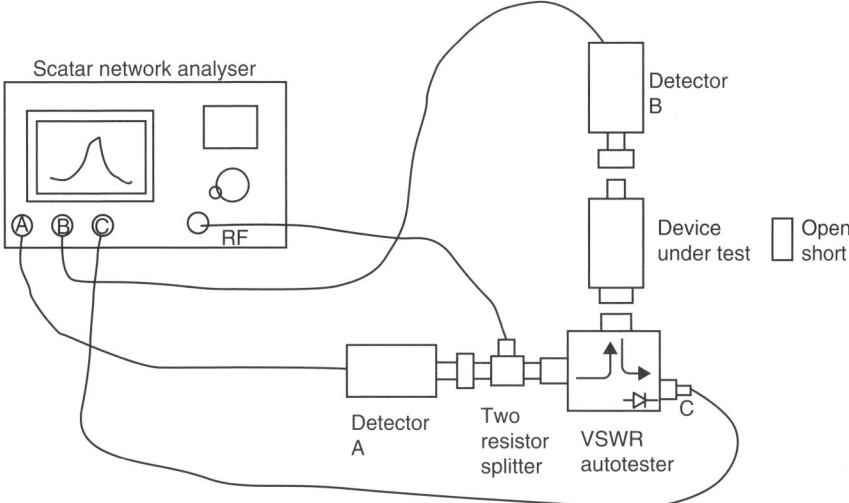

Figure 5.5   Scalar network analyser

as addition, subtraction, and averaging may be performed, making this set up very fast and versatile. The system has a claimed dynamic range of 70 dB (+20 dBm to −50 dBm), although this will be reduced if the source output is reduced by using a two-resistor splitter or padding attenuator.

Attenuation measurement uncertainty varies according to the frequency and applied power level to the detectors but is generally ±0.1 to ±1.5 dB from 10 to 50 dB. Although this may not be considered as being highly accurate, the system is fast and can identify resonances which may be hidden when using a stepped frequency measurement technique. It is also more useful when adjusting a device under test.

An important consideration for all of the above power ratio (homodyne) systems is that they make wide band measurements. When calibrating a narrow band device, such as a coupler or filter, it is important that the signal source be free of harmonic or spurious signals as they may pass through the device un-attenuated and be measured at the detector.

5.3.2   Voltage ratio method

Figure 5.6 represents a simple voltage attenuation measurement system, where a DVM (digital voltmeter) is used to measure the potential difference across a feed-through termination, first when it is connected directly to a matching attenuator, V_1, and then when the device under test has been inserted, V_2. Insertion loss may be calculated from:

$$L(\text{dB}) = 20 \log_{10} \frac{V_1}{V_2} \tag{5.11}$$

(Note that the generator and load impedance are matched.)

98 *Microwave measurements*

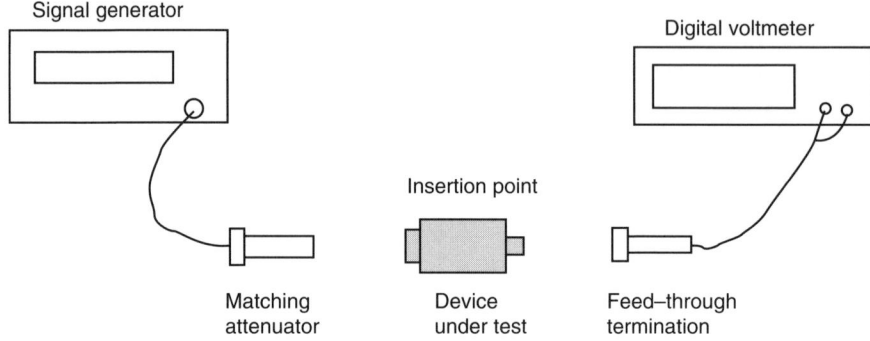

Figure 5.6 Voltage ratio

This simple system is limited by the frequency response and resolution of the DVM as well as variations in the output of the signal generator. The voltage coefficient of the device under test and resolution of the DVM will determine the range, typically 40–50 dB from dc to 100 kHz. A major contribution to the measurement uncertainty is the linearity of the DVM used, which may be typically 0.01 dB/10 dB for a good quality eight digit DVM. This may be measured using an inductive voltage divider, and corrections made.

5.3.3 The inductive voltage divider

Figure 5.7 is a simplified circuit diagram of an eight-decade IVD (inductive voltage divider), described by Hill and Miller [10]. This instrument is an extremely accurate variable attenuation standard operating over a nominal frequency range of 20 Hz to 10 kHz. (Special instruments have been constructed to operate at 50 kHz and 1 MHz.) It consists of a number of auto-transformers, the toroidal cores of which are constructed of wound insulated Supermalloy tape. (Supermalloy is an alloy that is 70 per cent Ni, 15 per cent Fe, 5 per cent Mo and 1 per cent other elements, and has a permeability $>100,000$ and hysteresis <1 J m^{-3} per Hz at 0.5 T).

The windings of a decade divider may be constructed by taking ten exactly equal lengths of insulated copper wire from the same reel and twisting them together to form a rope, which is wound around the core (double layer on the inner circumference and single layer on the outer circumference). Nine pairs of the ten strand cable are joined together, forming ten series-aiding coils. High-quality switches are required between the stages and care must be taken with the physical layout to avoid leakage and loading by the later decades.

The attenuation through a perfect IVD set to give a ratio D is given by

$$A = 20 \log_{10} \frac{V_{in}}{DV_{in}} = 10 \log_{10} \frac{1}{D} \qquad (5.12)$$

Attenuation measurement 99

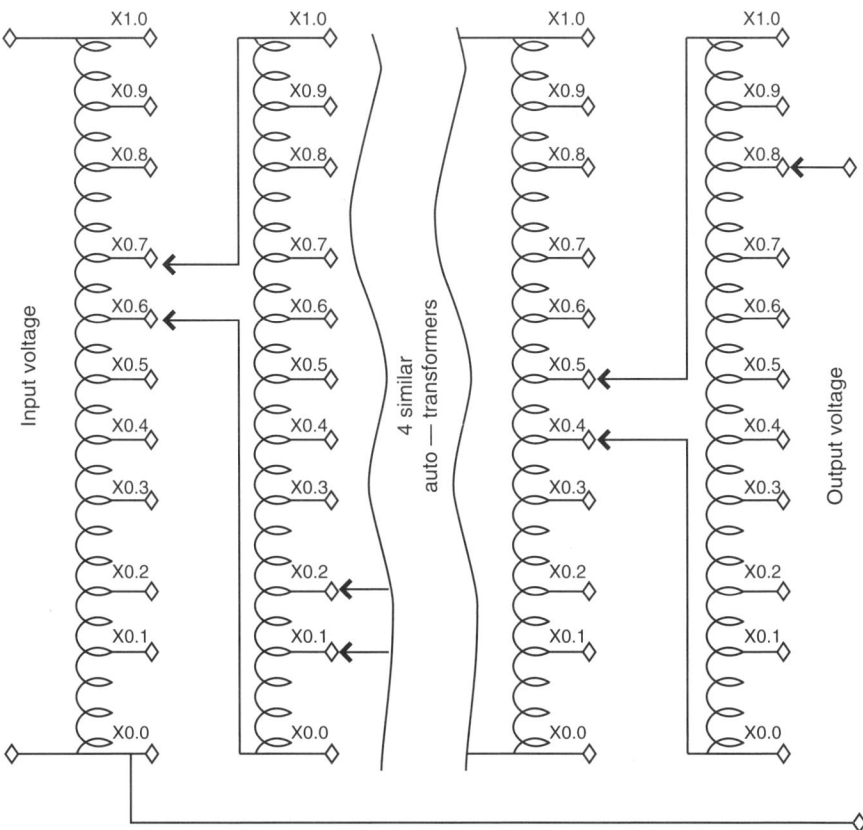

Figure 5.7 Inductive voltage divider

The overall error, ε, in an inductive voltage divider may be defined as

$$\varepsilon = \frac{V_{out} - DV_{in}}{V_{in}} \quad (5.13)$$

where V_{in} is the input voltage, D is the indicated ratio and V_{out} is the actual output voltage. $\varepsilon = \pm 4 \times 10^{-8}$ for a commercially available eight-decade IVD, or $\pm 10^{-6}$ to $\pm 10^{-7}$ for a seven-decade IVD.

There are three important voltage ratio attenuation measurement systems in use at the National Physical Laboratory (NPL), Teddington, UK.

Figure 5.8 is a computer controlled voltage ratio system described by Warner and co-workers [11–13] designed to calibrate the programmable rotary vane attenuators used at NPL in the Noise Standards Section. Two wide band synthesised signal generators are phase locked to a stable external reference frequency and their RF frequencies adjusted for a 50 kHz difference. RF from the signal generator is levelled and isolator and tuners used to match the system, presenting a generator and load

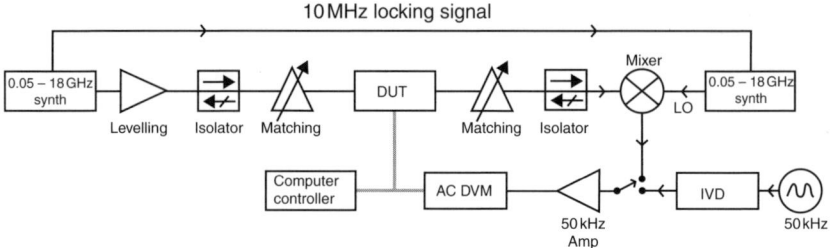

Figure 5.8 Automated voltage ratio system

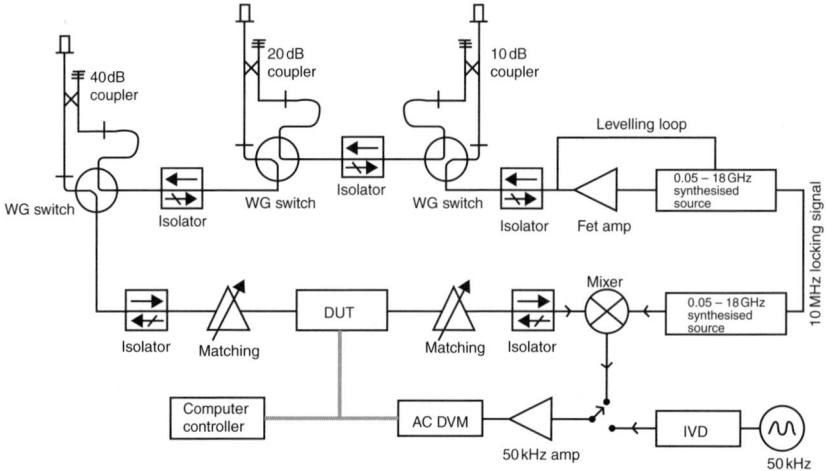

Figure 5.9 Gauge block voltage ratio system

reflection coefficient better than 0.005 to the device under test. The 50 kHz intermediate frequency (IF) signal is amplified and then measured on a precision AC DVM. The ac to dc conversion linearity of the DVM is periodically measured using a stable 50 kHz source and IVD. The non-corrected linearity is approximately 0.012/20 dB and the corrected linearity 0.0005/20 dB.

Figure 5.9 shows a ratio system described by Warner and Herman [14] similar to that shown in Figure 5.8 but with the addition of a gauge block attenuator, formed by three waveguide switched couplers in cascade. The gauge block may be switched from 0 to 70 dB in 10 dB steps, giving the system a 90 dB dynamic range. The standard deviation of ten repeated measurements at the same attenuation value is below 0.0005–40 dB, 0.001–70 dB and <0.002–90 dB. Attenuation may be found from

$$A_{\text{DUT}}(\text{dB}) = \left[20 \log_{10}\left(\frac{V_1}{V_2}\right)\right] + A_{\text{gb}} + C \tag{5.14}$$

where A_{gb} is the gauge block attenuation (dB) and C is the ac DVM correction dB.

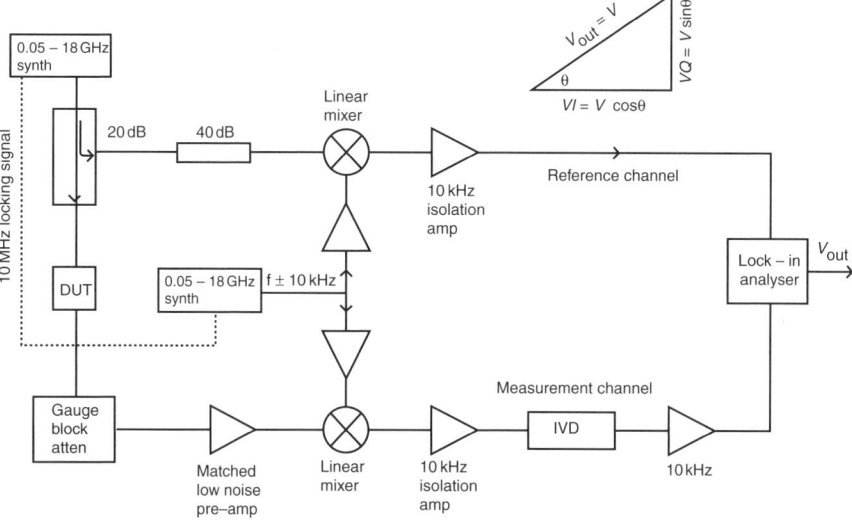

Figure 5.10   Dual channel voltage ratio system

Figure 5.10 shows a dual channel voltage ratio system developed at NPL by Kilby and co-workers [15,16] to provide extremely accurate attenuation measurements for the calibration of standard piston attenuators. It operates over the frequency range 0.5–100 MHz, is fully automatic and has a dynamic range of 160 dB without the need for noise balancing. The upper channel provides a reference signal for a lock-in analyser whilst the measurement process takes place in the lower channel. The lock-in analyser contains in-phase and quadrature phase sensitive detectors (PSDs), whose output ($VI = \cos\psi$ and $VQ = \sin\psi$) are combined in quadrature to yield a direct voltage given by:

$$V_{\text{out}} = (VI^2 + VQs)^{0.5}$$
$$= \left[(V\cos\psi)^2 + (V\sin\psi)^2\right]^{0.5} \quad (5.15)$$
$$= V$$

Thus, V_{out} is directly related to the peak value, V, of the 10 kHz signal emerging from the lower channel and is independent of the phase difference, ψ, between signals in the upper and lower channels.

Attenuation may be found from

$$A_{\text{DUT}} = \left[20\log_{10}\left(\frac{V_{\text{out1}}}{V_{\text{out2}}}\right)\right] A_{\text{gba}} + A_{\text{ivd}} \quad (5.16)$$

where V_{out1} and V_{out2} are the output voltages for the datum and calibration setting. A_{gba} and A_{ivd} are the attenuations in dB removed from the gauge block and IVD when the device under test is inserted.

102 Microwave measurements

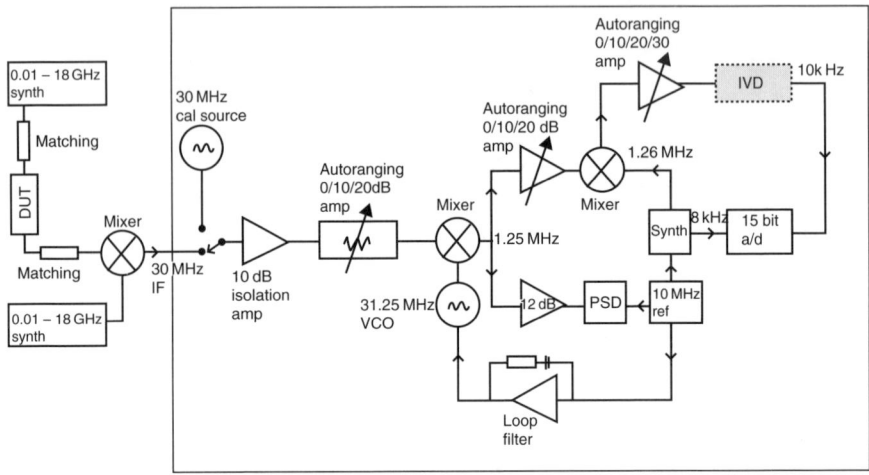

Figure 5.11 A commercial attenuator calibrator

This dual channel system also yields the phase changes that occur in the device under test. From the vector diagram in Figure 5.10 we have

$$\tan \psi = \frac{VQ}{VI}$$

Hence,

$$\psi = \arctan\left(\frac{VQ}{VI}\right) \tag{5.17}$$

Figure 5.11 is a simplified diagram of a commercially available voltage ratio attenuator calibrator manufactured by Lucas Weinschel of Gaithersberg, MD 2087, USA. The instrument is a microprocessor controlled 30 MHz triple conversion receiver and may be used at other frequencies with the addition of an RF generator, local oscillator and mixer.

The 30 MHz input signal is routed through a 10 dB isolation amplifier to a series of pin-switched attenuators, which are automatically selected to present the first mixer with a 30 MHz signal at the correct level for low noise linear mixing. A phase locked loop referenced to the instrument's internal 10 MHz oscillator provides the correction for the first voltage controlled local oscillator (31.25 MHz), resulting in a first IF of 1.25 MHz. The first IF is routed through a variable gain amplifier having switched gains of 0, 10 and 20 dB, to the second mixer. The signal is then mixed with a 1.26 MHz local oscillator derived from the 10 MHz reference, to form a second IF of 10 kHz. This 10 kHz signal is applied to a second variable gain amplifier having switched gains of 0, 10, 20 and 30 dB. The variable gain amplifiers and auto-ranging attenuator are microprocessor controlled in the auto-ranging sequence involved in making a measurement.

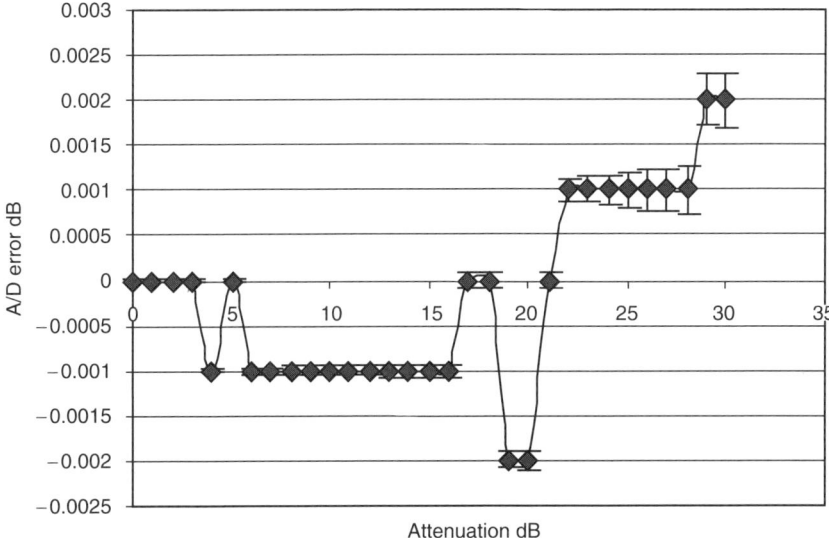

Figure 5.12   VM7 A/D calibration

The final conversion is made by a sample and hold amplifier within the 15 bit analogue to digital converter. The sample and hold amplifier is clocked at 8 kHz. The sampling of the 10 kHz signal every 8 kHz provides a quasi-mixing action that results in a 2 kHz signal with four samples per cycle. This analogue signal is then converted to digital form.

Before use, the instrument is calibrated by switching the 30 MHz receiver input to a 30 MHz internal oscillator. Each attenuation and gain block is automatically calibrated against the A/D converter and the calibration results stored in ram, to be used during the measurement process. The linearity of the A/D converter is paramount to the overall accuracy of the system and this may be checked by inserting a precision IVD into the 10 kHz path preceding the A/D.

This system has a sensitivity of −127 dBm and measurement accuracy of ±0.015 dB, +0.005 dB per 10 dB to 80 dB and ±0.1 dB per 10 dB from 80 dB to 105 dB.

Figure 5.12 shows the results of the A/D calibration using an eight-decade inductive voltage divider matched to the system. The 10 kHz signal is held between +14 to −2 dBm on all but the last range, where it may cover +14 to −12 dBm.

Figure 5.13 is a simplified diagram of an IFR 2309 FFT Signal Analyser. Input signals are down converted to a 10.7 MHz IF and then routed through signal conditioning circuits to a 1 bit sigma-delta analogue to digital converter (better known for its use in CD players, giving low noise and high linearity). The A to D converter contains a comparator and quantiser with a feed back loop, having a two-bit delay. The comparator compares the present input sample with the previous sample and outputs into a digital signal processor for complex analysis. The unit operates from

104 Microwave measurements

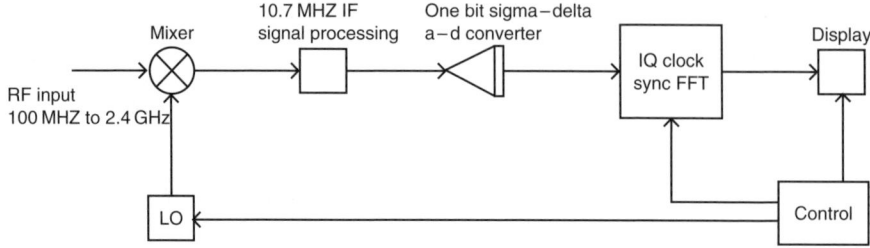

Figure 5.13   IFR 2309 FFT signal analyser

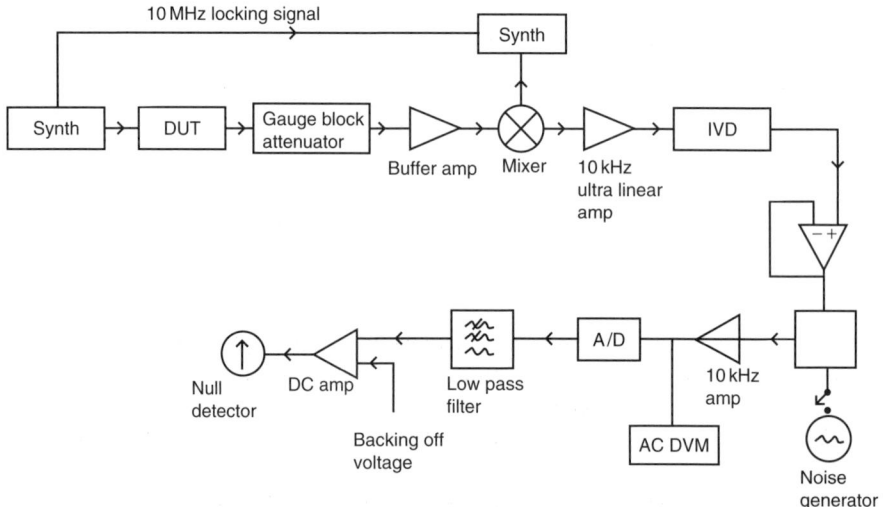

Figure 5.14   Audio frequency substitution system

100 MHz to 2.4 GHz and has a maximum sensitivity of −168 dBm, resolution of 0.0001 dB and a claimed linearity of ±0.01 dB per 10 dB.

5.3.4   AF substitution method

In an audio frequency substitution system, the attenuation through the device under test is measured by comparison with an audio frequency standard, usually a resistive or inductive voltage divider. Figure 5.14 is a simplified diagram of an AF substitution system designed and used at NPL and described by Warner [17]. It is used at national standards level over the frequency range 0.5–100 MHz and has a dynamic range of 150 dB.

An external 10 MHz reference frequency is used to lock the two synthesiser time bases and the local oscillator is operated at a frequency difference of 10 kHz to the RF source, giving a stable 10 kHz IF after mixing. The attenuation standard is a seven-decade precision IVD inserted between two 10 kHz tuned amplifiers.

The gauge block attenuator, which is a repeatable step attenuator, DUT and IVD are adjusted during the measurement so that the voltage to the null detector remains constant. The 10 kHz amplifiers are purpose built and are low noise, very stable and extremely linear (better than ±0.001 per cent). The IVD is driven from a low impedance source and its output is loaded with a high impedance. From 100 to 150 dB, noise must be injected into the system to compensate for the noise generated by the mixer. This is normally accomplished by setting the gauge block attenuator to zero, the IVD to unity and gain of the second 10 kHz amplifier for 2 V rms output. The backing off voltage is adjusted for a null meter reading. The signal source is now switched off and the output voltage caused by the noise alone is measured on the DVM. After this, the gauge block attenuator is switched to a value, A_{gba} and the DUT moved to its datum position, the IVD is set to zero, the noise generator is switched on and adjusted to give the same output reading as before. Finally, the signal source is switched back on and the IVD is adjusted to a ratio R, which gives a null reading on the output meter.

The attenuation change in the DUT is then given by

$$A_{\text{DUT}} = A_{gba} + 20 \log_{10}\left(\frac{1}{R}\right) \tag{5.18}$$

With great care being taken to reduce the effects of mismatch and leakage, the total uncertainty of measurement using this system is from ±0.0006 dB at 10 dB to ±0.01 dB at 100 dB, for 95 per cent confidence probability.

5.3.5 IF substitution method

An IF substitution attenuation system compares the attenuation through the device under test with an IF attenuation standard, which may be an IF piston attenuator, high-frequency IVD or a box of 'π' or 'T' type resistive attenuator networks.

Figure 5.15 shows a simplified piston attenuator arrangement normally used as an IF substitution standard. The tube acts as a waveguide below cut-off transmission line. An IF signal, usually 30 or 60 MHz, is launched into the tube from the fixed input coil. A metallic grid acts as a mode filter ensuring that only a single mode (H_{11}, E_{01}, H_{21}, E_{11} or H_{01}) is transmitted. The voltage in the output coil falls exponentially as the separation between the two coils increases. For perfectly conducting cylinder

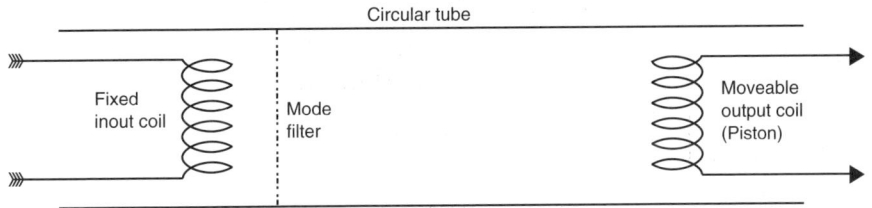

Figure 5.15 Piston attenuator

walls with a coil separation Z_1 to Z_2, the attenuation in dB may be found from:

$$\alpha p = 8.686 \times 2\pi (Z_2 - Z_1) \left[\left(\frac{Snm}{2\pi r} \right)^2 - \left(\frac{1}{\lambda^2} \right) \right]^{0.5} \quad (5.19)$$

where r is the cylinder radius, λ is the free space wavelength and Snm is a constant dependent upon the excitation mode.

Precision engineering is required in manufacturing the piston attenuator. A tolerance of ± 1 part in 10^4 on the internal radius is equivalent to ± 0.001 dB per 10 dB. The cylinder is sometimes temperature stabilised and a laser interferometer employed to measure the piston displacement.

When the piston attenuator is adjusted for a low attenuation setting, there is a systematic error resulting in non-linearity due to the interaction between the coils. This error reduces as the coils part, and is negligible at about 30 dB insertion loss.

A piston attenuator developed at NPL by Yell [18,19] is used as a national standard. The piston is mounted vertically and supported on air bearings to prevent contact with the cylinder wall. Displacements are measured with a laser interferometer. The cylinder is made of electroformed copper deposited on a stainless steel mandrel. This unit has a range of 120 dB, resolution of 0.0002 dB and a stated accuracy of 0.001 in 120 dB.

Figure 5.16 shows the basic layout of a parallel IF substitution system. Here, the 30 MHz IF input signal is compared with a 30 MHz reference signal, which may be adjusted in level by using a calibrated precision piston attenuator. These two signals are 100 per cent square wave modulated in counter phase and are fed to the IF amplifier. A tuned phase sensitive detector (PSD) is used to detect and display the difference between the two alternately received signals. This circuit is very effective and can detect a null in the presence of much noise. The system is insensitive to changes in IF amplifier gain, but an automatic gain control loop is provided, to automatically keep the detector sensitivity constant.

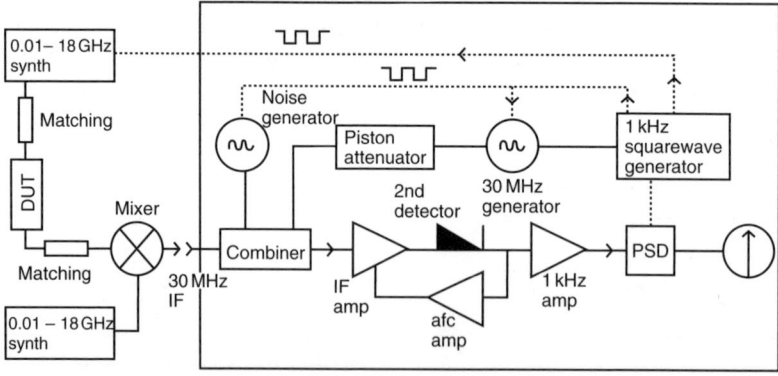

Figure 5.16   Parallel IF substitution system

In operation, the two signals are adjusted to be at the same amplitude, indicated by a null on the amplitude balance indicator. After the DUT is inserted, the system is again brought to a null by adjusting the precision piston attenuator. The difference in the two settings of the standard attenuator may be read directly and is a measure of the insertion loss of the unknown device.

For signal levels below -90 dBm, noise generated by the mixer in the signal channel may cause an error and it is necessary to balance this by introducing extra noise into the reference channel. The system described above has a specification of ± 0.01 dB per 10 dB up to 40 dB, ± 0.27 dB at 80 dB and ± 0.5 dB at 100 dB.

5.3.6 RF substitution method

With the RF substitution method, the attenuation through the device under test is compared with a standard microwave attenuator operating at the same frequency. The standard microwave attenuator is usually a precision waveguide rotary vane attenuator (rva) or microwave piston attenuator.

Figure 5.17 shows the basic parts of a waveguide rotary vane attenuator, described by Banning [20]. It consists of three metalised glass vanes: two end vanes fixed in a direction perpendicular to the incident electric vector and a third vane, which is lossy and able to rotate, set diametrically across the circular waveguide. When the central

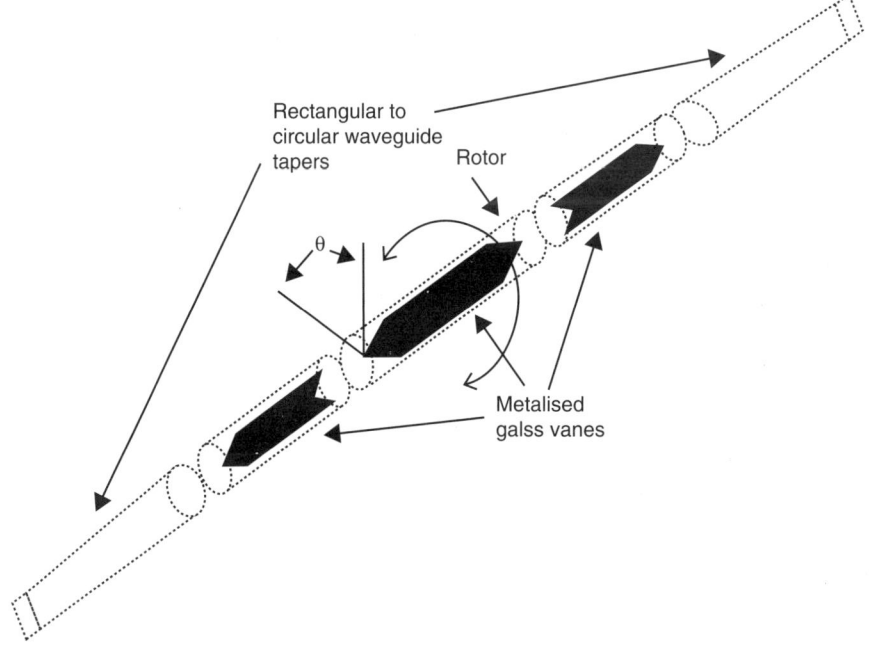

Figure 5.17   Rotary vane attenuator

108 Microwave measurements

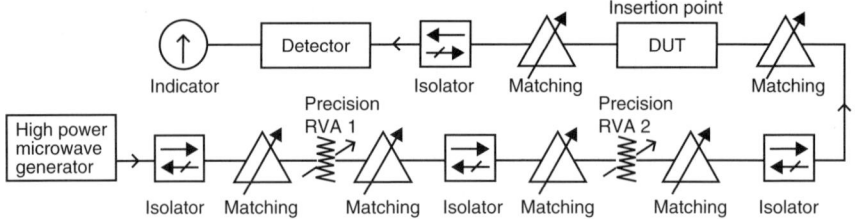

Figure 5.18 Series RF substitution

vane is at an angle θ relative to the two fixed vanes, the total attenuation is given by:

$$A_{\text{rva}} = 40 \log_{10}(\sec \theta) + A_0 \tag{5.20}$$

where A_0 is the residual attenuation when all three vanes lie in the same plane ($\theta = 0$).

The rotary vane attenuator is an extremely useful instrument as the attenuation is almost independent of frequency and there is little phase change as the attenuation is varied. As can be seen from (5.20), attenuation is not a linear function of the central vane angle. A change in θ from 0.000° to 0.615° = 0.001 dB, but a change in θ from 86.776° to 88.188° changes the attenuation from 50 to 60 dB. A high-quality rotary vane attenuator can have an attenuation range of 0–70 dB, dependent on the attenuation of the central vane and the vane alignment error. The accuracy of a commercially available rotary vane attenuator is in the order of ±2 per cent of reading or 0.1 dB whichever is the larger. Rotary vane attenuators have been made at NPL which have a display resolution of 0.0001°. They follow the $40 \log_{10}(\sec \theta)$ law to within ±0.006 dB up to 40 dB.

Figure 5.18 is a block diagram of a very simple series RF substitution system, employing two precision rotary vane attenuators. As attenuation through the DUT is increased, the attenuation in the precision rotary vane attenuator is adjusted from the datum setting to keep the detector output constant. The difference between the new rotary vane attenuator setting and the datum setting is equal to the loss through the DUT. This system could be enhanced by square wave modulating the signal source and using a synchronous detector. By using two precision rotary vane attenuators it is possible to obtain a dynamic range of 100 dB but a very stable high power signal generator would be necessary and the output power must be within the power coefficient limits of the microwave components used. Great care must be taken to reduce RF leakage.

5.3.7 The automatic network analyser

Figure 5.19 is a greatly simplified bock diagram of a commercially available microwave network analyser. The instrument is basically a microprocessor controlled two-channel superhetrodyne receiver, employing an ultra linear phase sensitive detector to detect magnitude and phase difference between the channels. The signal generator RF output is split into a reference channel and a measurement channel,

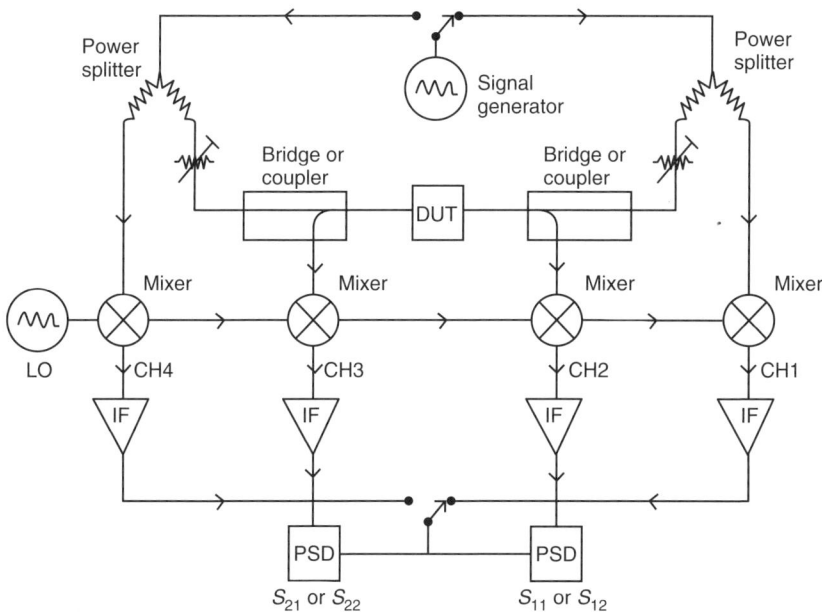

Figure 5.19 Vector network analyser

which consists of high-directivity directional couplers or RF Wheatstone bridges, capable of splitting the forward and reflected waves to and from the device under test.

With reference to Figure 5.18, when both switches are positioned to the right, channel 1 provides the reference signals for the phase sensitive detectors, channel 2 gives the real and imaginary parts of S_{11} and channel 3 yields the real and imaginary parts of S_{21}. With the switches set to the left, channel 4 provides the reference signals for the phase sensitive detectors, channel 2 gives the real and imaginary parts of S_{12} and channel 3 yields the real and imaginary parts of S_{22}. Thus, all four S-parameters may be determined without reversing the device under test.

In use, the instrument is treated as a black box, where known calibration standards, such as open and short circuits, air lines, sliding or fixed termination and known attenuators are measured at the test ports and the results stored and used to correct measurements made on the device under test. With the device under test removed and port 1 connected to port 2, the instrument will measure the complex impedance of its generator and detector. These data are stored and used in the equation for calculating attenuation.

These instruments are extremely fast and versatile and results may be displayed in frequency or time domain, using a fast Fourier transform process. They are very accurate when measuring reflection coefficient but have a limited dynamic range for attenuation measurement, due to available signal source level, harmonic mixing and channel crosstalk or leakage. The network analyser bridge or coupler sensitivity decreases with frequency (roll off); therefore, one may find an RF network analyser

110 Microwave measurements

Table 5.1 RF vector network analyser

Attenuation (dB)	300 kHz (dB)	3 GHz (dB)	6 GHz (dB)
0	±0.03	±0.04	±0.05
50	±0.11	±0.09	±0.12
75	±1.65	±1.34	±1.68

Table 5.2 Microwave vector network analyser

Attenuation (dB)	50 MHz (dB)	26.5 GHz (dB)
0	±0.03	±0.11
50	±0.05	±0.21
80	±0.60	±3.50

operating from 300 kHz to 6 GHz, a microwave network analyser operating from 50 MHz to 50 GHz and a millimetric wave network analyser operating from 75 to 110 GHz.

Typical measurement uncertainties are given in Tables 5.1 and 5.2.

5.4 Important considerations when making attenuation measurements

5.4.1 Mismatch uncertainty

Mismatch between the generator, detector and device under test is usually one of the most significant contributions to attenuation measurement uncertainty. With reference to (5.6) and (5.7), these parameters must be measured to calculate the attenuation of the DUT from the measured insertion loss. It may be possible to measure the device under test and the detector reflection coefficient by using a slotted line, RF bridge or scalar/network analyser. Measuring the signal generator output reflection coefficient is more difficult, as it may have an active levelling loop in its RF output stage.

If single frequency measurements are to be made then the generator and detector match may be improved by using ferrite isolators and tuners such as in Figure 5.8. The tuner must be adjusted for each frequency and this once tedious process has been improved by the introduction of computer-controlled motorised tuners. Tuners and isolators are narrow band devices and are not practical for low-RF frequencies (long wavelength) use due to their physical size. They are also prone to RF leakage and great care must be taken when making high-attenuation measurements with these devices in circuit.

For wide band measurements, the generator match may be improved by using a high-directivity directional coupler or a two-resistor splitter in a levelling loop circuit.

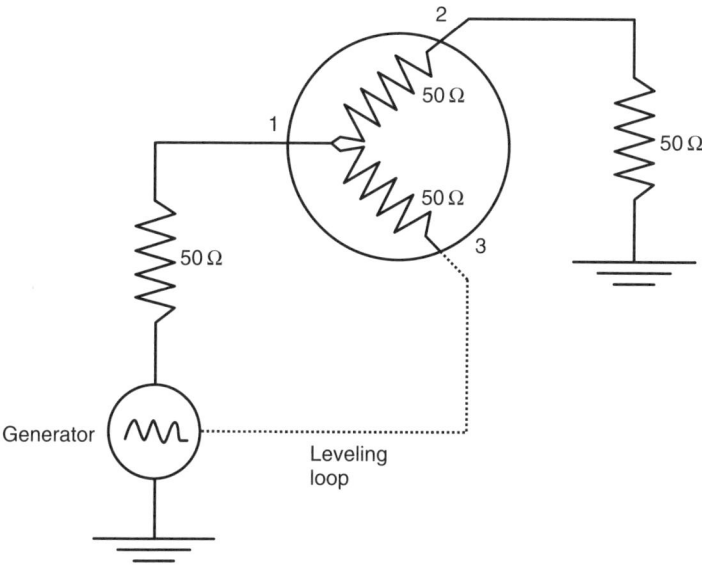

Figure 5.20   Two-resistor power splitter

A high-quality 'padding' attenuator placed before the detector will also improve the detector match but its attenuation may restrict the dynamic range of the system.

Figure 5.20 shows a two-resistor power splitter, which is constructed to have a 50 Ω resistor in series between port 1 and port 2, and an identical resistor between port 1 and port 3. This type of splitter should only ever be used in a generator levelling circuit configuration as in Figure 5.20 and must never be used as a power divider in a 50 Ω system. (If you connect a 50 Ω termination to port 1 and port 2 then the impedance seen at port 3 is approximately 83.33 Ω.)

This power splitter has been specifically designed to level the power from the signal source and to improve the generator source match. In Figure 5.19, if a detector is connected to port 3 and the output of the detector is compared with a stable reference and then connected to the signal generator amplitude control via a high-gain amplifier (levelling loop), then any change in the generator output at port 1 is seen across both 50 Ω series resistors and detected at port 3. This change is used to correct the generator output for a constant level at port 3. As port 2 is connected to port 1 by an identical resistor, port 2 is also held level.

If an imperfect termination is connected to port 2 then voltage will be reflected back into port 2 at some phase. This voltage will be seen across the 50 Ω resistor and the generator internal impedance. If the generator internal impedance is imperfect then some of the voltage will be re-reflected by the generator, causing a change in level at port 1. This change in level is detected at port 3 and used to correct the generator to maintain a constant output level at the junction of the two resistors. Port 1 is perceived as a virtual earth, thus when the levelling circuit is active, the effective impedance seen looking into port 2 will be the impedance of the 50 Ω series resistor.

112 *Microwave measurements*

Two-resistor splitters are characterised in terms of output tracking and equivalent output reflection coefficient. A typical specification for a splitter fitted with type N connectors would be ±0.2 dB tracking between ports at 18 GHz, and an effective reflection coefficient of ±0.025 to ±0.075 from DC to 18 GHz. The splitter has a nominal insertion loss of 3 dB between the input and output ports.

In keeping with the United Kingdom Accreditation Service (UKAS) document M3003 [21], the equation recommended for calculating mismatch uncertainty for a step attenuator is

$$M = \frac{8.686}{(2)^{0.5}} \left[|\Gamma_G|^2 \left(|S_{11a}|^2 + |S_{11b}|^2 \right) + |\Gamma_L|^2 \left(|S_{22a}|^2 + |S_{22b}|^2 \right) \right.$$
$$\left. + |\Gamma_G|^2 |\Gamma_L|^2 \left(|S_{21a}|^4 + |S_{21b}|^4 \right) \right]^{0.5} \tag{5.21}$$

where Γ_G and Γ_L are the source and load reflection coefficients. S_{11}, S_{22}, S_{21} are the scattering coefficients of the attenuator, a and b referring to the starting and finishing value.

The probability distribution for mismatch using this formula is considered to be normal, compared to the limit distribution of formulae (5.6) and (5.7).

5.4.2 RF leakage

In wide dynamic range measurements it is essential to check for RF leakage bypassing the device under test and entering the measurement system. This may be done by setting the system to the highest attenuation setting and moving a dielectric, such as a hand or metal object, over the equipment and cables. In a leaky system the detector output will vary as the leakage path is disturbed. If the measurement set up incorporates a levelling power meter, this can be used to produce a known repeatable step, which should be constant for any setting of the device under test.

RF leakage may be reduced by using good condition precision connectors and semi-rigid cables, wrapping connectors, isolators and other components in aluminium or copper foil, keeping the RF source a good distance from the detector and ensuring that no other laboratory work at the critical frequencies takes place. When the measured attenuation is much greater than the leakage, $A_1 \gg A_a$ then:

$$U_{LK} \text{ is approximately } \pm 8.686 \times 10^{-(A_1 - A_a)/20} \tag{5.22}$$

where A_1 = leakage path, A_a = DUT attenuator setting, U_{LK} = uncertainty due to leakage.

For example, if a leakage path 40 dB below the DUT attenuator setting is assumed, then error due to leakage is given in Table 5.3. This contribution is a limit having a rectangular probability distribution.

5.4.3 Detector linearity

This includes the linearity of the analogue to digital circuits of a digital voltmeter, square law of a diode or thermocouple detector, or mixer linearity of a superheterodyne receiver. The linearity may be determined by applying the same small repeatable

Table 5.3   Measured step and assumed leakage error

Measured step (dB)	Assumed leakage error (dB)
10	±0.000
20	±0.000
30	±0.000
40	±0.000
50	±0.000
60	±0.001
70	±0.003
80	±0.009
90	±0.027
100	±0.087
110	±0.275

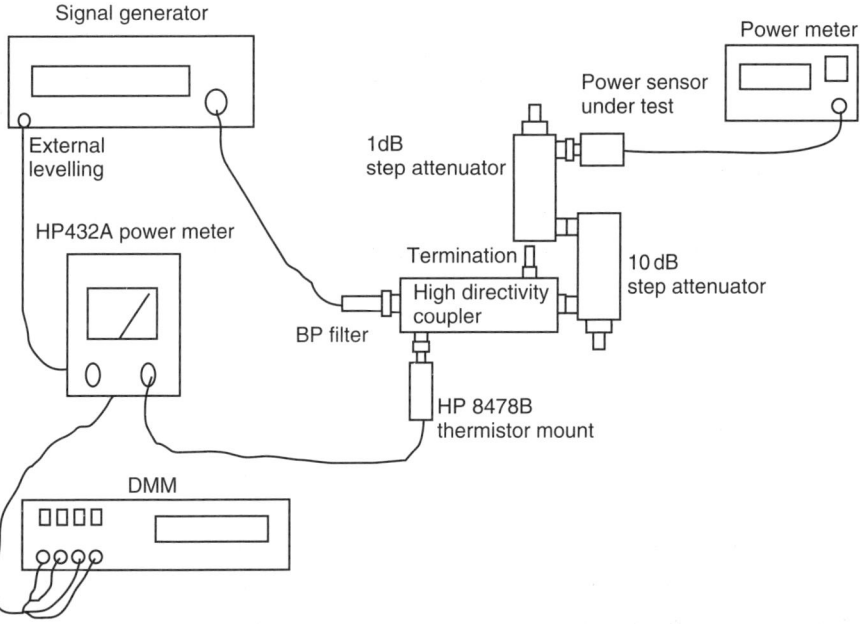

Figure 5.21   Detector linearity tests

level change to the detector over its entire dynamic range. This level change may be produced with a very repeatable switched attenuator operating in a matched system.

Figure 5.21 shows a variation of a technique described at the 22nd ARMMS Conference 1995 and at BEMC 1995. It involves applying a precise and repeatable 5 dB step, at various levels, to a power sensor, detector or receiver under test. If the signal generator is externally levelled from the power meter recorder output, as in

Figure 5.20, then switching to a consecutive range, say 3 mW to 1 mW, produces a nominal but highly repeatable 5 dB step. The power level applied to the device under test is adjusted using the step attenuators, such that the 5 dB step is applied over the instrument's operating range.

Measurements are best made starting with minimum power applied to the DUT and increasing the power to a maximum. The system may be automated using a GPIB controller setting the signal generator output and switching the step attenuator. The HP432A power meter bridge output voltages V_0, V_1, V_{comp} are measured using a long scale digital voltmeter.

5.4.4 Detector linearity measurement uncertainty budget

- Linearity of the 5 dB step due to the thermistor and power meter $= \pm 0.0005$ dB.
- Measurement repeatability $= \pm 0.01$ dB nominal (but should be determined by multiple measurements).
- Drift is determined by experiment and is dependent on temperature stability and the power meter range.
- Leakage is determined by experiment and is dependent on the step attenuator setting.
- *Mismatch.* The effective source match remains constant for the 5 dB step measurement; thus, uncertainty contributions due to mismatch between source and DUT cancel. Changes in load impedance are seen as contributing to the measured non-linearity.

Figure 5.22 shows a typical diode power sensor linearity response using the measurement method described above. Note that drift and noise effect the measurement at levels less than -50 dBm.

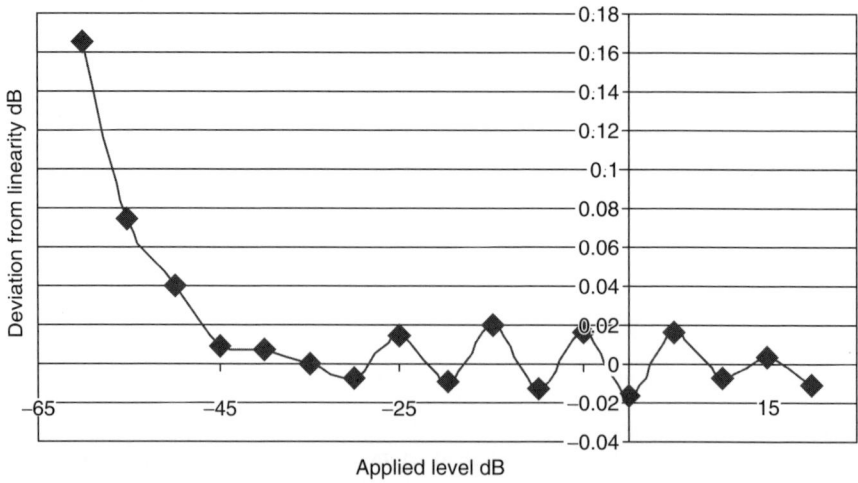

Figure 5.22 Power sensor linearity

5.4.5 System resolution

If the detector is a power meter or measurement receiver incorporating measurement ranges, then it is possible that the instrument resolution will depend on whether it is measuring at near full scale or at low scale. A power meter for instance, may resolve to three decimal places at full scale but only three decimal places at low scale. A digital voltmeter resolution is limited by the scale of its analogue to digital converter and an analogue meter may have a non-linear scale. The contribution due to resolution is taken as a limit, having a rectangular probability distribution.

5.4.6 System noise

The system noise is a measure of the receiver or detector sensitivity. As the measurement gets close to the sensitivity limit, thermal, shot and $1/f$ noise create a fluctuating reading. The detector output may be time averaged to present a more stable indication but there is a limit to applying averaging after which, this process becomes unhelpful and may obscure the correct answer. The system noise may be reduced by increasing the generator output level, using a phase sensitive detector, or using a low noise amplifier before the detector. The gauge block technique shown in Figure 5.9 is another way around this problem.

Noise is considered as a random (type A) uncertainty contribution, where multiple measurements using the same equipment set up will give different results. [For systematic (type B) uncertainties, results change if the system set up changes.]

5.4.7 Stability and drift

System stability and drift are particularly important uncertainty contributions for single channel measurement systems, where any drift in generator output or detector sensitivity will have a first-order effect on the measurement results. It is good measurement practice to allow test equipment to temperature stabilise before measurements commence and to measure system drift over the possible time required to make a measurement.

5.4.8 Repeatability

The repeatability of a particular measurement may only be derived by fully repeating that measurement a number of times. Statistics such as those described in UKAS document M3003 are used to determine the contribution to measurement uncertainty. If the device under test is a coaxial device inserted into the measurement system, it is usual to rotate the connector by 45° for each insertion. Uncertainty contributions due to the effect of flexing RF cables, mechanical vibrations, operator contribution, system noise and RF switch contacts, to list but a few, should be fully explored.

Semi-rigid RF cables give lower leakage and attenuation/phase change than screened coaxial cables when they are flexed. Chokeless waveguide flanges give better repeatability than choked joints. For precision measurements or where a microwave network analyser is employed it is useful to allow several people to make

116 *Microwave measurements*

the same measurement, in order to quantify operator uncertainty. Where an instrument or system contains mechanical microwave switches it has been found beneficial to exercise the switches several times, if they have not been in use for some hours (a small software routine can do this).

5.4.9 Calibration standard

Calibration standards such as the inductive voltage divider, IF piston attenuator, RF piston attenuator, rotary vane attenuator, switched coaxial attenuator and various power sensors have been covered previously in the text. These standards would normally be calibrated by the National Physical Laboratory or a UKAS Accredited Laboratory providing national traceability.

The National Standards are proven by exhaustive physical and scientific research and international intercomparison. The measurement uncertainties are usually calculated for approximately 95 per cent confidence probability.

When using the calibration results from the higher laboratory, the uncertainties must be increased to include the drift of the attenuator between calibrations and any interpolation between calibrated points. It is also important that the standard be used at the same temperature at which it was calibrated, particularly a piston or switched step attenuator. A 100 dB coaxial attenuator having a temperature coefficient of 0.0001 dB per °C may change by 0.01 dB per 1 °C change in temperature.

5.5 A worked example of a 30 dB attenuation measurement

The worked example in Table 5.4 is for a simple case of a 30 dB coaxial attenuator, measured using the dual channel power ratio system as shown in Figure 5.23. The measurement is repeated five times, from which the mean result and standard

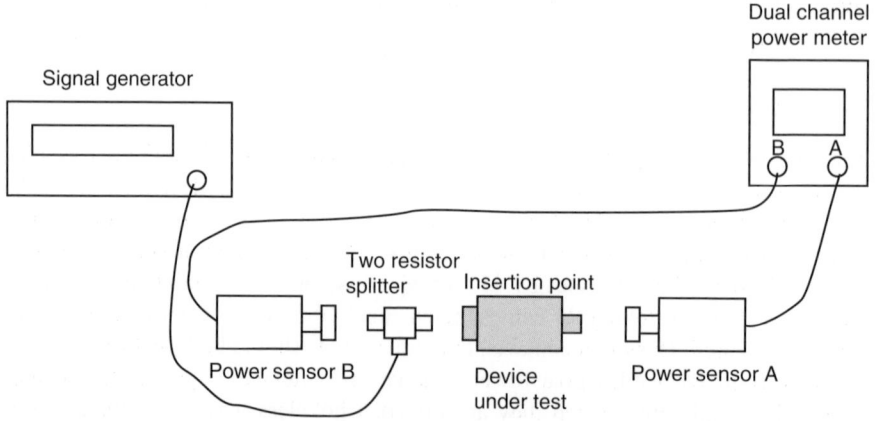

Figure 5.23   A simple dual channel power ratio system

Table 5.4   Measurements of dual channel power ratio system

P_1A (μW) without DUT	P_1B (μW) without DUT	P_2A (μW) with DUT inserted	P_2A (μW) with DUT inserted	Calculated insertion loss (dB)
10.001	10.1	10.01	10.2	30.039
10.002	10.1	10.01	10.2	30.039
9.993	10.1	9.99	10.1	30.047
9.997	10.0	9.99	10.2	30.089
9.995	10.0	9.98	10.1	30.050
Mean value of the five calculated samples:				30.053 dB.

deviation are calculated. This example follows the requirements of UKAS document M3003.

Insertion loss is calculated from

$$10 \log_{10} \left(\frac{P_1 A}{P_1 B}\right) \cdot \left(\frac{P_2 B}{P_2 A}\right) \tag{5.23}$$

5.5.1 Contributions to measurement uncertainty

Type A random contributions, U_{ran}. The estimated standard deviation of the uncorrected mean = 0.009 dB (normal probability distribution).

This may be calculated using

$$\frac{(\sigma n - 1)}{(n)^{0.5}} \tag{5.24}$$

where σ is the standard deviation of the population, n is the number of measurements, mismatch contribution, U_{mis}, where $\Gamma_G = 0.05$, $\Gamma_L = 0.02$, $S_{11} = 0.07$, $S_{22} = 0.05$, S_{12} and $S_{11} = 0.031$.

The measurement uncertainty of these values is taken as ± 0.02 and combining the measured value and its uncertainty in quadrature, we have $\Gamma_G = 0.054$, $\Gamma_L = 0.028$, $S_{11} = 0.073$ and $S_{22} = 0.054$. Putting these values into the mismatch formula (5.21) we arrive at an uncertainty of 0.026 dB (normal probability distribution).

Detector linearity, U_{lin}. The detector linearity was measured and found to be ± 0.02 dB over 30 dB range (limit distribution).

The measurement uncertainty of the detector linearity is in the order of ± 0.002 dB and is ignored, being an order less than other contributions.

Power meter resolution, U_{res}. The uncertainty due to power meter resolution was determined by experiment and found to be a maximum of ± 0.03 dB (limit distribution).

Leakage. The leakage was found to be less than 0.0001 dB and has not been included in the uncertainty calculations.

Table 5.5 Measurement uncertainty spreadsheet for a 30 dB coaxial attenuator at 10 GHz

Symbol	U_{lin}	U_{res}	U_{mis}	U_{ran}		v_{eff}	k	U_C (Atten$_x$)	Uncertainty
Source of uncertainty	Linearity of power sensor	Resolution	Mismatch	$u(xi) = \sigma_{est}(n)^{1/2}$					
Probability distribution	Rect	Rect	Normal	Normal			Normal	Normal ($k=2$)	
Divisor	$\sqrt{3}$	$\sqrt{3}$	1	1					
Sensitivity multiplier c_i	1	1	1	1	Measured step			Combined standard uncertainty ±dB	Expanded uncertainty ±dB
Contribution (dB)	0.02	0.03	0.026	0.009	30.053	1386	2	0.039	0.08

Measured attenuation: 30.53 dB ±0.08 dB

Power meter range and zero carry over are included in the power sensor linearity measurements and cannot be separated. Similarly, noise and drift cannot be separated from the random measurement of the five samples.

The spreadsheet in Table 5.5 is a convenient way to list the uncertainty contributions and calculate the measurement uncertainty.

The reported expanded uncertainty is based on a standard uncertainty multiplied by a coverage factor $k = 2$, providing a level of confidence of approximately 95 per cent. The uncertainty evaluation has been carried out in accordance with UKAS requirements.

References

1. Oldfield, L. C.: *Microwave Measurement*, IEE Electrical Measurement Series 3 (Peter Peregrinus, London, 1985), p. 107
2. Sinclair, M. W.: *Microwave Measurement*, IEE Electrical Measurement Series 3 (Peter Peregrinus, London, 1985), p. 202
3. Warner, F. L.: *Microwave Attenuation Measurement* (Peter Peregrinus, London, 1977), Chapter 2.9
4. Cherry, P., Oram, W., and Hjipieris, G.: 'A dynamic calibrator for detector non-linearity characterization', *Microwave Engineering Europe*, 1995
5. Orford, G. R., and Abbot, N. P.: 'Some recent measurements of linearity of thermistor power meters', *IEE Colloquium Digest*, 1981;**49**
6. Coster, A.: 'Calibration of the Lucas Weinschel attenuator and signal calibrator', *BEMC Conference Digest*, 1995
7. Coster, A.: 'Aspects of calibration – power splitters and detector linearity calibration', *27th ARMMS Conference Digest*, 1998
8. Stelzried, C. T., and Petty, S. M.: 'Microwave insertion loss test set', *IEEE Transactions on Microwave Theory and Techniques*, 1964;**MTT-12**:475–477
9. Stelzried, C. T., and Reid, M. S.: 'Precision DC potentiometer microwave insertion loss test set', *IEEE Transactions on Instrumentation and Measurement*, 1966;**LM-15**:98–104
10. Hill, J. J., and Miller, A. P.: 'A seven decade adjustable ratio inductively coupled voltage divider with 0.1 part per million accuracy', *Proc. IEEE*, 1962;**109B**:157–162
11. Warner, F. L.: *Microwave Measurement*, IEE Electrical Measurement Series 3 (Peter Peregrinus, London, 1985), Chapter 8
12. Bayer, H., Warner, F. L., and Yell, R. W.: 'Attenuation and ratio national standards', *Proc. IEEE*, 1986;**74**:46–59
13. Warner, F. L., Herman, P., and Cummings, P.: 'Recent improvements to the UK National microwave attenuation standards', *IEEE Transactions on Instrumentation and Measurement*, 1983;**LM-32**:33–37
14. Warner, F. L., and Herman, P.: 'Very precise measurement of attenuation over a 90 dB range using a voltage ratio plus gauge block technique', *IEE Colloquium Digest*, 1989;**53**:17/1–17/7

15 Kilby, G. J., and Warner, F. L.: 'The accurate measurement of attenuation and phase', *IEE Colloquium Digest*, 1994;**042**:5/1–5/4
16 Kilby, G. J., Smith, T. A., and Warner, F. L.: 'The accurate measurement of high attenuation at radio frequencies', *IEEE Transactions on Instrumentation and Measurement*, 1995;**IM-44**(2):308–311
17 Warner, F. L.: 'High accuracy 150 dB attenuation measurement system for traceability at RF', *IEE Colloquium Digest*, 1990;**174**:3/1–3/7
18 Yell, R. W.: 'Development of high precision waveguide beyond cut-off attenuator', *CPEM Conference Digest*, 1972, pp. 108–110
19 Yell, R. W.: 'Development in waveguide below cut-off attenuators at NPL', *IEE Colloquium Digest*, 1981;**49**:1/1–1/5
20 Banning, H. W.: *The measurement of Attenuation: a Practical Guide* (Weinschel Engineering Co. Inc., Gaithersburg, MD 20877, USA)
21 UKAS: *The expression of uncertainty and confidence in measurement*, M3003, 1st edn (HMSO, London, 1997)

Further reading

Warner, F. L.: *Microwave Attenuation Measurement* (Peter Peregrinus, London, 1977)

Chapter 6
RF voltage measurement
Paul C. A. Roberts

6.1 Introduction

The majority of signal amplitude or level measurements made at RF and microwave frequencies are measurements of power rather than voltage. Voltage measuring instruments such as RF millivoltmeters are still in frequent use but are nowadays less common than in the past. Probably, the most widespread test instrument is the oscilloscope, and modern oscilloscopes have bandwidths that extend well into the RF and microwave frequency ranges. There are a variety of tests and measurements of high-speed waveforms made with oscilloscopes and similar signal analysis instruments that rely on accurately probing and capturing signals at RF and microwave frequencies. This chapter overviews a variety of RF and microwave measuring instruments, including digitising and sampling oscilloscopes. It briefly describes their operating principles and characteristics, discusses probing and loading issues, and considers traceability and calibration for these instruments.

Figure 6.1 is a chart illustrating part of the electromagnetic spectrum and a number of typical applications. The range of the electromagnetic spectrum described by the term 'RF' (meaning radio frequency) is usually considered to begin at the frequency where AM broadcast radio transmission takes place (around 200 kHz). The frequency at which 'RF' becomes 'Microwave' is less clear, but is usually considered to be in the region of a few GHz. It is common in the calibration community to make the distinction between the 'DC and low frequency AC' and 'RF and microwave' fields at around 1 MHz. This chapter will consider the measurement of RF voltage in the frequency of 1 MHz to a few GHz.

122 *Microwave measurements*

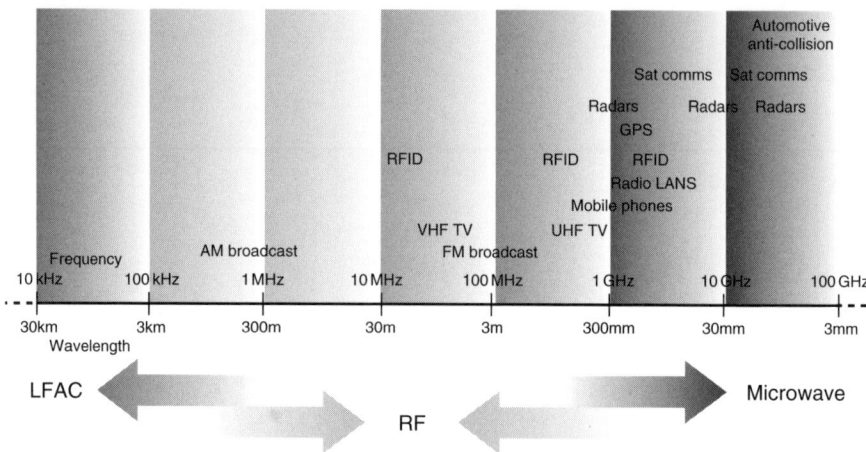

Figure 6.1   Chart showing frequency spectrum and typical usage

6.2   RF voltage measuring instruments

6.2.1   Wideband AC voltmeters

Wideband (or broadband) AC voltmeters are instruments capable of measuring voltages at frequencies up to 10 or 20 MHz. They are essentially low-frequency devices with an extended frequency range – either AC only or multimeter units – rather than dedicated high-frequency RF instruments. Internally, an analogue RMS measurement technique is employed with older instruments featuring an analogue display and modern instruments providing a digital readout. Generally input impedance is high, 1 MΩ or 10 MΩ with parallel capacitance of tens to hundreds of pF. Input connectors may be BNC coaxial or simple binding posts.

Few modern benchtop multimeter or voltmeter instruments are capable of measuring above 1 or 2 MHz. Among those that do have extended bandwidth (in the region of 10 MHz) are some precision systems digital multimeters (DMMs), which include signal sampling and digitising capability. Comprehensive triggering and acquisition controls together with sophisticated internal signal analysis features allow a variety of signal characteristics to be measured and displayed. Individual sample values can also be stored in internal memory for subsequent transfer to a controlling computer for analysis. There are also card-based digitising instruments that are capable of measuring signals at frequencies up to around 100 MHz, but these are not discussed here.

Older wideband voltmeter designs used matched thermal voltage converter (TVC) elements to perform an RMS to DC conversion. TVC is a device that responds to the heating effect produced by a voltage applied to a heater wire, which is sensed by a thermocouple junction attached to the heater wire. A modern alternative is a monolithic thermal sensor which uses a transistor as a temperature sensing device to sense the heating effect produced in a resistor – in this case, a pair of resistor/transistor sensors is manufactured on a silicon substrate. In common with the

RF voltage measurement 123

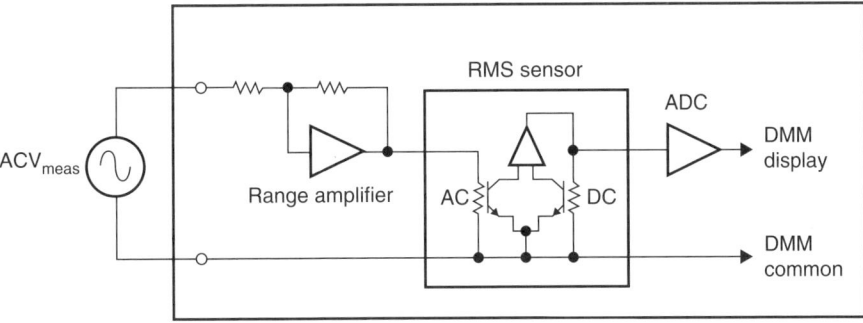

Figure 6.2 Block diagram of thermal RMS implementation

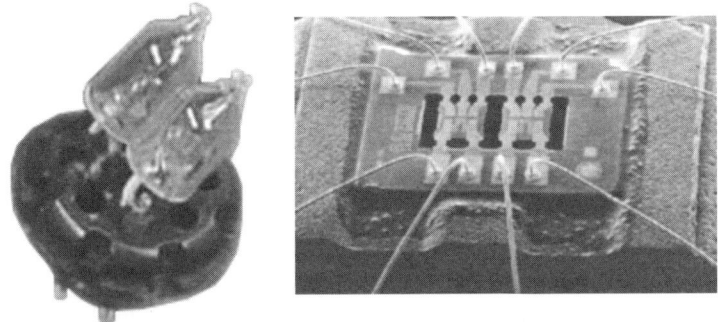

Figure 6.3 Thermal converter and monolithic thermal converter elements

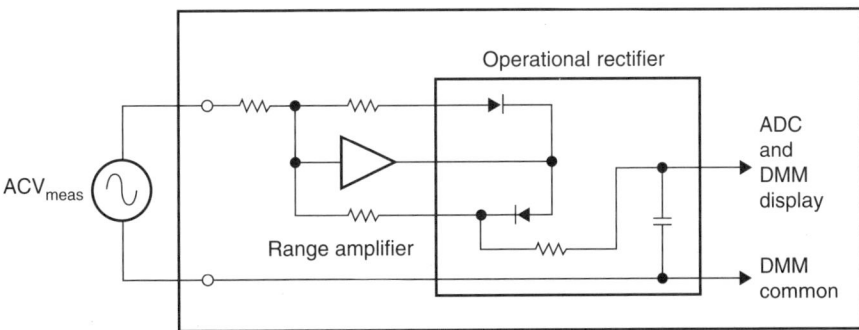

Figure 6.4 Block diagram of rectifier implementation

TVC-based designs, the sensors are used in a back-to-back feedback circuit which relies on the matching between devices to balance the heating effect from a DC voltage with that of the AC input, producing a DC output equal to the true RMS value of the input (Figures 6.2 and 6.3).

124 *Microwave measurements*

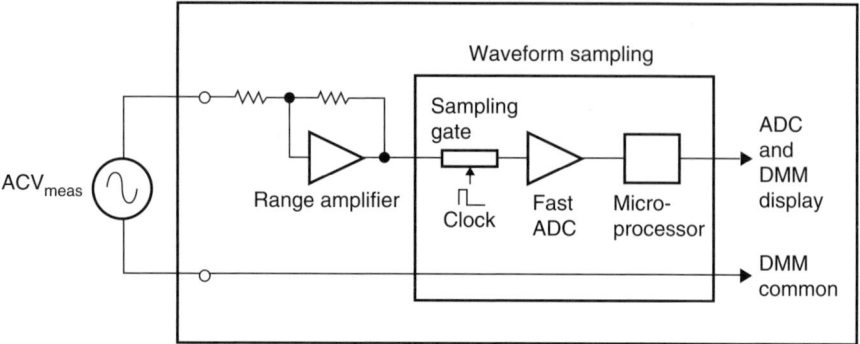

Figure 6.5   Block diagram of DMM sampling system

Other simple AC to DC conversion techniques can be employed, involving rectification and responding to either the peak or the average values of the signal. An example appears in Figure 6.4. The instrument is calibrated to respond in terms of the RMS value of a sinusoidal signal and will exhibit errors if the input is non-sinusoidal, has distortion or harmonic content.

Analogue RMS conversion techniques utilising log-feedback circuits implemented with bipolar transistor analogue multipliers (often referred to as Gilbert multiplier cells) are commonly used at lower frequencies to provide true RMS conversion. However, they tend to have limited bandwidth in precision applications and are generally not used in instrumentation above 1 or 2 MHz. Despite this, recently introduced RF signal level detector ICs operating up to 6 GHz for use in communications device and systems applications do employ Gilbert multiplier techniques.

6.2.2   Fast sampling and digitising DMMs

Although not specifically RF measuring instruments, some systems DMM and digitiser cards are capable of high-speed signal sampling and digitising of signals up to 10 MHz and above. Readouts can simply be the RMS or peak to peak value of the signal or can include more complex analysis and measurement of signal characteristics such as crest factor. Access may be provided to the individual sample values stored in an internal memory or transferred via a remote interface.

A sample and hold circuit acquires instantaneous samples of the input signal which are digitised by an analogue to digital converter (Figure 6.5). Alternatively, a fast 'flash' analogue to digital converter digitises the signal directly. Timing of the samples is critical, and a number of sampling schemes can be used.

Sampling schemes can take a number of approaches depending on whether the signal is a one-shot event or repetitive (Figure 6.6). In the simplest form, samples are taken at regular intervals to build up a complete representation of the waveform. For a repetitive signal the sampling may take place at a higher frequency than the signal, with many samples per signal cycle. Alternatively, synchronous subsampling takes samples at slightly later points on each subsequent waveform, building up

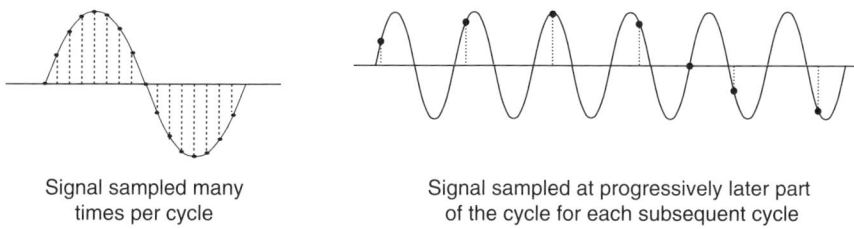

| Signal sampled many times per cycle | Signal sampled at progressively later part of the cycle for each subsequent cycle |

Figure 6.6   Sampling a repetitive signal

a complete representation of the waveform over a large number of cycles. This method is commonly used for high-frequency signals where the sampling frequency is much less than the input frequency. Random sampling can be used for non-repetitive signals such as noise. Care must be taken when choosing the sampling frequency to avoid aliasing (where the sampled output is not a true representation of the input signal). For sinusoidal signals consideration must be given to the input signal frequency and in the case of non-sinusoidal or distorted signals the highest frequency content present must be considered.

6.2.3   RF millivoltmeters

RF millivoltmeters are capable of measuring voltages from a few microvolts to several volts, at frequencies up to 1–2 GHz. Only a few manufacturers produce this type of instrument, but many of the popular instruments produced in the past are still in use today. The majority of instruments employ a high impedance probe which can be converted to 50 Ω or 75 Ω input with a variety of terminator and attenuator accessories. Insertion (or 'through') probes are available for some instruments, allowing the measurement of RF voltage in a matched coaxial system. Modern instruments are capable of accepting voltage detector probes and also power sensor probes allowing power or voltage measurements to be made with the same instrument.

Most instruments use diode detector probes that respond to the RMS value for low-level signals below about 30 mV and respond to the peak value for higher level signals. Their displays are calibrated in terms of the RMS value for a sinusoidal signal, so can indicate incorrectly for non-sinusoidal signals and signals with distortion and high harmonic content. Other instruments use an RF sampling technique (described later, not to be confused with the sampling/digitising features available on some precision systems DMMs or card-based digitisers), and measure the true RMS value of the signal.

A diode rectifier (usually full wave) within the meter probe converts the RF input into a DC signal for measurement and display in the main unit. At low levels (below around 30 mV) diode response is logarithmic, effectively measuring the RMS value of the input, even for non-sinusoidal waveforms. At higher levels the input signal peak value is sensed, and the meter is calibrated to indicate the equivalent RMS value of a sinewave. At these higher levels the reading will be in error if the input is non-sinusoidal, has distortion or harmonic content. Analogue linearisation may

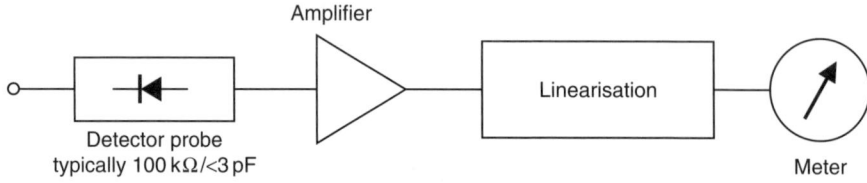

Figure 6.7 *Block diagram of RF millivoltmeter employing diode detector probe with analogue linearisation*

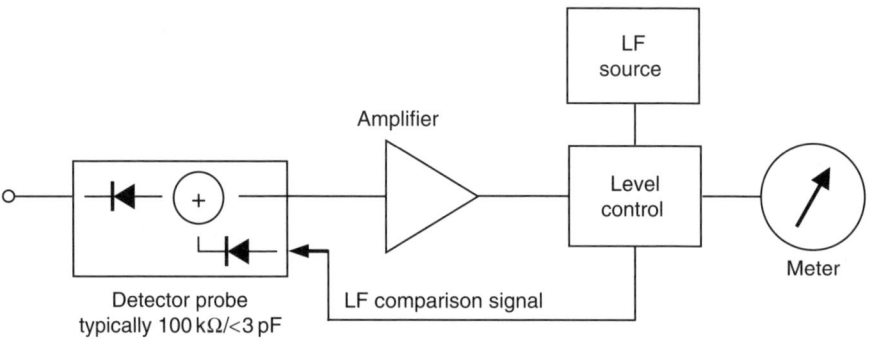

Figure 6.8 *Block diagram of RF millivoltmeter employing matched diode detector probe with LF feedback technique*

be applied to produce a linear scale over the entire amplitude range. Compensation is also applied for the temperature dependence of the rectifier diode, usually with another diode matched to device used for detection (Figure 6.7).

Other implementations have additional matched diodes in the probe to rectify a low-frequency comparison signal, with feedback to set the comparison signal amplitude equal to the RF input. The comparison signal is then measured and displayed, automatically producing a linear scale and compensating for the rectifier diode temperature dependency (Figure 6.8). Modern designs may incorporate digital compensation with linearisation, temperature compensation and frequency response correction information stored in a calibration memory within the probe itself.

All diode probes suffer from dependency of the input impedance on signal level and frequency. Input capacitance is usually small (<3 pF) but reduces at higher signal levels. Increased losses at higher frequencies within the diode capacitance can cause input resistance to fall as frequency increases. Manufacturers often provide data on input impedance variation with their products.

6.2.4 Sampling RF voltmeters

Sampling RF voltmeters use a fast sampling technique to measure repetitive waveforms. Sampling takes place using a high speed diode switch to charge a capacitor to the instantaneous value of the input waveform. Timing of the sampling is controlled

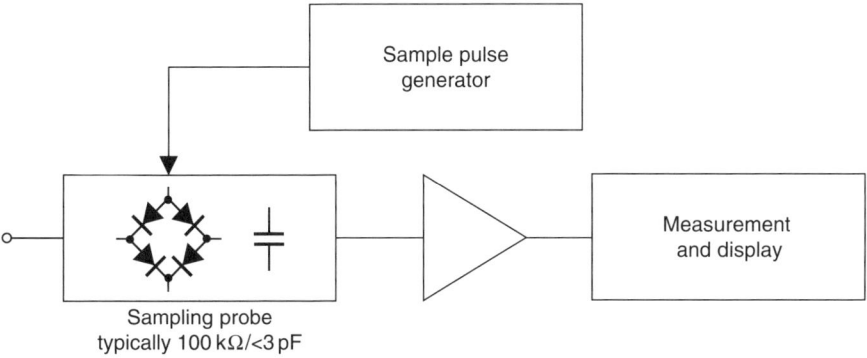

Figure 6.9   Block diagram of sampling voltmeter

by the sample pulse generator to sample the input at progressively later points of the waveform in subsequent cycles and create a much lower frequency representation of the input, preserving its RMS and peak values for measurement and display (Figure 6.9). Sampling voltmeters are much less common nowadays, but many of the older instruments employing this technique are still in use. A variety of implementations are used in commercial instruments, including dual and random sampling techniques capable of measuring sinusoidal, pulse and noise signals. Sampling probes typically have high-input impedance (100 kΩ/<3 pF), with 50 Ω terminator and attenuator accessories also available. Accuracies of 1–2 per cent are achieved at 100 MHz and around 10 per cent at 1 GHz.

Two channels of sampling using a coherent sampling technique which preserves signal amplitude and phase information enable the Vector Voltmeter to measure amplitude and phase of the RF input. A typical implementation employs a variable sampling frequency controlled by a phase lock loop to produce a sampled output waveform at the centre frequency of two identical bandpass filters, one in each channel. The filter outputs are then measured for amplitude and phase difference (Figure 6.10). Phase accuracy of 1° can be achieved at 100 MHz, 6° at 1 GHz and 12° at 2 GHz. Like the Sampling Voltmeter, Vector Voltmeter instruments are far less common than in the past; however, similar sampling techniques for measuring amplitude and phase relationship of two signals are employed in Vector Network Analysers.

6.2.5   Oscilloscopes

Modern oscilloscopes with bandwidths above 1 GHz are becoming more popular and are considered as general purpose test equipment rather than specialist instruments. Oscilloscopes display voltage waveforms, and can do so at RF and microwave frequencies but are seldom thought of as RF and microwave devices. Many oscilloscopes have sophisticated measurement and readout functions capable of displaying a variety of signal characteristics in the time and frequency domains. They will often have high impedance and 50 Ω inputs and be partnered with a variety of probes. Making accurate measurements depends on the appropriate choice of input and probe type.

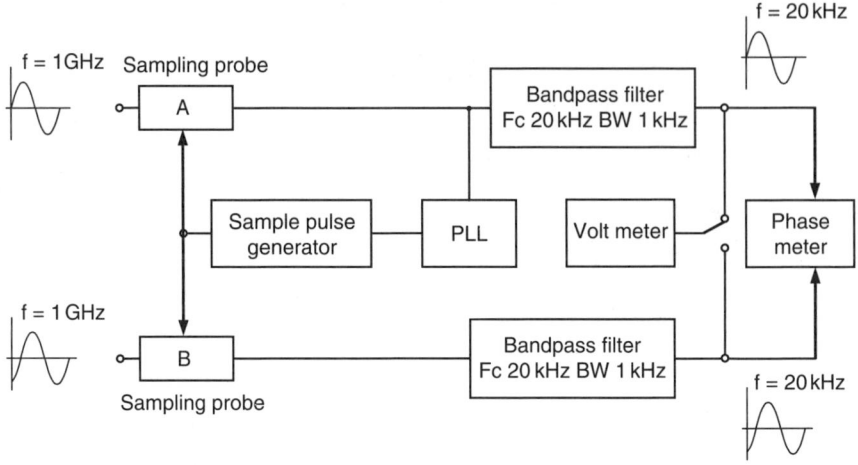

Figure 6.10   Block diagram of vector voltmeter

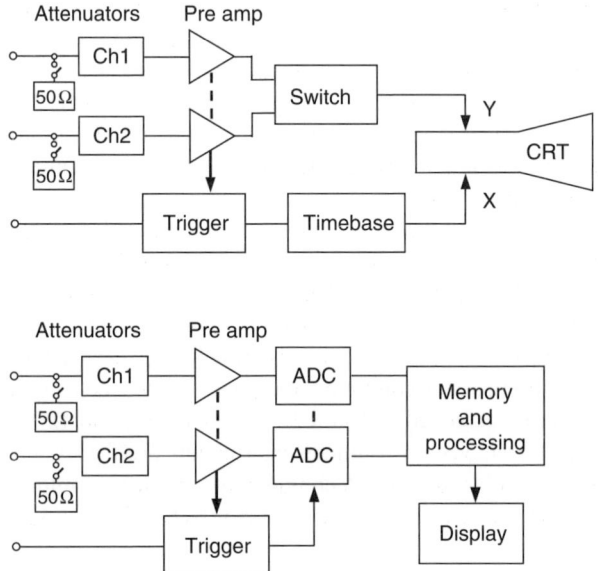

Figure 6.11   Block diagrams of analogue oscilloscope (top) and digital oscilloscope (bottom)

Along with increased bandwidth, sophisticated probing systems are now available to allow probing of high-speed and high-frequency circuits and signals, including wafer probe stations.

Analogue (or real time) oscilloscopes display the input signal directly on a cathode ray tube (CRT) screen (Figure 6.11). Bandwidths tend to be limited to around

RF voltage measurement 129

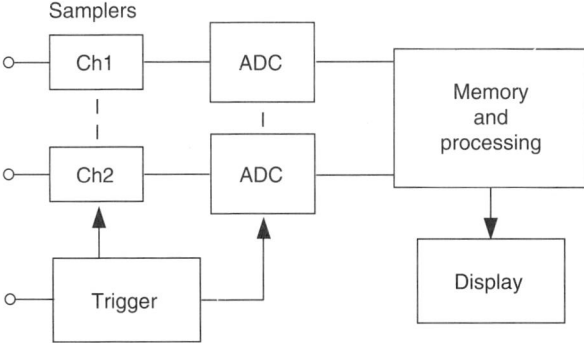

Figure 6.12   Block diagram of sampling oscilloscope

500 MHz, and these instruments will often have switched 1 MΩ/50 Ω inputs. Typical 1 MΩ inputs have up to 30 pF parallel capacitance from the input attenuator and preamplifier circuits. This capacitance will also appear across the internal 50 Ω terminating impedance and can adversely affect its voltage standing wave ratio (VSWR). Compensating networks are often employed which can also make the 50 Ω input appear slightly inductive at some frequencies.

Digital oscilloscopes use fast 'flash' analogue to digital converters to digitise and store the input signal for subsequent display and analysis (Figure 6.11). Bandwidths of up to more than 10 GHz are currently available, and these high-bandwidth instruments have dedicated 50 Ω inputs. Typical VSWR figures are <1.1 at 2 GHz, <1.3 at 4–6 GHz, with some instruments up to 2.0 at 4 GHz. VSWR worsens on the more sensitive ranges where there is less attenuation prior to the preamplifier. Lower bandwidth oscilloscopes have switched to 1 MΩ/50 Ω inputs and also tend to suffer the VSWR effects resulting from internal parallel capacitance described above. A typical 1 GHz oscilloscope input VSWR will be <1.5 at 1 GHz.

Sampling oscilloscopes offer bandwidths up to 70 GHz and sample the signal directly at their inputs using fast diode sampler techniques (Figure 6.12). Consequently the input range is typically limited to a few volts. As with all sampling systems, care should be taken when operating digital and sampling oscilloscopes to avoid aliasing between the sampling rate and signal frequency content, which may cause unexpected or erroneous results.

6.2.6 Switched input impedance oscilloscopes

Oscilloscopes with switched 1 MΩ and 50 Ω inputs are likely to have 50 Ω VSWR characteristics that suffer degradation at higher frequencies. These instruments tend to be the lower bandwidth devices, but 1 GHz oscilloscopes are available with switched inputs. Designing an oscilloscope to give a good 50 Ω input presents a challenge as the input circuitry typically has a reasonable amount of input capacitance, which will appear across the internal 50 Ω termination when it is switched in. There are usually compensating networks which counteract the effect (Figure 6.13), but inevitably result

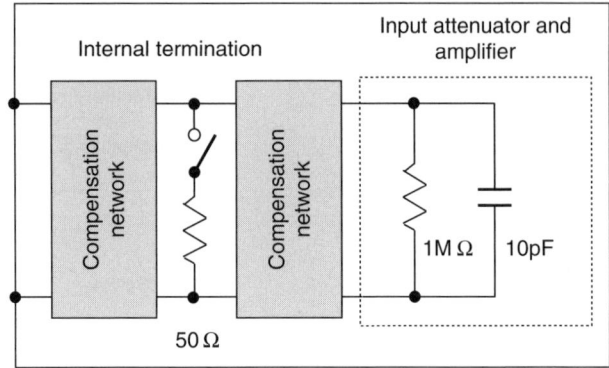

Figure 6.13 *Designs with switched input impedance typically employ compensating networks to improve 50 Ω input match*

in VSWR that worsens at high frequencies, and may even cause the input impedance to become inductive at some frequencies. Consequently some peaking or ringing may be seen with signals containing fast transients or high-speed edges, and there may be corresponding frequency response variations.

6.2.7 Instrument input impedance effects

The manufacturer's published specifications for instruments and probes generally include figures for nominal input impedance, but the actual input impedance can be dependent on input signal level and/or frequency. The instrument's input circuitry can be complex, modelled by a much simpler equivalent circuit representation for the specifications – often a parallel resistance/capacitance combination. Even with passive circuits, there can be variation with frequency, caused by such effects as capacitance dielectric loss increasing at higher frequency, appearing as an apparent input resistance drop. The presence of active devices can complicate matters even further. Where diodes are involved, such as diode detector or probes or sampling inputs, the reverse (depletion) capacitance of the diode(s) will have some voltage dependency, resulting in instrument or detector probe input capacitance varying with input level. Much less predictable variations can also occur, such as effective negative resistance being created by some input amplifier circuits. In some cases, manufacturers' specifications or instrument handbooks include information on input impedance variation, as illustrated by the examples in Figure 6.14.

Input impedance variation can produce unexpected voltage measurement results (higher or lower readings), where the input impedance of the probe or instrument form a voltage divider with the output impedance of the source. If the source is inductive, resonances can occur, and even if the resonant frequency is well away from the frequency of interest, high-precision measurements can be affected significantly.

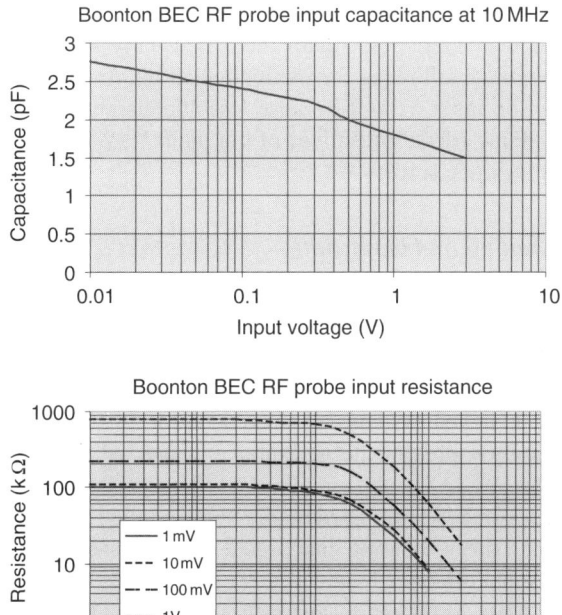

Figure 6.14 Example RF millivoltmeter probe input impedance characteristics

Variation of input impedance characteristics with frequency and level is not limited to instruments and probes with high-input impedances. Instruments and probes with matched inputs also exhibit variation of input impedance characteristics with frequency and level. It is common to see that these are variations described in specifications, for example, different VSWR figures for different frequency bands. The changes can be caused by frequency dependences in the input networks and the components themselves, such as frequency-dependent dielectric losses in capacitors, cables and printed circuit board materials. Active devices also contribute to variations, including the level dependent reverse capacitance of diodes used in rectifier and sampling probes.

Input attenuators are often employed to reduce input levels down to better suit the active circuitry. Input attenuators can also be used to provide some isolation (padding) of the input from variation in the input circuit active devices, input impedance or VSWR, and provide better overall performance. In multirange instruments the input attenuation is often range dependent, leading to different VSWR characteristics on different ranges. Often VSWR worsens on the more sensitive ranges where the amount of attenuation is less, reducing the padding effect and exposing more of the VSWR variation of the input stage active devices.

132 Microwave measurements

Oscilloscopes with dedicated 50 Ω inputs are also likely to have frequency-dependent VSWR characteristics. Again there is usually some small capacitance from the active circuitry following the input attenuators. At higher sensitivities this has greater impact on input VSWR because less attenuation is present at the input so there is less padding (de-sensitisation) of the input VSWR from variation of the active input circuitry impedance.

6.2.8 Source loading and bandwidth

If a 50 Ω source is measured by an instrument with a high-impedance probe or input, typically with $R_{in} > 100$ kΩ, the effect of the resistive loading is negligible (0.05 per cent). However, input capacitance can have a dramatic effect. For example, 25 pF produces an error of 3 dB (30.3 per cent) at 129 MHz. If the source is calibrated for operation in a matched system, terminated into a 50 Ω load, when measured without the termination present the measured voltage will be twice the expected value (Figure 6.15).

If the source is correctly terminated before being applied to the measuring instrument, the effect of the measuring instrument input capacitance is reduced. An alternative view is that effective bandwidth is increased, 25 pF will produce a 3 dB bandwidth of 257 MHz. If the input capacitance is reduced to 2.5 pF the bandwidth improves to 2.6 GHz. Figure 6.16 illustrates the effect in terms of frequency response.

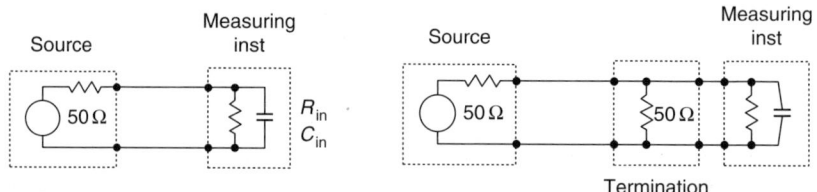

Figure 6.15 Source and measuring instrument: unterminated (left) and terminated (right)

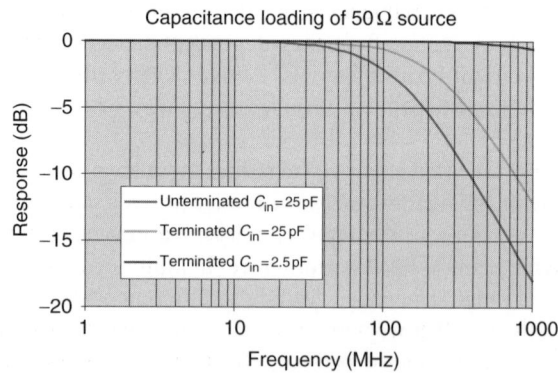

Figure 6.16 Frequency response effect of capacitive loading on 50 Ω source

A practical example is testing of an oscilloscope bandwidth with a high-frequency source. Most high-frequency sources have a 50 Ω output, so if the oscilloscope has a 1 MΩ input impedance best results will only be obtained with a 50 Ω termination at the oscilloscope input. However, results will be affected by the oscilloscope input capacitance appearing across the 50 Ω termination. Higher bandwidth oscilloscopes tend to have lower input capacitances.

6.3 AC and RF/microwave traceability

At lower frequencies, AC voltage traceability is established using thermal converters to compare the RMS value of the AC voltage to a known DC voltage. Up to 1 MHz extremely low uncertainties are achievable, and begin to rise above 1 MHz. Thermal converters are useable to several hundred MHz and 50 Ω thermal converters are available specifically for high-frequency use. At these higher frequencies in comparison with lower frequency, AC voltage is often used instead of DC, as a relative frequency response measurement rather than an absolute measurement.

At higher frequencies RF and microwave power traceability is established using microcalorimeter techniques, which compare RF and DC power. Microcalorimeter measurement times are extremely long, so practical measurements are made using thermistor, thermocouple or diode-based power sensors calibrated against these microcalorimeters.

There is overlap between the RF voltage and RF power traceability in the region from about 1 MHz to several hundred MHz, and it is also possible to obtain a measurement of voltage derived from a power measurement based on power traceability. To obtain a voltage measurement from power it is also necessary to have impedance traceability ($P = V^2/R$). Practically, this means knowledge of the power sensor VSWR when voltage measurement is required. Most calibration laboratories will use thermal converters or thermal transfer standards up to 1 MHz and use power sensors for higher frequencies (Figure 6.17).

6.3.1 Thermal converters and micropotentiometers

Thermal converters are used extensively to establish traceability for AC voltage at low frequencies (up to 1 MHz), but can also be used at higher frequencies up to several hundred MHz. These high-frequency thermal converters have matched inputs (50 Ω or 75 Ω). Construction is similar to the lower frequency devices, except for the addition of a terminating resistor (Figure 6.18). The heating effect in the resistive heater wire of the thermal converter element due to the applied RF voltage (and therefore its RMS value) can be compared to that of a known LF or DC voltage by measuring the thermocouple output. Thermal transfer instruments provide the comparison capability, and some older instruments have facility for external high-frequency TVC operation. However, use of thermal converters at RF is not very common, as RF power sensors provide a more useable alternative.

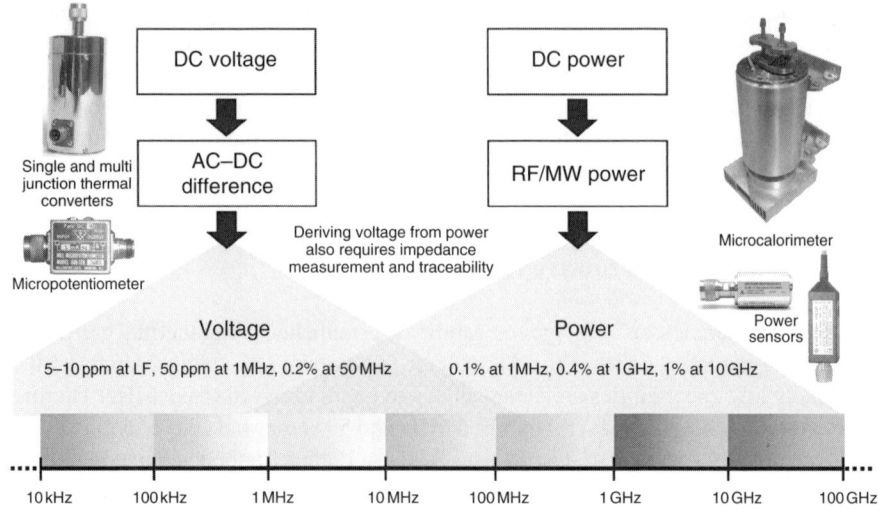

Figure 6.17 Traceability paths for the RF and microwave frequency ranges

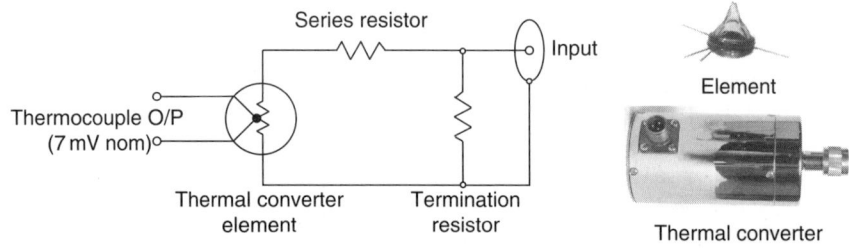

Figure 6.18 Terminated thermal converter. Photograph shows element (top) and a commercial coaxial thermal converter standard

Micropotentiometers provide a means of generating known voltages at microvolt and millivolt levels at frequencies up to several hundred MHz. In the micropot the TVC element is used as the upper arm of a voltage divider together with a coaxial disc resistor constructed within the output connector (Figure 6.19). This produces a voltage divider with extremely good high-frequency response and low output impedance (typically 20 Ω for 100 mV and 0.2 Ω for 1 mV). In use, the thermocouple output is measured with a nanovoltmeter as a means of comparing the RMS value of RF signals in the voltage divider with either a lower frequency or DC signal which can be accurately measured at the output. Thus, an RF to DC or RF to LF comparison can be made at microvolt and millivolt levels. Uncertainties around 1 per cent at 100 MHz and 5 per cent at 500 MHz are achievable, but avoiding signal switching and thermal stability errors can be difficult.

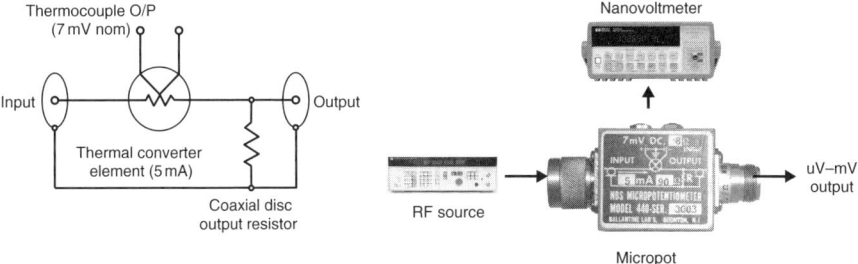

Figure 6.19   *Micropotentiometer (Micropot) used to generate known RF voltages from μV to mV*

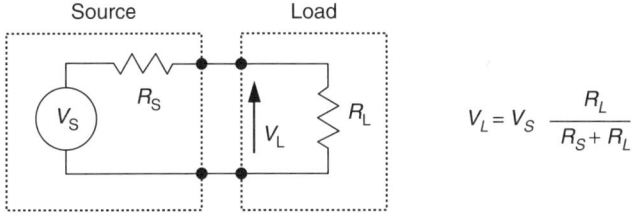

Figure 6.20   *Resistive source and load*

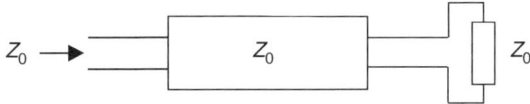

Figure 6.21   *Transmission line with characteristic impedance Z_0 terminated with Z_0*

6.4   Impedance matching and mismatch errors

If a source of voltage V_S and output impedance R_S is connected to a load of impedance R_L, the voltage developed across the load V_L is given by $V_L = V_S \cdot R_L/(R_S + R_L)$ (Figure 6.20).

Maximum power is transferred when $R_S = R_L$, which leads to the requirement to match load and source impedances. The full analysis at high frequency with complex impedances $(R + jX)$ becomes more involved, and also yields a requirement for impedance matching to ensure maximum power transfer.

$$V_L = V_S \frac{R_L}{R_S + R_L}$$

A high-frequency transmission line will have characteristic impedance Z_0 such that when terminated with Z_0 at its output it also presents Z_0 at its input (Figure 6.21). This impedance can be purely resistive, and practical transmission lines are typically designed to have an impedance of 50 Ω. Other values such as 75 Ω are also common.

In practice, the impedances are not purely resistive, particularly at high frequency, and are not expressed directly in terms of resistance, capacitance and inductance, but in terms of VSWR or return loss (which is related to VSWR). These parameters are high-frequency terms that are used to describe how well the actual impedance varies from the nominal (50 Ω) impedance, and relate the incident and reflected signals at a mismatch. If either the source impedance or the load impedance is extremely close to the nominal 50 Ω, the impact of difference from nominal (mismatch) of the other impedance is minimised.

The impact of mismatch on signal level accuracy can be assessed by considering the VSWR of source and load. If the magnitude and phase of the mismatch are known a correction could be applied for the mismatch error. Unless detailed measurements are made (e.g. with a vector network analyser), VSWR figures for typical instrument inputs and probes are worst case values and correction cannot be applied. Measuring source VSWR is even more complicated, and having phase information is extremely rare. Calculation of the measurement error due to mismatch can easily be made from the VSWR values, usually as an estimate based on the instrument specifications unless (scalar) measurements have been made, for example, using a return loss bridge.

If used, adapters will also introduce mismatch and attenuation errors. More complex multiple mismatch models can be used to evaluate their effects. As an alternative for simplicity, a single VSWR value can be determined by measurement of the load and adapter combination.

6.4.1 Uncertainty analysis considerations

When making voltage measurements, the mismatch error should be calculated as a voltage error. In most texts mismatch errors are treated as an error in the power delivered to the load, not the error in the voltage appearing at the load. Voltage mismatch error may be calculated using the expression below:

$$\text{Voltage error} = \left(1 - \frac{1}{(1 \pm |\Gamma_S| |\Gamma_L|)}\right) \times 100\%$$

where the magnitude of the reflection coefficient, Γ, is given by

$$|\Gamma| = \frac{\text{VSWR} - 1}{\text{VSWR} + 1}$$

and the return loss $= 20 \log |\Gamma|^{-1}$ dB.

As an alternative, if the errors are small, the voltage error will be half the power error. This is because voltage is proportional to the square root of power. More completely, in uncertainty analysis it is usual to calculate a sensitivity coefficient as the partial derivative of the expression describing the measured quantity with respect to the influence variable concerned – in this case the value is 0.5.

When performing an uncertainty analysis, it is necessary to combine the various uncertainty contributions at the same confidence level, known as standard uncertainty, equivalent to one standard deviation (1σ) or 68.3 per cent confidence level. Mismatch

RF voltage measurement 137

uncertainties are considered as one of the type B (systematic) contributions and are generally treated as having a U-shaped distribution. The estimate of voltage mismatch error calculated above is divided by $\sqrt{2}$ to convert it to standard uncertainty. More information on uncertainty analysis, including RF examples, can be found in UKAS publication M3003.

6.4.2 Example: Oscilloscope bandwidth test

As an example, consider the testing of an oscilloscope's bandwidth. The 50 Ω input of the oscilloscope is connected to the levelled sine output of an oscilloscope calibrator, which has a 50 Ω output impedance. The measurement procedure either determines the frequency at which the displayed response falls by 3 dB from a low-frequency value (usually at 50 kHz), or confirms that the displayed response at the nominal bandwidth frequency is within 3 dB of that at a low frequency.

The voltage amplitude displayed or measured by the oscilloscope will be subject to mismatch error, which may be estimated from the VSWR values involved. The following are VSWR figures for a popular oscilloscope calibrator and for some common high-bandwidth oscilloscopes:

Calibrator (source): VSWR <1.1 to 550 MHz, <1.2 550 MHz to 3 GHz, <1.35 3 GHz to 6 GHz
Oscilloscope (UUT):
Typical 1 GHz oscilloscope   VSWR <1.5 to 1 GHz
Typical 4–6 GHz oscilloscopes VSWR <1.1 to 2 GHz, <1.3 4 GHz to 6 GHz, some <2.0 at 4 GHz.

Oscilloscope VSWR can be worse on the more sensitive ranges. A popular 6 GHz oscilloscope quotes typical VSWR characteristics as 1.3 at 6 GHz for >100 mV per div and 2.5 at 6 GHz for <100 mV per div.

The chart in Figure 6.22 shows the mismatch error as a function of frequency for a number of UUT (load) VSWR values for this particular source and its specified VSWR figures. Note that the errors are not symmetrical – the amount by which a given mismatch can reduce the signal amplitude is slightly more than the amount by which it can increase the amplitude. The larger figure would usually be used as a worst case plus-or-minus contribution for uncertainty analysis purposes.

6.4.3 Harmonic content errors

Harmonic content causes waveform distortion which will lead to errors for measurements made with peak detecting instruments. Most instruments are calibrated in terms of the RMS value of a sinewave input, even when the measurement system detects the peak value of the waveform, such as a diode detector operating above the square-law region. The error depends on both the magnitude and phase of the harmonic signal, and the different harmonics will have different effects. Generally the effect becomes less sensitive to phase as the harmonic number increases, as there will be more cycles of harmonic per cycle of fundamental.

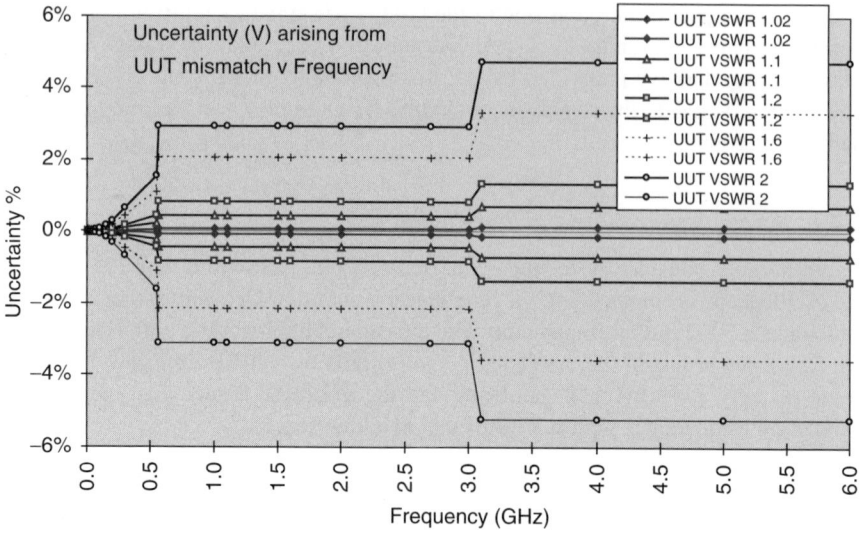

Figure 6.22 Chart showing voltage uncertainty due to mismatch

Determining the impact of harmonic content depends on the nature of the peak detector – whether it is a half or full wave rectifier, and whether the detector responds to the peak or average value. The example waveforms shown in Figure 6.23 illustrate the effects of third harmonic in phase with the fundamental, which flattens the positive and negative half cycles alike, and second harmonic at 45° which increases the peak value of positive half cycles and decreases the peak value of negative half cycles.

Usually there is no knowledge of phase relationships, so it is necessary to consider worst case conditions. The chart in Figure 6.24 shows the impact of second, third and fourth harmonics at various levels below the fundamental (as dBc). Often only the worst case harmonic level is known, not the harmonic number – for example, from harmonic content specifications of a signal source. In this case the worst case for any harmonic must be considered. The total RMS value is simply the Root-Sum-Square summation of the fundamental and harmonic voltages.

6.4.4 Example: Oscilloscope calibrator calibration

Consider the example of calibrating the levelled sine function of an oscilloscope calibrator. This calibrator feature produces a sinusoidal output with accurate peak to peak amplitude over a wide frequency range (6 GHz) for oscilloscope bandwidth testing.

The equipment used to calibrate the oscilloscope calibrator levelled sine output measures RMS voltage at low frequency (a precision AC voltmeter) and RF power at high frequency (a power meter and sensor). The set-up for high-frequency power meter measurements is shown in Figure 6.25. To perform the required calibration these measurements must yield a result in terms of peak to peak voltage. At low frequencies

RF voltage measurement 139

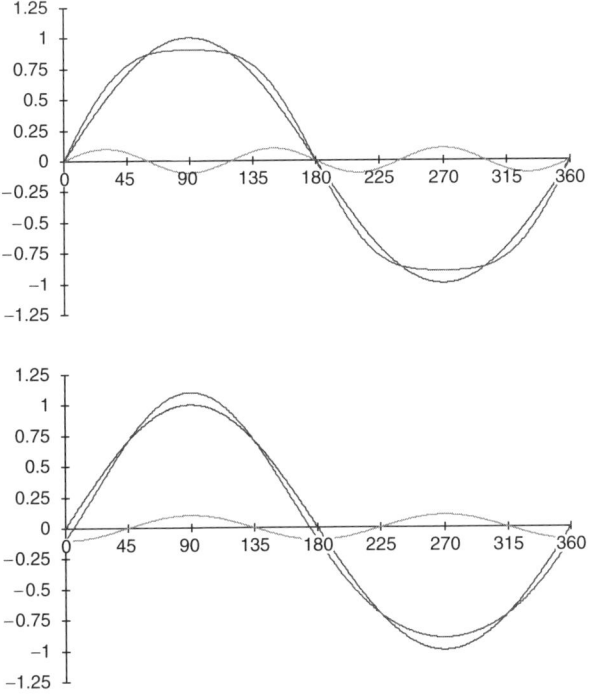

Figure 6.23 Examples of harmonic distortion: third harmonic at 0° (top) and second harmonic at 45° (bottom)

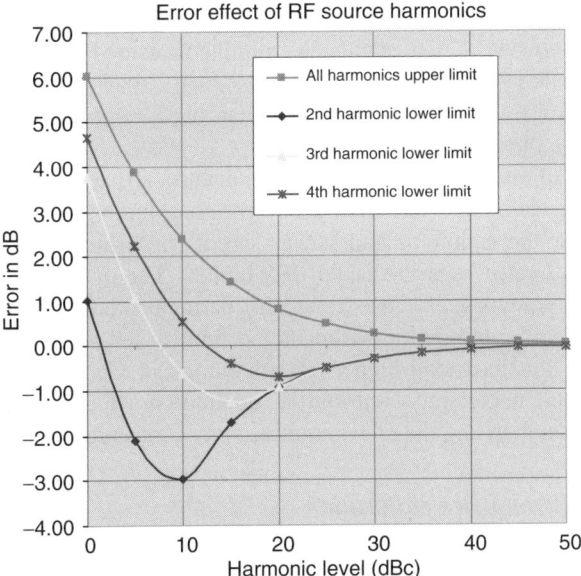

Figure 6.24 Chart showing error in peak to peak voltage due to harmonic distortion

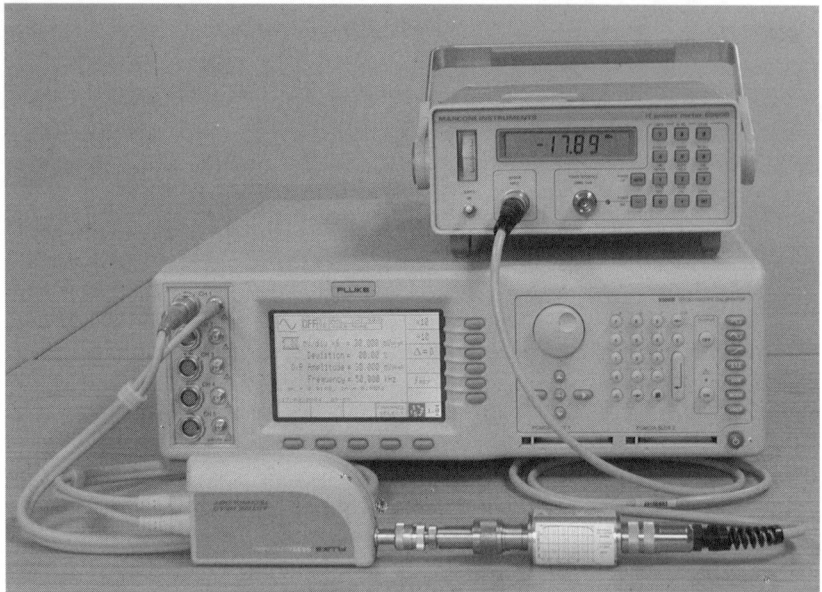

Figure 6.25   Photograph showing setup for levelled sine calibration up to 3.2 GHz against a power meter

the RMS voltage measurement is converted by multiplying by $2\sqrt{2}$, directly providing an absolute pk–pk voltage value. At high frequency, the measurements are made as flatness with respect to a reference frequency of 100 kHz (the lowest frequency for which the internally AC coupled power sensor presents a reasonable VSWR). The effective RMS voltage is first calculated from the measured power into 50 Ω, then RMS to pk–pk conversion is performed.

The calibrator 6 GHz output head is similar, but has an SMA connector and is calibrated with a different sensor.

The effects of mismatch and of waveform distortion (harmonic content) are key uncertainty considerations. The mismatch error contribution is calculated from the VSWR data for the calibrator and power sensor, in terms of voltage error (not power), and is treated as a U-shaped distribution. The distortion contribution is calculated from knowledge of the worst case harmonics taken from the calibrator specifications, confirmed by measurement (<−35 dBc for second-order harmonics and <−40 dBc for third- and higher-order harmonics). Connector repeatability is another applicable uncertainty contribution, common in RF calibration. Figure 6.26 shows a table summarising the relevant uncertainty contributions.

6.4.5   RF millivoltmeter calibration

RF millivoltmeters are usually calibrated against power sensors, as shown in Figure 6.27 (another example where RF voltage is derived from RF power). The

RF voltage measurement 141

Example flatness uncertainties for 10 mV to 5Vp/p at 1GHz					
Power meter (Pwr)	Cal factor at ref freq*	B	95% CL	0.6%
	Cal factor at test freq*	B	95% CL	0.5%
	1Yr Stability	B	Rect	0.3%
	±3°C tempco	B	95% CL	0.015%
	Linearity and resolution	B	Rect	0.4%
Mismatch (Pwr)		B	U	2.4%
Distortion (Vp/p)		B	Rect	0.5%
Combined noise (Pwr)		A	1σ	0.05%
Connector repeatability (Pwr)		A	1σ	0.05%
TOTAL (Expanded unc) (Vp/p)			95% CL	1.9%

Figure 6.26   Table showing individual type A and B uncertainty contributions. Total is expanded uncertainty after combination in accordance with UKAS M3003. Asterisk represents Power Meter Cal factor contribution, that is, power uncertainty, taken from its calibration certificate.

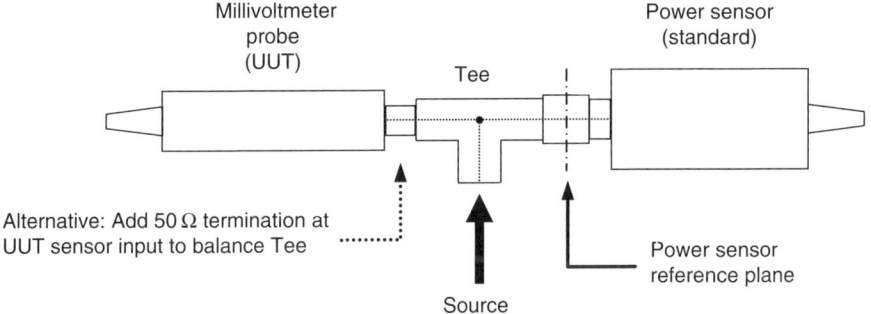

Figure 6.27   RF millivoltmeter calibration

millivoltmeter probe will usually present a high impedance (e.g. 100 kΩ/3 pF) and the power sensor will be a 50 Ω device.

A key issue is arranging that the voltage probe senses the voltage present at the reference plane of the power sensor. In practice, a Tee-piece must be used to make the physical interconnections, which will introduce a small length of transmission line between the power sensor reference plane and the voltage probe input. There will be a voltage standing wave in this short length of transmission line which produces a small difference between the voltage at the power sensor reference plane and the probe input, resulting in a measurement error. The length of line may be quite short, say 20–30 mm physical length for a precision N-series Tee (its effective electrical length depends on the dielectric properties of the insulation within the tee), but for

Estimated electrical length between connector and T reference plane						
	L_E(mm)	ε_R	Frequency for fractional λ separation			
			λ/500	λ/100	λ/20	λ/4
N-Type	43	2.2	16 MHz	70 MHz	350 MHz	1.75 GHz
BNC	24	2.2	25 MHz	125 MHz	625 MHz	3.12 GHz

Figure 6.28 Estimated transmission line length in the Tee-piece for typical connector types

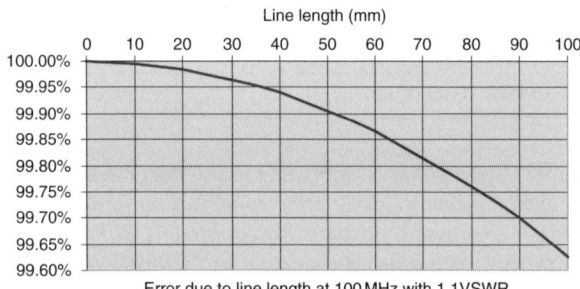

Figure 6.29 Change in voltage with electrical transmission line length

precision measurements the small error introduced could be significant (Figures 6.28 and 6.29). In theory a correction could be determined and applied, but in practice especially constructed Tee-pieces are employed to make the connections physically much closer, and minimise the error. Alternatively, a 50 Ω termination at the UUT sensor input will balance the Tee, and avoid the problem (assuming a 50 Ω power sensor).

Other error sources include mismatch (error from the nominal impedance of the sensor assumed for the power to voltage conversion), distortion of the source waveform and power meter uncertainties (calibration, linearity, resolution, stability, etc.) as discussed in previous examples.

RF voltage measurement 143

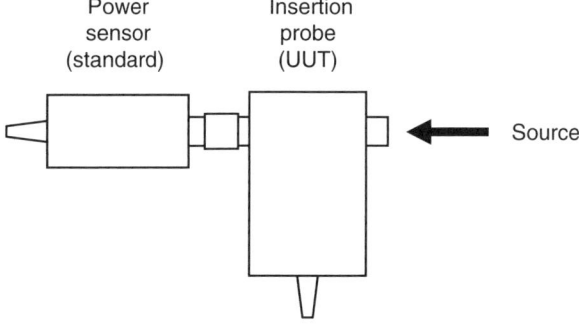

Figure 6.30   Insertion probe calibration

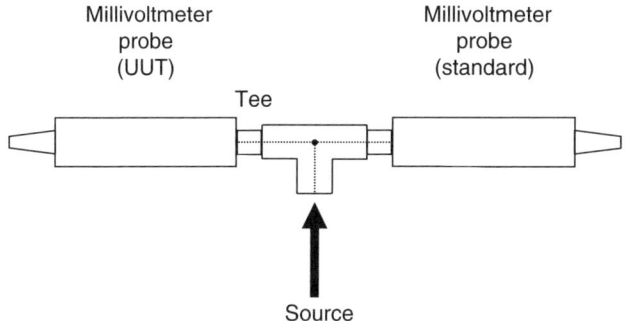

Figure 6.31   Intercomparing voltmeter probes having similar impedances

Insertion probes incorporate the voltage probe within a short transmission line, allowing a voltage measurement to be made in a matched system, and simplify the calibration issue. In this case the Tee-piece is effectively built into the insertion probe, allowing direct connection to the power sensor used as the reference standard as illustrated in Figure 6.30. When calibrating one voltmeter against another, they are usually connected either side of a Tee, as shown in Figure 6.31. If their impedances are similar and the Tee is symmetrical, no additional errors or corrections need to be considered.

Further reading

The popular test equipment manufacturers are useful sources for a variety of useful information, published in the form of application notes and guides, many available for download from their websites. The following are a few examples of particular relevance to the topics discussed in this chapter.

Agilent Technologies application note

Fundamentals of RF and Microwave Power Measurements (Parts 1, 2, 3 and 4), AN1449-1,2,3,4, literature number 5988-9213/4/5/6EN

Part 3 discusses theory and practice of expressing measurement uncertainty, mismatch considerations, signal flowgraphs, ISO 17025 and examples of typical calculations. Go to www.agilent.com and search for AN1449 to download.

Rohde & Schwarz application note

Voltage & Power Measurement: Fundamentals, Definitions, Products, PD 757.0835.23

Comprehensive document including definitions, discussion of reflections, standing waves, mismatches, sensors and instrumentation operating principles and characteristics, etc. Go to www.rohde-schwarz.com and search for 757.0835.23 to download.

Anritsu application note

Reflectometer Measurements – Revisited

Includes discussion of VSWR measurement, mismatch uncertainties and effects of adapters. Go to www.us.anritsu.com and search for reflectometer measurements to download.

Aeroflex booklet

RF Datamate, P/N 46891/883

Produced by the signal sources group. A 72-page guide to commonly used RF data, measurement methods, power measurement uncertainties, etc. Go to www.aeroflex.com and search for RF Datamate to order free copy online.

The IET publishes papers presented at conferences and other events in its Proceedings documents. UKAS (United Kingdom Accreditation Service) and the European Cooperation for Accreditation (EA) also provide downloadable guides. The following are a few examples of particular relevance to the topics discussed in this chapter.

Paper discussing mismatch uncertainties

Harris, I. A., and Warner, F. L.: 'Re-examination of mismatch uncertainty when measuring microwave power and attenuation', *IEE Proceedings* H, Microwaves Optics and Antennas, 1981;**128**: 35–41

UKAS Publication M3003

The Expression of Uncertainty and Confidence in Measurement

Produced by the United Kingdom Accreditation Service. An interpretation of the ISO Guide to the Uncertainty of Measurements (GUM) document, including measurement uncertainty estimation examples, for the use of ISO17025 Accredited laboratories. Go to www.ukas.com and select Information Centre, then Publications to download.

European Cooperation for Accreditation of Laboratories publication
EAL Guide EA-10/07 *Calibration of Oscilloscopes* (previously EAL-G30)
 Produced by EAL to harmonise oscilloscope calibration. Provides guidance to national accreditation bodies setting up minimum requirements for oscilloscope calibration and gives advice to calibration laboratories to establish practical procedures. Published June 1997. Go to www.euromet.org/docs/calguides/index.html and select document 'EA-10/07' to download.

Chapter 7

Structures and properties of transmission lines

R. J. Collier

7.1 Introduction

The number of different transmission lines has greatly increased in recent years. This range of lines enables the microwave circuit designer to choose particular features which meet the specifications. For instance, it may be that certain values of characteristic impedance, phase constant, dispersion or attenuation are required. Or it may be that the ease with which these lines can be used to couple to various solid-state devices is important. Or, finally, it may be that the line has unique properties in certain configurations, for example, gives good coupling to other circuits or radiates in a special way. As many microwave circuits are now completely integrated so that the largest dimension of the whole circuit could be less than a millimetre, special lines are required to couple signals into them.

Measuring microwave circuits is still as important as ever. Almost without exception, most microwave measurement equipment has remained with the same input and output transmission lines. For most microwave frequencies, a coaxial output is used. Above 20 GHz, a waveguide output is sometimes used. One of the major problems in circuit measurement is designing transitions from the standard coaxial and waveguide ports to numerous transmission lines that now exist in modern circuits. In the case of integrated circuits special surface probes are used that involve a tapered transmission line. As many microwave measurements of impedance, noise, gain, etc. involve using transitions from either coaxial or waveguide to these other transmission lines the properties of transitions are critical in the measurement. Finally, most impedance measurements consist of a comparison with a standard impedance. These standards are usually constructed out of either coaxial or waveguide transmission lines. Again, the properties of the coaxial or waveguide junctions can be the main limitation of these measurements, particularly at higher frequencies.

Most transmission lines are designed to operate with only one mode propagating. However, every transmission line will support higher-order modes if the frequency is high enough. Since these higher-order modes have separate velocities, it is not usually possible to do simple measurements when they are present. Hence, in the description that follows, the upper limit of mono-mode propagation is usually given. As a simple rule the transmission line has to get smaller as the wavelength gets smaller to avoid higher-order modes. This has a marked effect on the attenuation of transmission lines using metallic conductors. The attenuation will rise due to the skin effect by a factor of $f^{1/2}$. However, in addition as the structures get smaller, the increased current crowding means that the overall attenuation increases by a factor of $f^{3/2}$. Thus, a transmission line like coaxial cable is often made with a small diameter at high frequencies to ensure mono-mode propagation but the consequence is a large increase in attenuation which can greatly affect measurements.

Finally, dispersion has several causes. One of these is when the permittivity of any dielectric used in a transmission line changes with frequency. In practice this is often quite a small effect. This is called material or chromatic dispersion. In transmission lines where the electromagnetic waves propagate in only one dielectric and the mode has only transverse fields, for example, coaxial line, the material dispersion is the only type. For standards in coaxial lines, often an air-filled line is used to avoid even material dispersion. For transmission lines where the electromagnetic waves propagate in two or more dielectrics the modes are more complex. In general, as the frequency increases the energy concentrates in the dielectric with the highest dielectric constant. This dispersion is usually called waveguide dispersion and occurs in, for example, microstrip and mono-mode optical fibre. Waveguide dispersion also occurs where there is a firm cut-off frequency as in metallic rectangular waveguide. Modal dispersion occurs when there are many modes propagating which is the case in some forms of optical fibre. Since most transmission lines are designed to be mono-mode, modal dispersion is usually avoided.

7.2 Coaxial lines

A coaxial line is shown in Figure 7.1. The radius of the inner conductor is a and the inner radius of the outer conductor is b. At microwave frequencies the transmission line parameters are

$$L = \frac{\mu}{2\pi} \log_e \left(\frac{b}{a}\right) \tag{7.1}$$

$$C = \frac{2\pi \varepsilon}{\log_e (b/a)} \tag{7.2}$$

$$R = \frac{R_S}{2\pi} \left(\frac{1}{b} + \frac{1}{a}\right) \tag{7.3}$$

$$G = \frac{2\pi \sigma}{\log_e (b/a)} \tag{7.4}$$

Structures and properties of transmission lines 149

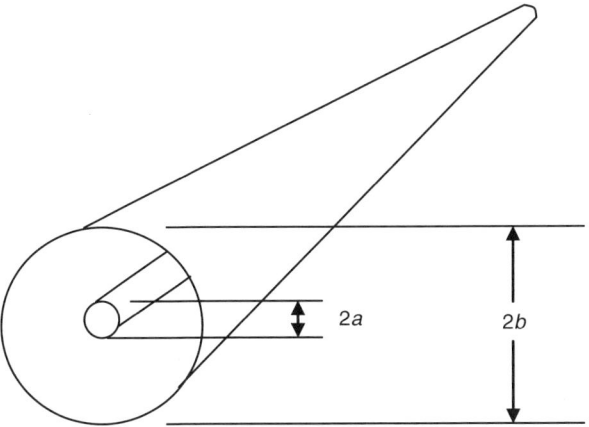

Figure 7.1   A coaxial line

This gives

$$Z_0 = \frac{1}{2\pi}\sqrt{\frac{\mu}{\varepsilon}}\log_e\left(\frac{b}{a}\right) \tag{7.5}$$

$$v = \frac{1}{\sqrt{\mu\varepsilon}} \quad \text{(see chapter 1 on transmission lines)} \tag{7.6}$$

$$\alpha = \frac{R}{2Z_0} + \frac{GZ_0}{2} \quad \text{(see chapter 1 on transmission lines)} \tag{7.7}$$

where R_S is the skin resistance of the conductors and is proportioned to $f^{1/2}$, and σ is the conductivity of the dielectric which is also a function of frequency.

Coaxial lines can be easily made with a range of characteristic impedances from 20 to 100 Ω. Their dispersion characteristics are good except at very low frequencies, where

$$Z_0 = \sqrt{\frac{R}{G}}, \quad \text{i.e. } \omega L \ll R; \omega C \ll G \tag{7.8}$$

and at very high frequencies when higher-order modes appear. These higher-order modes are discussed in Reference 1 and an approximate guide to their appearance is the condition

$$\lambda < 2\pi b$$

To maintain mono-mode propagation, the coaxial cable is usually made smaller at higher frequencies. Typical values for b are 7, 5, 3 and 1 mm. Unfortunately, as b gets smaller, the attenuation increases. Therefore, most cables are designed to be an acceptable compromise between attenuation and mono-mode propagation.

Coaxial lines are used as the input and output ports for most measurement equipment up to about 25 GHz with $2b = 7$ mm. Connectors with low insertion loss and good repeatability make high accuracy measurements possible. Transitions to other transmission lines exist for most types and in particular to rectangular waveguide and microstrip. Both these transitions have insertion loss and in the latter case the losses include radiation loss. As the microstrip transition has poor repeatability it is good measurement practice to measure at a coaxial junction where possible and use de-embedding techniques to find the circuit parameters.

7.3 Rectangular waveguides

Along with coaxial lines, metallic rectangular waveguides are used extensively in microwave measurements particularly above 25 GHz. Some of the properties of these guides were given in chapter 1 on transmission lines. It is worth repeating the comments about bandwidth. Take the TE_{10} mode in X Band waveguide as an example:

a	b	f_c	Usable frequency range	α
0.9"	0.4"	6.557 GHz	8.20–12.40 GHz	0.164 dB m^{-1}

A waveguide is not normally used near its cut-off frequency, f_c, and so an X Band waveguide is not used between 6.557 and 8.20 GHz. This is because all the properties of the guide are changing rapidly with frequency in this region. At 8.20 GHz most properties are within a factor of 1.66 of the free space values. To avoid higher-order modes which start at $2f_c$ or at 13.114 GHz, the usable frequency range ends at 12.4 GHz where the properties are within a factor of 1.18 of their free space values. Therefore, waveguide has a bandwidth of 4.2 GHz which is about two-thirds of the octave band theoretically available for mono-mode propagation. Even in this bandwidth the waveguide is still more dispersive than most other transmission lines. To cover a wide frequency range a series of waveguides are used with different a and b values. As with coaxial lines, the attenuation rises sharply as the waveguide size reduces and around 100 GHz is unacceptably high for many applications. One reason for this is that the surface roughness of the inner surface of the guide contributes significantly to the losses at these frequencies. Also at 100 GHz and above waveguide connectors do not have a satisfactory insertion loss or repeatability for accurate measurements.

At the lower microwave frequencies the waveguides have dimensions of several centimetres and are able to transmit high powers far better than any other transmission line. It is mainly for this reason that their use has continued at these frequencies.

7.4 Ridged waveguide

A technique for increasing the bandwidth of rectangular waveguide is to use a ridge in the centre of the guide as shown in Figure 7.2.

This ridge has a minimal effect on those waveguide modes with a null of electric field at the centre of the guide. Hence the cut-off frequency of the TE_{20} mode is almost

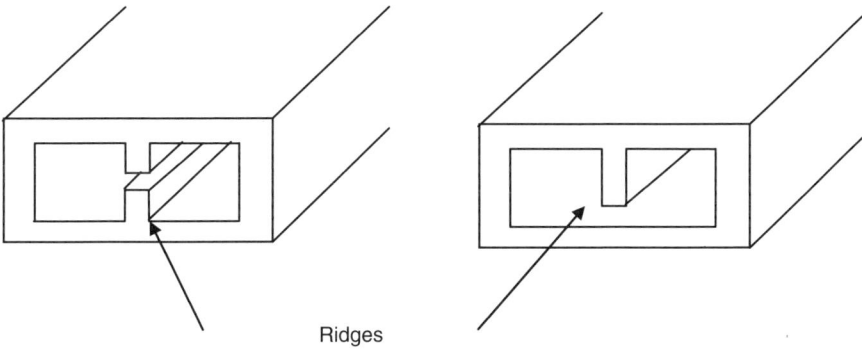

Figure 7.2 Ridged waveguides.

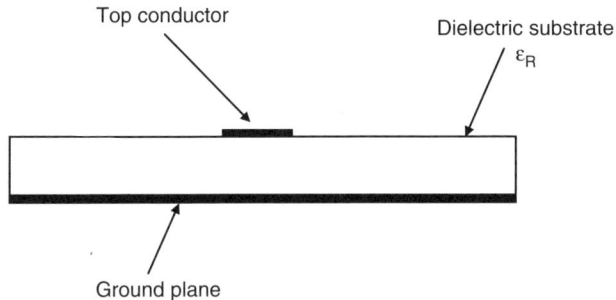

Figure 7.3 Cross section of a microstrip transmission line

unchanged. However, the cut-off frequency of the TE_{10} mode is greatly reduced, in some cases by as much as a factor of 4. For example, in an X Band waveguide this would lower the cut-off frequency to 1.64 GHz and the usable frequency range would be 2.054–12.4 GHz.

However, the concentration of fields in the gap also concentrates the currents in that region. This increases the attenuation and this factor limits the use of this transmission line to the lower microwave frequencies. However, the multioctave bandwidth with reduced dispersion is the feature that makes its use, particularly in wideband sources, quite common. Details of the guide's properties are in References 2 and 8.

7.5 Microstrip

Microstrip transmission line is one of the most common transmission lines used in microwave circuits (Figure 7.3).

It can be manufactured using conventional photolithographical techniques with great accuracy approaching ±50 nm. It has dimensions which make connections to solid-state components relatively easy and since the circuit is usually on one side of the substrate access to the input and output ports is also straightforward. With modern integrated circuit technology these microstrip circuits can often be made so small that the transmission line effects disappear and simple low-frequency circuit designs can be used.

Since the wave on a microstrip line moves partly in the air above the substrate and partly in the substrate itself the velocity, v, is in the range

$$\frac{3 \times 10^8 \text{ m s}^{-1}}{\sqrt{\varepsilon_R}} < v < 3 \times 10^8 \text{ m s}^{-1} \tag{7.9}$$

where ε_R is the relative permittivity of the substrate. The frequency range of microstrip is from 0 Hz to the cut-off frequency of the next higher-order mode. This can be found from λ_c which is approximately given by

$$\lambda_c = 2w$$

where w is the width of the top conductor.

At X Band, for example, typical microstrip parameters might be

Width	Thickness of substrate	v	ε_R	f_c
0.6 mm	0.6 mm	1.15×10^8 m s^{-1}	9.7	80 GHz

Dispersion in microstrip is much less than rectangular waveguide and the velocity and the characteristic impedance would typically change by only a few per cent over several octaves. However, the attenuation in microstrip is much greater than rectangular waveguide, by about a factor of 100. This is because the currents are far more concentrated in microstrip. Surprisingly, this is often not a critical factor as the circuits are usually only a wavelength in size and in integrated form often very much smaller. The range of characteristic impedance can be varied, usually by changing the width, and values in the range 5–150 Ω are possible. Transitions to microstrip are usually made with miniature coaxial lines at the edge of the substrate. 'On-wafer' probing is not possible with microstrip, but probing using electro-optic methods is possible. To a greater and lesser extent microstrip circuits radiate. Indeed, microstrip antennas are indistinguishable from some circuits. For this reason most microstrip circuits need to be enclosed to prevent radiation leaving or entering the circuits. The design of this enclosure is often a critical part of the whole circuit. Various alternatives to microstrip exist, including inverted microstrip, stripline and triplate [3–6].

7.6 Slot guide

Slot guide is shown in Figure 7.4 and is used in various forms of circuit.

The fields like microstrip are partly in the air and partly in the substrate, but the dispersion is less than microstrip. The characteristic impedance is mainly a function of the width of the slot and can have a range of values typically 50–200 Ω. There are some applications of couplers of slot line to microstrip line which have series rather

Structures and properties of transmission lines 153

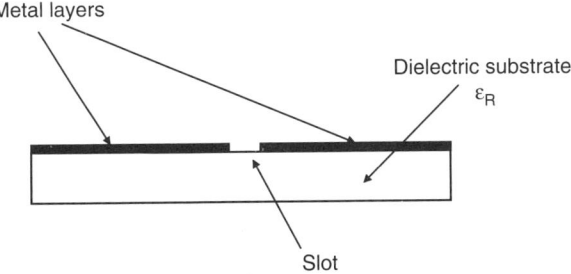

Figure 7.4 Slot guide

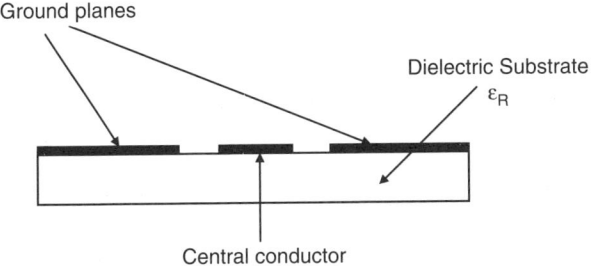

Figure 7.5 Coplanar waveguide

than the normal parallel configurations [6,7]. The main reason for including them in this list is to lead on to forms of slot line which are widely used, namely coplanar waveguide and finline.

7.7 Coplanar waveguide

Coplanar waveguide is shown in Figure 7.5 and is used in many integrated circuits as it can be measured on the chip using on-wafer probes. These probes are in the form of a tapered coplanar waveguide followed by a transition to either coaxial line or rectangular waveguide.

Since the guide consists of two slots in parallel, the low dispersion of slot guide is also present in coplanar waveguide. The range of characteristic impedances is typically 25–100 Ω depending on the width of the slots and the relative permittivity of the substrate. The velocity of coplanar waveguide, similar to that of slot line is approximately the average of the velocities of a TEM wave in air and in the substrate, that is

$$v = \frac{3 \times 10^8 \text{ m s}^{-1}}{((\varepsilon_R + 1)/2)^{1/2}} \tag{7.10}$$

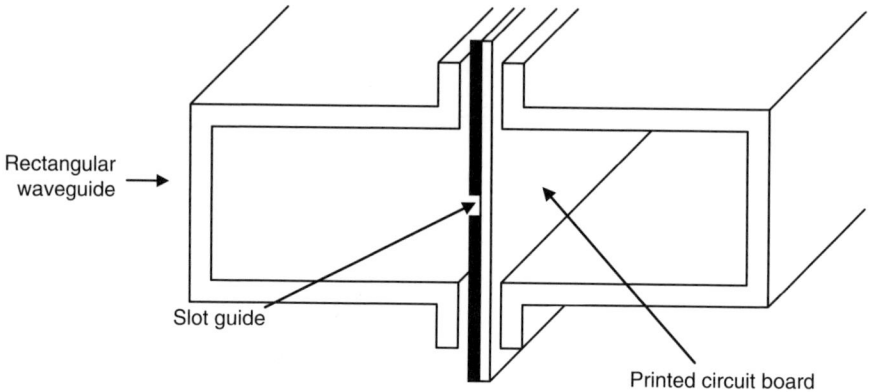

Figure 7.6   Finline

Similar to microstrip, the currents are concentrated and the attenuation of both slot line and coplanar waveguide is much higher than waveguide. However, if the circuit dimensions are much smaller than a wavelength this is not a limitation. The main advantage of the structure is that unlike microstrip the ground plane is easily accessible and connections to solid-state and other devices in series and parallel are possible. Higher-order modes do exist in these structures but one of the limitations is not these modes but box modes in the dielectric substrate and the enclosing box. The design of circuits to avoid losing energy to these modes requires special care [7].

7.8   Finline

Finline is shown in Figure 7.6. It is used in conjunction with conventional rectangular waveguide and enables circuits to be used with transitions to waveguide ports. It also avoids the use of an enclosure as the waveguide provides this. It is similar to ridged waveguide in that the printed circuit board provides the ridge [6, 8]. The advantage of finline is that solid-state components can be mounted on the substrate as in slot line thus avoiding the difficult problem of mounting such components in rectangular waveguide.

7.9   Dielectric waveguide

The increasing attenuation in all transmission lines using conductors makes their use less practical above 100 GHz. The dielectric waveguide shown in Figure 7.7 is a transmission line which overcomes these problems [9]. Its operation is similar to optical fibre in that the energy is trapped inside the waveguide by the principle of total internal reflection. The obvious difference between this waveguide and other lines described so far is that there is no easy way to connect many devices to it. However,

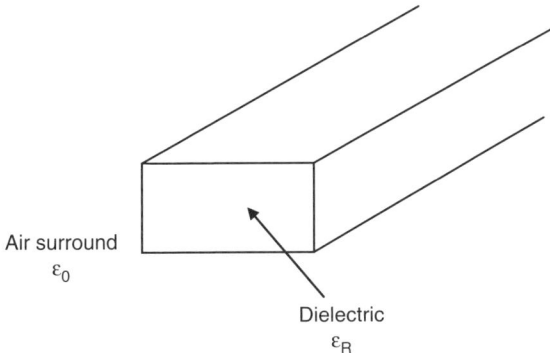

Figure 7.7   Dielectric waveguide

good transitions to rectangular metallic waveguide are available along with various circuit components.

The guide does have fields outside the structure which decay away rapidly in the directions transverse to the direction of propagation. These fields can be used for various circuit components. Supporting the guide requires a dielectric of lower dielectric constant and in the case of optical fibre this is made sufficiently large to ensure all the external fields have decayed to zero. For a dielectric guide at 100 GHz the dimensions might be 2×1 mm. The dispersion is less than metallic rectangular waveguide and bandwidths of an octave are possible. Recent research has shown that junctions between dielectric waveguides have superior repeatability and insertion loss to metallic waveguides. It is anticipated that dielectric waveguides will be commonly used in the future for frequencies above 100 GHz.

References

1 Marcuvitz, N.: *Waveguide Handbook*, IET Electromagnetic Waves Series 21 (IET, London, 1986)
2 Saad, T. S., Hansen, R. C., and Wheeler, G. J.: *Microwave Engineers' Handbook*, Vols. 1 & 2 (Artech House, Dedham, MA, 1971)
3 Edwards, T. C.: *Foundations for Microstrip Circuit Design*, 2nd edn (Wiley, Chichester, 1992)
4 Wolff, I.: *Microstrip – Bibliography 1948–1978* (V.H. Wolff, Aachen, Germany, 1979)
5 Chang, K.: *Microwave Solid-State Circuits and Applications* (Wiley, New York, 1994)
6 Wadell, B. C.: *Transmission Line Design Handbook* (Artech House, Norwood, MA, 1991)
7 Simons, R. N.: *Coplanar Waveguide Circuits, Components and Systems* (Wiley, New York, 2001)

8 Helszajn, J.: *Ridge Waveguides and Passive Microwave Components*, IET Electromagnetic Waves Series 49 (IET, London, 2000)
9 Rozzi, T., and Mongiardo, M.: *Open Electromagnetic Waveguides*, IET Electromagnetic Waves Series 43 (IET, London, 1997)

Further reading

1 Bhartia, P., and Bahl, I. J.: *Millimeter Wave Engineering and Applications* (Wiley, New York, 1984)
2 Smith, B. L., and Carpentier, M. H.: *The Microwave Engineering Handbook*, vol. 1 (Chapman Hall, London, 1993)
3 Hilberg, W.: *Electrical Characteristics of Transmission Lines* (Artech House, Dedham, MA, 1979)
4 Baden Fuller, A. J.: *Microwaves* (Pergamon, Oxford, 1969)
5 Chipman, R. A.: *Transmission Lines*, Schaum's Outline Series (McGraw-Hill, New York, 1968)

Chapter 8
Noise measurements
David Adamson

8.1 Introduction

In any classical system (i.e. non-quantum) the ultimate limit of sensitivity will be set either by interference or by random signals that are produced within the system. In this chapter we are not concerned with the situations where interference sets the ultimate limit and so henceforth, we will consider only situations where the ultimate sensitivity is set by random signals. The minimum possible value for these random signals is generally set by physical phenomena collectively called noise. If we are interested in determining the limit of sensitivity of a system then we need to measure the random signals which determine that limit. Alternatively, we may wish to design a system to reach a chosen level of sensitivity in which case we need to have methods to allow the calculation of the level of random signals that are to be anticipated in the system.

In this chapter, we are particularly interested in systems that are sensitive to electromagnetic signals, generally in the microwave and radiofrequency (RF) region of the spectrum. However, some of the principles apply at any frequency, or even to systems that are not concerned with electromagnetic signals.

Random signals produced in an electrical system are usually called electrical noise. The concept of noise is familiar to anyone who has tuned an AM radio to a point between stations where the loudspeaker will produce a hissing noise which is attributable to the electrical noise in the system. This example also illustrates an important general point about noise – the source of the noise may be either internal (caused by phenomena in the receiver in this case) or external (atmospheric and other sky noise in this case). Usually, we can attempt to choose our system components to bring the noise internal to the system to a level which is appropriate for that system whereas the external noise is often fixed by other phenomena and its level can only be controlled by careful design. For a system with no antenna, careful screening may ensure that the external noise is zero but a system with an antenna will always be susceptible to some external noise and the level can only be altered by careful design and even then, only within certain limits.

Sources of internal noise include various random fluctuations of electrons in the materials making up the electrical circuits. It is important to realise that in a classical system the level of these fluctuations cannot ever be zero except when the system is entirely at a temperature of absolute zero, 0 K. There are various ways of reducing the noise – choice of components and the temperature of the system are examples. In this chapter we are interested in methods of measuring the noise in an electrical system in the RF and microwave frequency range.

It is worth spending a little time considering what sort of random signal we are thinking about when we refer to noise. A noise signal will have arbitrary amplitude at any instant and the amplitude at another instant cannot be predicted by use of any historical data about previous amplitudes. As noise is a random signal the time averaged offset value will be zero and consequently we consider root mean square magnitude values. The root mean square value of the voltage is proportional to the average power of the noise signal. For theoretical reasons, it is often sensible to consider the amplitude of the signal as having a Gaussian probability density function. This is because one of the major sources of noise (thermal noise) gives a theoretical Gaussian distribution and because, if the noise has a Gaussian distribution, some analyses of noise are facilitated. In a practical situation, noise is unlikely to be truly Gaussian because there will be amplitude and bandwidth limitations which will prevent this from occurring. However, in a large majority of cases the assumption of a Gaussian distribution is sufficiently close to reality to make it a very satisfactory model. The term 'white noise' is often used by analogy with white light to describe a situation where the noise signal covers a very large (effectively infinite) bandwidth. Of course, white light from the sun is a noise signal in the optical band.

In almost all cases there is no correlation between sources of noise and so the noise is non-coherent. This means that, if we have several sources of noise in a system, the total noise can be found by summing the individual noise powers. In some cases there can be correlation between noise signals and in this case the analysis is more complex. A common example of this situation is where a noise signal generated within a system travels both towards the input and the output. If some of the noise signals are then reflected back towards the output from the input – due, for example, to a mismatch – there will be some degree of correlation between the reflected and original signals at the output.

8.2 Types of noise

8.2.1 Thermal noise

Thermal or Johnson [1] noise is the most fundamental source of noise and it is present in all systems. At any temperature above absolute zero, the electrons (and other charges) in the materials of the circuit will have a random motion caused by the temperature. This will occur in both active and passive components of the circuit.

Any movement of charges gives rise to a current and, in the presence of resistance, a voltage. The voltage will vary randomly in time and is described in terms of its mean square value. This was first done by Nyquist [2]

$$\overline{v^2} = 4kTBR \tag{8.1}$$

where k is the Boltzmann's constant (1.38×10^{-23} J K^{-1}), R represents resistance (Ω), T represents absolute temperature in Kelvin (K) and B represents system bandwidth (Hz).

Clearly, the available power associated with this mean square voltage is given by

$$P = \frac{\overline{v^2}}{4R} = kTB \text{ (Watts)} \tag{8.2}$$

The bandwidth, B, is a function of the system, not the noise source. Therefore, it can be useful to define a parameter that depends only on the noise source and not on the system. This is known as the available power spectral density

$$S = \frac{P}{B} = kT \text{ (Watts)} \tag{8.3}$$

In actual fact, these expressions are only approximate as full quantum mechanical analysis yields an equivalent expression for the power spectral density of

$$S = k\phi \tag{8.4}$$

$$\phi = T \cdot P(f) \tag{8.5}$$

$$P(f) = \left(\frac{hf}{kT}\right)\left(e^{hf/kT} - 1\right)^{-1} \tag{8.6}$$

where h is the Planck's constant (6.626×10^{-34} J s), f is the frequency (Hz) and ϕ has been referred to as the 'quantum noise temperature' [3].

Unless the temperature, T, is very low or the frequency, f, is very high, the factor $P(f)$ is close to unity and (8.3) and (8.4) become identical.

8.2.2 Shot noise

In an active device, the most important source of noise is shot noise [4] which arises from the fact that the charge carriers are discrete and are emitted randomly. This is most easily visualised in the context of a thermionic valve but applies equally to solid-state devices. Owing to this, the instantaneous current varies about the mean current in a random manner which superimposes a noise-like signal on the output of the device.

8.2.3 Flicker noise

Another cause of noise is flicker noise, also known as $1/f$ noise because the amplitude varies approximately inversely with frequency. As a consequence, it is rarely important at frequencies above a few kHz and it will not be considered further.

8.3 Definitions

The fundamental quantity measured when measuring noise is usually either a mean noise power or a mean square noise voltage. However, the relationship given in (8.2) allows us to express the noise as an equivalent noise temperature and it is very often convenient to do this. If the equivalent thermal noise power from a source is kT_eB then T_e is the equivalent available noise temperature of the source. Noise sources are very often specified as having a given value of the excess noise ratio or ENR, usually expressed in decibels. The ENR is defined as follows:

$$\text{ENR} = 10\log_{10}\left(\frac{T_e - T_0}{T_0}\right) \text{ dB} \tag{8.7}$$

where T_0 is the 'standard' temperature of 290 K (17 °C).

The definition given above is for available noise power into a conjugately matched load. In the past, some laboratories have measured noise power into a perfectly matched load giving an effective noise temperature T'_e and an effective value of ENR.

$$T'_e = T_e\left(1 - |\Gamma|^2\right) \tag{8.8}$$

$$\text{ENR}' = 10\log_{10}\left(\frac{T'_e - T_0}{T_0}\right) \text{ dB} \tag{8.9}$$

where Γ is the reflection coefficient of the noise source. The difference between ENR and ENR' is shown in Figure 8.1.

As can be seen, the error is small for reflection coefficients with small magnitude but becomes quite large as the reflection coefficient increases. Noise sources with very high values of ENR are often quite poorly matched so this could become important.

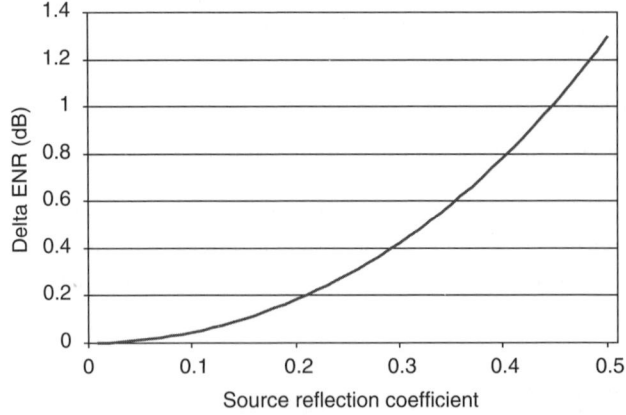

Figure 8.1 The difference between Available Noise Power and Effective Noise Power as a function of source reflection coefficient

The noise performance of a receiver is usually specified as a noise factor or noise figure. The definition of noise figure can be found in several places, for example, Reference 5, the Alliance for Telecommunications Industry Solutions. Excerpts from that definition state:

- It is determined by (1) measuring (determining) the ratio, usually expressed in dB, of the thermal noise power at the output, to that at the input; and (2) subtracting from that result, the gain, in dB, of the system.
- In some systems, for example, heterodyne systems, total output noise power includes noise from other than thermal sources, such as spurious contributions from image-frequency transformation, but noise from these sources is not considered in determining the noise figure. In this example, the noise figure is determined only with respect to the noise that appears in the output via the principal frequency transformation of the system, and excludes noise that appears via the image-frequency transformation.

In rare cases (most obviously radio astronomy) the principal frequency transformation may include both sidebands but usually only one sideband is considered. The terms 'noise figure' and 'noise factor' are normally considered to be synonymous although sometimes the term noise factor is used for the linear value (not in dB) whereas noise figure is used when the value is expressed logarithmically in dB. Adopting this distinction here we have

$$F = \frac{N_0}{GkT_0B} \tag{8.10}$$

where F is the noise factor, G is the gain and B is the bandwidth. Therefore, N_0, the total noise power from the output is

$$N_0 = GkT_0B \tag{8.11}$$

The input termination contributes an amount GkT_0B, and so N_r, the noise contribution from the receiver itself is

$$N_r = (F - 1)GkT_0B \tag{8.12}$$

In situations where there is very little noise (e.g. radio astronomy and satellite ground stations) it is more common to use the equivalent input noise temperature. A definition is given in Reference 6 and is as follows:

> At a pair of terminals, the temperature of a passive system having an available noise power per unit bandwidth at a specified frequency equal to that of the actual terminals of a network.

In most situations the pair of terminals chosen is at the input of the device so that all the noise at the output is referred back to the input and it is then imagined that all the noise is produced by a passive termination at temperature T_r. T_r is then the equivalent input noise temperature of the receiver.

The noise temperature and the noise factor can then be simply related. From (8.12) we have

$$N_i = (F - 1)kT_0B \qquad (8.13)$$

and from the definition of equivalent noise temperature we have

$$N_i = kT_rB \qquad (8.14)$$

and so

$$T_r = (F - 1)T_0 \qquad (8.15)$$

The total noise temperature at the input is often referred to as the operating noise temperature which is given by

$$T_{op} = T_s + T_r \qquad (8.16)$$

where T_s is the source temperature. Here we are assuming that there is no correlation between the source noise temperature and the receiver noise temperature and so the combined noise temperature is obtained by summing the noise temperatures.

8.4 Types of noise source

There are several types of noise source of which four are common. Two of these are particularly useful as primary standards of noise while the other two are more practical for general use.

8.4.1 Thermal noise sources

Thermal noise sources are very important because they are the type of noise source used throughout the world as primary standards of noise. To produce such a noise source, a microwave load is kept at a known temperature. In a perfect standard the transmission line between the non-ambient temperature and the ambient output would either have an infinitely sharp step change of temperature or it would have zero loss. If either of these idealisations occurred, calculation of the output noise temperature of the device would be trivial. However, in practice, the transmission line cannot fulfil either of these requirements. Along the length of the transmission line through the transition from non-ambient to ambient, each infinitesimal section of the lossy line will both produce noise power proportional to its local temperature and absorb power incident upon it. To calculate the effect of this, measurements of the loss of the line must be made and an integration along the length of the line performed.

In the United Kingdom, the majority of the primary standards used are hot standards operating at approximately 673 K and some are cold standards operating at 77 K. These are described in References 7–10, and are used at NPL to calibrate noise sources for customers worldwide. Commercial thermal noise sources have been available but the accuracy offered by these devices is limited by the attenuation measurements of the transition section and other factors and is not as good as what can be achieved from the devices at the National Standards Laboratories.

8.4.2 The temperature-limited diode

This is not a particularly common noise source. It is formed by using a thermionic diode in the temperature-limited regime, where all the electrons emitted from the cathode reach the anode. In this situation the current has noise which is determined by shot noise statistics and is calculable – in other words, it can be used as a primary standard. However, due to effects such as transit time and inter-electrode capacitances, these devices have previously only been used for relatively low frequencies up to perhaps 300 MHz. The device is described in Reference 11.

8.4.3 Gas discharge tubes

A gas discharge tube is an excellent broadband noise source. Tubes of this sort are described in Reference 12. The noise signal is produced by the random acceleration and deceleration of electrons in the discharge as they collide with atoms, ions or molecules in the gas. In general, the gas used is argon, neon or xenon at low pressure. The noise temperature is typically around 10,000 K. Commercially available waveguide devices can still be obtained and, for the higher frequencies, these are very good sources. The tube containing the gas is mounted at an angle across the waveguide and the waveguide is usually terminated with a good load at one end while the other is the mounting flange for the device. The match of the device is usually excellent and the variation in the reflection coefficient between the 'on' and the 'off' state is very small. In a practical measurement the 'on' state provides a high noise temperature whereas the 'off' state provides a noise temperature which is at the physical temperature of the device, that is, close to ambient. These devices are obtainable up to frequencies of 220 GHz [13]. These devices are not calculable and therefore require calibration before use. Once calibrated, they are relatively stable and will maintain their calibration for a considerable period if handled carefully.

8.4.4 Avalanche diode noise sources

A p–n junction which is reverse biased can produce noise through similar mechanisms to those in a gas discharge tube – that is, the random acceleration and deceleration of the charges in the material [14]. Provided the current is sufficiently high, the noise produced is almost independent of frequency as shown in Figure 8.2.

These devices are wideband, convenient, easy to use and are the preferred type of noise source for most applications. They are easily switched on and off rapidly, making measurements convenient. Recent developments include noise sources which contain a calibration table internally (on an EEPROM) and a temperature measuring device in the package so that the cold (or 'off') temperature can be measured *in situ* rather than assumed to be ambient. As for the discharge tube, the 'on' state has a high noise temperature while the 'off' state has a noise temperature which is close to ambient. In general, the diode will produce a high-ENR with a match which is poor and which has a considerable variation between the 'on' state and the 'off' state. To produce a lower value of ENR an attenuating pad is inserted which reduces the ENR and also improves the match and reduces its variation. These devices are

164 *Microwave measurements*

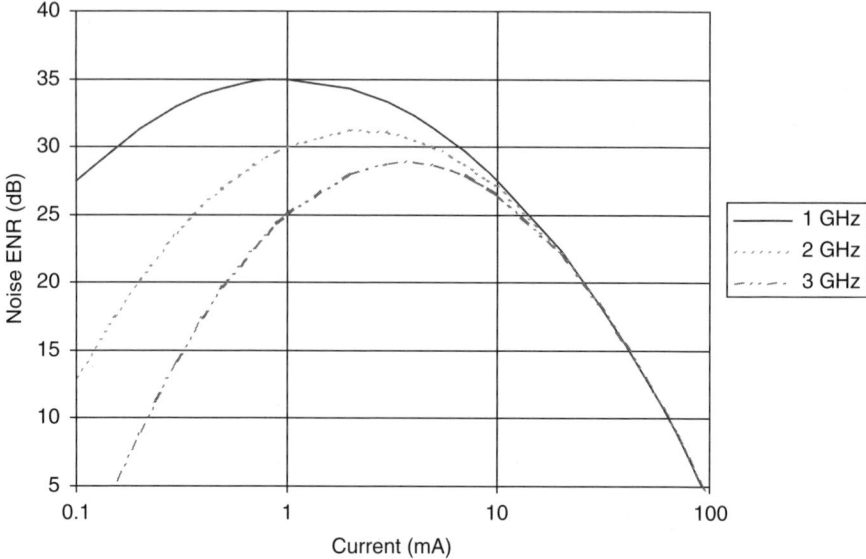

Figure 8.2   Avalanche diode noise as a function of current and frequency

not calculable and must be calibrated before use. They are reasonably stable although variations can occur. The condition of the connector is an important factor to consider when assessing the stability and repeatability of these devices.

8.5   Measuring noise

Instruments used to measure noise are classified under the general description of radiometers of which there are many types. Nowadays, a variety of instrument types can be configured to perform the measurement (e.g. noise figure analysers and spectrum analysers) but the actual operation is that of a radiometer and an understanding of the principle will allow the user to understand the way in which the measurement is made and the resulting limitations. There are many types of radiometer – the two most common are the total power radiometer and the Dicke (or switching) radiometer. In this chapter we will consider only the total power radiometer but those interested in the Dicke radiometer please refer to Reference 15.

8.5.1   The total power radiometer

The simple form of the total power radiometer is shown in Figure 8.3.

In the total power radiometer, the final output is a power reading which simply measures the total power coming from the input and from the noise generated in the receiver. We can assume that these are not correlated and that the total power is given by a simple sum of the individual powers.

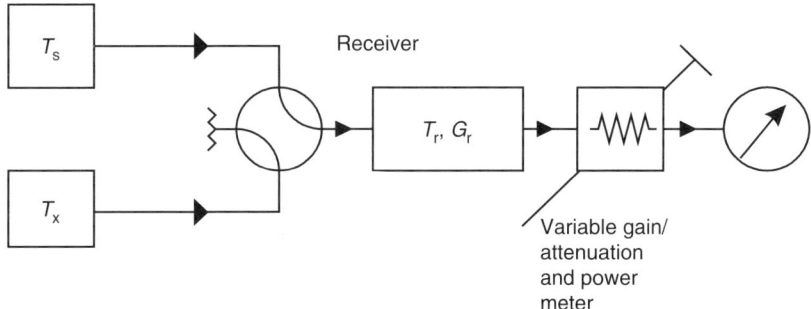

Figure 8.3 Simple block diagram of a Total Power Radiometer

When the first device, T_s (generally a standard), is attached to the input we have

$$(T_s + T_r)G_r = P_1 \tag{8.17}$$

and when the second device T_X (either an ambient standard or the unknown), is attached we have

$$(T_x + T_r)G_r = P_2 \tag{8.18}$$

The detection system is likely to have non-linearities and so it is good practice to include a calibrated attenuator in the system to ensure that the output power is held at a constant level and so

$$(T_s + T_r)G_r A_1 = (T_x + T_r)G_r A_2 \tag{8.19}$$

Commonly, the ratio of the attenuator settings, A_1/A_2 is called the Y-factor, Y and so, rearranging, we have

$$\left(\frac{T_x + T_r}{T_s + T_r}\right) = Y \tag{8.20}$$

The above equation can be rearranged to give either T_x or T_r depending on whether one is calibrating the receiver or the unknown.

$$T_x = YT_s + T_r(Y - 1) \tag{8.21}$$

or

$$T_r = \frac{(T_x - YT_s)}{(Y - 1)} \tag{8.22}$$

The receiver noise temperature must be obtained first through the use of two known noise temperatures, generally a 'hot' standard and, for convenience, an ambient load. Once this is done, the unknown noise temperature may be obtained using either the standard or an ambient device.

A usable total power radiometer requires very stable gain throughout the system since the gain must remain constant throughout the measurement. Very high values of gain (perhaps up to 100 dB) will be required since we are dealing with very small input powers; for example, a thermal noise source at 290 K has a power spectral density of −204 dBW Hz^{-1}. If the gain is not stable then measurement errors will result. In the past, this was difficult to achieve and was one of the motivations for the development of the Dicke radiometer which relies on a stable reference device. However, more recently, adequate stability has been achieved and more recent radiometers tend to be total power because the total power radiometer is more sensitive. Radiometer sensitivity is discussed in the next section.

8.5.2 Radiometer sensitivity

The sensitivity of the radiometer is limited by random fluctuations in the final output [16,17]. These fluctuations have an RMS value referred to the input given by

$$\Delta T_{\min} = \frac{aT_{\text{op}}}{\sqrt{B\tau}} \quad (8.23)$$

where B is the pre-detector bandwidth, τ is the post-detector time constant, T_{op} is as defined in (8.16), a is a constant that depends on the radiometer design and $\Delta T_{\min}$ is the minimum resolvable temperature difference.

The constant a will be unity for a total power radiometer and between 2 and 3 for a Dicke radiometer depending on the type of modulation and detection used. It is for this reason that a total power radiometer is to be preferred if adequate gain stability can be achieved.

8.6 Measurement accuracy

We have seen earlier that a total power radiometer can be used to measure an unknown noise temperature provided that two different standards of noise temperature are available (8.21) and (8.22). If we now assume that one of these sources is 'hot' (i.e. a calibrated noise source) and the other is 'cold' (i.e. an ambient temperature load) and denote these by T_h and T_c, respectively, we can re-write (8.22) as

$$T_r = \frac{(T_h - YT_c)}{(Y - 1)} \quad (8.24)$$

Wherever possible, the noise temperature being measured should be somewhere between the two standards. Rough guidelines for the choice of noise standards are given by

$$T_r = \sqrt{T_h T_c} \quad \text{or} \quad 4 \leqslant \frac{T_h}{T_c} \leqslant 10 \quad (8.25)$$

Commercial solid-state noise sources are typically either 5 dB ENR (about 1000 K) or 15 dB ENR (about 10,000 K) in the 'on' state. In the 'off' state, they have a temperature close to the ambient temperature and so the measurements can be made by connecting only one noise source to the device under test. For very low noise devices, a cold load might provide better measurement uncertainties. In order to see this, it is best to derive the uncertainties with respect to each input variable. This is done by partial differentiation

$$\Delta T_{r1} = \frac{1}{Y-1} \Delta T_h$$

$$\Delta T_{r2} = \frac{-1}{Y-1} \Delta T_c$$

$$\Delta T_{r3} = \frac{(T_c - T_h)}{(Y-1)^2} \Delta Y$$

These are the type 'B' uncertainties in the terminology defined in the appropriate guide [18] and the type 'A' uncertainties must be added to give an overall uncertainty. The type 'A' uncertainties are the uncertainties determined by statistical means. In the case of a noise radiometer, we can obtain an approximation to the magnitude of the type 'A' uncertainties from (8.23). This gives

$$\Delta T_{r4} = \left[\frac{(T_h + T_r)^2}{B\tau} + \frac{(T_c + T_r)^2}{B\tau} \right]^{1/2}$$

and the total uncertainty is then

$$\Delta T_{tot} = \left[\Delta T_{r1}^2 + \Delta T_{r2}^2 + \Delta T_{r3}^2 + \Delta T_{r4}^2 \right]^{1/2}$$

This relation has been used to derive the data shown in Figure 8.4 and Figure 8.5. In these figures, T_a denotes an ambient load used as one of the temperature references and T_v is a variable temperature noise source used as the other reference. ΔT_v, the uncertainty in T_v is assumed to be 2 per cent of T_v and ΔT_a, the uncertainty in T_a, is fixed at 0.5 K. The other parameters are

$$\Delta Y = 0.05 \text{ dB}$$

$$B = 2 \text{ MHz}$$

$$\tau = 1 \text{ s}$$

It is clear that the choice of suitable noise temperature references is very important if low uncertainties are desired, particularly for the measurement of very low noise temperatures.

When measuring a mixer or other device with frequency conversion, it is important to consider the image frequency. Noise sources are broadband devices; hence, in the absence of any image rejection, the noise source will provide an input signal to both sidebands. The definition of noise figure requires that only the principal frequency transformation is measured. If both sidebands are measured due to the lack of any image rejection, the measurement will be in error. When the loss

168 *Microwave measurements*

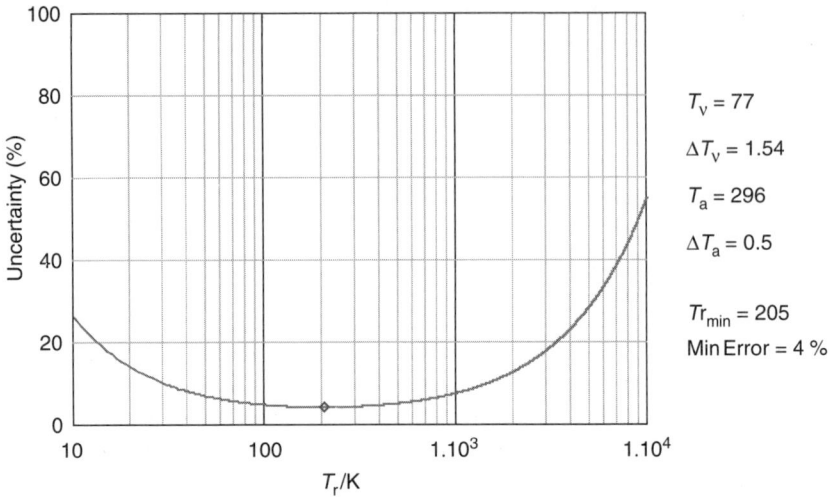

Figure 8.4 Uncertainty of measurement as a function of the noise temperature of the DUT for a radiometer using standards at 77 K and 296 K

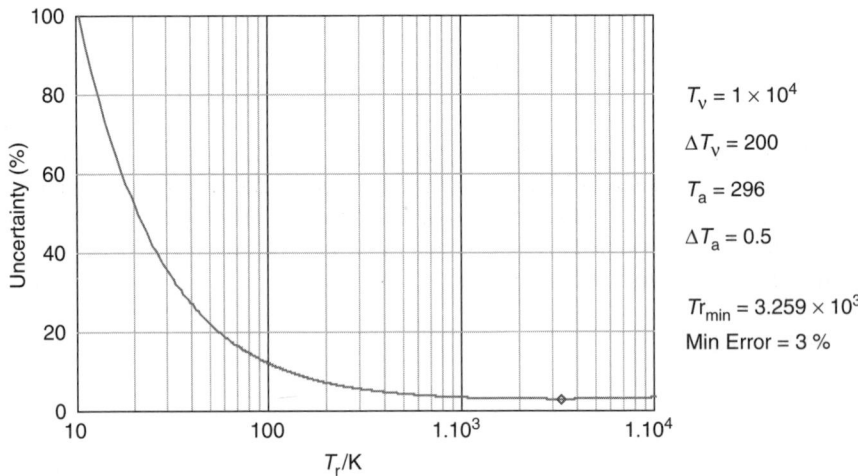

Figure 8.5 Uncertainty of measurement as a function of the noise temperature of the DUT for a radiometer using standards at 296 K and 10000 K

for both sidebands is equal this error will be a factor of 2, or 3 dB and so we have [19,20]

$$F_{ssb} = 2F_{dsb} \quad \text{when} \quad T_{ssb} = 2T_{dsb} + 290 \tag{8.26}$$

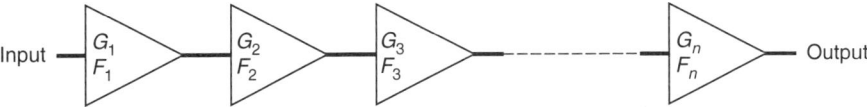

Figure 8.6   Cascaded receivers

but in many situations, this will not be the case and so the error will be large and difficult to quantify.

8.6.1 Cascaded receivers

If we have a cascade of receivers or amplifiers as shown in Figure 8.6, then the overall noise figure can be calculated using the expression

$$F_t = F_1 + \frac{(F_2 - 1)}{G_1} + \frac{(F_3 - 1)}{G_1 G_2} + \cdots + \frac{(F_n - 1)}{(G_1 G_2 \cdots G_{n-1})} \tag{8.27}$$

or in terms of noise temperature

$$T_t = T_1 + \frac{T_2}{G_1} + \frac{T_3}{G_1 G_2} + \cdots + \frac{T_n}{(G_1 G_2 \cdots G_{n-1})} \tag{8.28}$$

If G_1 is large then the higher-order terms can be ignored. It is also evident that the quality of this first amplifier has a large bearing on the performance of the overall system.

When we want to design a cascade with the lowest possible noise figure a parameter known as the noise measure is defined [21] as

$$M = \frac{(F - 1)}{\left(1 - \frac{1}{G}\right)} \tag{8.29}$$

The amplifier with the lowest value of noise measure should come first in the cascade.

8.6.2 Noise from passive two-ports

Any real system will contain passive devices which will have some loss. These devices will be noise sources with a noise temperature equivalent to their physical temperature. Consider the arrangement in Figure 8.7.

Here the transmission coefficient of the two-port is denoted by α and hence its loss is $(1 - \alpha)$. It will therefore have a noise temperature of $(1 - \alpha)T_2$. The incident noise from the source on the input will be attenuated by the two-port to a value of $T_1 \alpha$ and so the total noise temperature (assuming no correlation) is

$$T_{\text{out}} = T_1 \alpha + (1 - \alpha)T_2 \tag{8.30}$$

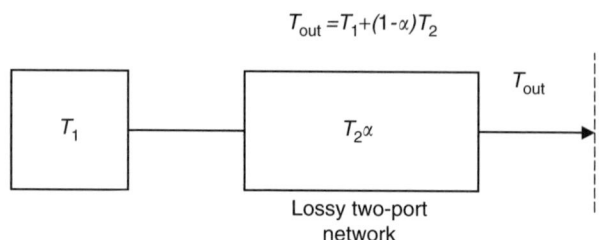

Figure 8.7 Lossy two-port

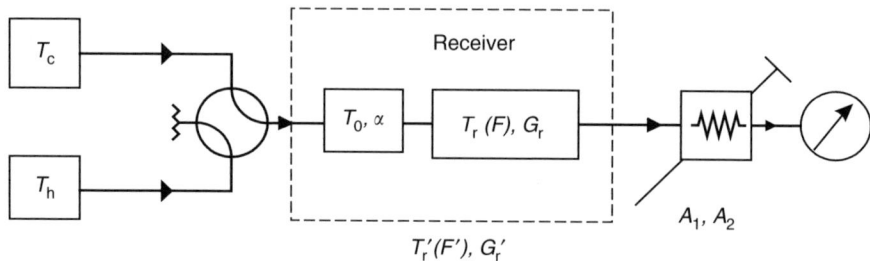

Figure 8.8 Radiometer preceded by a lossy two-port

If this is applied to the case in Figure 8.8, where a radiometer is preceded by a lossy two-port then we can use (8.19) and (8.30) to write

$$[T_c\alpha + (1-\alpha)T_0 + T_r]G_r A_1 = [T_h\alpha + (1-\alpha)T_0 + T_r]G_r A_2$$

Now, letting $Y = A_1/A_2$ and rearranging

$$T_r = \left[\alpha\frac{(T_h - YT_c)}{(Y-1)}\right] - (1-\alpha)T_0 \tag{8.31}$$

If we now treat the lossy network and the radiometer as a single unit (i.e. outside the dotted box in Figure 8.8)

$$T'_r = \frac{(T_h - YT_c)}{(Y-1)} \tag{8.32}$$

Using $F = (T_r/T_0) + 1$ and (8.31) we obtain

$$F = \alpha\left[\frac{(T_h - YT_c)}{(Y-1)T_0} + 1\right] \tag{8.33}$$

From (8.32) we have

$$M = \frac{(1-|\Gamma_S|)(1-|\Gamma_S|)}{|1-\Gamma_S\Gamma_L|} \tag{8.34}$$

and, hence,

$$F = \alpha < 1 \quad \text{and} \quad F'(\text{dB}) = F(\text{dB}) + \alpha(\text{dB}) \tag{8.35}$$

When measured in dB, any losses in front of the radiometer add directly to the noise temperature of the radiometer. This is only strictly true if the lossy two-port is at the standard temperature of 290 K. If it is at any other temperature a new expression for F in (8.33) must be derived using the same procedure.

8.7 Mismatch effects

Up to now, the effect of mismatch has been ignored in the analyses. In reality, there will usually be mismatches and these must be considered (Figure 8.9).

$$\text{Mismatch factor} = \frac{\text{Power delivered to load}}{\text{Power available from source}}$$

$$M = \frac{\left(1 - |\Gamma_S|^2\right)\left(1 - |\Gamma_L|^2\right)}{|1 - \Gamma_S \Gamma_L|^2}$$

Noise is affected in two ways by mismatches. First, in common with all other microwave signals, there will be mismatch loss [22]. Second, and more subtly, the noise temperature of a receiver is affected by the input impedance. This is because noise emanating from the first active device in the direction of the input will be correlated with the noise emanating from it in the direction of the output. When this noise is reflected off the input mismatch it will still be partially correlated with the noise going towards the output and so the two noise powers cannot be simply summed; a more complex analysis is required. However, simple steps can be taken to reduce or eliminate this effect. In the past, tuners were frequently used but this is less common since manual tuners are slow and therefore expensive in operator time and automatic tuners are expensive in capital cost. Therefore the use of isolators is more common now (Figure 8.10).

In this situation the radiometer sees a constant input impedance for both switch positions and so is not affected by any variation in the reflection coefficients of the two noise sources. There is still mismatch loss at the input port and so accurate noise

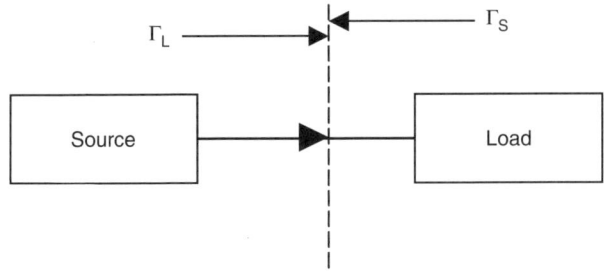

Figure 8.9   Mismatch

172   Microwave measurements

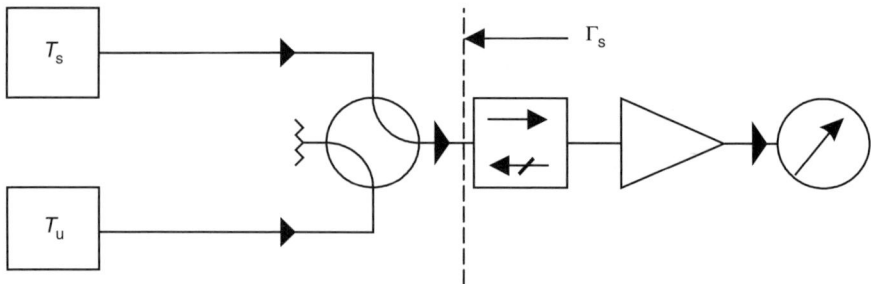

Figure 8.10   The use of an Isolator on the radiometer input

measurements require the measurement of the complex reflection coefficients of both the isolator input and of the noise sources. Some accurate noise systems incorporate instrumentation to allow this to be done *in situ* [23].

8.7.1   Measurement of receivers and amplifiers

For many people, measurement of noise means measurement of the noise figure of an amplifier. It is important to realise that this figure is not a unique parameter – the noise figure is an insertion measurement which depends on the source impedance the amplifier sees when it is measured. A measurement made with a different source impedance will yield a different result. Fortunately, the effect is often quite small, but it should not be overlooked. Two options exist:

- Measure the amplifier in a defined impedance environment and inform the user what the measurement conditions are.
- Provide the full complex noise parameters so that the user can calculate the noise for any impedance configuration.

In the past, the first option was often adopted and the amplifier was measured in a perfectly matched environment. However, increasingly, users wish to obtain the very best performance from their amplifiers and, to do this, full knowledge of the complex amplifier noise parameters is required.

There have been several different representations of the complex amplifier noise parameters. These all provide the same information and so one set can readily be converted to another. The most common (because they are the most useful to the practising engineer) are the parameters defined by Rothe and Dahlke [24]. The most familiar form is

$$F = F_{\min} + \frac{R_n}{G_s}|Y_s - Y_{opt}|^2 \tag{8.36}$$

in terms of noise factor and admittances. It can be written in terms of noise temperatures and reflection coefficients as

$$T_r = T_{min} + \frac{4T_0 R_n}{Z_0}\left[\frac{|\Gamma_s - \Gamma_{opt}|^2}{|1 + \Gamma_{opt}|^2 (1 - |\Gamma_s|^2)}\right] \tag{8.37}$$

In the above equation the noise temperature, T_r, will reach its minimum value, T_{min}, when the reflection coefficient at the source, Γ_s, is at its optimum value, Γ_{opt}. Similarly, in (8.36) the noise factor, F, will reach its minimum value F_{min} when the source admittance, Y_s is at its optimum value, Y_{opt}. In both equations R_n is the noise resistance and determines how rapidly the noise increases as the source admittance or reflection moves away from optimum. In (8.37) T_0 is the usual standard noise temperature 290 K, Z_0 is the characteristic impedance of the transmission line and in (8.36) G_s is the conductance of the transmission line.

The issue of correlated noise has been touched upon several times already. The following description [25] may make this clearer.

Referring to Figure 8.11, the amplifier (enclosed within the dotted box) can conceptually be split into a perfect, noise free two-port with a two-port noise source. The latter may be on the input or the output; here it is assumed to be on the input. Noise waves, a_{1n} and b_{1n}, are produced and propagate in each direction. As these are produced in the same place, in the same way, they are correlated. After b_{1n} is reflected from the source, there will still be some degree of correlation with a_{1n} and so the output noise temperature, T_t cannot be correctly evaluated without accounting for this. The amount of correlation depends on both the magnitude and the phase of the source reflection coefficient. It should be noted that the apparent noise temperature at the input, T_B, is different and can be well below ambient [26–28].

The total noise incident on the two-port (remembering the noise generated within the two-port is referred to the input) will be

$$N_t = N_s + N(a_{1n}) + \Gamma_s N(b_{1n})$$

where N_t is the noise at the output of the receiver, $N(a_{1n})$ is the noise due to the forward going noise wave, $N(b_{1n})$ is the noise due to the reverse going noise wave and Γ_s is the source reflection coefficient.

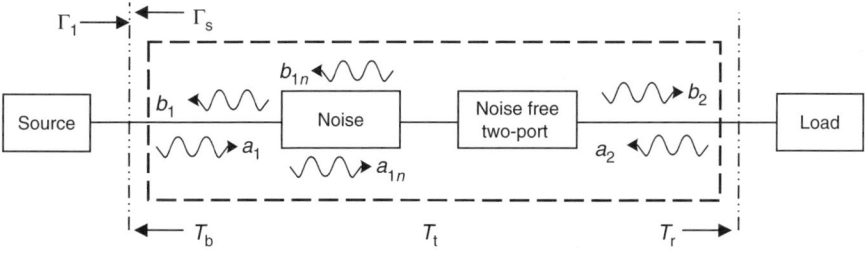

Figure 8.11 Conceptual split of an amplifier into a perfect, noise free two-port and a two-port noise source

174 Microwave measurements

If we assume no correlation between the noise source on the input and the noise generated by the receiver then

$$\overline{|N_t|^2} = \overline{|N_s|^2} + \overline{|N(a_{1n})|^2} + |\Gamma_s|^2 \overline{|N(b_{1n})|^2} + 2\text{Re}\overline{[\Gamma_s N(a_{1n})^* N(b_{1n})]}$$

where, for example, $\overline{|N_t|^2}$ denotes the mean square value of N_t and the asterisk denotes the complex conjugate.

Recalling the earlier definitions we can say

$$\overline{|N_t|^2} = kT_t B$$

$$\overline{|N_s|^2} = kT_s B$$

$$\overline{|N(a_{1n})|^2} = kT_r B$$

$$\overline{|N(b_{1n})|^2} = kT_b B$$

Now define a new parameter Γ' as

$$\Gamma' = \frac{(\Gamma_s - \Gamma_1^*)}{(1 - \Gamma_s \Gamma_1)} \tag{8.38}$$

which has the property $M_s = 1 - |\Gamma'_s|^2$, where M_s is the mismatch factor at the input of the amplifier.

Dividing throughout by kB and introducing a complex parameter T_c which represents the degree of correlation between T_r and T_b gives

$$T_t = T_s + T_r + |\Gamma'|^2 T_b + 2\text{Re}(T_c \Gamma') \tag{8.39}$$

The terms T_r, T_b and T_c are the parameters defined by Meys [29].

Measurement of the noise parameters of an amplifier expressed in any of the various forms can be done relatively easily by measuring the total noise power output with a variety of input terminations – at least one of which must be at a different noise temperature to the others. Usually, this is done by using a noise source which provides a hot noise temperature (on) and ambient noise temperature (off) at a reflection close to a match and a set of mismatches which provide ambient terminations away from a match. Choice of the values for the mismatches is not trivial if a low uncertainty measurement is to be achieved with the minimum of mismatches [30,31]. Obviously, whatever input sources are used, their reflection coefficient must be measured and used to calculate the noise parameters and this, alone, is enough to make the whole measurement much more time consuming, complex and expensive in terms of equipment.

8.8 Automated noise measurements

The great majority of noise measurements are made using some sort of automated system, most commonly a noise figure analyser. The first widely available instruments

Noise measurements 175

of this kind were introduced more than two decades ago and, although there have been many improvements in usability and in accuracy since then, the basic principles have not changed and are, indeed, those of the total power radiometer already described.

8.8.1 Noise figure meters or analysers

The basic block diagram of a noise figure analyser is shown in Figure 8.12.

Modern instruments will cover a wide bandwidth (e.g. 10 MHz to 26.5 GHz) in a single unit and will have inbuilt filtering to avoid image problems. The most modern instruments have selectable measurement bandwidth (achieved by digital signal processing). The calculation of the receiver noise temperature is performed by the processor and then used to correct the DUT measurements. Often a variety of parameters may be displayed, for example, gain and noise figure, but it is important to remember that the only measurement which is actually made by the instrument is the Y-factor. The sources of uncertainty in the measurement are the same as those made by other methods described earlier.

The block diagram (Figure 8.12) shows an isolator on the input. This component is not, in fact, generally part of the instrument but should be added externally for best uncertainty for reasons described earlier. The mixer shown is an internal component, but external mixers may be added to extend the frequency range. If this is done then the user may have to be concerned with image rejection.

8.8.2 On-wafer measurements

Increasingly, noise measurements are being performed on-wafer. There are many difficulties with this, in common with all on-wafer measurements. The main interest is in the measurement of the full complex noise parameters and so variable reflection coefficients are required. These are usually achieved using tuners, either solid state or mechanical. The majority of these systems use off-wafer noise sources, off-wafer automatic tuners and off-wafer instrumentation (noise figure analyser and network analyser). A probing station is used to link all these to the on-wafer devices [32]. This is a complex and error prone method of measurement. Discussion of the intricacies

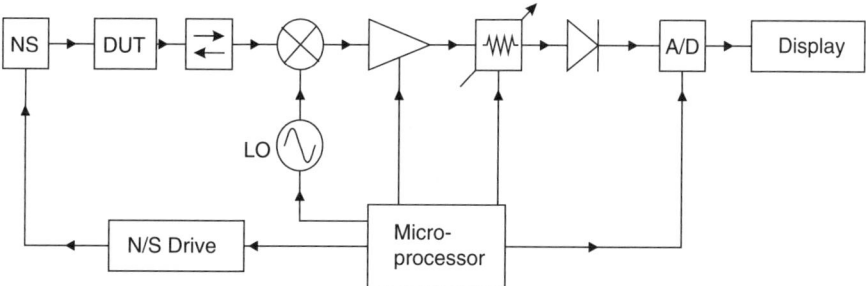

Figure 8.12   Simple block diagram of a Noise Figure Analyser

of on-wafer measurement is not within the scope of this document and so will not be considered further here.

Comparative measurements on-wafer are much easier to perform and these are fairly routinely performed. The measurements can be checked by including passive devices such as an attenuator on the wafer [33]. Other workers have proposed a passive device based upon a Lange coupler which also has calculable noise characteristics and in addition it is designed in such a way that its scattering coefficients are similar to the FET structures often being investigated [34].

8.9 Conclusion

This chapter has attempted to give an overview of noise metrology from primary standards to practical systems. The view is, of necessity, partial and brief. There are other works on measurements which also include a discussion of noise metrology. The interested reader may refer in particular to References 35 and 36.

Acknowledgements

The author wishes to acknowledge the contribution made to the content of this chapter by his co-workers and predecessors in the field of noise metrology in the United Kingdom. Particular mention must be made of Stephen Protheroe, Malcolm Sinclair (author of the previous version of this chapter) and Gareth Williams.

© Crown copyright 2005
Reproduced with the permission of the Controller of HMSO
and Queen's Printer for Scotland

References

1. Johnson, J. B.: 'Thermal agitation of electricity in conductors', *The Physical Review*, 1928;**32**:97–109
2. Nyquist, H.: 'Thermal agitation of electric charge in conductors', *The Physical Review*, 1928;**32**:110–13
3. 'Joint service review and recommendations on noise generators', Joint Service Specification REMC/30/FR, June 1972, UK
4. Schottky, W.: 'Spontaneous current fluctuations in various conductors', (German) *Annalen der Physik*, 1918;**57**:541–67
5. ATIS Committee T1A1 (2001) *Noise figure* [online], available from: http://www.atis.org/tg2k/_noise_figure.html [Accessed December 2006]
6. ATIS Committee T1A1 (2001) *Noise temperature* [online], available from: http://www.atis.org/tg2k/_noise_temperature.html, [Accessed December 2006]

7 Blundell, D. J., Houghton, E. W., and Sinclair M. W.: 'Microwave noise standards in the United Kingdom', *IEEE Transactions on Instrumentation and Measurement*, 1972;**IM-21** (4):484–88
8 Sinclair, M. W.: 'A review of the UK national noise standard facilities', *IEE Colloquium on Electrical Noise Standards and Noise Measurements*, Digest no. 1982/30, Paper no. 1, pp. 1/1–1/16, March 1982
9 Sinclair, M. W., and Wallace, A. M.: 'A new national electrical noise standard in X-band', *IEE Proc. A, Phys. Sci. Meas. Instrum. Manage. Educ. Rev.* 1986;**133** (5):272–74
10 Sinclair, M. W., Wallace, A. M., and Thornley, B.: 'A new UK national standard of electrical noise at 77 K in WG15', *IEE Proc. A, Phys. Sci. Meas. Instrum. Manage. Educ. Rev.* 1986;**133** (9): 587–95
11 Harris, I. A.: 'The design of a noise generator for measurements in the frequency range 30–1250 MHz', *Proc. Inst. Electr. Eng.*, 1961;**108** (42):651–58
12 Hart, P. A. H.: 'Standard noise sources', *Philips Technical Review*, 1962;**23** (10):293–309
13 CP Clare Corporation, *Microwave Noise Tubes and Noise Sources TD/TN Series* [online], available from: http://www/ortodoxism.ro/datasheets/clare/TD-77.pdf [Accessed 10 January 2007]
14 Haitz, R. H., and Voltmer, F. W.: 'Noise of a self-sustaining avalanche discharge in silicon: studies at microwave frequencies', *Journal of Applied Physics*, 1968;**39** (7):3379–84
15 Dicke, R. H.: 'The measurement of thermal radiation at microwave frequencies', *The Review of Scientific Instruments*, 1946;**17**:268–75
16 Kelly, E. J., Lyons, D. H., and Root, W. L.: *The theory of the radiometer*, MIT Lincoln Lab, Report no. 47.16, 1958
17 Tiuri, M. E.: 'Radio astronomy receivers', *IEEE Trans. Mil. Electron*, 1964; **MIL-8**:264–72
18 United Kingdom Accreditation Service: 'The expression of uncertainty and confidence in measurement', NAMAS publication M3003, December 1997
19 Pastori, W. E.: 'Image and second-stage corrections resolve noise figure measurement confusion', *Microwave Systems News*, May 1983, pp. 67–86
20 Bailey, A. E. (ed.): *Microwave Measurements*, 2nd edn (Peter Peregrinus, London, 1989)
21 Haus, H. A., and Adler, R. B.: *Circuit Theory of Linear Noisy Networks* (Wiley, New York, 1959)
22 Kerns, D. M., and Beatty, R. W.: *Basic Theory of Waveguide Junctions and Introductory Microwave Network Analysis* (Pergamon Press, New York, 1967)
23 Sinclair, M. W.: 'Untuned systems for the calibration of electrical noise sources', *IEE Colloquium Digest* no. 1990/174, 1990, pp. 7/1–7/5
24 Rothe, H., and Dahlke, W.: 'Theory of noisy fourpoles', *Proceedings of the Institute of Radio Engineers*, 1956;**44**:811–18
25 Williams, G. L.: 'Source mismatch effects in coaxial noise source calibration', *Measurement Science and Technology*, 1989;**2**:751–56

26 Frater, R. H., and Williams, D. R.: 'An active "cold" noise source', *Transactions on Microwave Theory and Techniques*, 1981;**29** (4):344–47
27 Forward, R. L., and Cisco, T. C.: 'Electronically cold microwave artificial resistors', *Transactions on Microwave Theory and Techniques*, 1983;**31**(1):45–50
28 Randa, J., Dunleavy, L. P., and Terrell, L. A.: 'Stability measurements on noise sources', *IEEE Transactions on Instrumentation and Measurement*, 2001;**50** (2):368–72
29 Meys, R. P.: 'A wave approach to the noise properties of linear microwave devices', *IEEE Transactions on Microwave Theory and Techniques*, 1978; **MTT-26**(1):34–37
30 Van den Bosch, S., and Martens, L.: 'Improved impedance-pattern generation for automatic noise-parameter determination', *IEEE Transactions on Microwave Theory and Techniques*, 1998;**46** (11):1673–78
31 Van den Bosch, S., and Martens, L.: 'Experimental verification of pattern selection for noise characterization', *IEEE Transactions on Microwave Theory and Techniques*, 2000:**48** (1):156–58
32 Hewlett Packard Product Note 8510-6: 'On-wafer measurements using the HP8510 network analyser and cascade microtech probes', pp. 1-16, May 1986, Hewlett Packard, Palo Alto, http://www.home.agilent.com/upload/cmc_upload/All/6C065954-1579.PDF [Accessed 03 January 2007]
33 Fraser, A. et al.: 'Repeatability and verification of on-wafer noise parameter measurements', *Microwave Journal*, 1988;**31** (11): 172–176
34 Boudiaf, A., Dubon-Chevallier, C., and Pasquet, D.: 'An original passive device for on-wafer noise parameter measurement verification', *CPEM Digest*, Paper WE3B-1, pp. 250–51, June 1994
35 Bryant, G. H.: *Principles of Microwave Measurements* (Peter Peregrinus, London, 1988)
36 Engen, G. F.: *Microwave Circuit Theory and Foundations of Microwave Metrology* (Peter Peregrinus, London, 1992)

Chapter 9
Connectors, air lines and RF impedance
N. M. Ridler

9.1 Introduction

This chapter gives information concerning some 'impedance' considerations that can be useful when making transmission line measurements at radio frequency (RF) and microwave frequencies. The subject matter is divided into the following three areas:

(1) connectors – the mechanisms used to join together two or more transmission lines;
(2) air lines – components used to define certain characteristics of transmission lines (such as impedance and phase change); and
(3) RF impedance – special considerations needed at lower microwave frequencies (typically, below 1 GHz) when defining the impedance of electrical components and networks.

The treatment of connectors deals only with coaxial connectors used to perform precision transmission line measurements (e.g. of power, attenuation, impedance and noise)[1]. Similarly, only air lines used to realise standards of impedance for these connector types are considered. For both connectors and air lines, only the 50 Ω variety is dealt with in any detail. Finally, the electromagnetic properties (such as characteristic impedance and propagation constant) of these lines at lower microwave and radio frequencies are considered.

---

[1] *Note*: 7/16 connectors are not referred to explicitly in these notes. However, some information is given on this type of connector in the appendix towards the end of these notes.

180 *Microwave measurements*

9.2 Historical perspective

The use of high-frequency electromagnetic signals dates back to the late nineteenth century and the experiments of Hertz [1] validating the theory of electromagnetic radiation proposed by Maxwell [2]. For a majority of these experiments, Hertz chose to use guided electromagnetic waves, including a primitive form of coaxial line, and this essentially gave birth to the science of guided-wave RF and microwave measurements. However, a period of approximately 40 years passed before commercial needs for this new science began to emerge and new technologies (e.g. reliable signal sources) became available.

9.2.1 Coaxial connectors

During the 1940s, work began on developing coaxial connectors that were suitable for high-frequency applications [3], and this led to the introduction of the Type N connector, which is still used extensively today throughout the industry. Other connector types followed. Many of these connectors are still in use today (such as BNC, TNC and SMA connectors), however, many others have since become obsolete. By the late 1950s, a general awareness began to emerge concerning the need for precision coaxial connectors to enable accurate measurements of transmission line quantities to be made. To address this need, committees were established during the early 1960s (including an IEEE committee on precision coaxial connectors) and these produced standards [4] for the 14 and 7 mm precision connectors. These connectors were manufactured by General Radio and Amphenol, respectively, and hence became known colloquially as the GR900 and APC-7 connectors (GR900 being the 900 series General Radio connector and APC-7 being the 7 mm Amphenol Precision Connector). Around the same time, a precision version of the Type N connector was also introduced.

During the 1970s and 1980s, additional 'precision' connectors were introduced, generally of smaller size[2] to accommodate a wider frequency range of operation. These included the 3.5 mm [5], 2.92 mm (or K connector)[3] [7], 2.4 mm [8] and 1.85 mm (or V connector) [9] connectors. In recent years, devices and measuring instruments have been manufactured fitted with 1 mm connectors [10]. Table 9.1 shows the approximate dates for the introduction of all these connectors (as well as, for reference, the BNC, TNC and SMA connectors, although these are not precision connectors).

The precision connectors discussed above are used nowadays in most laboratories involved in high-precision RF and microwave measurements. This warrants a closer look at the properties of these connectors (as given in Section 9.3).

---

[2] The 'size' of a coaxial connector refers to the internal diameter of the outer conductor constituting the coaxial line section of the connector containing air as the dielectric.

[3] It is interesting to note that a version of the 2.92 mm connector was actually introduced back in 1974 [6], but this was not a commercial success. This was probably due to compatibility problems with other connector types existing at the time.

Table 9.1 *Approximate dates at which some coaxial connectors were introduced*

Decade of introduction	Connector
1940s	Type N
	BNC
1950s	TNC
	SMA
1960s	7 mm (APC-7)
	14 mm (GR900)
1970s	Precision Type N
	3.5 mm
1980s	2.92 mm
	2.4 mm
1990s	1.85 mm
	1 mm

9.2.2 Coaxial air lines

The use of precision coaxial lines as primary impedance standards also dates back to the early 1960s [11,12]. These lines use air as the dielectric medium due to the simple, and predictable, electromagnetic properties (i.e. permeability and permittivity) of air at RF and microwave frequency. The subsequent development of these lines has closely followed the development of the precision coaxial connectors, mentioned above, with smaller diameter line sizes being produced to interface with the various different types of connector. Air lines are now commercially available in the 14, 7, 3.5, 2.92, 2.4 and 1.85 mm line sizes. It is only for the 1 mm line size that air lines are not presently available. This is presumably due to the difficulties involved in accurately machining such small diameter conductors (bearing in mind that in order to achieve a characteristic impedance of 50 Ω, the 1 mm line size would require a centre conductor with a diameter of less than 0.5 mm). The air lines currently used as impedance standards are discussed in Section 9.4.

9.2.3 RF impedance

Closely following the evolution of air lines as absolute impedance standards at microwave frequencies has been a consideration of the problems involved in utilising these, and similar, standards for impedance at RF[4]. However, the physical phenomena

---

[4] The term RF is used in these notes to indicate frequencies ranging typically from 1 MHz to 1 GHz. In general, the term RF does not define a specific frequency region. Therefore, this term might be used in other texts to indicate different frequency regions.

affecting the characteristics of these standards at these frequencies have been known about for many years. Indeed, the discovery of the so-called 'skin effect', which affects the use of air lines as standards at these frequencies, was made by Maxwell and other eminent workers in this field during the late nineteenth century (e.g. Rayleigh [13]). Subsequent work during the early-to-mid twentieth century established expressions for the series resistance and inductance of conductors due to the skin effect [14] and this led to formulas being developed [15,16] for various forms of transmission line, including coaxial lines.

During the 1950s and 1960s, precision near-matched terminations were developed [17,18] as alternative impedance standards, especially for use at lower microwave frequencies. Recent work has used a combination of air lines and terminations for RF impedance standardisation [19,20]. Some of the considerations involved in RF impedance measurement and standardisation are given in Section 9.5.

9.3 Connectors

It is often the case that in many RF and microwave measurement applications, the role played by the connectors is overlooked. This is presumably because many connectors, particularly coaxial connectors, appear to be simple devices that are mechanically robust. In fact, the performance of any device, system or measuring instrument can only be as good as the connector used to form its output. A greater awareness of connector performance is therefore a great asset for individuals involved in RF and microwave applications involving connectors.

There are several useful documents giving detailed information on various aspects of coaxial connectors. Particularly recommended is [21], which provides up-to-date information on the use, care and maintenance of coaxial connectors as well as performance and specification figures. Related information can also be found in [22,23], although these documents are now obsolete and hence difficult to obtain. Excellent reviews of coaxial connector technology, both past and present, have been given in [24,25]. The following information is based on the above documents and other similar sources.

9.3.1 Types of coaxial connector

There are several ways of categorising coaxial connectors; for example, as either sexed or sexless or as precision or non-precision. These categories are explained below, along with the GPC and LPC categories used for precision connectors.

9.3.1.1 Sexless and sexed connectors

Sexless connectors

A connector is said to be 'sexless' when both halves of a mated pair are nominally identical (both electrically and mechanically) and hence look the same. In recent times, the two most common sexless connectors are the 14 and 7 mm connectors. However, during the evolution of coaxial connector designs in the 1960s [26], other

Connectors, air lines and RF impedance 183

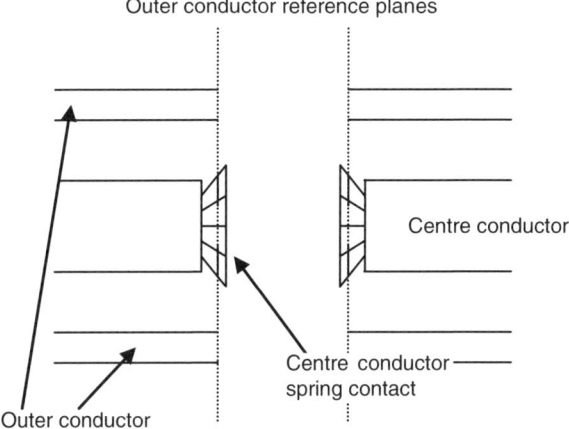

Figure 9.1 Schematic diagram of a generic sexless connector pair

sexless connectors were produced (e.g. see [27]), but these are less common today. A schematic diagram of a generic sexless connector pair is shown in Figure 9.1. This diagram shows the principle components of the connector, that is, the centre and outer conductors, the centre conductor's spring contact and the outer conductor reference plane. Notice that the position of the centre conductor is recessed with respect to the outer conductor reference plane. The spring contact thus ensures that good electrical contact is made between the centre conductors of a mated pair of sexless connectors.

Sexed connectors

The vast majority of connectors in current use are of the so-called 'sexed' variety. These connectors use a male and female (i.e. pin and socket) arrangement to produce a mated pair. The most common examples of these connectors are the SMA, BNC and Type N. However, these connector types are not generally of the precision variety (see Section 9.3.1.2). A schematic diagram of a generic sexed connector pair is shown in Figure 9.2. This diagram is similar to Figure 9.1, except that the centre conductor's spring contacts are replaced by the pin (male) and socket (female) arrangements used for making the contact between the centre conductors of a mated pair of sexed connectors.

9.3.1.2 Precision and non-precision connectors

Precision connectors

The term 'precision' was used originally to describe only the sexless variety of connectors [28]. However, this category was later modified to include sexed connectors exhibiting very good electrical performance and coincident mechanical and electrical reference planes [29] (e.g. the 3.5 mm connector). These days, the term is generally used for high-quality connectors having air as the dielectric at the connector's reference planes.

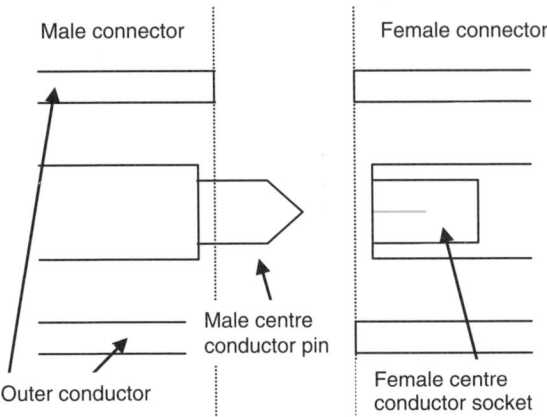

Figure 9.2 Schematic diagram of a generic sexed connector pair

The two precision sexless connectors are the 7 and 14 mm connectors mentioned above. Precision sexed connectors include the 3.5, 2.92, 2.4, 1.85 and 1 mm connectors. The 2.92 and 1.85 mm connectors are often called the K and V connectors, respectively. In addition, there is a precision version of the Type N connector[5] although the reference planes for the male and female sexes of this connector are not coincident.

Another major development in the acceptance of the above sexed connectors as being suitable for use as precision connectors has been the introduction of the slotless female contact. Conventional female connectors have longitudinal slots cut into the centre conductor that enable the pin of the male connector to be held tightly during mating. This provides a secure connection but also produces electrical discontinuities due to the slotted gaps in the female centre conductor and diameter variations, also in the female centre conductor, due to the diameter variations of male pins. The slotless female contact [30] avoids these two problems by placing the female's grasping mechanism inside the female centre conductor socket. The removal of the electrical discontinuities caused by the slotted arrangements means that slotless sexed connectors with a very high performance can be achieved (i.e. rivalling the performance of the sexless connectors).

However, devices fitted with slotless female contacts are generally more fragile, and more expensive, than the equivalent slotted devices and are therefore generally only used in applications requiring the very highest levels of measurement accuracy. Slotless female contacts can be found on Type N, 3.5 and 2.4 mm precision connectors. There is also a special version of the 3.5 mm connector (called WSMA [31]), which

---

[5] In fact, there are a considerable number of different qualities of Type N connector of which most would not be classified as 'precision'. Therefore, care should be taken when using different varieties of Type N connectors so that the overall desired performance of a measurement is not compromised.

uses a slotless female contact, designed specifically to achieve good quality mating with SMA connectors.

Non-precision connectors

There are a very large number and variety of non-precision coaxial connectors in use today. Some of the more popular non-precision connectors include the SMA, BNC, Type N[6], UHF, TNC, SMC and SMB connectors. The scope of this chapter does not include non-precision connectors, so further details are not given for these connectors.

9.3.1.3 GPC and LPC terminologies

The precision connectors, discussed above, can be further classified in terms of either GPC (General Precision Connector) or LPC (Laboratory Precision Connector) versions. GPCs include a self-contained solid dielectric element (often called a 'bead') to support the centre conductor of the connector, whereas LPCs use only air as the dielectric throughout. The LPC therefore requires that the centre conductor of the connector is held in place by some other means. For example, the centre conductor of a reference air line fitted with LPCs can be held in place by the test ports of a measuring instrument to which the line is connected. LPCs are used where the very highest levels of accuracy are required (e.g. at national standard level). LPCs and GPCs for the same line size are mechanically compatible (i.e. they are of nominally the same cross-sectional dimensions). The use of the terms GPC and LPC also avoids any confusion caused by using manufacturers' names to identify specific connectors. For example, using the term APC-7 implies a connector manufacturer (e.g. Amphenol) but does not necessarily indicate whether the connector is an LPC-7 or GPC-7 version.

9.3.2 Mechanical characteristics

Two very important mechanical characteristics of a coaxial connector are its size (i.e. the diameters of the two coaxial line conductors) and mating compatibility with other connectors. Table 9.2 gives the nominal sizes of the precision connectors discussed previously.

Several of the above connector types have been designed to be mechanically compatible with each other, meaning that they can mate with other types of connector without causing damage. Specifically, 3.5 mm connectors can mate with K connectors, and 2.4 mm connectors can mate with V connectors. This is achieved by using the same diameter for the male pin of the connector (as shown in Table 9.3). However, since the diameters of the coaxial line sections in the connectors are of different sizes (as shown in Table 9.2) there will be an electrical discontinuity at the interface between a mated pair of these mechanically compatible connectors. This discontinuity is caused by the step changes in the diameters of both the centre and outer conductors of the coaxial

---

[6] The Type N connector is included in lists of both precision and non-precision connectors due to the wide range of performance for this connector (depending on the connector specification), as mentioned previously.

Table 9.2   Line diameters of precision

Connector name	Line size, i.e. the outer conductor internal diameter (mm)	Centre conductor diameter (mm)
14 mm (e.g. GR900)*	14.2875	6.204
7 mm (e.g. APC-7)	7.000	3.040
Type N	7.000	3.040
3.5 mm	3.500	1.520
2.92 mm (K connector)	2.920	1.268
2.4 mm	2.400	1.042
1.85 mm (V connector)	1.850	0.803
1 mm	1.000	0.434

* This connector actually has an outer diameter of 14.2875 mm, and not 14 mm as its name implies. This is because the original connector design was based around an outer diameter of 9/16". When the connector was standardised, it was decided to keep the diameter as 14.2875 mm (9/16") but to use a less precise name (i.e. 14 mm).

Table 9.3   Electrical discontinuities caused by joining mechanically compatible connectors

Connector pair	Centre conductor pin diameter for both connectors (mm)	Equivalent discontinuity capacitance (fF)	Maximum linear reflection coefficient magnitude
3.5 mm and K connector	0.927	8	0.04 (at 33 GHz)
2.4 mm and V connector	0.511	10	0.08 (at 50 GHz)

line and can be represented electrically as a single shunt capacitance at the reference plane of the connector pair [32]. This discontinuity capacitance produces a reflection at the connector interface that varies with frequency. This effect has been investigated in [33], and typical maximum values for this reflection are given in Table 9.3.

In addition to the mechanical compatibility of the above precision connectors, the 3.5 mm and K connectors are also mechanically compatible with the SMA connector. In this case, the presence of a solid dielectric (e.g. Teflon) at the reference plane of the SMA connector causes an additional discontinuity capacitance (this time due to the dielectric) leading to even larger electrical reflections than those produced from a 3.5 mm to K-connection. However, as mentioned previously, the WSMA precision 3.5 mm connector was designed specifically to produce high-performance mating with SMA connectors [31]. This is achieved by deliberately setting back the position of the centre conductor pin by a prescribed amount, and hence introducing an amount of inductance to compensate for the additional capacitance caused by the SMA's dielectric [34].

9.3.3 Electrical characteristics

Two very important electrical characteristics of a coaxial connector are the nominal characteristic impedance and the maximum recommended operating frequency to ensure a stable, and repeatable, measurement. The characteristic impedance of coaxial air lines is discussed in detail in Section 9.4 of this chapter. This discussion is also applicable to the precision coaxial connectors used with these air lines.

The maximum recommended operating frequency for a coaxial line is usually chosen so that only a single electromagnetic mode of propagation is likely to be present in the coaxial line at a given frequency. This is the dominant transverse electromagnetic (or TEM) mode and operates exclusively from DC to the maximum recommended operating frequency. Above this frequency, other higher-order modes[7] can also propagate to some extent.

The maximum recommended operating frequency is often called the 'cut-off frequency' as it corresponds to the lower frequency cut-off for these higher-order waveguide modes. The first higher-order mode in 50 Ω coaxial line is the TE_{11} mode (also known as the H_{11} mode in some references). The cut-off frequency is given by [35]

$$f_c = \frac{c}{\lambda_c \sqrt{\mu_r \varepsilon_r}} \qquad (9.1)$$

It has been shown in [36] that the approximate cut-off wavelength for the TE_{11} mode is given by

$$\lambda_c \approx \pi(a + b) \qquad (9.2)$$

which corresponds to the average circumference of the line's conductors. More precise expressions for the cut-off wavelength can be obtained from [37] and these produce the theoretical upper frequency limits (i.e. the cut-off frequencies) for each line size shown in Table 9.4.

Table 9.4 also gives recommended usable upper frequency limits for each line size. These are lower than the theoretical upper frequency limits and this is due to potential higher-order mode resonances (again, the TE_{11} mode being the most likely) caused by solid material dielectric (e.g. Teflon) being present between the two conductors of the coaxial line. These resonances are most problematic when they occur in the vicinity of the transitions from air to solid dielectric, such as when a dielectric bead is

---

[7] These modes are often called 'waveguide modes' since they are similar to the modes found in hollow waveguide. These modes are either transverse electric (TE) or transverse magnetic (TM) and have a longitudinal component to their propagation. It should be noted that the TEM mode can continue to propagate at frequencies where TE and TM modes are also possible, since the TEM mode does not actually have an upper frequency limit.

Table 9.4 Theoretical and recommended upper frequency limits for coaxial connectors

Connector name	Theoretical upper frequency limit (GHz)	Recommended usable upper frequency limit (GHz)
14 mm (e.g. GR900)	9.5	8.5
7 mm (e.g. APC-7)	19.4	18.0
Type N	19.4	18.0
3.5 mm	38.8	33.0
2.92 mm (K connector)	46.5	40.0
2.4 mm	56.5	50.0
1.85 mm (V connector)	73.3	65.0
1 mm	135.7	110.0

used to support the centre conductor of the coaxial line (as in GPCs). Such resonances can occur in single connector beads as well as in a mated connector pair containing two dielectric beads.

These higher-order mode resonances can cause significant electromagnetic changes in both the reflection and transmission properties of the coaxial line. (In general, these changes cause the reflection coefficient of the line to increase whereas the transmission coefficient decreases.) These resonances are highly unpredictable and can be initiated by subtle asymmetries, eccentricities or other irregularities that may be present in the line – as can be the case at connector interfaces. For example, if dielectric beads form part of the connector interface (as in GPCs), these electromagnetic changes can vary according to the orientation of the connectors each time a connection is made. Under these conditions, even pristine precision connectors can exhibit very poor repeatability of connection.

The presence of bead resonances in precision coaxial connectors has been investigated in [38], while [39] presents some methods proposed to reduce the likelihood of excitation of these modes (e.g. through connector bead design). In any case, care should be taken when performing measurements near the upper frequency limits of coaxial connectors – even the recommended usable upper frequency limits, given in Table 9.4. Acute changes in the reflection and transmission coefficients (or a lack of repeatability of these coefficients) may indicate the presence of a higher-order mode resonance.

9.4 Air lines

Precision air-dielectric coaxial transmission lines (or, air lines, for short) can be used as reference devices, or standards, for impedance measurements at RF and microwave frequencies. (The term impedance is used here to imply a wide range of electrical quantities, such as *S*-parameters, impedance and admittance parameters, VSWR, and return loss.) This includes the use of air lines as calibration and verification standards for measuring instruments such as vector network analysers (VNAs) [40]. For example, VNA calibration schemes, such as Thru-Reflect-Line (TRL) [41] and Line-Reflect-Line (LRL) [42], use air lines as standards to achieve very high accuracy impedance measurement capabilities. This is the method currently used to realise the UK primary national standard for impedance quantities [43] at RF and microwave frequencies (typically, from 45 MHz and above). Similarly, verification schemes determining the residual systematic errors in a calibrated VNA [44] use air lines as the reference devices, and these methods are currently endorsed by organisations involved in the accreditation of measurement, such as the European co-operation for Accreditation (EA) [45].

This section describes the different types of air line that are available and reviews their use as standards of characteristic impedance and/or phase change. Consideration is also given for the effects caused by imperfections in the conductors used to realise these air lines.

9.4.1 Types of precision air line

There are basically three types of air line depending on the number of dielectric beads used to support the centre conductor of the line. These beads are usually to aid in the connection of the line during measurement.

9.4.1.1 Unsupported air lines

These lines do not contain any support beads and therefore the connector interfaces conform to the LPC category. The ends of the centre conductor are usually fitted with spring-loaded contacting tips to facilitate connecting the line to other connectors. The line's centre conductor is held in place by the test ports of a measuring instrument (or whatever else is being connected to the line). The centre and outer conductors of these lines come in two separate parts and are assembled during connection. These lines (which are of a calculable geometry) are used where the very highest levels of accuracy are required. Therefore, such lines are often found in VNA calibration kits used to realise TRL and LRL calibration schemes.

9.4.1.2 Partially supported air lines

These lines contain a support bead at only one end of the line. This design is often used for relatively long lengths of line that may be difficult to connect if they were not supported in some way. The unsupported end of the line is usually connected first – this being the more difficult of the two connections – followed by the supported end (which connects like a conventional connector). Such a line therefore has connections that are LPC at one end and GPC at the other. The centre and outer conductors of these lines often come as two separate components, although fully assembled versions also exist where the centre conductor is held in place by the bead in the air line's GPC. Semi-supported lines are often found in VNA verification kits where a calculable geometry is not required (although a high-electrical performance is still necessary).

These lines can also be used in applications where minor reflections from one end of the line do not cause problems (e.g. some applications of the 'ripple' technique [44]).

9.4.1.3 Fully supported air lines

These lines contain support beads at both ends of the line. This is equivalent to GPCs being present at both ends of the line thus making it relatively easy to connect. These lines come fully assembled with the centre conductor being held in place by both beads in the air line's GPCs. Such lines find application where only relatively modest levels of accuracy are required or where only a part of the length of a line needs to be of a known, or calculable, impedance (e.g. when calibrating time-domain reflectometers). In such applications, the minor reflections and discontinuities caused by the presence of the beads will be inconsequential.

9.4.2 Air line standards

In the above applications, the air lines are used as references of either characteristic impedance or phase change, or both. These two applications are discussed in the following subsections.

9.4.2.1 Characteristic impedance

In general, the characteristic impedance of a particular electromagnetic mode supported by a coaxial line is a complex function of the dimensions and alignment of the conductors, the physical properties of the materials of the line, and the presence of discontinuities such as connectors. However, for a uniform line with lossless conductors and air between the centre and outer conductors, the characteristic impedance of the TEM mode can be approximated by

$$Z_0 = \frac{1}{2\pi}\sqrt{\frac{\mu}{\varepsilon}}\log_e\left(\frac{b}{a}\right) \approx 59.93904 \times \log_e\left(\frac{b}{a}\right) \tag{9.3}$$

From the above expression, it is clear that the characteristic impedance of a line can be found from measurements of the diameters of the line's conductors. Such measurements are often made using air gauging techniques [46] that enable measurements to be made continuously along the entire lengths, and at all possible orientations, of both conductors. This is a very useful technique since the determination of an air line's characteristic impedance can be made with direct traceability to the SI base unit of length (i.e. the metre).

Similarly, it is also clear that, from the above expression, values of characteristic impedance can be established by using different diameters for a line's conductors. This is evident from the diameter values presented in Table 9.2 that show a range of diameter values for coaxial air lines each with a nominal characteristic impedance of 50 Ω. Similarly, Table 9.5 gives diameter values that achieve a nominal characteristic impedance of 75 Ω for the 14 and 7 mm line sizes, mentioned previously. These

Table 9.5   Line diameters for 75 Ω line sizes

Connector name	Line size, i.e. the internal diameter of the outer conductor (mm)	Centre conductor diameter (mm)
GR900	14.2875	4.088
Type N	7.000	2.003

diameters are used to realise 75 Ω versions of the GR900 and Type N connectors, respectively[8].

Having established that a wide range of characteristic impedance values can be achieved simply by choosing different diameters for the centre and outer conductors, this raises the question 'Why is 50 Ω a preferred value for the characteristic impedance of coaxial lines?' The answer appears to be that it was chosen as a compromise in performance between the theoretical characteristic impedance needed to obtain minimum attenuation in a line (which occurs at nominally 77.5 Ω[9]) and the theoretical characteristic impedance needed to obtain the maximum power transfer along a line (which occurs nominally at 30 Ω). The average of these two values is 53.75 Ω, which rounds to 50 Ω (to one significant figure). Hence, 50 Ω is a good compromise value for the characteristic impedance of lines used in many and diverse applications.

9.4.2.2   Phase change

Air lines can also be used as standards of phase change since a lossless line will only introduce a phase change to a signal, which relates directly to the line's length. The phase change is given by

$$\Delta\varphi = 2\pi \frac{\sqrt{\varepsilon_r}}{c} f\, l \text{ (radians)}$$

or

$$\Delta\varphi = 360 \frac{\sqrt{\varepsilon_r}}{c} f\, l \text{ (degrees)}$$

Air lines have been used successfully as phase change standards to calibrate reflectometers and VNAs at the very highest levels of accuracy (e.g. see [47,48]). These techniques use the lines in conjunction with high reflecting terminations (such as short-circuits and open-circuits) to produce a known phase change at the instrument

---

[8] Caution! Great care should be taken when performing measurements where both 75 and 50 Ω versions of the same connector type are available. For the Type N connector, damage will occur to a 75 Ω female connector if an attempt is made to mate it with a 50 Ω male connector. This is due to the substantial difference in diameters of the male pin and the female socket. (Note that the same situation occurs with 50 and 75 Ω versions of BNC connectors!).

[9] This may also explain why 75 Ω is also often used in some applications (such as in certain areas of the communications industry).

test port. In recent years, the use of such techniques is beginning to re-emerge in applications where it is not practical to use unsupported air lines primarily as standards of characteristic impedance (e.g. in calibration schemes such as TRL and LRL). For example, a kit currently available for VNA calibrations in the 1 mm coaxial line size [10] uses short-circuits offset by different lengths of line to achieve calibrations from around 50 to 110 GHz.

An important consideration when using air lines in conjunction with high reflecting terminations (e.g. as offset short-circuits) is that the effective electrical length of the offset line is actually double the mechanical length. This is because the electrical signal has to make a 'there-and-back' journey along the length of the line having been reflected back from the termination at the end of the line.

9.4.3 Conductor imperfections

In the above discussion concerning using air lines as standards of characteristic impedance and phase change, it has been assumed that the line's conductors are made up of lossless material (i.e. the conductors are perfectly conducting or, in other words, possess infinite conductivity). However, in practice, conductors are not perfectly conducting and therefore possess finite conductivity (or loss). This causes problems for the electrical properties of lines especially at low frequencies when the conductivity at the surface of the conductors becomes important. Manufacturers attempt to minimise these problems by producing lines made up of high-conductivity materials (such as alloys of copper) or by applying a plated layer of high-conductivity material (such as silver) to the surface of the conductors in the coaxial line.

Even so, as frequency decreases, the finite conductivity of a line causes the propagating wave to penetrate the walls of the conductors to some extent. The attenuation constant associated with the wave propagating into the walls of the conductors[10] is considerably higher than for the wave propagating in the dielectric between the conductors, and therefore the wave attenuates rapidly as it penetrates the walls of the conductors. The reciprocal of this attenuation constant is called the skin depth and is defined as the distance travelled into the walls of the conductors by the wave before being attenuated by one neper (≈ 8.686 dB).

The skin depth is given by

$$\delta_s = \sqrt{\frac{1}{\pi f \sigma \mu}} \qquad (9.4)$$

This indicates that skin depth increases as the frequency decreases. The skin depth will also be larger for a line with a lower value of conductivity. To illustrate this, values for skin depth are given in Table 9.6, for conductors made up of silver, brass and beryllium copper (BeCu), with assumed conductivities of 62, 16 and 13 MS m^{-1}, respectively, as these are materials often used to fabricate precision air lines. Further detailed discussions on skin depth effects can be found in [49].

---

[10] The wave decays exponentially as it penetrates the walls of the conductors.

Table 9.6   Skin depth values as a function of frequency

Frequency (MHz)	Skin depth (μm)		
	Silver ($\sigma = 62\,\text{MS m}^{-1}$)	Brass ($\sigma = 16\,\text{MS m}^{-1}$)	BeCu ($\sigma = 13\,\text{MS m}^{-1}$)
1	64	126	140
10	20	40	44
100	6	13	14
1000	2	4	4

It is generally only necessary to accurately determine the conductivity of a line's conductors at RF (typically, between 1 MHz and 1 GHz) in order to determine the line's characteristics. This is because lines are rarely used as impedance standards below these frequencies and skin depth becomes less of a problem at higher frequencies. This requirement, however, is not trivial. If the line's constitution is known then a value may be obtained from tables of physical data (e.g. from sources such as [50]). However values specified in tables usually refer to bulk material samples. These values are often different from actual values for the same material after it has been subject to manufacturing processes, as is the case for air lines (e.g. see [51,52]).

An additional problem in determining a value for the conductivity of a line is caused by plating layers that may be applied by manufacturers either to increase conductivity (e.g. silver plating) or increase longevity (e.g. gold 'flashing'). Several studies have been carried out evaluating effects of plating on the effective conductivity of conductors [53–55] but these assume prior knowledge of the material of each layer and ignore additional complications caused by impurities which will doubtlessly be present. A recent validation of theoretical predictions based on assumed conductivity values has been performed by comparison with precision attenuation measurements [56].

Finally, another consideration concerning the characteristics of a line relates to the non-uniformity of the conductor's surfaces caused either by changes in the longitudinal dimensions of the line [57,58] or surface roughness [59]. In both cases, these effects will cause the properties of the line to depart significantly from ideal values.

9.5   RF impedance

The measurement of impedance, and impedance-related quantities, requires special consideration when the measurement frequency is in the RF region (i.e. from 1 MHz to 1 GHz). This is generally due to techniques used at the higher frequencies becoming inappropriate at these longer wavelengths. Similarly, low-frequency techniques, used

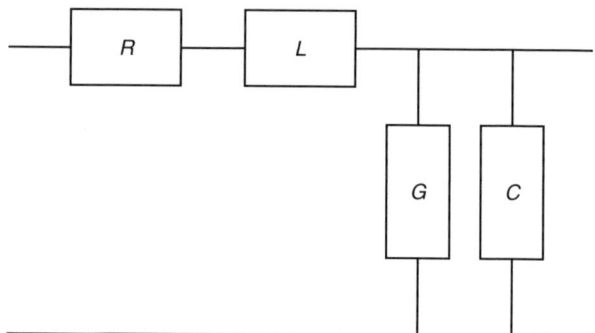

Figure 9.3 Distributed circuit model for a section of coaxial line

below 1 MHz, are also unsuitable – for example, because the connector configurations are often different (e.g. four-terminal pair connections). Information concerning use of air lines and terminations (i.e. one-port devices) as impedance standards at RF is given below – for example, to calibrate a VNA. More detailed information can be found in [60].

9.5.1 Air lines

Air lines can be used in conjunction with terminations as calibration items for reflectometers (or VNA one-port calibrations). In this configuration, one end of the air line is connected to the instrument test port while the other end is connected to the termination. Lines can also be used for VNA two-port calibrations (such as TRL and LRL, where they act as the Line standard) and are connected between the two test ports during calibration. In either application, the accuracy achieved using modern VNAs requires that the electrical characteristics of the air lines are defined very precisely, as shown in Figure 9.3.

A coaxial line can be characterised using the distributed circuit model given in Figure 9.3, where R, L, G and C are the series resistance and inductance, and the shunt conductance and capacitance, respectively, per unit length of line.

Expressions for the four line elements R, L, G and C can be used to obtain further expressions for two fundamental line parameters – the characteristic impedance and the propagation constant – which are defined as follows:

$$Z = \sqrt{\frac{(R+j\omega L)}{(G+j\omega C)}} \tag{9.5}$$

$$\gamma = \alpha + j\beta = \sqrt{(R+j\omega L)(G+j\omega C)} \tag{9.6}$$

9.5.1.1 Lossless lines

For a lossless line (i.e. with conductors of infinite conductivity) both the series resistance and the shunt conductance are zero. The series inductance and the shunt capacitance have fixed values independent of frequency and are given by

$$L_0 = \frac{\mu \log_e (b/a)}{2\pi} \tag{9.7}$$

$$C_0 = \frac{2\pi \varepsilon}{\log_e (b/a)} \tag{9.8}$$

The characteristic impedance of the lossless line is therefore (as before)

$$Z_0 = \sqrt{\frac{L_0}{C_0}} = \frac{1}{2\pi}\sqrt{\frac{\mu}{\varepsilon}} \log_e \left(\frac{b}{a}\right) \tag{9.9}$$

This shows that the line's characteristic impedance is a purely real quantity (i.e. containing no imaginary component), is independent of frequency and determined by the ratio (b/a). For example, to achieve a characteristic impedance of 50 Ω this ratio is approximately 2.3. The propagation constant of the lossless line is

$$\gamma_0 = j\beta = j\omega\sqrt{L_0 C_0} = j\omega\sqrt{\mu\varepsilon} = j\frac{\omega}{v} = j\frac{2\pi}{\lambda} \text{ (rad m}^{-1}\text{)} \tag{9.10}$$

This shows that the line's propagation constant is purely imaginary (i.e. containing no real component) and is determined only by the wavelength (or equivalent) of the propagating wave. The attenuation constant is zero which is consistent with a line having no loss. The phase constant is a linear function of frequency, indicating a non-dispersive line.

9.5.1.2 Lossy lines

As mentioned previously, metallic air-filled coaxial lines are not lossless. An important part of line characterisation at RF is a determination of the effects due to line loss. An attempt at dealing with this problem for RF impedance standardisation has been given in [61]. Further work has since been presented in [62], giving expressions for all four line elements – R, L, C and G – containing frequency-dependent terms for each element. Additional work has also solved this problem for frequencies below the RF region, obtaining exact field equations for lossy coaxial lines [63].

The expressions derived in [62] for the four line elements at RF are as follows:

$$R = 2\omega L_0 d_0 \left(1 - \frac{k^2 a^2 F_0}{2}\right) \tag{9.11}$$

$$L = L_0 \left[1 + 2d_0 \left(1 - \frac{k^2 a^2 F_0}{2}\right)\right] \tag{9.12}$$

$$G = \omega C_0 d_0 k^2 a^2 F_0 \tag{9.13}$$

$$C = C_0 (1 + d_0 k^2 a^2 F_0) \tag{9.14}$$

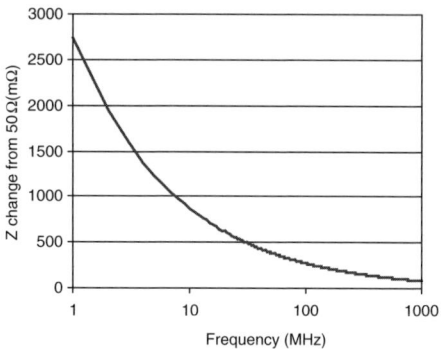

Figure 9.4 Change in characteristic impedance magnitude for a 7 mm BeCu line

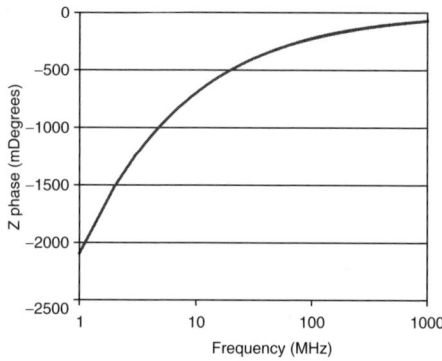

Figure 9.5 Characteristic impedance phase angle for a 7 mm BeCu line

where

$$F_0 = \frac{(b^2/a^2) - 1}{2 \log_e(b/a)} - \frac{(b/a) \log_e(b/a)}{(b/a) + 1} - \frac{1}{2}\left[\frac{b}{a} + 1\right] \quad (9.15)$$

$$d_0 = \frac{\delta_s(1 + (b/a))}{4b \log_e(b/a)} \quad (9.16)$$

These expressions can be used to calculate the characteristic impedance, which, for a line with finite conductivity, is clearly a complex quantity, material dependent and a function of frequency. Figures 9.4 and 9.5 illustrate the effect on the characteristic impedance of a nominal 50 Ω 7 mm air line made up of BeCu with an assumed conductivity of 13 MS m^{-1}.

The deviation in the characteristic impedance causes a problem for impedance measurements (such as S-parameters) since they are usually specified with respect to the lossless line value (e.g. 50 Ω). Measurements made on instruments calibrated with lines of different material will vary systematically since the impedance parameters will be measured with respect to different characteristic impedances. This problem is

Connectors, air lines and RF impedance 197

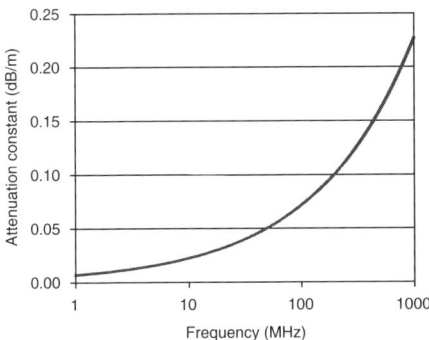

Figure 9.6   Attenuation constant for a 7 mm BeCu line

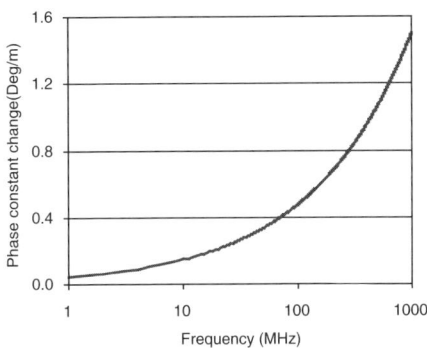

Figure 9.7   Change in phase constant for a 7 mm BeCu line

overcome by transforming from the actual line characteristic impedance to the defined lossless value (e.g. 50 Ω for the 50 Ω line size). Further information on impedance transformations of this type is given in [64].

The above expressions can also be used to calculate the propagation constant, which, for a line with finite conductivity has both real and imaginary parts and is non-linear with frequency. Figures 9.6 and 9.7 illustrate the effect on the propagation constant for a nominal 50 Ω 7 mm air line made up of BeCu. The attenuation constant is non-zero (Figure 9.6), which is consistent with a line containing loss. The increase in the phase constant from its lossless value indicates that the line's electrical length is longer than its physical length – this discrepancy varying as a function of frequency. The line is therefore dispersive and imparts group delay to broadband signals.

A comparison of parameters characterising lossless and lossy lines reveals that only one extra term is included to allow for the loss effects, that is, the conductor's conductivity. If the conductivity is assumed to be infinite, the skin depth becomes zero and the term d_0 in the expressions for the four lossy line elements vanishes. This causes R and G to become zero and L and C to revert to their lossless values (i.e. L_0 and C_0). The finite conductivity (and hence non-zero skin depth) of the conductors is therefore solely responsible for departures from the lossless line conditions. The expression given earlier for skin depth also contains a $1/\sqrt{f}$ term indicating that skin depth increases as frequency decreases, causing a subsequent increase in the values for all four line elements.

9.5.2 Terminations

It is often very convenient to use terminations (i.e. one-port devices) as calibration standards for reflectometers and VNAs. These terminations can be used in both one-port and two-port VNA calibration schemes. The terminations can be connected directly to the instrument test port or separated by a length of air line called an 'offset'. The air line section can be an integral part of the item or connected separately. The three most common terminations used for this purpose are short-circuits, open-circuits and near-matched terminations (including so-called sliding loads). Mismatched terminations (and capacitors) can also be used, particularly at lower frequencies.

9.5.2.1 Short-circuits

A coaxial line short-circuit is simply a flat metallic disc connected normally to the line's centre and outer conductors. Its radius must exceed the internal radius of the outer conductor and be of sufficient thickness to form an effective shield for the electromagnetic wave propagating in the line. The disc is usually made up of a similar material as the line's conductors. Short-circuits can be connected directly to an instrument test port or via a length of line producing an offset short-circuit.

Short-circuits provide a good approximation to the lossless condition at RF (i.e. with both series resistance and inductive reactance being close to zero). This produces a reflection coefficient with real and imaginary parts of -1 and 0, respectively. Loss due to skin depth and surface finish of the disc can be considered for high-precision metrology applications. Such losses have been considered in [65] by analysing the effects of a TEM wave incident normally to a conducting plane.

9.5.2.2 Open-circuits

In principle, a coaxial open-circuit is produced by having nothing connected to the instrument test port. However, this produces a poorly defined standard for two reasons: (1) it will radiate energy producing a reflected signal dependent on obstacles in the vicinity of the test port and (2) the test port connector's mating mechanism affects the established measurement reference plane which limits accurate characterisation as a standard.

The first of these problems can be overcome by extending the line's outer conductor sufficiently beyond the position of the open-circuited centre conductor so that the evanescent radiating field decays to zero within the outer conductor shield – the extended outer conductor acting as an effectively infinite length of circular waveguide below cut-off.

The second problem can be overcome either by depressing the mating mechanism using a dielectric plug or attaching a length of line to the centre conductor, terminated in an abrupt truncation. The dielectric plug technique is used as a standard with numerous VNA calibration kits. The abruptly truncated line technique has been used to realise primary national impedance standards [47,66]. In both cases, the open-circuit behaves as a frequency-dependent 'fringing' capacitance. Calculations for the capacitance of an abruptly truncated coaxial line can be found in the literature (i.e. [67–69]). These values have been verified for RF impedance applications using a computer-intensive equivalent circuit technique [70].

Coaxial open-circuits have a reflection coefficient of nominally unity magnitude and a phase angle dependent on the fringing capacitance and the length of any line used to fabricate the device. They can therefore be very useful as standards for calibrating reflectometers and VNAs.

9.5.2.3 Near-matched terminations

A low-reflection (or near-matched) termination can be produced by mounting a cylindrical thin-film resistive load in the centre conductor of a line with a tractorially shaped outer conductor. A parabolic transition between the conventional coaxial line and the tractorial section transforms incident plane wave fronts to spherical wave fronts required to propagate in the tractorial section of the termination. This produces near-uniform power dissipation along the length of the resistive load element with minimal frequency dependence. This design of low reflecting termination has been discussed in [71].

Low-reflection terminations are usually assumed to have zero reflection during a reflectometer, or VNA, calibration (and are therefore often called 'matched' loads). Alternatively, low reflecting load elements can be used to 'synthesise' the performance of a matched termination, using sliding load techniques. This is achieved by measuring the response of a load element at several positions along a variable length of precision air line. The characteristics of a 'perfectly' matched termination can then be computed by fitting a circle to the measured reflection values (the centre of the fitted circle being the point in the complex reflection coefficient plane corresponding to a perfect match, i.e. zero reflection). However, problems due to imperfections in the air line section and inadequate phase differences produced by realisable lengths of air line make this technique of limited use at RF.

9.5.2.4 Mismatched terminations

In principle, mismatched terminations (and capacitors) can be very useful devices for providing values of reflection that are significantly different from those achieved using short-circuit, open-circuit and near-matched terminations. Such reflection values could be used in certain calibration applications (e.g. as alternatives to the short-open-load values used during conventional VNA calibration schemes). However, devices used for calibration (i.e. standards) are usually assumed to have 'known' values based on either a calculated and/or measured performance[11]. In general, it is not possible to calculate, to any degree of accuracy, the performance of a mismatched termination. Indeed, the same can be said of near-matched terminations where an assumed value (i.e. zero) is often used for calibration purposes.

There have been several attempts recently at characterising near-matched terminations using measurement data at DC and RF. Some work in the 1990s [72] used equivalent circuit models for characterising these devices at lower RF (300 kHz to 30 MHz) based on measurement data at higher RF. More recent work [73,74] has concentrated on implementing interpolation schemes for characterising these devices. The interpolation schemes have the advantage that very few assumptions need to be made concerning the characteristics of the device. In principle, such schemes can be extended to characterise 'any' device (e.g. mismatch terminations) without requiring detailed knowledge concerning the physical (i.e. calculable) properties of the device. This is leading to the development of generalised techniques for VNA calibrations [75] that do not need to rely on the classical assumptions implicit in the short-open-load calibration schemes. Such techniques are expected to greatly enhance our knowledge of calibration devices and instruments used traditionally to perform RF impedance measurements.

9.6 Future developments

Coaxial connectors and coaxial transmission lines continue to play a crucial role in the realisation of the majority of measurements made at radio and microwave frequencies. This chapter has presented some of the important issues relating to the various types of coaxial connector currently available for making high-precision measurements. Even so, the connector itself can still be the limiting factor for the accuracy achieved by today's measurement systems.

Similarly, coaxial air lines provide very useful standard reference artefacts for realising impedance quantities for these connector types and the associated transmission lines. These devices are simple structures with well-defined electromagnetic properties. But once again, the precision at which today's instruments can operate means that these standards will need to be defined to an even greater level of precision. This is particularly true at lower RF (and, indeed, at extremely high frequencies) where the line's characteristics depart substantially from their idealised values.

It is unlikely that future requirements for these technologies will be less demanding than they are at present. Indeed, it can be expected that most measurement applications will require broader bandwidths, improved electrical capabilities (including repeatability, insertion loss and lower passive inter-modulation) and higher levels of accuracy. These demands are likely to continue to drive developments in precision coaxial connectors, air lines and other impedance standards for the foreseeable future.

---

[11] For example, the characteristics of unsupported air lines can be calculated based on the measured values of the diameters of the line's conductors.

Appendix: 7/16 connectors

The 7/16 connector was developed during the 1960s primarily for high-performance military applications. In recent years, it has become a popular choice for certain applications in the mobile communications industry, such as in base stations and antenna feed lines. This is due to its suitability for uses involving high power levels, low receiver noise levels and where there are requirements for low passive intermodulation (PIM).

The 7/16 connector is a sexed connector with a nominal characteristic impedance of 50 Ω. It is available in both GPC and LPC versions – LPCs are found on 7/16 unsupported air lines used in VNA calibration kits to realise calibration schemes such as TRL and LRL. Terminations are also available which can be used for Short-Open-Load calibration schemes. The nominal diameters of the centre and outer conductors are 7 and 16 mm, respectively, and this yields a recommended usable upper frequency limit of approximately 7.5 GHz.

Primary national standards of impedance for 7/16 connectors have recently been introduced at the UK's National Physical Laboratory.

Symbols

a = Radius of coaxial line centre conductor (m).
α = Attenuation constant (Np m^{-1}).
b = Radius of coaxial line outer conductor (m).
β = Phase constant (rad m^{-1}).
γ = Propagation constant for coaxial line containing conductor loss. This is generally a complex-valued quantity (m^{-1}).
γ_0 = Propagation constant of lossless coaxial line. This is an imaginary-valued quantity (m^{-1}).
C = Shunt capacitance, per unit length, of coaxial line including conductor loss (F m^{-1}).
C_0 = Shunt capacitance, per unit length, of lossless coaxial line (F m^{-1}).
c = Speed of light in vacuum (defined exactly as 299,792,458 m s^{-1}).
δ_s = Skin depth of air line conductors (m).
$\Delta\phi$ = Phase change (in degrees or radians) introduced by a length of line, l.
e = 2.718281828…(base of Naperian logarithms).
ε = Permittivity, $\varepsilon = \varepsilon_0 \varepsilon_r$ (F m^{-1}).
ε_r = Relative permittivity of an air line's dielectric (e.g. $\varepsilon_r = 1.000649$ for 'standard' air at 23 °C, 50 per cent relative humidity and 1013.25 hPa atmospheric pressure).
ε_0 = Permittivity of free space (defined exactly as $(c^2\mu_0)^{-1} = 8.854187817\ldots \times 10^{-12}$ F m^{-1}).
f = Frequency (Hz).
f_c = Cut-off frequency for the TEM mode (Hz).

G = Shunt conductance, per unit length, for a coaxial line including conductor loss (S m^{-1}).
$j = \sqrt{-1}$.
k = Angular wave number, $k = 2\pi/\lambda$ (rad m^{-1}).
l = Length of air line (m).
L = Series inductance, per unit length, for a coaxial line including conductor loss (H m^{-1}).
L_0 = Series inductance, per unit length, for a lossless coaxial line (H m^{-1}).
λ = Wavelength = v/f (m).
λ_c = Cut-off wavelength for the TEM mode (m).
μ = Permeability, $\mu = \mu_0 \mu_r$ (H m^{-1}).
μ_r = Relative permeability of an air line's dielectric (e.g. $\mu_r = 1$ for 'standard' air, to six decimal places).
μ_0 = Permeability of free space (defined exactly as $4\pi \times 10^{-7}$ H m^{-1}).
π = 3.141592653...
R = Series resistance, per unit length, for a coaxial line including conductor loss (Ω m^{-1}).
σ = Conductivity of an air line's conductors (S m^{-1}).
v = Speed of the electromagnetic wave in the air line [$v = c/\sqrt{\varepsilon_r}$ (m s^{-1})].
ω = Angular frequency, $\omega = 2\pi f$ (rad s^{-1}).
Z = Characteristic impedance of a coaxial line containing conductor loss. This is generally a complex-valued quantity (Ω).
Z_0 = Characteristic impedance of a coaxial line with lossless conductors. This is a real-valued quantity (Ω).

References

1 Hertz, H.: *Electric waves, being researches on the propagation of electric action with finite velocity through space*, Trans. Jones, D. E. (Dover Publications Inc, New York, 1962), Chapter 10
2 Maxwell, J. C.: *A treatise of electricity and magnetism*, 3rd edn, vol. 2 (Oxford University Press, London, 1892)
3 Bryant, J. H.: 'Coaxial transmission lines, related two-conductor transmission lines, connectors, and components: A US historical perspective', *IEEE Transactions on Microwave Theory and Techniques*, 1984;**32** (9):970–83
4 G-IM Subcommittee on Precision Coaxial Connectors: 'IEEE standard for precision coaxial connectors', *IEEE Transactions on Instrumentation and Measurement*, 1968;**17** (3):204–18
5 Adam, S. F., Kirkpatrick, G. R., Sladek, N. J., and Bruno, S. T.: 'A high performance 3.5 mm connector to 34 GHz', *Microwave Journal*, 1976;**19** (7):50–4
6 Maury, M. A., and Wambach, W. A.: 'A new 40 GHz coaxial connector', *Millimeter Waves Techniques Conference Digest* (NELC, San Diego, CA, 1974)
7 Browne, J.: 'Precision coaxial cables and connectors reach 45 GHz', *Microwaves & RF*, Sep 1983;131–6

8 Kachigan, K., Botka, J., and Watson, P.: 'The 2.4 mm connector vital to the future of 50 GHz coax', *Microwave Systems News*, 1986;**16** (2):90–4
9 Manz, B.: 'Coaxial technology vies for emerging V-band applications', *Microwaves & RF*, Jul 1989;35–41
10 Howell, K., and Wong, K.: 'DC to 110 GHz measurements in coax using the 1 mm connector', *Microwave Journal*, 1999;**42** (7):22–34
11 Weinschel, B. O.: 'Air-filled coaxial lines as absolute impedance standards', *Microwave Journal*, Apr 1964;47–50
12 Harris, I. A., and Spinney, R. E.: 'The realization of high-frequency impedance standards using air spaced coaxial lines', *IEEE Transactions on Instrumentation and Measurement*, 1964;**13**:265–72
13 Rayleigh, L.: 'On the self-inductance and resistance of straight conductors', *Philosophical Magazine S5*, 1886;**21** (132):381–94
14 Russell, A.: 'The effective resistance and inductance of a concentric main, and methods of computing the Ber and Bei and allied functions', *Philosophical Magazine*, 1909;**17**:524–52
15 Wheeler, H. A.: 'Formulas for the skin effect', *Proceedings of the Institute of Radio Engineers*, 1942;**30**:412–24
16 Stratton, J. A.: *Electromagnetic theory* (McGraw-Hill Book Company Inc, New York and London, 1941), Chapter 9
17 Harris, I. A.: 'The theory and design of coaxial resistor mounts for the frequency band 0-4000 Mc/s', *Proc. Inst. Electr. Eng.*, 1956;**103** Part C(3):1–10
18 MacKenzie, T. E., and Sanderson, A. E.: 'Some fundamental design principles for the development of precision coaxial standards and components', *IEEE Transactions on Microwave Theory and Techniques*, 1966;**14** (1):29–39
19 Ridler, N. M., and Medley, J. C.: 'Improvements to traceability for impedance measurements at RF in the UK', *IEE Engineering, Science and Education Journal*, 1997;**6** (1):17–24
20 Ridler, N. M., and Medley, J. C.: 'Improving the traceability of coaxial impedance measurements at lower RF in the UK', *IEE Proceedings Science Measurement and Technology*, 1996;**143** (4):241–45
21 Skinner, A. D.: *ANAMET connector guide*, ANAMET Report 032, 2001 (Available at: www.npl.co.uk/anamet)
22 'Coaxial Systems. Principles of microwave connector care (for higher reliability and better measurements)', Hewlett Packard Application Note 326, July 1986
23 'Coaxial connectors in radio frequency and microwave measurements', NAMAS Information Sheet 4303, edn 1, December 1991
24 Maury, M. A.: 'Microwave coaxial connector technology: a continuing evolution', *Microwave Journal (State of the Art Preference Supplement)*, Sep 1990; 39–59
25 Weinschel, B. O.: 'Coaxial connectors: a look to the past and future', *Microwave Systems News*, 1990;**20** (2):24–31
26 Anderson, T. N.: 'Evolution of precision coaxial connectors', *Microwave Journal*, Jan 1968;18–28

27 Huber, F. R., and Neubauer, H.: 'The Dezifix connector – a sexless precision connector for microwave techniques', *Microwave Journal*, Jun 1963;79–85
28 Weinschel, B. O.: 'Standardization of precision coaxial connectors', *Proceedings of the IEEE*, 1967;**55** (6):923–32
29 Sladek, N. J., and Jesch, R. L.: 'Standardization of coaxial connectors in the IEC', *Proceedings of the IEEE*, 1986;**74** (1):14–18
30 Botka, J.: 'Major improvement in measurement accuracy using precision slotless connectors', *Microwave Journal*, 1988;**31** (3):221–26
31 'Connector relieves nagging SMA measurement problems', *Microwaves*, Jan 1979;97–9
32 Whinnery, J. R., Jamieson, H. W., and Robbins, T. E.: 'Coaxial line discontinuities', *Proceedings of the Institute of Radio Engineers*, 1944;**32**:695–709
33 Ide, J. P.: 'Estimating the electrical compatibility of mechanically compatible connectors', *Microwave Engineering Europe*, 1994;**43**:39–40
34 Oldfield, W. W.: 'Comparing miniature coaxial connectors', *Microwaves and RF*, 1985;**24** (9):171–74
35 Dimitrios, J.: 'Exact cutoff frequencies of precision coax', *Microwaves*, Jun 1965; 28–31
36 Ramo, S., and Whinnery, J.: *Fields and waves in modern radio* (John Wiley & Sons, New York, 1959)
37 Marcuvitz, N.: *Waveguide handbook*, MIT Radiation Laboratory Series 10 (McGraw-Hill Book Company, New York, 1951), pp. 72–80
38 Gilmore, J. F.: 'TE_{11}-mode resonances in precision coaxial connectors', *GR Experimenter*, 1966;**40** (8):10–13
39 Neubauer, H., and Huber, R. F.: 'Higher modes in coaxial RF lines', *Microwave Journal*, 1969;**12** (6):57–66
40 Wong, K. H.: 'Using precision coaxial air dielectric transmission lines as calibration and verification standards', *Microwave Journal*, Dec 1998; 83–92
41 Engen, G. F., and Hoer, C. A.: 'Thru-Reflect-Line: an improved technique for calibrating the dual six-port automatic network analyzer', *IEEE Transactions on Microwave Theory Techniques*, 1979;**MTT-27** (12):987–93
42 Hoer, C. A., and Engen, G. F.: 'On-line accuracy assessment for the dual six-port ANA: extension to nonmating connectors', *IEEE Transactions on Instrumentation and Measurement*, 1987;**IM-36** (2):524–29
43 Ridler, N. M.: *A review of existing national measurement standards for RF and microwave impedance parameters in the UK*, IEE Colloquium Digest No 99/008, 1999, pp. 6/1–6/6
44 Baxter, W., and Dunwoodie, D.: *An easy-to-use method for measuring small SWRs to better than computer-aided accuracy levels*, Wiltron Technical Review, No 8, 1978
45 EA: *Guidelines on the evaluation of Vector Network Analysers (VNA)*, EA-10/12, 2000 (Available at www.european-accreditation.org)
46 Ide, J. P.: 'Traceability for radio frequency coaxial line standards', NPL Report DES 114, 1992

47 Ridler, N. M., and Medley, J. C.: *An uncertainty budget for VHF and UHF reflectometers*, NPL Report DES 120, 1992
48 Ridler, N. M.: 'Improved RF calibration techniques for network analyzers and reflectometers', *Microwave Engineering Europe*, Oct 1993;35–39
49 Zorzy, J.: 'Skin-effect corrections in immittance and scattering coefficient standards employing precision air-dielectric coaxial lines', *IEEE Transactions on Instrumentation and Measurement*, 1966;**IM-15** (4):358–64
50 Kaye, G. W. C., and Laby, T. H.: *Tables of physical and chemical constants*, 15th edn (Longman, London and New York, 1986), pp. 117–20
51 Gray, D. A.: *Handbook of coaxial microwave measurements* (General Radio Company, Massachusetts, USA, 1968), Chapter 1
52 Weinschel, B. O.: 'Errors in coaxial air line standards due to skin effect', *Microwave Journal*, 1990;**33** (11):131–43
53 Faraday Proctor, R.: 'High-frequency resistance of plated conductors', *Wireless Engineer*, 1943;**20**:56–65
54 von Baeyer, H. C.: 'The effect of silver plating on attenuation at microwave frequencies', *Microwave Journal*, 1960;**3** (4):47–50
55 Somlo, P. I.: *The computation of the surface impedance of multi-layer cylindrical conductors*, CSIRO National Standards Laboratory (Australia), Report No APR 12, 1966
56 Kilby, G. J., and Ridler, N. M.: 'Comparison of theoretical and measured values for attenuation of precision coaxial lines', *IEE Electronics Letters*, 1992;**28** (21):1992–94
57 Hill, D. A.: 'Reflection coefficient of a waveguide with slightly uneven walls', *IEEE Transactions on Microwave Theory Techniques*, 1989;**MTT-37** (1):244–52
58 Holt, D. R.: 'Scattering parameters representing imperfections in precision coaxial air lines', *Journal of Research of NIST (USA)*, 1989;**94** (2):117–33
59 Sanderson, A. E.: 'Effect of surface roughness on propagation of the TEM mode', in Young, L. (ed.), *Advances in Microwaves*, vol. 7 (Academic Press Inc, New York, 1971), pp. 1–57
60 Ridler, N. M.: *VHF impedance – a review*, NPL Report DES 127, 1993
61 Nelson, R. E., and Coryell, M. R.: 'Electrical parameters of precision, coaxial, air-dielectric transmission lines', *NBS Monograph 96*, National Bureau of Standards (USA), 1966
62 Daywitt, W. C.: 'First-order symmetric modes for a slightly lossy coaxial transmission line', *IEEE Transactions on Microwave Theory Techniques*, 1990;**MTT-38** (11):1644–51
63 Daywitt, W. C.: 'Exact principal mode field for a lossy coaxial line', *IEEE Transactions on Microwave Theory Techniques*, 1991;**MTT-39** (8):1313–22
64 Woods, D.: 'Relevance of complex normalisation in precision reflectometry', *IEE Electronics Letters*, 1983;**19** (15):596–98
65 Collin, R. E.: *Foundations for microwave engineering*, (McGraw-Hill Book Company, New York, 1966)
66 Ridler, N. M., and Medley, J. C.: 'Calibration technique using new calculable standard for RF reflectometers fitted with GPC-7 connectors', *Conference on*

Precision Electromagnetic Measurements (CPEM) Digest, Boulder, CO, 1994, pp. 117–18

67 Somlo, P. I.: 'The computation of coaxial line step capacitances', *IEEE Transactions on Microwave Theory Techniques*, 1967;**MTT-15** (1):48–53

68 Razaz, M., and Davies, J. B.: 'Capacitance of the abrupt transition from coaxial-to-circular waveguide', *IEEE Transactions on Microwave Theory Techniques*, 1979;**MTT-27** (6):564–69

69 Bianco, B., Corana, A., Gogioso, L., and Ridella, S.: 'Open-circuited coaxial lines as standards for microwave measurements', *IEE Electronics Letters*, 1980;**16** (10):373–74

70 Ridler, N. M., Medley, J. C., Baden Fuller, A. J., and Runham, M.: 'Computer generated equivalent circuit models for coaxial-line offset open circuits', *IEE Proceedings A, Science, Measurement and Technology*, 1992;**139** (5): 229–31

71 Fantom, A. E.: *Radio frequency and microwave power measurement* (Peter Perigrinus Ltd, London, 1990), Appendix A

72 Ridler, N. M., and Medley, J. C.: *Traceable reflection coefficient measurements in coaxial line at MF and HF*, IEE Colloquium Digest No 1994/042, 1994, pp. 8/1–8/4

73 Cox, M. G., Dainton, M. P., and Ridler, N. M.: 'An interpolation scheme for precision reflection coefficient measurements at intermediate frequencies. Part 1: theoretical development', *IMTC'2001 Proceedings of the 18th IEEE Instrumentation and Measurement Technology Conference*, Budapest, Hungary, 21–23 May 2001, pp. 1720–25

74 Ridler, N. M., Salter, M. J., and Young, P. R.: 'An interpolation scheme for precision reflection coefficient measurements at intermediate frequencies. Part 2: practical implementation', *IMTC'2001 Proceedings of the 18th IEEE Instrumentation and Measurement Technology Conference*, Budapest, Hungary, 21–23 May 2001, pp. 1731–35

75 Morgan, A. G., Ridler, N. M., and Salter, M. J.: 'Generalised calibration schemes for RF vector network analysers', *IMTC'2002 Proceedings of the 18th IEEE Instrumentation and Measurement Technology Conference*, Anchorage, AL, 21–23 May 2002

Chapter 10
Microwave network analysers
Roger D. Pollard

10.1 Introduction

This chapter is intended to cover the basic principles of measuring microwave networks by using a network analyser. The objectives are to discuss the kind of measurements which can be made and the major components in a network analyser covering the basic block diagram, the elements and the advantages and disadvantages of different hardware approaches. Material on error correction is the subject of another chapter. The fundamental concept of microwave network analysis involves incident, reflected and transmitted waves travelling along a transmission line. It must be appreciated, at the outset, that measurement in terms of impedance, which is the ratio of voltage to current, implies knowledge of the characteristic impedance Z_0 which describes the mode of propagation in the transmission line. Microwave network analysis is concerned with measuring accurately the incident, reflected and transmitted signals associated with a linear component in a transmission line environment. It is important to appreciate that the same quantities may be defined as different values, for example, return loss, reflection coefficient, VSWR, S_{11}, impedance and admittance are all ways of describing reflection coefficient, and, similarly, gain, insertion loss, transmission, group delay and insertion phase are all ways of describing transmission coefficient.

It is also necessary to understand the fundamental difference between a network analyser and a spectrum analyser. Network analysers are used to measure components, devices and circuits, but a network analyser is always looking at a known signal in terms of frequency and is described as a stimulus–response system. With a network analyser, for example, it is very hard to get an accurate trace on the display, for reasons which will be explained later, but very easy to interpret the results using vector error correction. A network analyser can provide much higher accuracy than a spectrum analyser. Spectrum analysers on the other hand are used to measure signal characteristics on unknown signals. They are usually a single channel receiver without a source and have a much wider range of IF bandwidths than a network analyser. With a spectrum analyser it is easy to get a trace on the display, but interpreting the results can often be much more difficult than with a network analyser.

10.2 Reference plane

The measurements under consideration are those which characterise travelling waves on a uniform transmission line and the (usually voltage) ratios which are detected are functions of position on the lines. Furthermore, any change in the cross section of the transmission line will give rise to a reflection and the launch of evanescent modes. It is therefore necessary to be able to specify a *reference plane* which is appropriately located in a sufficient length of uniform transmission line. The reference plane often, but not necessarily, is the plane of contact of the outer conductors of a mating pair of coaxial connectors or a pair of waveguide flanges.

10.2.1 Elements of a microwave network analyser

Figure 10.1 shows the general block diagram of a network analyser showing the major signal processing parts.

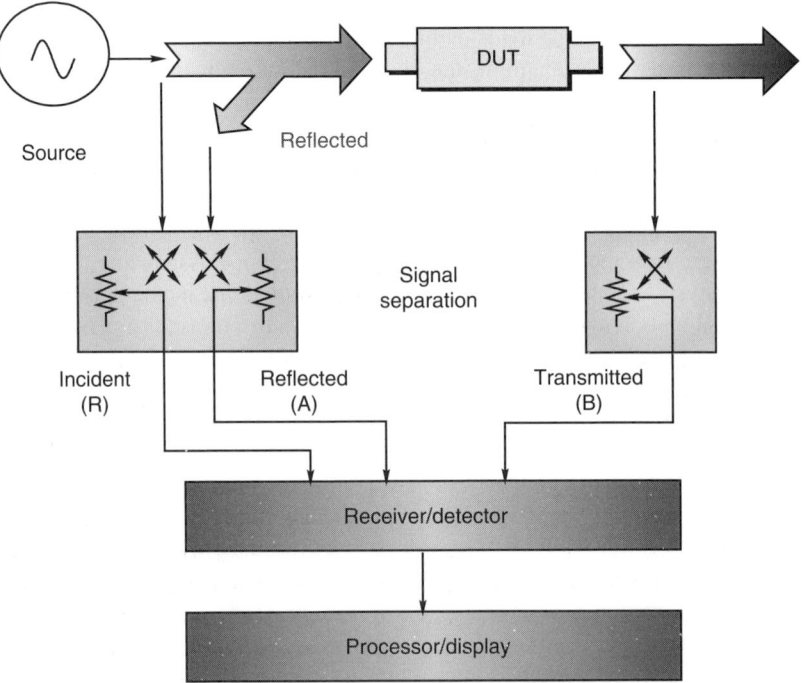

Figure 10.1   General block diagram of a network analyser

Microwave network analysers 209

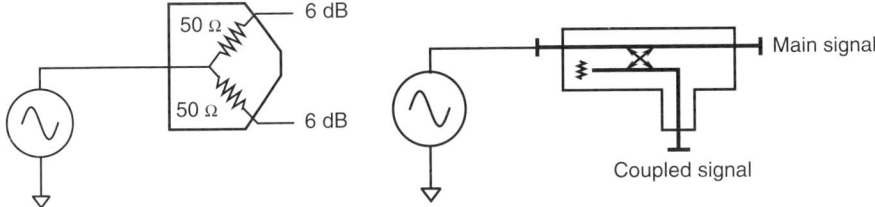

Figure 10.2 Separation of reference signal using power splitter or directional coupler

Four elements are present: (1) source to provide a stimulus, (2) signal separation devices, (3) a receiver for detecting the signals, and (4) a processor and display for calculating and showing the results.

10.2.1.1 Source
The signal source supplies the stimulus for the test system and can either sweep the frequency of the source or its power level. Traditionally, most network analysers had a separate source but nowadays the source is often a built-in part of the instrument. The source may be either a voltage-controlled oscillator or a synthesised sweeper.

10.2.1.2 Signal separation
This is normally described as the test set, which can be a separate box or integrated into a network analyser. The signal separation hardware must provide two functions. The first is to separate a portion of the incident signal to provide the reference signalling for ratioing. This can be done with a power splitter or a directional coupler (Figure 10.2).

Power splitters are usually resistive, non-directional devices and can be very broadband; the trade-off is that they have some loss (usually 6 dB or more) in each port. Directional couplers can be built to have very low loss through the main arm and offer good isolation and directivity, but it is difficult to make them operate at very low frequencies.

The second function is to separate the incident and reflected travelling waves at the input to the device under test (DUT). Directional couplers are ideal because they have the necessary directional properties, low loss in the main arm and good reverse isolation. However, owing to the difficulty of making very broadband couplers, directional bridges are often used. Bridges can operate over a very wide range of frequency but exhibit more loss to the transmitted signal resulting in less power delivered to the DUT.

A directional coupler is a device that separates a component of the signal travelling in one direction only. In the diagrams in Figure 10.3, the signal flowing through the main arm is shown as a solid line, the coupled signal as a dotted line. Note that the fourth port of the coupler is terminated with a matched load. The signal appearing at the coupled port is a fraction of the input signal; this fraction is the *coupling factor*. In the example in Figure 10.3, the coupling factor is 20 dB and therefore when 1 mW (0 dBm) is supplied to the input port, 0.01 mW (−20 dBm) will appear at the coupled

210 *Microwave measurements*

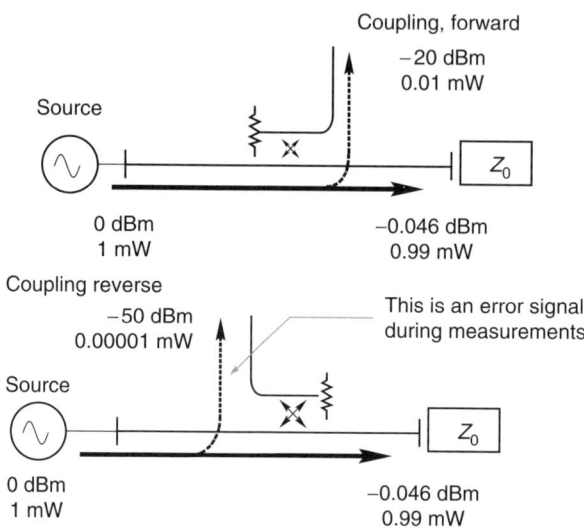

Figure 10.3   *Directional coupler: coupling and directivity*

port. Note that as a result there is a small loss through the main arm. The coupling factor is rarely constant with frequency and the frequency response can become a significant measurement error term.

In an ideal coupler, there will be no component of a signal travelling in the reverse direction at the coupled port, but in practice a coupler has finite isolation and some energy will leak in the reverse direction. In the example in Figure 10.3, the coupler is reversed and the isolation measured at -50 dB.

The most important single parameter for a directional coupler is its directivity, which is a measure of a coupler's ability to separate signals flowing in opposite directions. It can be thought of as the dynamic range for reflection measurements. By definition, directivity is the ratio between the reverse coupling factor (isolation) and the forward coupling factor. In the example of Figure 10.3, the coupler has a directivity of 30 dB. During a reflection measurement the error signal can be, at best, the directivity below the desired signal. The better the match of the DUT the greater measurement error the directivity error will cause. Directivity error is the main reason that will be seen as a large ripple pattern in many measurements of return loss. At the peak of the ripple, directivity is added in phase with the signal reflected from the device. In other cases the directivity will cancel the DUT reflection, resulting in a sharp dip in the response (Figure 10.4).

The directional bridge is similar in operation to the Wheatstone bridge. If all four arms have equal resistance and 50 Ω is connected to the test port then a voltage null will be measured at the detector and the bridge is balanced. If the load at the test port is not 50 Ω then the voltage across the detector is proportional to the mismatch presented by the DUT. If both magnitude and phase are measured at the detector, the complex impedance of the test port can be calculated. A bridge also has an

Microwave network analysers 211

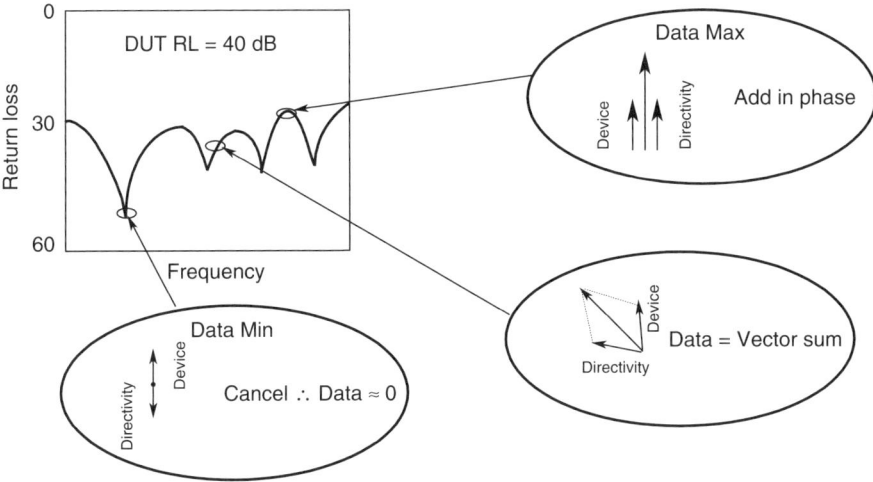

Figure 10.4   Return loss ripple caused by coupler directivity

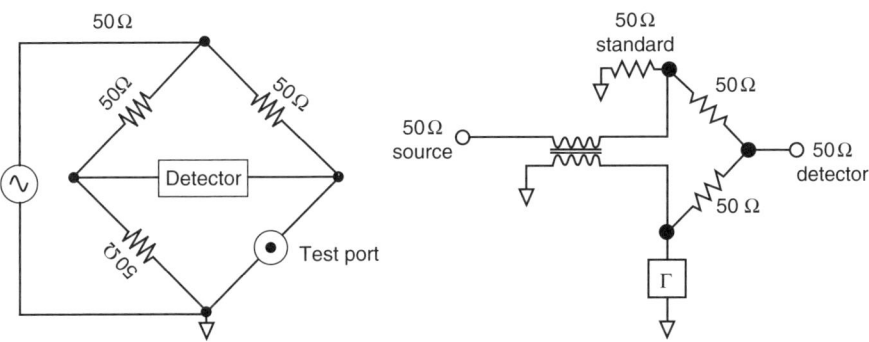

Figure 10.5   Directional bridge: theoretical and actual circuit

equivalent directivity that is the ratio between the best balance measuring a perfect load and the worst balance measuring an open circuit or a short circuit. The effect of bridge directivity on measurement accuracy is exactly the same as for a directional coupler. The basic arrangement of a directional bridge is shown in Figure 10.5. Notice that in a microwave system there is generally a requirement that one terminal of each component is connectable to ground; the key therefore to designing a successful broadband directional bridge to operate at microwave frequencies is the provision of a suitable balun as shown in Figure 10.5.

10.2.1.3   Detectors and receivers

There are two basic ways of providing detection in network analysers – diode detectors, which simply convert the RF to a proportional DC level, or tuned receivers. Diode

212 *Microwave measurements*

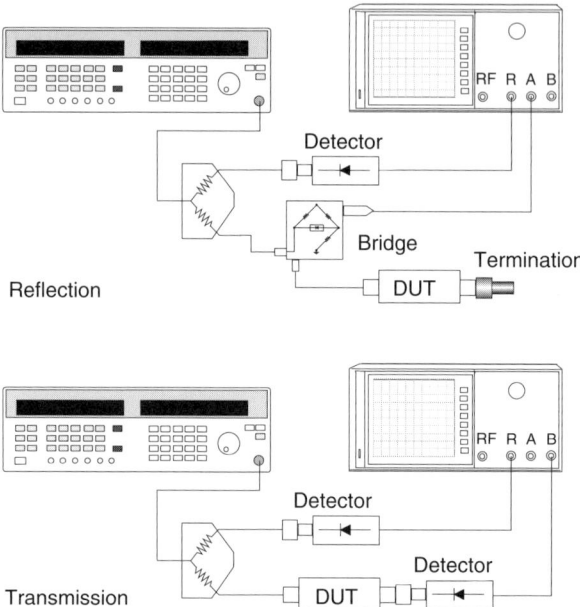

Figure 10.6   Scalar network analyser measurements using diode detectors

detection is inherently scalar and loses phase information. The main advantages of diode detection are low cost and broadband frequency range which is a significant benefit when measuring frequency translating devices (Figure 10.6). Offset against this is the limited sensitivity and dynamic range and susceptibility to source harmonics and spurious signals. Drift in a diode detector, a major source of measurement error, can be eliminated by the use of AC detection that also reduces noise and susceptibility to unwanted signals. However, the necessary modulation of the RF signal can affect the measurements of some devices (e.g. amplifiers with AGC).

The tuned receiver uses a local oscillator (LO) to mix the RF down to an intermediate frequency (IF). The LO is locked either to the RF or to the IF so that the receiver in the network analyser is always correctly tuned to the RF present at the input (Figure 10.7). The IF signal is filtered, which narrows the receiver bandwidth, allows large amounts of gain and greatly improves the sensitivity and the dynamic range. A modern network analyser uses an analogue-to-digital converter (ADC) and digital signal processing to extract the magnitude and phase information from the IF signal.

Tuned receivers not only provide the best sensitivity and dynamic range but also provide harmonic and spurious signal rejection. The narrow band IF filter produces a considerably lower noise floor resulting in significant improvement in sensitivity and dynamic range. For example, a microwave network analyser might have a 3 KHz IF bandwidth and an achievable dynamic range, better than 100 dB. The dynamic range can be improved by increasing the input power by decreasing the IF bandwidth

Microwave network analysers 213

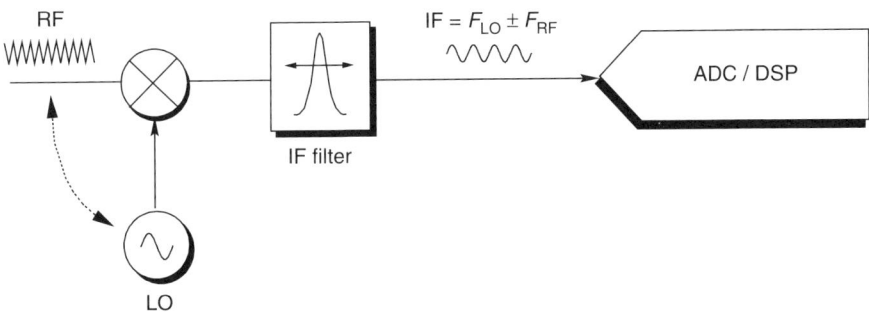

Figure 10.7   Downconverting tuned receiver

or by averaging. This provides a trade-off between noise floor and measurement speed. Averaging reduces the noise floor of the network analyser because complex data are being averaged. Without phase information, as in, for example, a spectrum analyser, averaging only reduces the noise amplitude and does not improve sensitivity. Also because the RF signal is downconverted and filtered before it is measured, any harmonics associated with the source appear at frequencies outside the IF bandwidth and are removed. This eliminates response to harmonics and spurious signals and results in increased dynamic range.

A tuned receiver can be implemented with a mixer or a sampler based front-end. It is often cheaper and easier to make wide band front-ends using samplers instead of mixers. The sampler uses diodes to sample very short time slices of the incoming RF signal. Conceptually the sampler can be thought of as a mixer with an internal pulse generator. The pulse generator creates a broadband frequency spectrum (often known as a 'comb') composed of harmonics of a local oscillator. The RF signal mixes with one of the spectral lines (or 'comb-tooth') to produce the desired IF.

Figure 10.8 shows the block diagram of a sampler system. The local oscillator is tuneable and drives a harmonic generator. The output from the harmonic generator drives a diode that can be thought of simply as a switch. In terms of frequency behaviour the output from the harmonic generator provides a comb of harmonics of the local oscillator and by tuning the local oscillator to the right frequency, the difference between the incoming RF and one of the comb-teeth will be exactly the IF frequency which can pass through the IF filter. A phase lock loop will ensure that the local oscillator is always correctly tuned as the source frequency changes. In most modern designs the local oscillator is pre-tuned to ensure that the same comb-tooth of the local oscillator is used every time that the same frequency is input. Compared to a mixer-based network analyser the LO in a sampler-based front-end covers a much smaller frequency range and a broadband mixer is no longer needed. The trade-off is that the phase lock algorithms for locking the various comb-teeth are much more complex. Sampler based front-ends also have a somewhat lower dynamic range than those based on mixers and fundamental local oscillators because the additional noise is converted into the IF from all of the comb-teeth. Nonetheless, network analysers

214 *Microwave measurements*

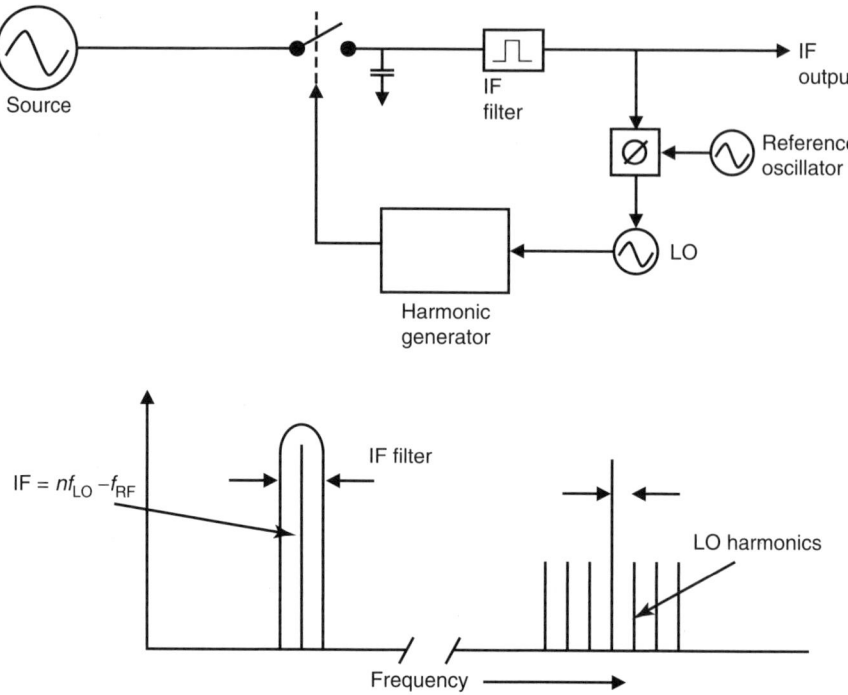

Figure 10.8 *Principle of operation of a sampling receiver in a network analyser*

with narrow band detection based on samplers still have far greater dynamic range than analysers based on diode detection. Dynamic range is usually defined as the maximum power the receiver can measure accurately minus the receiver noise floor. There are many applications requiring large dynamic range, the most common being filter applications. Also the presence of harmonics from the source may create a false response which will be removed by a tuned receiver.

10.3 Network analyser block diagram

Figure 10.9 shows the general schematic of an S-parameter measurement system whilst Figure 10.10 is the block diagram of a modern microwave vector network analyser.

The schematic diagram shown in Figure 10.10 is a RF system which has an integrated source and a tuned receiver based on samplers (labelled S). The system can be configured with a three-channel or four-channel receiver and consequently the test set can be either a transmission/reflection type or capable of full S-parameters.

There are two basic types of test set that are used with network analysers for transmission/reflection (TR) test sets. The RF power always comes out of test port 1 and test port 2 is always connected to a receiver. To measure reverse transmission

Microwave network analysers 215

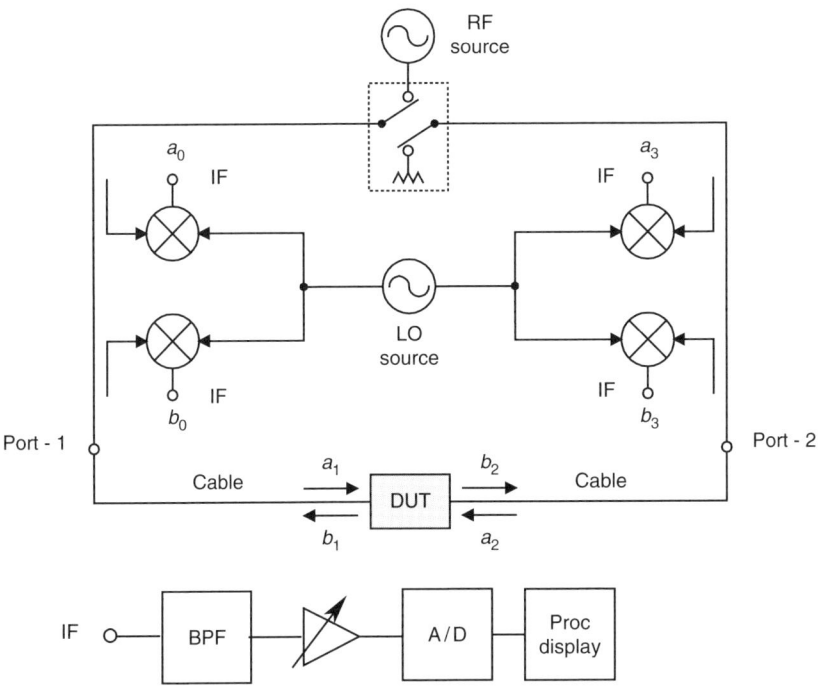

Figure 10.9 Schematic of an S-parameter measurement system

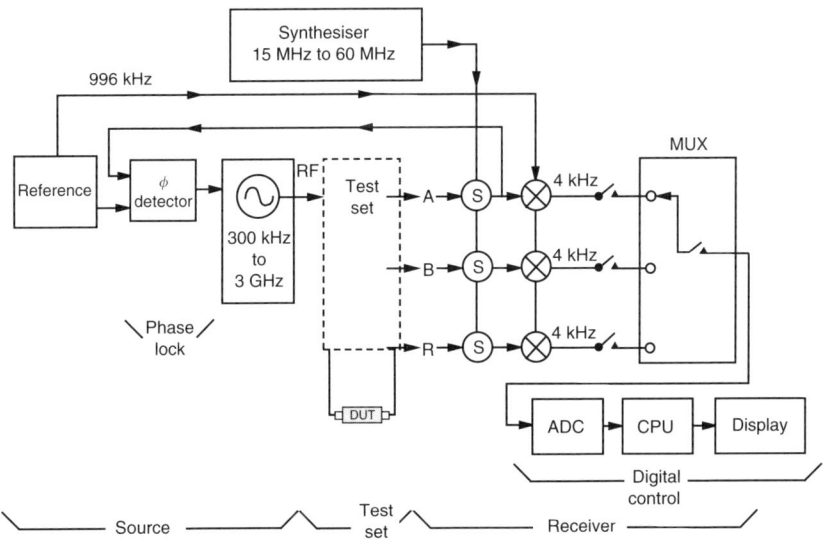

Figure 10.10 Block diagram of an RF network analyser

or output reflection the device must be disconnected, turned around and reconnected again. TR-based network analysers offer only response and one-port calibration so measurement accuracy is not as good as the one that can be achieved using S-parameter test sets. An S-parameter test set allows both forward and reverse measurements without reconnection and allows characterisation of all four S-parameters. RF power can come out of either test port 1 or test port 2 and either test port can be connected to a receiver. The internals rearrangement is carried out by switches inside the test set. These are usually solid-state switches which are fast and do not wear out. Although it is possible to configure an S-parameter test set with only three samplers or mixers the architecture provides fewer choices for calibration as does a four receiver architecture.

The display and processor section allows in current systems is usually an in-built, full-featured PC that not only the reflection and transmission data to be formatted in many ways to allow for easy display, comparison and interpretation but also supports algorithms for calibration, data storage and various other features.

Further reading

Warner, F. L.: 'Microwave vector network analysers' in Bailey, A. E. (ed.), *Microwave Measurements*, 2nd edn (Peter Peregrinus Ltd, London, 1989), Chapter 11

Chapter 11

RFIC and MMIC measurement techniques

Stepan Lucyszyn

11.1 Introduction

All electronic sub-systems are made up of devices and networks. In order to simulate the overall performance of a sub-system under development, all the components that make up the sub-system must be accurately characterised. To this end, precision measurement techniques must be employed at component level. Not only do precision measurements enable a manufacturer to check whether devices are within their target specifications, and to monitor variations in parameter tolerances due to process variations, they also allow more accurate empirical models to be extracted from the measurements and help new modelling techniques to be validated. Also, the operation and performance of some experimental devices can often only be understood from accurate measurements and subsequent modelling. Conversely, poor measurements could result in the needless, and therefore expensive, redesign of high-performance components or sub-systems.

Devices and networks are traditionally characterised using Z, Y or h-parameters. To measure these parameters directly, ideal open and short circuit terminations are required. These impedances can be easily realised at low frequencies. However, at microwave frequencies such impedances can only be achieved over narrow bandwidths (when tuned circuits are employed) and can also result in circuits that are conditionally stable (when embedded within a 'matched load' reference impedance environment) becoming unstable. Fortunately, scattering- (or S)- parameters can be determined at any frequency. To perform such measurements, the device under test (DUT) is terminated with matched loads. This enables extremely wideband measurements to be made and also greatly reduces the risk of instability; however, only when the DUT is terminated with near ideal matched loads (this is irrespective of whether the measurement system is calibrated or not). S-parameter measurements also offer the following advantages:

(1) Any movement in a measurement reference plane along an ideal transmission line will vary the phase angle only.
(2) For a linear device or network, voltage or current and measured power are related through the measurement reference impedance (normally 75 or 50 Ω for coaxial lines and 1 Ω for rectangular waveguides).
(3) With some passive and reciprocal structures, ideal S-parameters can be deduced from spatial considerations, enabling the measurements of the structure to be checked intuitively.

By applying a known incident wave to the DUT and then measuring the reflected and transmitted wave amplitudes, S-parameters can be calculated from the resulting wave amplitude ratios. The equipment most commonly used to perform this measurement is called a vector network analyser (VNA) [1]. The DUT can now be characterised using complete S-parameter measurements (along with DC measurements). The element values associated with the small-signal equivalent circuit model of the DUT can be determined using direct calculations, iterative optimisation and intuitive tuning. This process is referred to as parameter extraction.

With a radio frequency integrated circuit (RFIC), also known as a monolithic microwave integrated circuit (MMIC), either a test fixture or probe station is employed to secure the MMIC in place and to provide a stable means of electrically connecting the MMIC to the measurement system [2]. In this chapter, the use of test fixtures and probe stations at ambient room temperature is reviewed and their role at thermal and cryogenic temperatures is discussed. Finally, with the increasing need for performing non-invasive (or non-contacting) measurements, experimental field probing technologies are introduced.

11.2 Test fixture measurements

Although probe stations result in much more accurate and reproducible measurements, test fixtures are still widely used. The principal reasons are that they are very much cheaper than probe stations and they offer a greater degree of flexibility, such as facilitating larger numbers of RF ports and enabling DC bias circuitry and any off-chip resonators to be located next to the chip. Also, the heat dissipation required when testing monolithic power amplifiers can be easily provided with test fixtures. In addition, test fixtures are ideally suited when RF measurements are required during temperature-cycling and when cryogenic device characterisation is required [3–5].

An illustration of a basic two-port test fixture is shown in Figure 11.1. Most test fixtures are, in principle, based on this generic design, typically consisting of four different components: (1) a detachable metal chip carrier with high-permittivity substrate, (2) a rigid metal housing, (3) connector/launchers and (4) bond wires. The MMIC is permanently attached to the chip carrier with either conductive epoxy glue or solder. The metal housing is employed to hold the chip carrier and the connector/launchers in place. The launcher is basically an extension of the coaxial

RFIC and MMIC measurement techniques 219

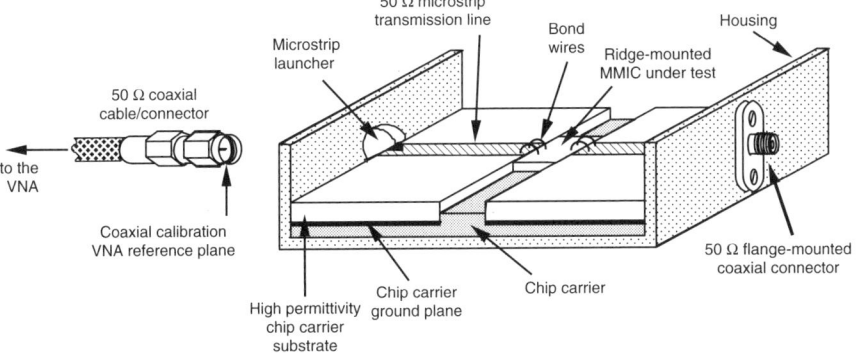

Figure 11.1   *Generic design of a two-port test fixture*

connector's centre conductor, which passes through the housing wall to make electrical contact with the associated chip carrier's microstrip transmission line. Bond wires or straps are used to connect the other end of the microstrip line to the MMIC under test. The parasitic element values associated with a test fixture are typically an order of magnitude greater than those of the MMIC under test.

Before any accurate measurements can be performed, measurement systems must first be calibrated, in order to correct for the systematic errors resulting from the numerous reflection and transmission losses within the measurement system. A calibration kit is required to perform this calibration procedure. This 'cal. kit' has a number of electrical reference standards and software that must be downloaded into the VNA's non-volatile memory or associated PC/workstation controller. For a two-port measurement system, the calibration standards must:

(1) define the primary reference planes;
(2) remove any phase ambiguity using open circuit and/or short circuit reflection standard(s); and
(3) define the reference impedance using delay line, matched load or attenuator impedance standard(s).

Some of the various combinations of different standards that can be employed in a two-port calibration procedure are listed in Table 11.1. The software should contain accurate models for the associated standards and the algorithms required to implement the chosen calibration method. The accuracy of subsequent measurements ultimately depends on how well all the standards remain characterised. Any deviation in the electrical parameters of the standards will degrade the magnitudes of the effective directivity and source match for the measurement system. As a result, great care must be taken to look after these calibration standards.

The non-idealities of a measurement system are characterised using mathematical error correction models, represented by flow diagrams (also known as error adapters or boxes). The function of the calibration procedure is to solve for the error coefficients in these models by applying the raw, uncorrected, S-parameter measurements of the

Table 11.1 Common calibration methods

Method	Calibration standard					
	Through		Reflect	Reference impedance		
	$L = 0$	$L \neq 0$	$\rho_1 = \rho_2$	Line	Match	Atten.
TRL	•		•	•		
LRL		•	•	•		
TRM	•		•		•	
LRM		•	•		•	
TRA	•		•			•
LRA		•	•			•
TSD	•		$\rho = -1$	•		

standards to a set of independent linear equations. The basic two-port calibration procedures have an eight-term error model (four terms associated with each port) and require only three standards. These error terms should correspond directly to the raw hardware performance, including the directivity, source match and frequency tracking. A more accurate 12-term error model, as used in two-port coaxial calibration, takes crosstalk and the effects of impedance mismatches at the RF switches within the VNA's test-set into account. Once the calibration procedure has been performed, it can be verified by measuring separate verification standards.

11.2.1 Two-tier calibration

One method of calibrating the measurement system is to split the process into two tiers [6]. Initially, a coaxial calibration is performed, where the VNA reference planes are located at the end of its cable connectors. Historically, the VNA was calibrated using short-open-load-through (SOLT) standards. These lumped-element standards can give high-quality coaxial calibrations across an ultra-broad bandwidth (e.g. DC to 50 GHz), so long as all the standards remain accurately characterised across the entire bandwidth. Since test fixtures are far from ideal, a second process is required to shift the initial VNA reference planes to the MMIC under test, in order to eliminate the effect of the test fixture. This second process is known as de-embedding or de-convolution [7]. To perform de-embedding it is necessary to accurately characterise the test fixture [8].

Another reason why you may need to characterise a test fixture is when multiple RF port MMICs are to be measured using a two-port VNA [9]. Here, power reflected from impedance-mismatched loads on the auxiliary ports of the MMIC can result in significant measurement errors. These errors will increase as the mismatch losses increase and/or the number of RF ports increases. As a result, MMICs that have more RF ports than the VNA require all the loads to be individually characterised, and a further process of matrix renormalisation [9–13] in order to remove the effects of the mismatched loads on the auxiliary ports. Three methods that can, in principle, be employed to characterise a test fixture are (1) time-domain (T-D) gating, (2) in-fixture calibration and (3) equivalent circuit modelling.

11.2.1.1 Time-domain gating

Some VNAs can be upgraded with a synthetic-pulse T-D reflectometry (TDR) option [14–20]. Here, the discrete form of the inverse Fourier transform (IFT) is applied to a real sequence of harmonically related frequency-domain (F-D) measurements; in our case, of the MMIC embedded within its test fixture. This is directly equivalent to mathematically generating synthetic unity-amplitude impulses (or unity-amplitude steps), which are then 'applied' to the embedded MMIC. The resulting T-D reflection and transmission responses can then be analysed to provide information about the MMIC and test fixture discontinuities. In reflection measurements, it is possible to remove the effects of unwanted impedance mismatches or else isolate and view the response of an individual feature. With a multiple port test fixture, transmission measurements can give the propagation delay and insertion loss of signals travelling through a particular path by removing the responses from the unwanted paths.

With an MMIC fed with transmission lines that only support a pure TEM mode of propagation, time and actual physical distance are simply related:

$$\text{Physical distance} = \begin{cases} c\Delta t\zeta/2 & \text{with reflection measurements} \\ c\Delta t\zeta & \text{with transmission measurements} \end{cases}$$

where c is the speed of light in free space; Δt is the time difference, relative to a reference (e.g. $t = 0$); and $\zeta = 1/\sqrt{\varepsilon_r}$ is the velocity factor.

Also, F-D nulls in $|S_{11}|$ are at frequency harmonics of $1/\Delta t$, where Δt is the time difference between two reflected impulses.

If the feed lines are non-TEM, and therefore dispersive, impulse spreading will occur, which could significantly distort the impulse shape (in time and amplitude). If the dispersive nature is known, the frequency sweep can be pre-warped [20].

With either a banded VNA (which may cover just one of the main waveguide bands), or a broadband VNA, the *band-pass* T-D mode can be selected, where only synthetic impulses are generated. This is useful for band-limited guided-wave structures (e.g. rectangular waveguides). In general, in this mode, only the magnitudes of the individual reflection and transmission coefficients are available. As a result, the exact nature of any discontinuity (e.g. resistive, inductive and capacitive) cannot be identified. However, it is still possible to extract some information about the nature of a defect in *band-pass* mode with a phasor impulse.

With a broadband VNA, a *low-pass* T-D mode is also available where both synthetic impulses and synthetic steps can be generated. The *low-pass* mode is used to emulate a real-pulse TDR measurement system. This allows the user to identify the nature of any discontinuity. F-D measurements are taken from the *start frequency*, f_1, to the *stop frequency*, f_2. When compared with *band-pass*, for the same bandwidth (i.e. *frequency-span*) $B = f_2 - f_1$, the *low-pass* mode offers twice the *response resolution* in the T-D. However, with the *low-pass* mode, the F-D measurements must be harmonically related, from DC to f_2, such that $f_2 = n_{fd} f_1$, where n_{fd} is the number of points in the F-D (e.g. 51, 101, 201, 401 and 801). The DC data point is extrapolated from the f_1 measurement. However, if the measurement at f_1 is noisy, the T-D trace will be unstable and difficult to interpret.

In TDR, the *width* of a band-limited unit impulse (or window function) is defined as the interval between its two half-amplitude (i.e. -6 dB power) points. The corresponding *response resolution* is defined as the interval between two impulses that are just distinguishable from each other as separate peaks. With equal amplitude impulses, the *response resolution* is equal to the 6 dB *impulse width*. With no window function applied to the F-D measurements:

$$6 \text{ dB Impulse width} = \begin{cases} \dfrac{1.2}{B} & \text{for band-pass} \\ \dfrac{0.6}{B} & \text{for low-pass} \end{cases}$$

$$\text{Main Lobe's null-to-null width} = \begin{cases} \dfrac{2}{B} & \text{for band-pass} \\ \dfrac{1}{B} & \text{for low-pass} \end{cases}$$

The time *range* is the length of time that measurements can be made without encountering a repetition of the same response. The *range* must be set longer than the furthest discontinuity, otherwise aliasing will occur, where out-of-range discontinuities will fold-over and appear in-range at (*two range – target position*)

$$\text{Range} = \frac{1}{\Delta f}$$

where $\Delta f = B/(n_{fd} - 1)$.

If a feature lies exactly midway between two T-D points then the energy associated with the discontinuity will be distributed between the two points, resulting in the displayed amplitude being reduced by almost 4 dB [20]. Therefore, care must be taken to ensure there is sufficient *range resolution* (or *point spacing*) in the T-D.

$$\text{Range resolution} = \frac{\text{Range}}{n_{td}}$$

where n_{td} is the number of points in the T-D.

The point spacing can be reduced to any desired level, at the expense of processing time, by using a chirp-Z fast Fourier transform algorithm. This allows range to be replaced by an arbitrary display time-span in the above range resolution equation. It is worth noting that with range and range resolution, either the one-way time or round-trip time may be quoted, depending on the manufacturer.

De-embedding using synthetic-pulse TDR is not de-embedding in the true sense. It is specifically T-D gating, which can isolate a time feature and emphasise its frequency response. With time-gating, a mathematical window (called a gate or time filter) is used to isolate the embedded MMIC, so that only the MMIC's frequency

RFIC and MMIC measurement techniques 223

response can be emphasised. When the gate is switched on, all reflections outside the gate are set to zero. This is equivalent to terminating the MMIC with the complex conjugate of its respective port impedance(s).

The synthetic-pulse TDR option can be a very useful tool, although it can suffer from a number of sources of errors [15,19,20]; some of these are listed as follows:

(1) Noise errors [15].
   (a) *Sweep mode*: The VNA's synthesised source can be operated in either the *ramp-sweep* or *step-sweep* mode. With the former, small non-linearities and phase discontinuities generate low-level noise sidebands on the T-D impulse and step stimuli. However, with the *step-sweep* mode, the improved source stability eliminates these noise sidebands and improves the T-D's dynamic range by as much as 30 dB. Moreover, to reduce the noise floor of the T-D measurements further, the *step-sweep* mode enables more averaging of the F-D measurements, compared with the *ramp-sweep* mode, without greatly increasing the sweep time.
   (b) *Bandwidth*: The noise floor in the T-D response is directly related to noise in the F-D data. Therefore, the number of F-D data points taken at, or below, the system's noise floor can be minimised by reducing the *frequency-span* to the bandwidth of the MMIC.
   (c) *Test-set*: If the test-set does not have a flat response down to the *start frequency* then the reduction in the F-D's dynamic range towards f_1 will cause an increase in the T-D's noise floor, the resulting trace bounce, in the *low-pass* mode, can be improved by turning on T-D trace averaging.
(2) Frequency-domain window errors.
   There is usually a choice of F-D window functions (e.g. Kaiser–Bessel) that can be applied prior to the IFT, for example, *minimum* (0*th order*), *normal* (6*th order*) and *maximum* (13*th order*). The *minimum* window has a rectangular function that produces the sin x/x impulse shape, having the minimum 6 dB impulse width and also the maximum sidelobe levels (with a minimum sidelobe suppression of only 13 dB in its power response). The other two window functions reduce the sidelobe levels (with a minimum suppression of 44 and 98 dB, respectively) at the expense of a wider impulse (by a factor of 1.6 and 2.4, respectively). It will be seen that a trade-off has to be made when choosing the F-D's windowing function, between the desired resolution and dynamic range in the T-D. Note that this windowing function does not affect the displayed F-D response.
   (a) *Time resolution errors*: With narrow bandwidth VNAs, the impulse may be too wide. As a result, it may be difficult to resolve the MMIC and the connector/launchers features, down to the baseline, when the associated discontinuities are too close to one another. In practice, the MMIC should be separated by at least two 6 dB impulse widths from the connector/launcher.

(b) *Dynamic range errors*: Impulse sidelobes limit the dynamic range of the T-D responses, since the sidelobes from a large impulse can hide a small adjacent target impulse.

(c) *Moding errors*: If the bandwidth of the VNA is too high, such that overmoding in transmission lines or box mode resonances occur in the test jig, the T-D responses become un-interpretable.

(d) *Out-of-band response*: The amplitude of the impulses represents the average value over the entire *frequency-span*. Therefore, the displayed amplitude of an impulse can be different from the expected value if the *frequency-span* includes an MMIC with a non-flat frequency response; for example, having highly abrupt out-of-band characteristics.

(3) Discontinuity errors.

(a) *Masking errors*: If the target discontinuity is preceded by other discontinuities that either reflect or absorb energy, then these other discontinuities may remove some of the energy travelling to and emanating from the target discontinuity. The trailing edge of earlier features can also obscure the target feature.

(b) *Multi-reflection aliasing errors*: Multiple reflections between discontinuities can cause aliasing errors. For example, if a two-port MMIC is positioned midway between two connector/launchers, reflection from the furthest connector/launcher will be corrupted by multiple reflections between the MMIC and the nearest connector/launcher.

(4) Time-domain window errors.

In practice, a time filter having a non-rectangular response is used for gating, otherwise the sin x/x weighting would be conveyed to the F-D. There is usually a choice of T-D window functions that can be applied before the Fourier transform: *minimum, normal, wide* and *maximum*. The *minimum* window has the fastest roll-off and largest sidelobes, while the *maximum* window has the slowest roll-off and smallest sidelobes.

(a) *Baseline errors*: The *gate-start* and *gate-stop* times, which define the -6 dB gate-span of the filter, must be set at the baseline if low frequency distortion in the F-D is to be minimised [14].

(b) *Truncation errors*: A limited gate width may truncate lengthy target features. To minimise truncation error, the wider gates are preferred.

(c) *Sidelobe errors*: The time filter sidelobes may 'see' earlier or subsequent features. This could significantly corrupt the F-D response of the target feature. To minimise sidelobe errors, the wider gates are preferred.

(d) *Gate offset errors*: F-D distortion can occur if the *gate-centre* is offset from the centre of the target feature(s). This is because a near-symmetrical target response may lose its symmetry when applied to a time-offset gate that has significant in-gate attenuation. To minimise gate offset errors, the wider gate shapes are preferred.

(e) *Minimum gate-span errors*: The gate-span must be set wider than the minimum value, otherwise the gate will have no passband and may have high sidelobe levels.

(f) *Attenuation errors*: For a fixed gate-span, the level and duration of in-gate attenuation may be excessive with wider gates.

(g) *Reflection/transmission switching errors*: If gating is performed on a voltage reflection coefficient response then the associated return loss F-D measurement is valid. If the same gating times are applied to the voltage transmission coefficient response(s) then this may not be appropriate. For example, when a two-port MMIC is not placed midway between connectors, the transmission pulse may not be fully enclosed within the reflection response's gate. The resulting insertion loss F-D measurement will not represent accurate de-embedding.

As an example, a gallium arsenide (GaAs) MMIC with a 2.9 mm length of 55 Ω microstrip through-line was placed at the centre of a 25.4 mm alumina chip carrier. An Agilent Technologies 8510B VNA was calibrated with a 20 GHz bandwidth and 401 frequency points. The *band-pass* mode was selected with a *minimum* F-D windowing function. This combination provides a minimum *response resolution* and maximum *range* values of 60 ps and 20 ns, respectively. The F-D power responses are shown in Figure 11.2a. The corresponding T-D response of the input port's voltage reflection coefficient is shown in Figure 11.2b. Here, the first and last peaks correspond to the impedance mismatches associated with the coaxial-to-microstrip transitions of the input and output ports, respectively. The two centre peaks correspond to the reflections associated with the microstrip-to-MMIC transitions. It will be apparent from Figure 11.2b that accurate de-embedding would not be possible using T-D gating. This is because the unwanted reflections cannot be resolved down to the baseline. If de-embedding was attempted in the above example then the ripples in the F-D responses would be smoothed out, as one would expect, although this would not constitute accurate de-embedded measurements. In order to achieve accurate de-embedded measurements, a VNA with more bandwidth, or alternatively, a real-pulse TDR system having ultra-short impulses, can be used.

11.2.1.2 In-fixture calibration

In general, a quality test fixture is much cheaper to buy than a probe station. Suitably designed quality test fixtures can be accurately characterised using in-fixture calibration techniques. As with coaxial calibration, the most appropriate algorithms use a combination of through, reflection and delay line standards, with common methods being through-reflect-match (TRL), TSD and line-reflect-line (LRL). The main reason for employing these types of calibration is that only one discrete impedance standard is required, such as an open or short, which is relatively easy to implement. The matched load is avoided; this is advantageous, as it is more difficult to fabricate non-planar 50 Ω loads to the same level of accuracy that can be achieved with low dispersion transmission lines. However, there are still significant disadvantages with in-fixture calibration:

(1) Multiple delay lines may be required for wideband calibration (any one line must introduce between about 20 and 160 of electrical delay to avoid phase ambiguity, limiting the bandwidth contribution of each line to an 8:1 frequency range).
(2) The use of multiple lines can add uncertainty to the measurements, since the launchers are continually being disturbed during calibration, although freely available software (called *MultiCal*™) can eliminate the effects of non-repeatability, by measuring either the same line a number of times or different lengths of line, in order to reduce the uncertainty [21].
(3) A frequency-invariant measurement reference impedance must be taken from the characteristic impedance, Z_0, of the delay lines, however, frequency dispersion in microstrip lines may not always be corrected for. In practice, the Z_0 of the lines can be determined using TRL calibration [22,23] and then subsequent measurements can be renormalised to any measurement reference impedance.
(4) The high level of accuracy is immediately lost with test fixtures that employ poor quality components and/or non-precision assembly.
(5) The calibration substrates dictate and, therefore, restrict the location of the RF ports.
(6) For devices with more than two ports the calibration procedure must be significantly extended and all the results from this routine must be easily stored and retrieved.
(7) The microstrip-to-MMIC transition is not taken into account.

11.2.1.3 Equivalent circuit modelling

Test fixtures made in-house tend to be simple in design, such as the type shown in Figure 11.1, and cost only a small fraction of the price of a good quality commercial test fixture. Unfortunately, these non-ideal test fixtures suffer from unwanted resonances [24], poor grounding [25] and poor measurement repeatability. The problem of unwanted resonances can be clearly seen in the F-D responses of Figure 11.2a. Here, the resonances at 3 and 12 GHz are attributed to the production grade coaxial connectors used in the test fixture. Because of poor repeatability, employing elaborate and expensive calibration techniques to characterise such fixtures would appear unjustified, because significant measurement degradation is inherent. As an alternative, equivalent circuit models (ECMs) can provide a crude but effective means of de-embedding. This 'stripping' process results in about the same level of degradation as would be found if in-fixture calibration was used with a non-ideal test fixture, but with minimal expense and greater flexibility. Also, ECMs based on the physical structure of the fixture have demonstrated a wide bandwidth performance. The ECMs can be easily incorporated into conventional F-D simulation software packages. They can also be employed to simulate packaged MMICs. An example of an ECM for a test fixture similar to the one in Figure 11.1 is shown in Figure 11.3. This model has

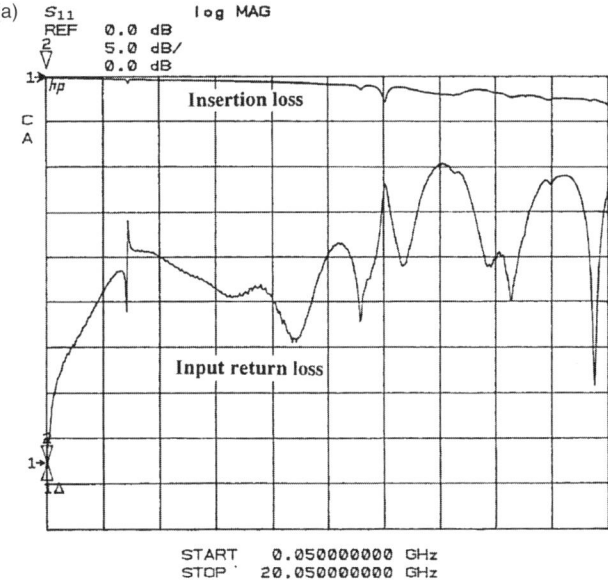

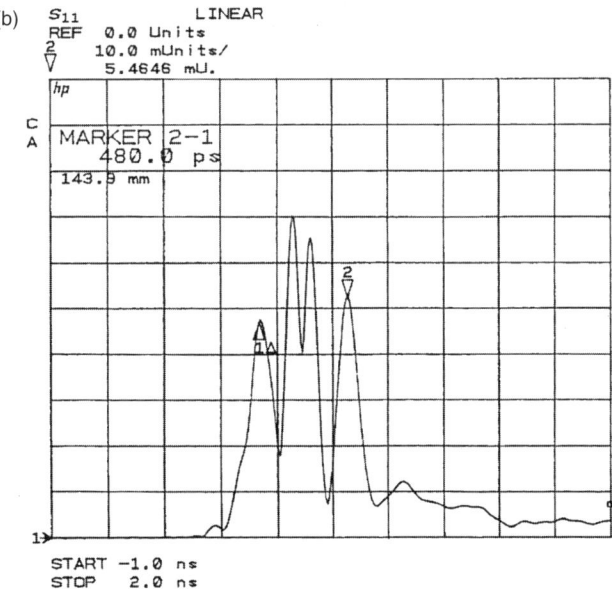

Figure 11.2 Embedded 55 Ω MMIC through-line: (a) frequency-domain power responses and (b) corresponding time-domain response for the input voltage reflection coefficient

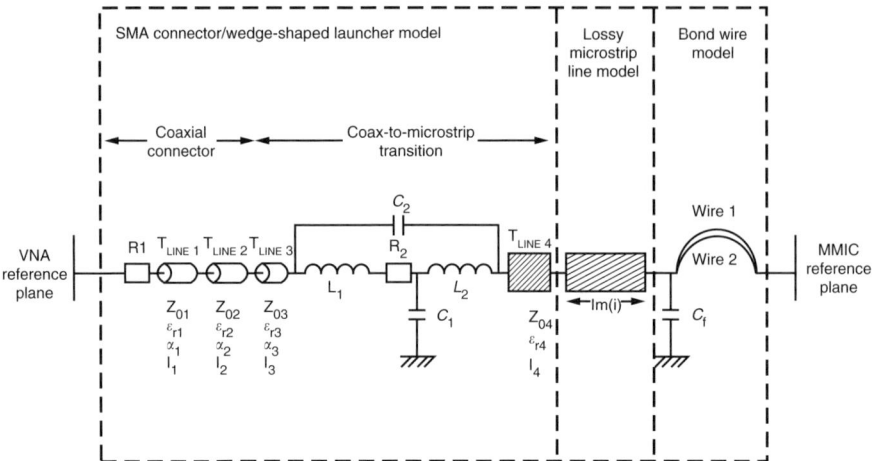

Figure 11.3   Equivalent circuit model of a microstrip test fixture

demonstrated a sufficient degree of accuracy from DC to 19 GHz for the popular Omni-Spectra SMA connector/wedge-shaped launcher [9], which is similar to the more popular SMA printed circuit board socket.

The exact nature of the ECM, the element values and the microstrip parameter data are extracted from through-line measurements of the test fixture. Both a direct microstrip through-line and an MMIC through-line should be used in order to provide more information for the parameter extraction process, and to make it possible to model the microstrip-to-MMIC transition accurately. De-embedding can be carried out with most F-D CAD packages by converting the ECM into a series of negative elements connected onto the ports of the measured data. Some CAD packages provide a 'negation' function that allows the ECM sub-circuit to be directly stripped from the measured data. With either method, the order of the node numbers is critical, and the de-embedding routine should be verified.

In addition to those already mentioned, de-embedding using equivalent circuit modelling has the following advantages:

(1) dispersion in the microstrip lines does not have to be corrected for in the VNA's calibration;
(2) there is no restriction by the calibration procedure on the location of the RF ports;
(3) systematic errors resulting from variations in the characteristic impedance of the chip carrier's microstrip lines, due to relaxed fabrication tolerances, can easily be corrected for;
(4) bond wires [26] and the microstrip-to-MMIC transition can be modelled [27];
(5) resonant mode coupling between circuit components, due to a package resonance, can also be modelled [24]. Better still, package resonances can, in some instances, be removed altogether [28,29].

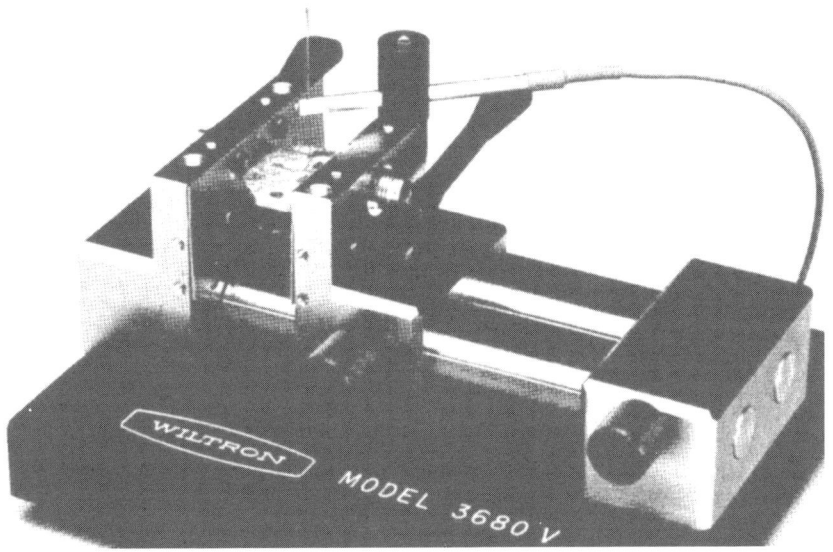

Figure 11.4   Photograph of the Anritsu 3680V universal test fixture

11.2.2 One-tier calibration

Improved contact repeatability and prolonged contact lifetime are two considerations that favour the two-tier process [6], as they are only assembled once with T-D gating and ECMs. In practice, however, to achieve the best performance, in-fixture TRL or line–network–network [30] calibration is applied directly to a quality test fixture, without the need for the two-tier coaxial calibration/de-embedding process. This one-tier calibration procedure gives more accurate measurements than the two-tier method, since de-embedding is inherently prone to errors, and the propagation of measurement errors is reduced [6]. Using this approach, the Anritsu 3680 V universal test fixture, shown in Figure 11.4, can perform repeatable measurements up to 60 GHz. At the time of writing, a number of other companies produce test fixtures for accurate in-fixture calibration, including Agilent Technologies, Intercontinental Microwave, Argumens and Design Techniques. They are either split-block fixtures, with a removable centre section, or they use launchers attached to sliding carriages.

With the high levels of accuracy that can be achieved using quality test fixtures, the poor characterisation of bond wires, due to the poor repeatability of conventional manually operated wire-bonding machines, becomes significant. Improvements in the modelling accuracy and physical repeatability of the microstrip-to-MMIC transition when using automatic wire-bonding assembly techniques have been reported [27]. In addition, flip-chip technology (also known as solder-bump technology) is now well established [31–39]. Here, a tiny bead of solder is placed on all the MMIC bond pads and the MMIC is placed upside down directly onto the chip carrier. When heated to the appropriate temperature, the solder flows evenly and a near perfect connection is made between the MMIC pad and its associated chip carrier pad. The advantages of this technology over bond wire technology, for the purposes of measurements, are its ultra-broad bandwidth, superior contact repeatability and high characterisation accuracy of the carrier's transmission line-to-MMIC transition.

11.2.3 Test fixture design considerations

The following guidelines are useful when selecting, designing or using a test fixture:

(1) Split-block test fixtures [5,40] are ideal for two-port in-fixture TRL calibration, since they can provide good repeatability. Here, a short circuit standard is preferred, since significant energy may be radiated with an open circuit standard.
(2) Side walls can form a waveguide or resonant cavity. The size of the waveguide/cavity should be made small enough so that the dominant mode resonant frequency is well above the maximum measurement frequency. Carefully placed tuning screws and/or multiple RF absorbing pads can eliminate or suppress unwanted modes [28,29].
(3) Poor grounding, due to excessively long ground paths and ground path discontinuities, must be avoided.
(4) Avoid thick chip carrier substrates, wide transmission lines (sometimes used for off-chip RF de-coupling) and discontinuities, in order to minimise the effects of surface wave propagation and transverse resonances at millimetric frequencies. Transverse currents can be suppressed by introducing narrow longitudinal slits into the low impedance lines.
(5) Use substrates with a high dielectric constant to avoid excessive radiation losses and to minimise unwanted RF coupling effects.
(6) New precision connector/launchers should be used whenever possible, and measurements should be performed below the connector's dominant TEM mode cut-off frequency.
(7) Launchers should be separated from the DUT by at least three or four times the substrate thickness, so that any higher-order evanescent modes, generated by the non-ideal coax-to-microstrip transition, are sufficiently attenuated at the DUT.

11.3 Probe station measurements

Until relatively recently, the electrical performance of an MMIC was almost always measured using test fixtures. Nowadays, extremely accurate MMIC measurements can be achieved using probe stations. Such techniques were first suggested for use at microwave frequencies in 1980 [41], demonstrated experimentally in 1982 [42], and introduced commercially by Cascade Microtech in 1983. During the past two decades there have been rapid developments in probe station measurement techniques.

Today, the partnership between Cascade Microtech and Agilent Technologies provides a total solution for on-wafer probing, which can perform repeatable

RFIC and MMIC measurement techniques 231

F-D measurements at frequencies as high as 220 GHz [43], although single-sweep measurements from 45 MHz to 110 GHz are routinely undertaken. When compared with test fixtures, commercial probe station measurements have the following advantages:

(1) they are available in a single-sweep system from DC to 110 GHz;
(2) they are more accurate and much more repeatable, since they introduce much smaller systematic errors;
(3) they have a simpler calibration procedure, which can be automated with on-wafer calibration and verification standards [12,44];
(4) they enable the VNA measurement reference planes to be located at the probe tips or at some distance along the MMIC's transmission line; in the latter case, transition effects can be removed altogether;
(5) they provide a fast, non-destructive means of testing the MMIC, thus allowing chip selection prior to dicing and packaging; and
(6) banded measurements are possible up to 220 GHz.

Overall, the microwave probe station can provide the most cost effective way of measuring MMICs when all costs are taken into account.

11.3.1 Passive microwave probe design

At frequencies greater than a few hundred megahertz, DC probe needles suffer from parasitic reactance components, due to the excessive series inductance of long thin needles and shunt fringing capacitances. If the needles are replaced by ordinary coaxial probes that are sufficiently grounded, measurements up to a few gigahertz can be achieved. The upper frequency is ultimately limited by the poor coax-to-MMIC transition. A tapered coplanar waveguide (CPW) probe provides a smooth transition with low crosstalk.

Cascade Microtech have developed tapered CPW probes and microstrip hybrid probes (Infinity) that enable measurement to be made from DC to 110 GHz with a single coaxial input. With waveguide input, 50–75 GHz (V-band) or 75–110 GHz (W-band) [43] probes are available in both the tapered waveguide and Infinity versions, as shown in Figure 11.5. The Infinity probes are also available for 90–140 GHz (F-band), 110–170 GHz (D-band) and 140–220 GHz (G-band) operation.

The maximum frequency limit for coaxial-input probes is imposed by the onset of higher-order modes propagating in the conventional coaxial cables and connectors. For W-band operation, Agilent Technologies developed a coaxial cable and connector that has an outer screening conductor diameter of only 1 mm, while Anritsu have their own 1.1 mm coaxial technology.

A photograph illustrating the use of Agilent's 1 mm coaxial technology to give state-of-the-art performance up to 110 GHz, with a Cascade Microtech Summit 12000 probe station, is shown in Figure 11.6. This arrangement uses the latest Agilent N5250 110 GHz VNA. Fully automatic calibration of the probing system can be performed up to 110 GHz. The D-band version is shown in Figure 11.7.

232 *Microwave measurements*

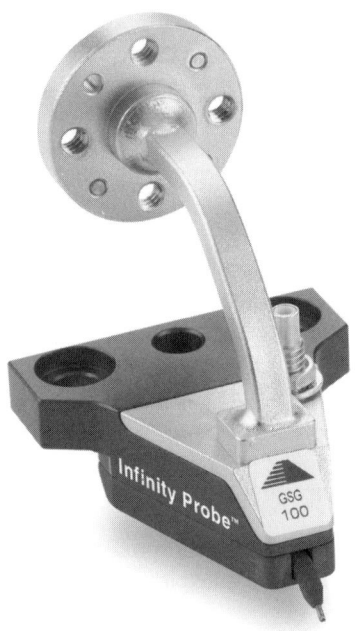

Figure 11.5 Photograph of a waveguide input Infinity probe

In the past, the tapered coplanar waveguide probe was made from an alumina substrate or an ultra-low-loss quartz substrate. The probe tips that made the electrical contacts consisted of hard metal bumps that were electroplated over small cushions of metal, allowing individual compliance for each contact. As the probes were over-travelled (in the vertical plane) the probe contacts wiped or 'skated' the MMICs' probe pads (in the horizontal plane). One of the major limitations of these tapered CPW probes was their short lifetime, since the substrate had limited compliance and the probe contacts could wear down quite quickly. As a result, the more the probe was used, the more over-travel had to be applied to them. Eventually, either the probe substrate begins to crack or the probe tips fall apart.

For this reason, GGB Industries developed the *Picoprobe*™. This coaxial probe is more compliant and can achieve operation between DC and 120 GHz, with a coaxial input, and between 75 and 120 GHz with a waveguide input [45]. From DC to 40 GHz, this probe has demonstrated an insertion loss of less than 1.0 dB and a return loss better than 18 dB. However, one potential disadvantage of coaxial probes is that the isolation between probes may be limited when operating above V-band.

For even better compliancy, durability, ruggedness and flexibility, Cascade Microtech developed the *Air Coplanar*™ tipped coaxial probe [46]. This probe has demonstrated an insertion loss of less than 1.0 dB from DC to 110 GHz and can operate at temperatures from -65 to $+200\,°C$. A cross-sectional view and photograph can be seen in Figure 11.8.

RFIC and MMIC measurement techniques 233

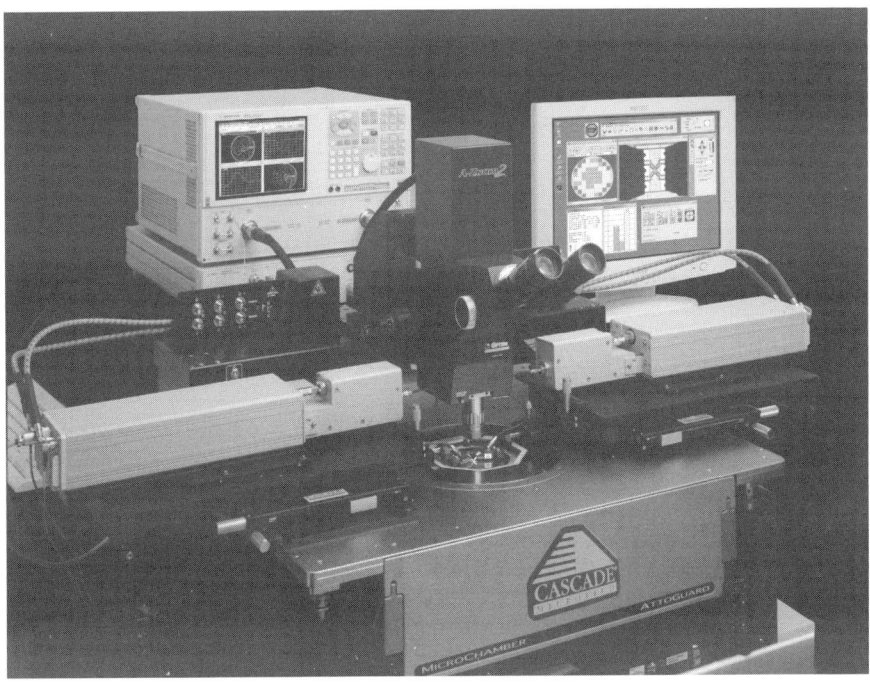

Figure 11.6 Single-sweep, 10 MHz to 110 GHz, on-wafer probing system with Agilent's N5250 110 GHz PNA series network analyser and the Summit 12971 probe

Cascade Microtech still produce the ACP probe, as it is useful for applications that require high power/bias use (above 500 mA), poor contact planarity, large pitches (above 250 μm) or temperatures above 125 °C.

Cascade Microtech's latest generation of Infinity probes, shown in Figure 11.9, was initially designed to improve the contact resistance characteristics of probing onto aluminium pads but it also has significant advantages for probing onto gold pads. Figure 11.10 shows a comparison between the resistance characteristics of the tungsten ACP probe and Infinity probe, when probing onto aluminium.

Inherent to the design is a coaxial-to-microstrip transition. This, in turn, uses vias to connect to extremely small contacts. The microstrip construction ensures vastly improved isolation between the underside of the probe and the measurement of the substrate underneath, allowing adjacent devices to be placed closer to the test structure. This design also dramatically improves the calibration and crosstalk characteristics. Moreover, as a result of the reduced contact size, as shown in Figure 11.11, the damage to the contact pads is also greatly reduced; this is very useful for tests that require multiple tests or with applications where very little pad damage is allowed.

234 *Microwave measurements*

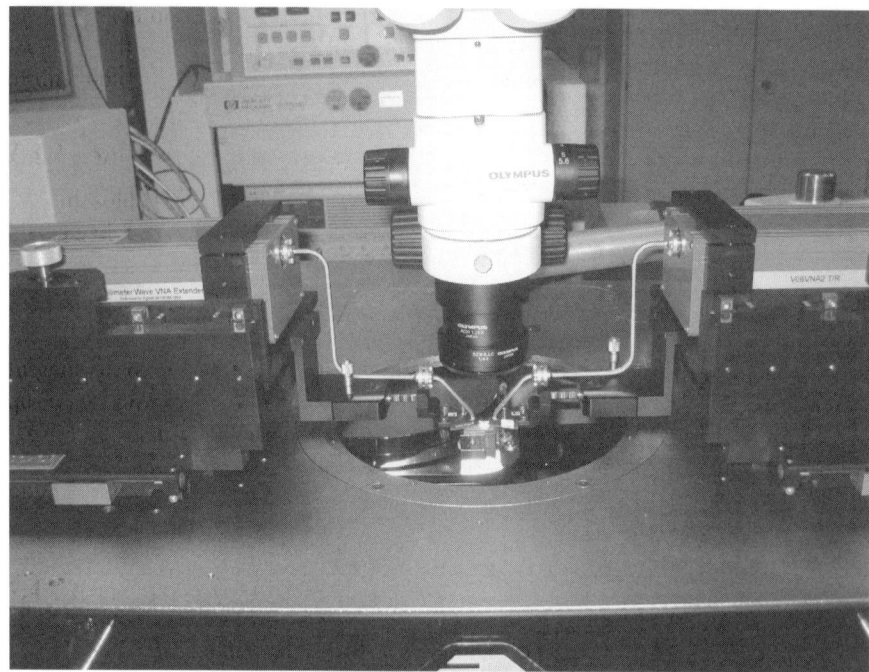

Figure 11.7 Photograph of the banded, 110–170 GHz, on-wafer probing system

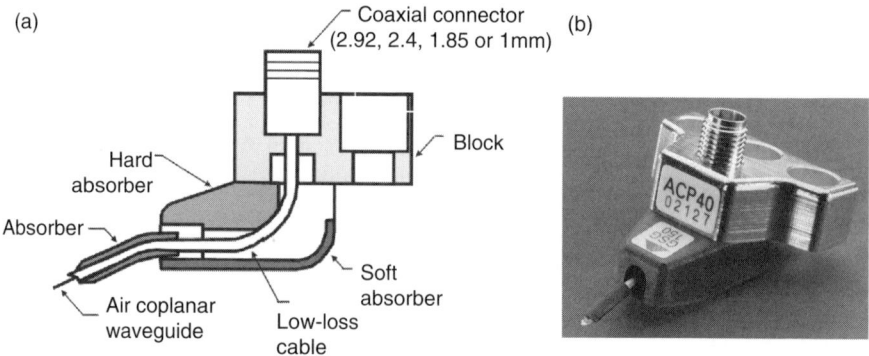

Figure 11.8 APC probe: (a) cross-sectional view of construction and (b) photograph

When selecting the type of microwave probe required, it is necessary to supply the vendor with the following specifications:

(1) *Footprint*: Ground–signal–ground (GSG) is the most common for MMICs, although ground–signal (GS) probes are used below 10 GHz.
(2) *Probe tip contact pitch (i.e. distance between the mid-points of adjacent contacts)*: For microwave applications, 200 μm is very common, although

RFIC and MMIC measurement techniques 235

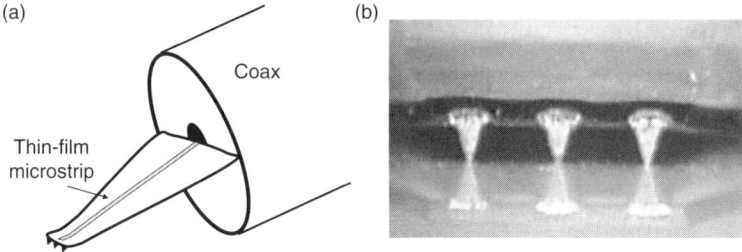

Figure 11.9   Infinity tip: (a) illustration and (b) contact bumps in contact with wafer

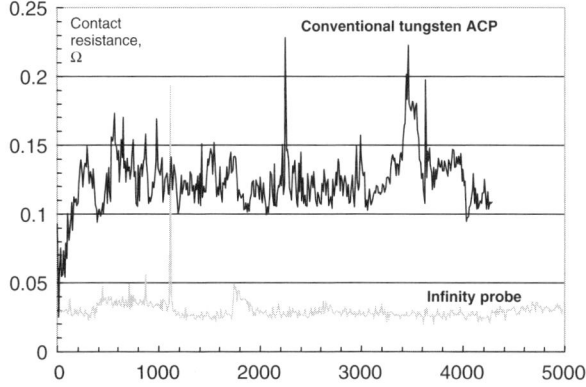

Figure 11.10   Variation of contact resistance with touchdowns for conventional tungsten ACP and Infinity probes

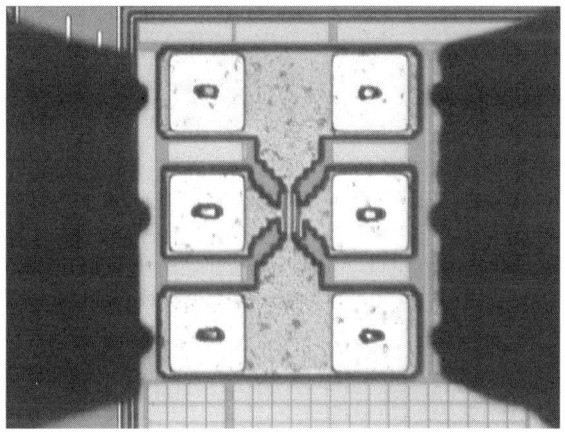

Figure 11.11   Contact damage from Infinity probes, typically $12 \times 25 \mu m^2$

probes are commercially available with pitches ranging from 50 to 1250 μm. Smaller pads result in smaller extrinsic launcher parasitics. A 100 μm pitch is commonly used from applications in the 40–120 GHz frequency range, while 75 μm is used above 120 GHz.

(3) *Probe tip contact width*: 40 and 25 μm are typical for operation up to 65 and 110 GHz, respectively.
(4) *Probe tip contact metal-plating*: BeCu is optimised for GaAs chips (having gold pads) and tungsten is optimised for silicon and SiGe chips (having aluminium pads).
(5) Launch angle, ϕ.
(6) *Coaxial connector type*: The 3.5 mm Amphenol Precision Connector (APC3.5) is used for operation to 26.5 GHz; the Anritsu K-connector (2.92 mm), for single-mode operation to 46 GHz, is compatible with 3.5 mm connectors; the APC2.4 can be used for measurements up to 50 GHz, while the Anritsu V-connector (1.85 mm), for single-mode operation to 67 GHz, is compatible with 2.4 mm connectors; the Agilent Technologies 1 mm connector is used for operation up to 120 GHz, while the Anritsu W-connector (1.1 mm) has a cut-off frequency of either 110 or 116 GHz, depending on the coaxial dielectric used.

Cascade Microtech now sells probes that are capable of making on-wafer measurements in dual configurations, such as GSGSG. In the case of the dual infinity, this allows dual measurements up to 67 GHz. Such probes may be used in conjunction with modern four-port VNAs, such as Agilent's N5230A PNA-L.

If the launch angle is too small, unwanted coupling between the probe and adjacent on-wafer components may occur. For this reason, it is recommended that adjacent components have at least 600 μm of separation for 110 GHz measurements. On the other hand, if the angle is too large there will not be enough skate on the probe pads. It has been found analytically and empirically that the best angle occurs when the horizontal components of the phase velocity for the probe and MMIC transmission lines match one other [44]. Therefore

$$\phi = \cos^{-1}\left\{\frac{\varepsilon_{\text{eff. probe}}}{\varepsilon_{\text{eff. MMIC}}}\right\}$$

where $\varepsilon_{\text{eff. probe}}$ is the effective permittivity of probe line and $\varepsilon_{\text{eff. MMIC}}$ is the effective permittivity of MMIC line.

For example, a CPW line on GaAs has $\varepsilon_{\text{eff. MMIC}} \approx 6.9$ at 76.5 GHz and, therefore, $\phi = 68°$ with *Air Coplanar*™ probes, since $\varepsilon_{\text{eff. probe}} = 1$. However, in practice, the launch angle is approximately 20°. This may raise questions as to the possibility of launching unwanted parasitic modes, due to uncompensated velocity mismatches at the RF probe tip, and also fringe fields coupling from the RF probe tip into the wafer.

11.3.2 Probe calibration

During the placement of probes onto an MMIC, there are two mechanisms by which the probe tips become soiled. First, since the probe tip contact's metal-plating is designed to be much harder than the MMIC probe pad ohmic contact's metal, particles of either gold or aluminium will be deposited onto the respective BeCu or tungsten contacts. Second, it is not uncommon for the probe tip contacts to overshoot the unpassivated probe pads and scratch off some of the Si_2N_3 (silicon nitride) passivation material surrounding the pads. Without regular cleaning, a build-up of gold/aluminium and Si_2N_3 particles can form around the probe tip contacts. This build-up is likely to degrade the performance of measurements at millimetric frequencies. Therefore, prior to calibrating the measurement system, it is recommended that the probe tips be very gently cleaned. Here, forced-air can be blown onto the probe tip – in a direction parallel to the tip and towards its open contact end – in order to remove any particles. For more stubborn objects, a lint-free cotton bud, soaked in isopropanol (IPA), can be carefully brushed in a direction parallel to the tip and towards its open contact end.

After the probe tips have been inspected for any signs of damage and cleaned, a planarity check must be made between the probes and the ultra-flat surface of the wafer chuck. A contact substrate, consisting of a polished alumina wafer with defined areas of patterned gold, is used to test that all three of the probe tip contacts (e.g. ground–signal–ground) make clear and even markings in the gold. Once this procedure is complete, the probe tip contacts can be cleaned of any residual gold by simply probing onto the exposed, un-metallised, areas of alumina. This is particularly important for tungsten contacts, because tungsten oxidises, and therefore the contact resistance would otherwise increase. However, this is not the case for BeCu contacts as they do not oxidise.

Probe stations use a one-tier calibration procedure, with the standards located either on an impedance standard substrate (ISS) or on the test wafer. With a precision ISS, the standards can be fabricated to much tighter tolerances. For example, a pair of 100 Ω resistors are used to implement the CPW 50 Ω load reference impedance. Here, these resistors can be laser-trimmed to achieve an almost exact value of 50 Ω, but at DC only. For D- and G-band operation, in order to reduce the effects of moding from the underside of the calibration substrate, Cascade Microtech produce a 250 μm thin ISS that, when used in conjunction with their ISS absorber blocks, drastically reduces the effects of substrate moding.

It should be noted, however, that if a calibration is performed using a 635 μm thick alumina ISS and the verification is performed using 200 μm thick GaAs on-wafer standards (which is a realistic measurement scenario), then problems may be encountered at millimetric frequencies. This is because the probe-to-ISS interface is electromagnetically different from that of the probe-to-wafer interface. As a result, even though the specifications for corresponding calibration and verification standards may be identical, their measured characteristics may differ significantly. For this reason, the use of on-wafer standards is by far the best choice. This is because the probe-to-wafer interface can be electromagnetically the same for calibration, verification and all subsequent measurements. Moreover, on-chip launch transition discontinuities (e.g. probe pads and their transmission lines) can be treated as part of the overall measurement system to be calibrated. Ideally, the reference planes within the on-wafer standards should have the same line geometries as those at the on-chip

DUT. The UK National Physical Laboratory (NPL) and the US National Institute of Standards and Technology (NIST) have developed GaAs ISS wafers with calibration standards and verification components of certified quality [47,48].

There are a number of calibration techniques that are used for on-wafer measurements [44,47–52]. The SOLT technique is not used at upper-microwave frequencies due to the poor quality of planar open standards. For TRL, the reflect standards (either an open or short circuit) must be identical at both ports, but they can be non-ideal and unknown. The TRL technique also requires a minimum of two transmission lines. The reference impedance is taken from the characteristic impedance, Z_0, of these lines [53]. Since a 50 Ω load is not required for TRL calibration, only transmission line standards are needed, and these are easily realisable on-wafer. In practice, in order to cover a useful frequency range, it is necessary to employ a number of different delay line lengths to overcome phase ambiguity at all the measurement frequencies. This means that the probe separation has to be adjusted during the calibration procedure. For many applications such as automated test systems this is a major limitation, and for these applications the line-reflect-match (LRM) calibration [49] is preferred to TRL. The multiple CPW delay lines required with the TRL calibration are effectively replaced by the CPW 50 Ω load, to theoretically represent an infinitely long delay line. This results in the following advantages:

(1) an ultra-wideband calibration can be achieved (e.g. DC to 120 GHz),
(2) the probes can be set in a fixed position,
(3) automatic calibration routines can be applied,
(4) reflections and unwanted modes in long CPW delay lines are avoided and
(5) a considerable saving of wafer/ISS area can be made.

With SOLT and LRM, the accuracy to which the load is known directly determines the accuracy of the measurement. In other words, perfect models are required for the load impedances. These loads inevitably have some parasitic shunt capacitance (which is equivalent to having negative series inductance), and furthermore, have frequency-dependent resistance due to the 'skin effect'. In addition, with microstrip technology there will be significant series inductance associated with the short and load standards. Cascade's line-reflect-reflect-match (LRRM) calibration is a more accurate version of the standard LRM calibration, in which load-inductance correction is incorporated by including an extra reflection standard.

NIST recently released some public domain software on the worldwide web called *MultiCal*™. This software provides a new method for the accurate calibration of VNAs [21–23]. Here, multiple and redundant standards are used to minimise the effects of random errors caused by imperfect contact repeatability. Moreover, with split-band methods (e.g. LRL and TRL), the calibration discontinuities at the frequency break points can be eliminated. With *MultiCal*™-TRL, only the physical lengths of the standards and the DC measurement of the line resistance per unit length (by applying a least-square error fit to the multiple shorted line lengths) are required. The Z_0 of the lines can then be determined and subsequent measurements can be renormalised to the 50 Ω measurement reference impedance.

For the ultimate in ultra-wideband calibration, verification and measurement accuracy, there is strong support for having *MultiCal*™-TRL calibration for frequencies above a few gigahertz (say 1 GHz), combined with LRM for the frequencies below 1 GHz, using on-wafer standards. The LRM's standards should be characterised at DC and at 1 GHz (using *MultiCal*™-TRL); conventional modelling techniques can be used to interpolate the results. A recent comparison was made between the calibration coefficients obtained from a NIST multiline calibration and those obtained from an assortment of other techniques; the results are shown in Figure 11.12.

A two-port probe station traditionally uses a 12-term error model, although a 16-term error model has been introduced that requires five two-port calibration standards [54]. This more accurate model can correct for poor grounding and the additional leakage paths and coupling effects encountered with open-air probing. With the extremely high levels of accuracy that are possible with modern probe stations, the effects of calibration errors become more noticeable. Calibration errors can result from the following (in the order of greatest significance):

(1) probe placement errors – position, pressure and planarity variations;
(2) degradation with use in the probe tips and the standards' probe pads surface wave effects on calibrations [55]; and
(3) ISS manufacturing variations.

It has been found that the effects of probe misplacement are greatly reduced when calibration is carried out on an automated probe. Cascade Microtech produce an automated calibration package, called WinCal, which allows full automation of the calibration on a Cascade semi-automatic probe station, such as the Summit 12000 or the 300 mm S300. Manual calibration is also possible. WinCal incorporates all the main family of calibrations (e.g. TRL, SOLT, LRM and also has LRM/LRRM

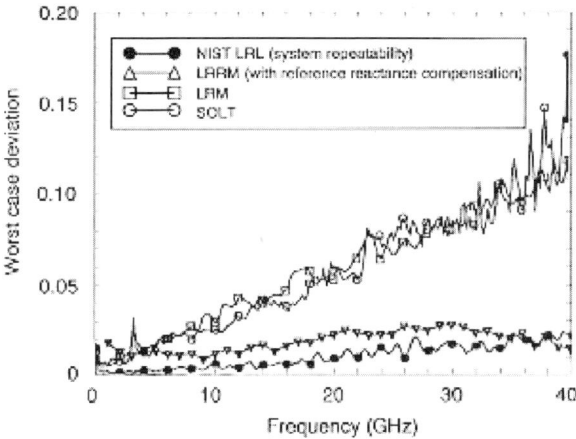

Figure 11.12 Comparison of calibration coefficients obtained from LRRM, LRM, SOLT and NIST multiline

240 Microwave measurements

with auto load inductance compensation). Another routine, called Short Open Load Reflect, is included that allows accurate calibration to be conducted with non-ideal through-line standards. Such situations are almost unavoidable when device ports are orthogonal in nature. WinCal has the ability to measure, record and display S-parameters in a variety of formats and also carry out compensation to remove the effects of pad parasitics. A stability checker is also provided in order to determine the validity of the calibration at any given moment.

With the extremely high level of measurement accuracy that can be achieved, the effects of on-chip launch transition discontinuities can be significant above a few gigahertz. So far, it has been assumed that the effects of probe pads and their associated transmission lines have been calibrated out. Here, the on-wafer calibration standards would have the same launch transition discontinuities as the on-chip DUT. However, effective de-embedding techniques can still be performed within the MMIC. If ECMs are to be employed, the foundry that fabricates the MMIC should provide very accurate models for probe pads and transmission lines. The metrologist must use these foundry-specific models to determine the actual measurements of the on-chip DUT. When de-embedding is performed using equivalent circuit modelling, these foundry-specific models can be easily incorporated into conventional F-D simulation software packages.

11.3.3 Measurement errors

Even when the system has been successfully calibrated, measurement errors (or uncertainty) can still occur. Some of the more common sources of errors are as follows:

(1) probe placement errors,
(2) temperature variation between calibration and measurement,
(3) cable-shift induced phase errors between calibration and measurement,
(4) radiation impedance changes due to the probes/wafer chuck moving,
(5) matrix renormalisation not being performed with multiple port MMICs,
(6) resonant coupling of the probes into adjacent structures [56],
(7) low frequency changes in the characteristic impedance and effective permittivity of both microstrip and CPW transmission lines [56] and
(8) optically induced measurement anomalies associated with voltage-tunable analogue-controlled MMICs [57].

11.3.4 DC biasing

Depending on the nature and complexity of the device or circuit under test, DC bias can be applied to an MMIC in a number of ways:

(1) through the RF probes, via bias-tees in the VNA's test set;
(2) through single DC needles mounted on probe station positioners; and
(3) with multiple DC needles attached to a DC probe card, which may in turn be mounted on a positioner.

The DC probe needle has significant inductance, and as a result, provides RF de-coupling for the bias lines that helps to prevent stability problems. However, additional off-chip de-coupling capacitors and resistors can usually be added to the card to further minimise the risk of unwanted oscillations. With bias-tees and DC needles, the maximum DC bias voltage and current are approximately 40 V and 500 mA, respectively. With multiple DC needles, standard in-house DC footprints are recommended wherever possible, in order to provide card re-use. This will reduce measurement costs considerably. There is a limit to the maximum number of needles per card, but ten is typical. One needle is normally required to provide a ground reference.

11.3.5 MMIC layout considerations

The foundry's design guidelines will define a minimum distance between the centres of probe pad vias and the minimum distance from the vias to the edge of the MMIC's active area. Generally, a particular company or institute may standardise on a certain pad size and pitch for a particular probe tip specification. In order to save expensive chip area, probing directly onto via-hole grounds is tempting. However, the probe tip contacts may puncture the gold pads on top of the via-holes, which could damage the probe tips and destroy the MMIC. While on-via probing can be used, in principle, it is likely that the chip would fail a subsequent QA inspection. As a result, when designing MMICs for on-wafer probed measurement, it is important to consult the foundry design guidelines for the probe pad specifications.

The location and orientation of the probe pads must also be considered. If the pads associated with one port are too close to those of another port, the very fragile probe tips are at risk of severe damage if they accidentally touch one another during the probe alignment procedure. The minimum separation distance between probe tips is determined by the design rule on probe pad spacing (typically 250 μm with vias or 200 μm without vias, depending on the thickness of the chip). Moreover, if the spacing between port pads is less than 200 μm, there could be significant measurement errors due to RF crosstalk effects between probes. Finally, if three or four RF probe positioners are attached to the probe station then they will be oriented orthogonal to one another. As a result, the RF probe pads for a three- or four-port MMIC must also be orthogonal to one another.

On the MMIC, launch transitions are required to interface between the probes and the DUT. In many cases, the DUT is in the microstrip medium, and so transitions from CPW-to-microstrip must be employed before and after the DUT. With reference to Figure 11.13a, microstrip launchers require through-GaAs vias to provide a low inductance earth path from the probe to the MMIC's backside metallisation layer. A microstrip launcher should be long enough for the higher-order evanescent modes, resulting from the CPW-to-microstrip transition, to be sufficiently attenuated and have minimum interaction with the DUT. As a rule of thumb, the microstrip launchers should ideally have a length of three to four times the substrate thickness. With reference to Figure 11.13b, when the DUT is in the CPW medium, through-GaAs vias are not required and a matched taper from the probe pads to the DUT is used. Even though this taper is very short, if the 50 Ω characteristic impedance is not maintained throughout the transition, significant parasitic capacitance or inductance

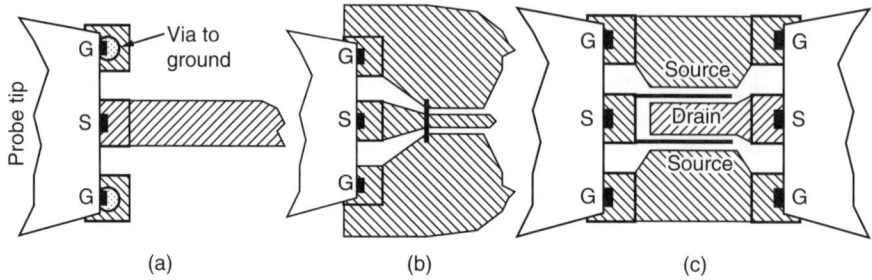

Figure 11.13 Common launcher techniques: (a) microstrip, (b) coplanar waveguide and (c) direct probing onto a FET device

can be introduced. In special cases, launchers are not required at all for some devices. One example of this is with a simple FET structure, as shown in Figure 11.13c, where two GSG probes are placed directly onto the source–gate–source and source–drain–source pads. This approach eliminates the need for de-embedding the effects of launchers from the measurements, but the effects of the bond pads should still be considered. At this point, it is important to note that for frequencies above a few gigahertz, the equivalent circuit model of a device that has been characterised in one medium (e.g. microstrip or CPW) should only be used in circuits designed in the same medium.

Single devices such as transistors and diodes can be biased through the bias-tees of the network analyser. However, in order to test a complete circuit using a probe station, special consideration has to be given to the layout of the DC bias pads and the design of the bias networks. When using DC needles to bias a circuit, the following points should be considered:

(1) The foundry may impose minimum pad sizes and centre-to-centre pitch.
(2) For ease of DC probe card fabrication and probe alignment, the DC probe pads should be arranged in a linear array along the edge of the chip's active area, and should be kept away from the RF pads. A common method is to have the RF probe pads on the east and west edges of the chip, and the DC bias pads on the north and/or south edges. If layout constraints suggest that orthogonal RF inputs and outputs would be more convenient, first check that suitable positioners are available.
(3) The bias networks of the circuit should be modelled separately to ensure that oscillations will not occur. Off-chip de-coupling capacitors cannot always be placed as near to the chip as they can be in a test fixture.
(4) High-value resistors can be added on-chip to prevent RF leakage and catastrophic failure resulting from excess forward biasing of diodes and transistors. With varactor diodes, cold-FETs and switching-FETs, a trade-off may have to be made in the value of these bias resistors. If the resistance is too small there may not be enough RF isolation. If the resistance is too high the maximum switching speed may not be reached, due to an excessive R-C time

constant. In practice, a minimum resistance value of approximately 300 Ω should suffice for most applications.

11.3.6 Low-cost multiple DC biasing technique

Conventional DC probe cards may need to be replaced for every new MMIC design, unless standard DC probe footprints can be used. This throwaway approach is very costly, especially when the DC probe cards are supplied by a commercial vendor (as automated and precision manufacturing techniques generally have to be used for aligning multiple needles). Moreover, the cost of the cards increases with the number of needles, as the individual needles are themselves precision-made components.

A flexible, low-cost technique has been developed for providing an experimental active filter with multiple DC bias connections [58]. The MMIC is attached to a gold-plated chip carrier using conductive epoxy glue. An array of single-layer microwave capacitors is then attached to the chip carrier in close proximity to the MMIC. BAR-CAPS™, made by Dielectric Labs Inc., are ideal for this purpose since they are available as single-chip strips of three, four or six 100 pF shunt capacitors, each having a probeable area of approximately 650×325 μm^2 and separated by approximately 170 μm. A gold bond wire is then used to connect the MMIC's DC probe pad to its off-chip capacitor.

As an example of this technique, a microphotograph of the experimental MMIC, requiring 15 DC bias lines, is shown in Figure 11.14.

It has been found that this low-cost solution has a number of important advantages for use in the R & D laboratory.

(1) The high-inductance bond wires and off-chip de-coupling capacitors minimise the risk of unwanted oscillations.
(2) When designing the MMIC layout, the DC probe pads do not need to be arranged in a linear array along the edges of the chip. This provides greater design layout flexibility.
(3) The linear array of off-chip capacitors automatically provides a standard in-house DC footprint, reducing long-term measurement costs considerably.
(4) The probeable area of the off-chip capacitors is approximately 15 times larger than that of the MMIC probe pads and the capacitors can withstand greater mechanical forces. As a result, in-house DC probe cards can be made by hand because of the relaxation in manufacturing precision, reducing short-term costs considerably.

11.3.7 Upper-millimetre-wave measurements

The past few years have seen considerable developments in the proposed uses of the millimetric frequency range above 75 GHz for new civil applications; for example, collision avoidance radar at 77 GHz. Also, the 94 GHz band is no longer dominated by military applications. High-resolution radiometric imaging at 94 and 140 GHz has a number of important applications, including aircraft landing systems, finding victims trapped in fires and locating concealed weapons without the use of X-rays. Ultra-high

244 *Microwave measurements*

Figure 11.14 Microphotograph of an experimental MMIC with multiple DC biasing using the low-cost technique [58]

data rate optical communications – using a 'radio-fibre' system at 180 GHz – could transform the way domestic computer networks are distributed. Future EC directives on environmental air pollution monitoring will require cheap high-performance terahertz sensors to be mass-produced. Sensors for sub-cellular probing are opening up new areas of medical research. Finally, passive tagging/identification systems are possible, which are both easy to conceal and extremely difficult to forge. With most (if not all) of these applications, monolithic technology will be sought. To this end, there have been major advances in both high electron mobility transistor (HEMT) and heterojunction bipolar transistor (HBT) technologies, both of which have attained values of f_{max} greater than 500 GHz [59,60].

Today, VNAs are commercially available that can operate in either broadband or banded configurations up to 110 GHz. The Agilent Technologies N5250 and the Anritsu ME7808B are examples of two broadband VNAs that are able to measure small-signal S-parameters from about 45 MHz to 110 GHz in a single-sweep. Both systems use coaxial cables between the test-sets and the probes.

As frequency increases, the combined losses of all the components between the test-sets' reflectometers and the MMIC under test (e.g. test-set combiners, transmission lines, probes, transitions and connectors) also increase. As a result, the overall system suffers from a reduction in both accuracy and stability [61]. The Anritsu 360B employs two test-sets: one rack-mounted (operating from 40 MHz to 67 GHz) and the other mounted on the probe station (operating from 67 to 110 GHz). Here, a test-set combiner (or forward wave MUX coupler) is used to combine the signals from both test-sets. The drawback with this approach is the considerable losses associated with test-set combiners, which will degrade the effective directivity, source match and frequency tracking of the system at W-band. Ultimately, this will have an impact on the quality of calibrations and the system's ability to hold a calibration in the presence of drift. The Agilent Technologies 8510XF minimises this problem by removing the need for a test-set combiner. Here, ultra-broadband (45 MHz to 110 GHz) directional couplers are utilised to create a single test-set [61].

In order to minimise the losses between the test-set's reflectometer and the MMIC under test, a banded VNA is preferred. This can utilise coaxial cables up to W-band and metal-pipe rectangular waveguides at and/or above W-band. The UK's National Physical Laboratory has recently established a new primary national standard measurement facility for S-parameters with rectangular waveguide operating over the frequency range of 75–110 GHz, using such a banded VNA system [62]. This facility represents a significant extension to the existing UK national standards for S-parameter and impedance measurements [63].

To date, there are still no traceable standards for on-wafer measurements above 75 GHz, from either NPL or NIST. This is due to a multitude of issues (e.g. mechanical precision, multi-moding, radiation effects, dielectric and surface wave propagation, ohmic losses in the dielectric and anomalous skin-effect losses in the conductors) associated with accurate calibration and verification measurements using non-ideal standards. However, there is a great deal of experimental work being undertaken to find the optimum calibration strategy for W-band [64–66].

With the ever-increasing interest in performing on-wafer measurements above 110 GHz, Oleson Microwave Laboratories Inc. can now supply frequency extension modules for the commercial market to include the following waveguide bands: WR-8 for F-band (90–140 GHz) [67]; WR-5 for G-band (140–220 GHz); and WR-3 for H-band (220–325 GHz). Cascade Microtech and GGB Industries supply the Infinity and *Picoprobe*™ on-wafer probes, respectively, for frequencies up to 220 GHz.

In addition to these commercial systems, the University of Kent has developed an experimental passive on-wafer probing system. Here, ultra-low loss PTFE dielectric waveguides are used to avoid the problem of the skin-effect altogether [68–72]. The dielectric waveguide has been used to implement the multistate reflectometer, interconnecting transmission lines, and even the on-wafer probes. In principle, this system can operate from 118 to 178 GHz [72]. However, the ultimate challenge is to remove all the losses between the test-set's reflectometer and MMIC under test. In an experimental set-up, a full two-port VNA has been implemented with active probes, enabling S-parameter measurements to be made from DC up to 120 GHz [73]. Here, high-speed non-linear transmission line (NLTL)-gated directional T-D reflectometers (which are essentially directional samplers) were realised using GaAs MMIC technology [74]. More recently, a 70–230 GHz VNA has been demonstrated that also employs MMIC reflectometers located on the on-wafer probes [75,76]. The NLTL-based active probes serve as S-parameter test-sets for the Agilent Technologies 8510 VNA. Using the Agilent Technologies 8510XF system, good agreement has been demonstrated from 70 to 120 GHz [75].

11.4 Thermal and cryogenic measurements

11.4.1 Thermal measurements

In real-life applications, microwave circuits can be exposed to temperatures other than ambient room temperature (i.e. 23 °C or approximately 296 K). For example, some components in geostationary orbiting satellites (e.g. within the antenna sub-system) may be periodically exposed to temperatures ranging from −150 to +80 °C, depending on the amount of visible sunlight, the levels of localised heat generated within the satellite and the effectiveness of the thermal control sub-system. Also, Gunn diodes can have junction temperatures in excess of +200 °C. At the other extreme, cryogenically cooled LNAs can operate at −196 °C, with a liquid nitrogen cryogen having a boiling point temperature of 77 K.

During the development of a sub-system, the levels of performance degradation while operating over a predefined temperature range must be known. Therefore, the temperature-dependent characteristics of all the MMIC components that make up a sub-system must be determined. Once the complete sub-system has been assembled, temperature-cycling is performed so that the measured levels of performance degradation can be compared with those predicted during simulation.

The Cascade Microtech Summit S300-863 semi-automatic probing system, in conjunction with the *Microchamber*™ enclosure, enables very fast set-up and measurements to be performed up to 110 GHz in a dark, temperature controlled and electromagnetic interference-isolated environment. The Summit S300-973 thermal probing system [77] can be seen in Figure 11.15.

The MMIC under test sits on a temperature controlled wafer chuck, which can be subjected to temperatures ranging from −65 to +200 °C or from 0 to +300 °C. Across these temperature ranges, the parameter values within, say, a FET's equivalent circuit model exhibit a linear temperature dependency. Here, all the resistive and capacitive elements have a positive temperature coefficient, while I_{ds}, g_m and f_T have negative temperature coefficients. Also, as the temperature drops, the gain of an active device can increase significantly. Therefore, to ensure linear operation, and thus avoid oscillation, the input RF power levels need to be reduced accordingly. Also, if the RF probes and cables exhibit large temperature gradients, significant phase changes will be found, even at low microwave frequencies. As a result, an air flow purge is introduced into the chamber in order to minimise the thermal coupling between the chuck and the probe/connector/cables. The air-flow purge also creates a dry, frost-free environment. The system is calibrated for every new wafer chuck temperature setting. An LRRM calibration is used, with the ISS located on a separate thermally isolated stage. The at-temperature calibration procedure can be performed 15 min after the chuck temperature has been changed. This short wait corresponds to approximately three thermal time constants for the probe/connector/cable assembly. Since all but the matched load impedance standards are insensitive to temperature, the ISS chuck temperature can be set at −5 °C, for a wafer chuck temperature of −65 °C. This approach results in less than a 1 per cent error in measurements between DC and 65 GHz.

Figure 11.15   Photograph of the Summit S300-973 thermal probing system, capable of over temperature measurements from −65 to 200 °C

As a wafer chuck changes temperature it expands or contracts. For example, the total chuck expansion, from −65 to +200 °C, can be about 230 μm. As a result, probe placement errors will become significant. Therefore, at each temperature, the overtravel of the probe tips may need to be adjusted. In addition, as the wafer diameter changes with temperature, there will be small changes in the spacing between devices. Cascade Microtech's Summit series of semi-automatic thermal probe stations include control software that automatically compensates for such changes. This minimises the impact of measurement accuracy.

Cascade Microtech now has a new microscopy system called Evue. This enables the contact height to be adjusted dynamically to ensure that the chuck is maintained at a constant height. This has the potential to enable fully automatic over temperature probing. Moreover, the technology employed in this system allows for an extremely large field of view that can be zoomed into a far smaller field of view at a single software command.

11.4.2 Cryogenic measurements

Cryogenic hybrid MICs, employing high-performance active semiconductor and passive superconductor components, are being more widely used in applications ranging from radio astronomy, to space communications, to medical nuclear magnetic resonance scanners. Therefore, it is important to be able to determine the cryogenic temperature characteristics of these components [3–5,78–82]. At cryogenic temperatures, the noise figures of conventional GaAs transistors are reduced dramatically from their ambient room temperature values. For example, at 10 GHz the measured noise figure of a typical 0.6 × 100 μm MESFET is 0.8 dB at 300 K and only 0.4 dB at 35 K [80]. With HEMT technology, electron mobility can increase by a factor of 5 when the lattice temperature is reduced from 300 to 77 K [80], resulting in a considerable improvement in gain and noise performance. Furthermore, measurements made at temperatures as low as 10 K may provide information that can give a unique insight into the physics of experimental devices. Also, in addition to the advances being made in new semiconductor devices, there is considerable interest in the developments of ultra-low loss high temperature superconducting microwave components that currently have to be refrigerated below around 100 K.

The first microwave test fixture to be used in cryogenic measurements was reported in 1976 [3]. The fixture was designed to be immersed in liquid nitrogen (LN_2), which has a boiling point of 77 K. This approach suffers from the problems of poor accuracy and poor repeatability due to the changing temperature gradients exhibited by the cable/connector/launcher assembly, and requires a complicated calibration procedure. Accurate measurements have been reported using a TRL calibrated split-block test fixture mounted on the cold-head of an RMC *Cryosystems*™ LTS-22-IR helium refrigerator [5]. This approach enables small-signal S-parameter measurements to be made at 300 and 77 K.

Cryogenic probe stations have either the MMIC under test and the probes immersed in liquid nitrogen or a liquid cryogen-cooled copper stage with a dry nitrogen vapour curtain. The former approach suffers from poor repeatability (due to varying amounts of LN_2), a short measurement duration (in order to limit the build-up of ice formation) and a limited lifetime due to the degradation of the probes in contact with the LN_2. With the latter approach, accuracy is limited by mechanical stress, caused by the large thermal gradients between the microwave hardware and the MMIC under test. Also, reliability is limited by moisture and the build-up of ice, which increases the wear and tear on manipulators and requires extensive re-planarisation of the mechanical apparatus. Researchers at the University of Illinois have, however, demonstrated the design and operation of a cryogenic vacuum microwave probe station, for the measurement of S-parameters from DC to 65 GHz, which minimises the problems of limited accuracy and repeatability [80]. Within a vacuum chamber, the vacuum probe station has high-frequency CPW probes connected to cable feeds via a custom bellows and manipulator system. A liquid helium cryogen, with a boiling point temperature of 4.2 K, enables measurements to be performed at temperatures as low as 20 K. The copper stage is continually fed with liquid cryogen, and the system is then left to stand for 15–20 min in order to achieve temperature equilibrium. Once the at-temperature calibration has been performed, the actual device measurements can be taken for up to 4 h before having to recalibrate.

Today, complete on-wafer cryogenic characterisation (from 20 to 300 K) can be performed for S-parameters, noise parameters and load-pull measurements [82].

11.5 Experimental field probing techniques

So far only invasive MMIC measurement techniques have been discussed, which generally do not perform internal function and failure analysis. However, one simple technique that can perform such tasks is to realise a coaxial probe with a high-impedance tip. Here, a 500 Ω resistor is used to create a potential divider with the 50 Ω oscilloscope. The internal node voltage can be measured without perturbing the operation of the circuit. This technique has been demonstrated on an MMIC power amplifier [83]. Alternatively, non-contacting methods also exist. Again, all the RF ports of the MMIC under test are terminated with matched loads. An RF signal is injected into the MMIC's input port and a micron-level probing system is used to detect the internal signal strength. In the case of non-contacting techniques, different types of field are detected along transmission lines and at discontinuities. Field probing can detect current crowding, standing waves and unwanted modes of propagation, and S-parameters can be determined from T-D network analysis measurements.

11.5.1 Electromagnetic-field probing

The simplest method of field detection uses a semiconductor diode. At microwave frequencies, however, it becomes difficult to match the diode because its impedance varies with power level. At low power levels, bolometers are traditionally employed for use above 1 GHz. The device is similar to a thin-film resistor, where a high-resistivity bismuth film is evaporated onto metallic electrodes. When exposed to microwave radiation, the bolometer absorbs the electromagnetic energy and converts it into heat energy. As the film heats up, its resistivity decreases. Since the bolometer is inherently a square law detector, the measured voltage change across the device is proportional to the change in incident RF power. In practice, however, since the signal levels are so small, the incident microwave signal must be pulsed. This causes the resistance of the bolometer to change at the pulse repetition frequency, which is usually below 100 kHz. With a DC bias current applied, the low-frequency voltage signal across the bolometer is applied to a lock-in amplifier that acts as a coherent detector. This technique exhibits a high degree of sensitivity; as an example, a 4 × 5μm device with a noise equivalent power of 160 pW/Hz$^{1/2}$ has been reported [84]. With the use of conventional probe microfabrication techniques, microbolometers can be employed to detect power levels as low as a few nanowatts along MMIC transmission lines. A microbolometer probe that can be used for microstrip and CPW transmission lines is illustrated in Figure 11.16. With a perfectly symmetrical probe positioned directly above a CPW line, the wanted CPW (or even) mode will be detected and the unwanted slotline (or odd) mode will not. As well as their simple fabrication and calibration, microbolometer probes can be designed to operate in the terahertz frequency range. Unfortunately, the attainable stability and uniformity of the resistive film does not yet appear to be sufficient for the commercial production of these probes.

A more recent development uses a dielectric rod probe, with a thin copper strip at its end face that helps to pick up the electromagnetic field and couple it to the dielectric

250 Microwave measurements

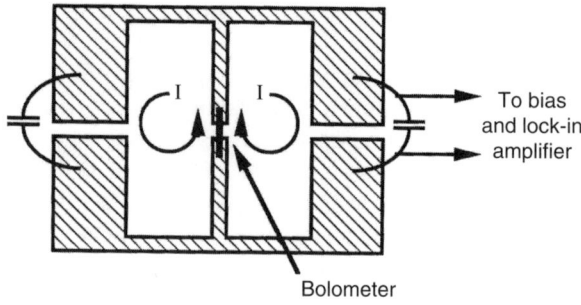

Figure 11.16   Illustration of an electromagnetic-field probe

waveguide [85]. Using this technique, measured results have been demonstrated between 200 and 220 GHz to show standing wave patterns on a mismatched dielectric waveguide [85].

11.5.2 Magnetic-field probing

The simplest magnetic-field probing technique is to connect a conventional spectrum analyser to a magnetic-field probe. Using wafer probe microfabrication techniques, a miniature magnetic quadrupole antenna can be configured to match the magnetic fields associated with microstrip and CPW transmission lines, as illustrated in Figure 11.17. Placed directly above the transmission line, the lines of magnetic flux will come up through one loop and back down through the other loop. As a result, the induced signals add. From a distance, the probe sees a near uniform magnetic field which induces signals that tend to cancel each other out. In addition to amplitude, phase measurements can also be measured. A reference signal at the same frequency, with a variable amplitude and phase, is combined with the measured signal. The measured phase is equal to the reference phase when the amplitude displayed on the spectrum analyser is at its peak. Therefore, the probe can be used to measure the amplitude and phase of currents at any node within an MMIC.

An experimental system has been reported that can operate in the 26.5–40 GHz frequency range [86]. Here, a 25–50 μm separation distance provides sufficient coupling and discrimination, while providing a negligible effect on the MMIC under test. One of the major sources of error is electrostatic pickup. Increasing the width of the loops increases the ratio of magnetic to electric coupling, but it also increases the random radiation picked up from other circuit elements. Reducing the width of the metal conductors reduces capacitive pickup, but increases the conductor's resistance and self-inductance. In practice, an effective method of limiting the errors due to electrostatic pickup is to rotate the probe and average the measurements. This problem can be avoided by having just a single-loop probe [87].

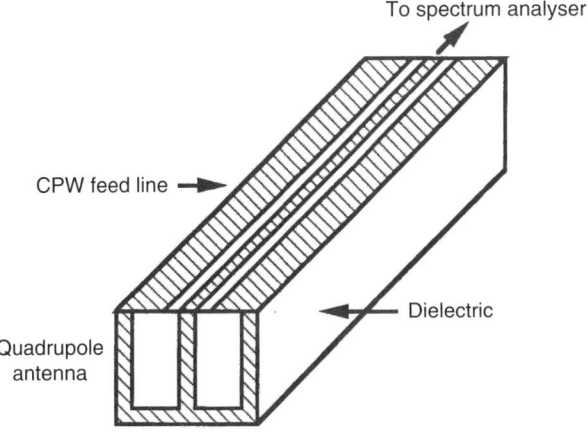

Figure 11.17   Illustration of a magnetic-field probe

11.5.3 Electric-field probing

The simplest electric-field probing technique is to connect a conventional spectrum analyser to a near electric-field (i.e. capacitive) probe. This technique was first demonstrated on MICs in 1979 [88], but it is still being used today [89]. The probe can be simply realised by removing a small section of the outer screening conductor and dielectric from the end of the analyser's coaxial feed line. Unfortunately, these probes have significant unwanted parasitic reactances at high microwave frequencies, which can severely perturb the operation of the circuit under test, thus causing measurement errors. However, micromachining techniques can be adopted to limit this problem, to realise dipole and monopole antennas [90]. In practice, this technique is only accurate when used with shielded transmission lines. As a result, it is unsuitable for micron-level features found in MMICs.

Over the past decade, a number of alternative electric-field probing techniques have been investigated, with varying degrees of success.

11.5.3.1 Electron beam probing

The voltage-contrast scanning electron microscope (SEM) was developed in the late 1960s for detecting voltages on the conductor tracks of integrated circuits. A pulsed electron beam stimulates secondary electron emissions from the irradiated surface of metals. For conductors at a negative potential, the secondary electrons have more energy than for conductors at a more positive potential. Commercial SEMs suffer from a poor millivolt potential sensitivity and limited bandwidths of only a few gigahertz [91], although larger bandwidths have been reported [92]. Also, apart from its very high complexity and cost, the electron beam may affect the operation of GaAs MMICs due to charging of deep levels in the GaAs substrate. However, the major advantage of this technique is that the attainable spatial resolution that can be achieved is in the order of a few angstroms.

11.5.3.2 Photo-emissive sampling

Instead of using an electron beam to stimulate secondary electron emissions, another approach uses a high-intensity pulsed laser beam to illuminate the surface of the metals [91]. This T-D sampling technique offers an improved potential sensitivity and a greatly extended bandwidth. However, as with the SEM, the performance of GaAs MESFETs may be affected by charging of deep level traps.

11.5.3.3 Opto-electronic sampling

Time-domain network analysis can be performed using opto-electronic sampling techniques. Here, electrical pulses can be generated on an MMIC by illuminating DC biased photoconductive switches with a pulsed laser beam. The optical excitation of a photoconductive switch can also perform signal sampling. By comparing the Fourier transforms of the sampled incident and reflected or transmitted waveforms, the complex two-port S-parameters can be determined for the DUT [91,93–99]. Sub-picosecond electrical pulse generation with a photoconductive switch has been reported, enabling terahertz measurement bandwidth [100]. This T-D opto-electronic sampling technique (also known as photoconductive sampling) requires the DUT to be embedded in a single-chip GaAs test fixture. Each RF port of the DUT is connected to a test structure consisting of a 50 Ω matched load termination, photoconductive switches, DC bias lines and a length of transmission line. These test components are not only wasteful of expensive chip space, but they must also be de-embedded from the measurements. In addition, the fabrication process of the photoconductive switches must be compatible with that of the MMIC under test. However, a DC to 500 GHz measurement system has been demonstrated [99] using this technique.

11.5.3.4 Electro-optic sampling

The most promising electric-field probing technique is electro-optic sampling. A variety of non-centrosymmetric crystals, such as gallium arsenide and indium phosphide, exhibit Pockel's electro-optic effect. The presence of an electric field will induce small anisotropic variations in the crystal's dielectric constant, and therefore, its refractive index. If a laser beam passes through this material it will experience a voltage-induced perturbation in its polarisation, which is directly proportional to the change in the electric-field strength. As a result, this linear electro-optic effect can be used to provide a non-invasive means of detecting electric fields [91,93,94,101–111].

With internal (or direct) electro-optic probing the laser beam penetrates the GaAs MMIC in a reflection mode, as illustrated in Figure 11.18a, giving good beam access and requiring only a single focusing lens [91,93,94,101–103,107–111]. However, optical polishing of the MMIC substrate is required for best results. With front-side probing, the beam is reflected off the back-side ground plane metallisation, adjacent to the circuit conductor. With back-side probing, the beam is reflected off the back of the circuit conductor itself, making this scheme ideal for conventional CPW or coplanar strip lines and slotlines. Today, internal electro-optic sampling can achieve a spatial resolution down to less than 0.5 μm [110].

Centrosymmetric crystals, such as silicon and germanium, do not exhibit the linear electro-optic effect. Therefore, silicon MMICs must employ external (or indirect)

RFIC and MMIC measurement techniques 253

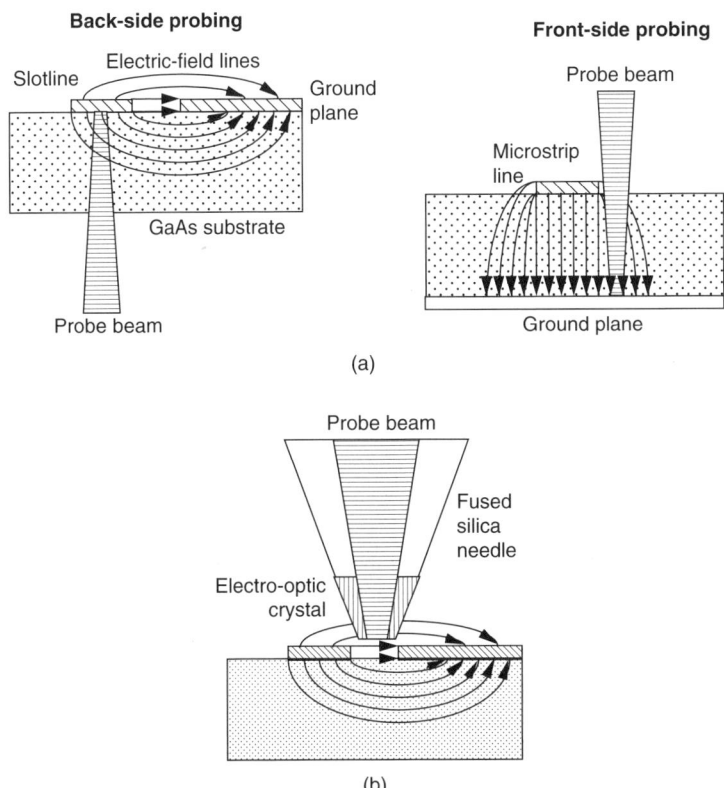

Figure 11.18 Illustration of electric-field probe: (a) internal and (b) external

electro-optic probing [91,93,104–106]. This technique uses an extremely small electric field sensor, consisting of a 40 × 40 μm^2 electro-optic crystal (lithium tantalate) at the end of a fused silica needle, placed in close proximity to the circuit conductor, as shown in Figure 11.18b. Sending a laser beam down the needle and measuring the induced change in the refractive index of the crystal from the returning beam can detect the conductor's fringing fields. Since the beam can be focused down to a spot size of 3–5 μm in diameter, excellent spatial resolution is achieved. Also, there is no need for MMIC substrate polishing.

With electro-optic probing, picosecond optical pulses (generated by a laser with an output power level that is lower than the band-gap energy of the MMIC's semiconductor) pass through the electric fields associated with the MMIC's circuit conductors. After being passed through a common beam splitter, the incident and return beams are combined, before being passed through a polarising beam splitter. Two photodiodes detect the intensity of the orthogonally polarised components and lock-in amplifiers are then used to determine the electric-field vectors. As a result, internal node voltage measurements can be determined and impressive two-dimensional mappings of the amplitude [105,107–109] and phase angles [109] of microwave fields within the MMIC can be obtained. T-D network analysis can also be performed using electro-optic sampling. Here, picosecond electrical pulses are applied to the input port of the MMIC under test, with the generator connected to the MMIC using traditional invasive techniques. By comparing the Fourier transform of the detected incident and reflected or transmitted waveforms, the complex two-port S-parameters can be determined. To date, a 50–300 GHz network analyser has been demonstrated using this technique [106]. A European consortium (which includes NPL and the Fraunhofer Institute for Applied Solid State Physics) has developed the first optical instrument capable of testing terahertz circuits and tracing the measurements back to international standards [111].

11.5.3.5 Electrical sampling scanning-force microscopy

A number of non-invasive measurement techniques have been introduced that can perform internal function and failure analysis of MMICs. The electron beam probing technique is well established and has excellent spatial resolution, but the temporal resolution is limited because of electron transit time effects. Optical probing techniques have a superior temporal resolution, but because of the micron-beam diameters they have a limited spatial resolution. Scanning-force microscopy, in the electrical sampling mode, is a relatively new non-contacting measurement technique that has high spatial, temporal and voltage resolutions [112,113]. Here, an atomically sharp needle is mounted on one end of a cantilever. When the needle is placed at a fixed working distance of between 0.1 and 0.5 μm above the MMIC, it will be subjected to attraction or repulsion forces, causing a detectable bending of the cantilever. This very experimental technique has so far demonstrated a spatial resolution of 0.5 μm and a bandwidth of 40 GHz [112].

11.6 Summary

A wide range of techniques has been briefly introduced for the measurement of MMICs. A summary of the main features associated with the most practical invasive techniques is given in Table 11.2. In general, the level of accuracy and repeatability

Table 11.2   Comparison of the invasive measurement technologies

	In-house test fixture		Commercial test fixture	On-wafer probe station
Calibration	2-tier with ECM de-embedding	1-tier	1-tier	1-tier
Accuracy	Moderate	High	High	Very high
Repeatability	Moderate	Moderate	High	Very high
Bandwidth	Wideband	Wideband	Wideband	Ultra-wideband
Flexibility	Excellent	Poor	Poor	Poor
Cost	Very low	Low	High	Very high

obtainable is proportional to the initial investment costs of the measurement system.

Compared with traditional invasive on-wafer measurement techniques, optical systems have so far demonstrated a lower dynamic range and inferior frequency resolution. In addition, optical techniques have complicated and lengthy calibration procedures. However, with its excellent spatial resolution and extremely wide bandwidth capabilities, electro-optic probing may become commonplace in the not too distant future.

References

1 Lorch, P.: 'Applications drive the evolution of network analyzers', *Microwaves & RF*, 1994;79–84
2 Lucyszyn, S., Stewart, C., Robertson, I. D., and Aghvami, A. H.: 'Measurement techniques for monolithic microwave integrated circuits', *IEE Electronics and Communication Engineering Journal*, 1994; 69–76
3 Liechti, C. A., and Larrick, R. B.: 'Performance of GaAs MESFET's at low temperatures', *IEEE Transactions on Microwave Theory and Techniques*, 1976;**MTT-24**:376–81
4 Smuk, J. W., Stubbs, M. G., and Wight, J. S.: 'Vector measurements of microwave devices at cryogenic temperatures', *IEEE MTT-S International Microwave Symposium Digest*, 1989, pp. 1195–8
5 Smuk, J. W., Stubbs, M. G., and Wight, J. S.: 'S-parameter characterization and modeling of three-terminal semiconductive devices at cryogenic temperatures', *IEEE Microwave and Guided Wave Letters*, 1992;**2** (3):111–13
6 Lane, R.: 'De-embedding device scattering parameters', *Microwave Journal*, 1984; 149–56
7 Riad, S. M.: 'The deconvolution problem: an overview', *Proceedings of the IEEE*, 1986;**74**(1):82–5
8 Romanofsky, R. R., and Shalkhauser, K. A.: 'Fixture provides accurate device characterization', *Microwaves & RF*, 1991; 139–48
9 Lucyszyn, S., Magnier, V., Reader, H. C., and Robertson, I. D.: 'Ultrawideband measurement of multiple-port MMICs using non-ideal test fixtures and a 2-port ANA', *IEE Proceedings A Science, Measurement and Technology*, 1992;**139** (5):235–42
10 Tippet, J. C., and Spaciale, R. A.: 'A rigorous technique for measuring the scattering matrix of a multiport device with a 2-port network analyzer', *IEEE Transactions on Microwave Theory and Techniques*, 1982;**30**:661–6
11 Dropkin, H.: 'Comments on – a rigorous technique for measuring the scattering matrix of a multiport device with a two-port network analyzer', *IEEE Transactions on Microwave Theory and Techniques*, 1983;**MTT-31** (1):79–81
12 Selmi, L., and Estreich, D. B.: 'An accurate system for automated on-wafer characterization of three-port devices', *IEEE Gallium Arsenide Integrated Circuit Symposium Digest*, 1990, pp. 343–6

13 Goldberg, S. B., Steer, M. B., and Franzon, P. D.: 'Accurate experimental characterization of three-ports', *IEEE MTT-S International Microwave Symposium Digest*, 1991, pp. 241–4
14 Stinehelfer, H. E.: 'Discussion of de-embedding techniques using the time domain analysis', *Proceedings of IEEE*, 1986;**74**
15 'Introduction to time domain measurements', *HP 8510C Network Analyser: Operating Manual*, (Hewlett Packard, USA, 1991)
16 Gronau, G., and Wolff, I.: 'A simple broad-band device de-embedding method using an automatic network analyzer with time-domain option', *IEEE Transactions on Microwave Theory and Techniques*, 1989;**MTT-37** (3):479–83
17 Dumbell, K. D.: 'TDR for microwave circuits', *Proceedings of IEEE Asia-Pacific Microwave Conference*, 1992, pp. 361–4
18 Gronau, G.: 'Scattering-parameter measurement of microstrip devices', *Microwave Journal*, 1992; 82–92
19 Lu, K., and Brazil, T. J.: 'A systematic error analysis of HP8510 time-domain gating techniques with experimental verification', *IEEE MTT-S International Microwave Symposium Digest*, 1993, pp. 1259–62
20 Hjipieris, G.: 'Time and frequency domain measurements', *Proceedings of 11th IEE Training Course on Microwave Measurements*, Great Malvern, UK, May 2000
21 Marks, R. B.: 'A multiline method of network analyser calibration', *IEEE Transactions on Microwave Theory and Techniques*, 1991;**MTT-39** (7): 1205–15
22 Marks, R. B., and Williams, D. F.: 'Characteristic impedance determination using propagation constant measurements', *IEEE Microwave and Guided Wave Letters*, 1991;**1** (6):141–3
23 Williams, D. F., and Marks, R. B.: 'Transmission line capacitance measurements', *IEEE Microwave and Guided Wave Letters*, 1991;**1** (9):243–5
24 Burke, J. J., and Jackson, R. W.: 'A simple circuit model for resonant mode coupling in packaged MMICs', *IEEE MTT-S International Microwave Symposium Digest*, 1991, pp. 1221–4
25 Swanson, D., Baker, D., and O'Mahoney, M.: 'Connecting MMIC chips to ground in a microstrip environment', *Microwave Journal*, 1993; 58–64
26 March, S. L.: 'Simple equations characterize bond wires', *Microwaves & RF*, 1991; 105–10
27 Nelson, S., Youngblood, M., Pavio, J., Larson, B., and Kottman, R.: 'Optimum microstrip interconnects', *IEEE MTT-S International Microwave Symposium Digest*, 1991, pp. 1071–4
28 Iwasaki, N., Katsura, K., and Kukutsu, N.: 'Wideband package using an electromagnetic absorber', *Electronics Letters*, 1993;**29** (10):875–6
29 Williams, D. F., and Paananen, D. W.: 'Suppression of resonant modes in microwave packages', *IEEE MTT-S International Microwave Symposium Digest*, 1989, pp. 1263–5
30 Heuermann, H., and Schiek, B.: 'The in-fixture calibration procedure line-network-network-LNN', *Proceedings of 23rd European Microwave Conference*, 1993, pp. 500–3

31 Munns, A.: 'Flip-chip solder bonding for microelectronic applications', *Metals and Materials*, 1989; 22–5
32 Warner, D. J., Pickering, K. L., Pedder, D. J., Buck, B. J., and Pike, S. J.: 'Flip chip-bonded GaAs MMICs compatible with foundry manufacture', *IEE Proceedings H, Microwaves, Antennas and Propagation*, 1991; **138** (1):74–8
33 Felton, L. M.: 'High yield GaAs flip-chip MMICs lead to low cost T/R modules', *IEEE MTT-S International Microwave Symposium Digest*, 1994, pp. 1707–10
34 Jin, H., Vahldieck, R., Minkus, H., and Huang, J.: 'Rigorous field theory analysis of flip-chip interconnections in MMICs using the FDTLM method', *IEEE MTT-S International Microwave Symposium Digest*, 1994, pp. 1711–14
35 Sakai, H., Ota, Y., Inoue, K. *et al.*: 'A novel millimeter-wave IC on Si Substrate using flip-chip bonding technology', *IEEE MTT-S International Microwave Symposium Digest*, 1994, pp. 1763–66
36 Lau, J. H. (ed.): *Flip Chip Technologies* (McGraw-Hill, New York, 1996)
37 Baumann, G., Ferling, D., and Richter, H.: 'Comparison of flip chip and wire bond interconnections and the technology evaluation on 51 GHz transceiver modules', *Proceedings of 26th European Microwave Conference*, 1996, pp. 98–100
38 Kim, J., and Itoh, T.: 'A novel microstrip to coplanar waveguide transition for flip-chip interconnection using electromagnetic coupling', *Proceedings of 28th European Microwave Conference*, 1998, pp. 236–40
39 Spiegel S. J., and Madjar, A.: 'Characterization of flip chip bump interconnects', *Proceedings of 28th European Microwave Conference*, 1998, pp. 524–28
40 'Network analysis: applying the HP 8510B TRL calibration for non-coaxial measurements', Hewlett-Packard Production Note 8510-8, 1987
41 Strid, E., and Gleason, K.: 'A microstrip probe for microwave measurements on GaAs FET and IC wafers',*IEEE Gallium Arsenide Integrated Circuit Symposium*, Las Vegas, Paper 31, 1980
42 Strid, E., and Gleason, K.: 'A DC-12GHz monolithic GaAs FET distributed amplifier', *IEEE Transactions on Microwave Theory and Techniques*, 1982;**MTT-30**:969–75
43 Godshalk, E. M.: 'A W-band wafer probe', *IEEE MTT-S International Microwave Symposium Digest*, 1993, pp. 171–174
44 Bahl, I., Lewis, G., and Jorgenson, J.: 'Automatic testing of MMIC wafers', *International Journal of Microwave and Millimeter-wave Computer-aided Engineering*, 1991;**1**(1):77–89
45 Liu, J. S. M., and Boll, G. G.: 'A new probe for W-band on-wafer measurements', *IEEE MTT-S International Microwave Symposium Digest*, 1993, pp. 1335–38
46 Godshalk, E. M., Burr, J., and Williams, J.: 'An air coplanar wave probe, *Proceedings of 24th European Microwave Conference*, 1994, pp. 1380–85
47 Bannister, D. J., and Smith, D. I.: 'Traceability for on-wafer CPW S-parameter measurements', *IEE Colloquium Digest on Analysis, Design and Applications of Coplanar Waveguide*, 1993, pp. 7/1–5

48 Pence, J. E.: 'Technique verifies LRRM calibrations on GaAs substrates', *Microwaves & RF*, 1994, 505–07
49 Eul, H. J., and Schiek, B.: 'Thru-match-reflect: one results of a rigorous theory for de-embedding and network analyzer calibration', *Proceedings of 18th European Microwave Conference*, 1988
50 Pradell, L., Caceres, M., and Purroy, F.: 'Development of self-calibration techniques for on-wafer and fixtured measurements: a novel approach', *Proceedings of 22nd European Microwave Conference*, 1992, pp. 919–24
51 Ferrero, A., and Pisani, U.: 'Two-port network analyzer calibration using an unknown 'thru', *IEEE Microwave and Guided Wave Letters*, 1992;**2** (12):505–7
52 Purroy, F., and Pradell, L.: Comparison of on-wafer calibrations using the concept of reference impedance', *Proceedings of 23rd European Microwave Conference*, 1993, pp. 857–859
53 Marks, R. B., and Williams, D. F.: 'A general waveguide curcuit theory', *Journal of Research of the National Institute of Standards and Technology*, 1992;**97** (5):533–62
54 Silvonen, K. J.: 'Calibration of 16-term error model', *Electronics Letters*, 1993;**29** (17):1544–5
55 Godshalk, E. M.: 'Wafer probing issues at millimeter wave frequencies', *Proceedings of 22nd European Microwave Conference*, 1992, pp. 925–30
56 Miers, T. H., Cangellaris, A., Williams, D., and Marks, R.: 'Anomalies observed in wafer level microwave testing', *IEEE MTT-S International Microwave Symposium Digest*, 1991, pp. 1121–24
57 Lucyszyn, S., and Robertson, I. D.: 'Optically induced measurement anomalies with voltage-tunable analog-control MMIC's', *IEEE Transactions on Microwave Theory and Techniques*, 1998;**MTT-46** (8):1105–14
58 Lucyszyn, S., and Robertson, I. D.: 'Monolithic narrow-band filter using ultrahigh-Q tunable active inductors', *IEEE Transactions on Microwave Theory and Techniques*, 1994;**MTT-42** (12):2617–22
59 Smith, P. M., Liu, S.-M. J., Kao, and M.-Y. et al.: 'W-band high efficiency InP-based power HEMT with 600 GHz fmax', *IEEE Microwave and Guided Wave Letters*, 1995;**5**:230–2
60 Lee, Q., Martin, S. C., Mensa, D. et al.: 'Deep submicron transferred-substrate heterojunction bipolar transistors', *Proceedings of Device Research Conference*, 1998
61 Gibson, J.: 'New capabilities for enhancing mm-wave network measurements', *Microwave Journal*, 1998; 86–94
62 French, G., and Ridler, N.: 'A primary national standard millimetric waveguide S-parameter measurements', *Microwave Engineering Europe*, 1999; 29–32
63 Ridler, N. M.: 'A review of existing national measurement standards for RF and microwave impedance parameters in the UK', *IEE Colloquium Digest*, 1999, no. 99/008, pp. 6/1–6
64 Kok, Y.-L., DuFault, M., Huang, T.-W., and Wang, H., 'A calibration procedure for W-band on-wafer testing', *IEEE MTT-S International Microwave Symposium Digest*, 1997, pp. 1663–66

65 Marks, R. B.: 'On-wafer millimeter-wave characterization', *Gallium Arsenide and its Applications Symposium Digest*, 1998, pp. 21–6
66 Edgar, D. L., Elgaid, K., Williamson, F. *et al*.: 'W-band on wafer measurements of active and passive devices', *IEE Colloquium Digest*, 1999, pp. 2/1–6
67 Anritsu Co.: '140 GHz extender modules for vector network analyzers', *Microwave Journal*, 1998; 148–50
68 Collier, R. J., and Boese, I. M.: 'Microwave measurements above 100 GHz', *Proceedings of Microwaves and RF Conference*, 1995, pp. 147–151
69 Boese, I. M., and Collier, R. J.: 'Novel measurement system within 110–170 GHz using a dielectric multistate reflectometer', *Proceedings of 26th European Microwave Conference*, 1996, pp. 806–10
70 Boese, I. M., Collier, R. J., Jastrzebski, A. K., Ahmed, H., Cleaver, J. R., and Hasko, D.: 'An on wafer probe for measurements at 140 GHz', *IEE Colloquium Digest*, 1997, pp. 9/1–7
71 Collier, R. J.: 'Measurements of impedance above 110 GHz', *IEE Colloquium Digest*, 1998, pp. 1/1–6
72 Boese, I. M., and Collier, R. J.: 'Measurements on millimeter wave circuits at 140 GHz', *IEE Proceedings – Science, Measurement and Technology*, 1998;**145** (4):171–6
73 Yu, R., Reddy, M., Pusl, J., Allen, S., Case, M., and Rodwell, M.: 'Full two-port on-wafer vector network analysis to 120 GHz using active probes', *IEEE MTT-S International Microwave Symposium Digest*, 1993, pp. 1339–42
74 Rodwell, M., Allen, S., Case, M., Yu, R., Bhattacharya, U., and Reddy, M.: 'GaAs nonlinear transmission-lines for picosecond and millimeter-wave applications', *Proceedings of 23rd European Microwave Conference*, 1993, pp. 8–10
75 Wohlgemuth, O., Rodwell, M. J. W., Reuter, R., Braunstein, J., and Schlechtweg, M.: 'Active probes for network analysis within 70–230 GHz', *IEEE Transactions on Microwave Theory and Techniques*; 1999; **47** (12):2591–8
76 Wohlgemuth, O., Agarwal, B., Pullela, R. *et al*.: 'A NLTL-based integrated circuit for a 70–200 GHz VNA system', *Microwave Engineering Europe*, 1999; 35–39
77 D'Almeida, D., and Anholt, R.: 'Device characterization with an integrated on-wafer thermal probing system', *Microwave Journal*, 1993; 94–105
78 Laskar, J., and Kolodzey, J.: 'Cryogenic vacuum high frequency probe station', *Journal of Vacuum Science Technology*, 1990; 1161–5
79 Meschede, H., Reuter, R., Albers, J. *et al*.: 'On-wafer microwave measurement setup for investigations on HEMT's and high Tc superconductors at cryogenic temperatures down to 20 K', *IEEE Transactions on Microwave Theory and Techniques*, 1992;**MTT-40** (12):2325–31
80 Laskar, J., and Feng, M.: 'An on-wafer cryogenic microwave probing system for advanced transistor and superconductor applications', *Microwave Journal*, 1993, 104–14
81 Laskar, J., Lai, R., Bautista, J. J. *et al*.: 'Enhanced cryogenic on-wafer techniques for accurate InxGa1-xAs HEMT device models', *IEEE MTT-S International Microwave Symposium Digest*, 1994, pp. 1485–88

82 Laskar, J., Murti, M. R., Yoo, S. Y., Gebara, E., and Harris, H. M.: 'Development of complete on-wafer cryogenic characterization: S-parameters, noise-parameter and load-pull', *Gallium Arsenide and its Applications Symposium Digest*, 1998, pp. 33–38

83 Wei, C.-J., Tkachenko, Y. A., Hwang, J. C. M., Smith, K. R., and Peake, A. H.: 'Internal-node waveform analysis of MMIC power amplifiers', *IEEE Transactions on Microwave Theory and Techniques*, 1995;**MTT-43** (12):3037–42

84 Schwarz, S. E., and Turner, C. W.: 'Measurement techniques for planar high-frequency circuits', *IEEE Transactions on Microwave Theory and Techniques*, 1986;**MTT-34** (4):463–7

85 Basu, A., and Itoh, T.: 'A new field-probing technique for millimeter-wave components', *IEEE MTT-S International Microwave Symposium Digest*, 1997, pp. 1667–70

86 Osofsky, S. S., and Schwarz, S. E.: 'Design and performance of a non-contacting probe for measurements on high-frequency planar circuits', *IEEE Transactions on Microwave Theory and Techniques*, 1992;**MTT-40** (8):1701–8

87 Gao, Y., and Wolff, I.: 'A new miniature magnetic field probe for measuring three-dimensional fields in planar high-frequency circuits', *IEEE Transactions on Microwave Theory and Techniques*, 1996;**MTT-44** (6):911–18

88 Dahele, J. S., and Cullen, A. L.: 'Electric probe measurements on microstrip', *IEEE Transactions on Microwave Theory and Techniques*, 1980;**MTT-28** (7):752–5

89 Gao, Y., and Wolff, I.: 'Electric field investigations on active microwave circuits', *Proceedings of 26th European Microwave Conference*, 1996, pp. 662–4

90 Budka, T. P., Waclawik, S. D., and Rebeiz, G. M.: 'A coaxial 0.5–18 GHz near electric field measurement system for planar microwave circuits using integrated probes', *IEEE Transactions on Microwave Theory and Techniques*, 1996;**MTT-44** (12):2174–82

91 Bloom, D. M., Weingarten, K. J., and Rodwell, M. J. W.: 'Probing the limits of traditional MMIC test equipment', *Microwaves & RF*, 1987; 101–06

92 Kubalek, E., and Fehr, J.: 'Electron beam test system for GHz-waveform measurements on transmission-lines within MMIC', *Proceedings of 22nd European Microwave Conference*, 1992, pp. 163–8

93 Bierman, H.: 'Improved on-wafer techniques evolve for MMIC testing', *Microwave Journal*, 1990; 44–58

94 Lee, T. T., Smith, T., Huang, H. C., Chauchard, E., and Lee, C. H.: 'Optical techniques for on-wafer measurements of MMICs', *Microwave Journal*, 1990; 91–102

95 Huang, S.-L. L., Chauchard, E. A., Lee, C. H., Hung, H.-L. A., Lee, T. T., and Joseph, T.: 'On-wafer photoconductive sampling of MMICs', *IEEE Transactions on Microwave Theory and Techniques*, 1992;**MTT-40** (12):2312–20

96 Kim, J., Son, J., Wakana, S. *et al.*: 'Time-domain network analysis of mm-wave circuits based on a photoconductive probe sampling technique', *IEEE MTT-S International Microwave Symposium Digest*, 1993, pp. 1359–61

97 Golob, L. P., Huang, S. L., Lee, C. H. et al.: 'Picosecond photoconductive switches designed for on-wafer characterization of high frequency interconnects', *IEEE MTT-S International Microwave Symposium Digest*, 1993, pp. 1395–98

98 Armengaud, L., Gerbe, V., Lalande, M., Lajzererowicz, J., Cuzin, M., and Jecko, B.: 'Electromagnetic study of an electronic sampler for picosecond pulse measurements', *Proceedings of 23rd European Microwave Conference*, 1993, pp. 751–4

99 Frankel, M. Y.: '500-GHz characterization of an optoelectronic S-parameter test structure', *IEEE Microwave and Guided Wave Letters*, 1994;**4** (4):118–20

100 Valdmanis, J. A., and Mourou, G.: 'Subpicosecond electrooptic sampling: principles and applications', *IEEE Journal of Quantum Electronics*, 1986;**QE-22**: 69–78

101 Bloom, D. M., Weingarten, K. J., and Rodwell, M. J. W.: 'Electrooptic sampling measures MMICs with polarized light', *Microwaves & RF*, 1987; 74–80

102 Mertin, W., Bohm, C., Balk, L. J., and Kubalek, E.: 'Two-dimensional field mapping in MMIC-substrates by electro-optic sampling technique', *IEEE MTT-S International Microwave Symposium Digest*, 1992, pp. 1443–6

103 Lee, C. H., Li, M. G., Hung, H.-L. A., and Huang, H. C.: 'On-wafer probing and control of microwave by picosecond optical beam', *Proceedings of IEEE Asia-Pacific Microwave Conference*, 1992, pp. 367–370

104 Wu, X., Conn, D., Song, J., and Nickerson, K.: 'Calibration of external electro-optic sampling using field simulation and system transfer function analysis', *IEEE MTT-S International Microwave Symposium Digest*, 1993, pp. 221–4

105 Mertin, W., Roths, C., Taenzler, F., and Kubalek, E.: 'Probe tip invasiveness at indirect electro-optic sampling of MMIC', *IEEE MTT-S International Microwave Symposium Digest*, 1993, pp. 1351–54

106 Cheng, H., and Whitaker, J. F.: '300-GHz-bandwidth network analysis using time-domain electro-optic sampling', *IEEE MTT-S International Microwave Symposium Digest*, 1993, pp. 1355–58

107 Hjelme, D. R., Yadlowsky, M. J., and Mickelson, A. R.: 'Two-dimensional mapping of the microwave potential on MMIC's using electrooptic sampling', *IEEE Transactions on Microwave Theory and Techniques*, 1993;**MTT-41** (6/7):1149–58

108 David, G., Redlich, S., Mertin, W. et al.: 'Two-dimensional direct electro-optic field mapping in a monolithic integrated GaAs amplifier', *Proceedings of 23rd European Microwave Conference*, 1993, pp. 497–99

109 Mertin, W., Leyk, A., David, G. et al.: 'Two-dimensional mapping of amplitude and phase of microwave fields inside a MMIC using the direct electro-optic sampling technique', *IEEE MTT-S International Microwave Symposium Digest*, San Diego, 1994, vol. 3, pp. 1597–1600

110 David, G., Tempel, R., Wolff, I., and Jager, D.: 'Analysis of microwave propagation effects using 2D electro-optic field mapping techniques', *Optical and Quantum Electronics*, 1996;**28**:919–31

111 'Maps of electric fields traced back to standards', *Optics and Laser Europe (OLE) Magazine*, 1997; 31–2
112 Bohm, C., Roths, C., and Kubalek, E.: 'Contactless electrical characterization of MMICs by device internal electrical sampling scanning-force microscopy', *IEEE MTT-S International Microwave Symposium Digest*, 1994, pp. 1605–8
113 Mueller, U., Boehm, C., Sprengepiel, J., Roths, C., Kubalek, E., and Beyer, A.: 'Geometrical and voltage resolution of electrical sampling scanning force microscopy', *IEEE MTT-S International Microwave Symposium Digest*, 1994, pp. 1005–8

Chapter 12

Calibration of automatic network analysers

Ian Instone

12.1 Introduction

Network analysers are very complex instruments so it is important to define terms such as calibration to avoid confusion. The two dictionary definitions of calibration that can be applied to network analysers are 'to mark (a gauge) with a scale of readings' [1], and 'to correlate the readings of (an instrument, etc.) with a standard to find the calibre of' [1]. Unfortunately neither of these expressions defines the term calibration as it is applied to network analysers, instead they relate better to verification which is the process where the network analyser's measurements are compared with those performed in a higher level laboratory.

12.2 Definition of calibration

Calibration in the network analyser sense is the process by which the errors within the instrument are compensated for, whereas verification checks that the resultant corrections have been properly assessed and applied. The extent of calibration used will depend on the desired measurement accuracy and the type of network analyser employed. To a large extent the available time will influence the type of calibration. There are two basic types of network analyser, both of them having their own advantages and limitations.

12.3 Scalar network analysers

The scalar network analyser usually consists of a source, display/processor and a transducer. Earlier scalar network analysers rarely included a receiver, instead they

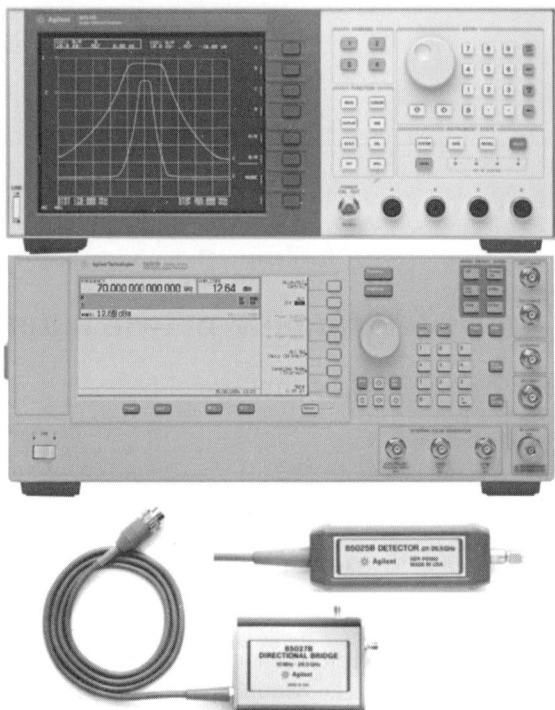

Figure 12.1   Photograph of a typical wideband detector based scalar network analyser and accessories

normally employ wide band diode detectors that have the advantage of being able to make measurements over a very wide frequency range at high speed (Figure 12.1). Because this type operates over such a wide range the noise floor usually limits their low amplitude response to around −70 dBm. Diode detectors do not have a linear response to amplitude so the display/processor will also include a table of corrections (within the memory) that are applied to the measured values before being displayed. A very useful application of the scalar network analyser is its ability to characterise the transmission properties of mixers where the incident signal will be at a different frequency to the output signal. Filters might need to be selected to reject any unwanted signals generated by the mixer.

More modern scalar network analysers are based on spectrum analysers (with one or more inputs) with a tracking generator (or two) included (Figure 12.2). With the rapidly decreasing costs of electronic equipment both the sources and receiver sections of these instruments are usually synthesised. A scalar network analyser of this design will be similar in complexity to its vector cousin, although it will lack many of the useful features (due to it being unable to measure the phase component of any signal). It will often have the advantage that it can be used as a standalone source or spectrum analyser, in some cases making it a more cost-effective solution.

Calibration of automatic network analysers 265

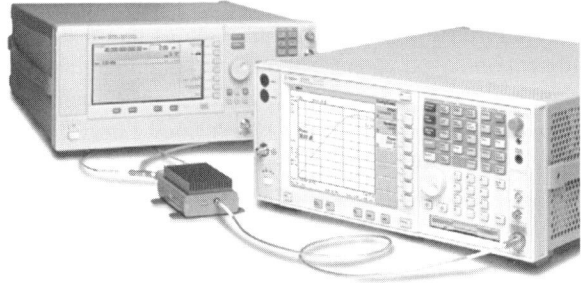

Figure 12.2   Photograph of a high-performance spectrum analyser based scalar network analyzer, which uses a high-performance external source as the tracking generator

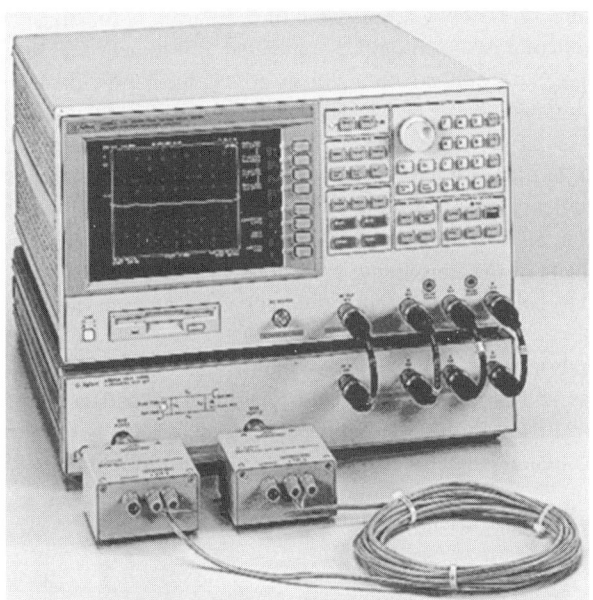

Figure 12.3   Network, spectrum, impedance analyser combined with a test-set used for making a wide range of RF and LF measurements

Due to it using a spectrum analyser as the detector this type of scalar network analyser will usually have a very large dynamic range, and depending on the quality of the included spectrum analyser, will often have a good linearity characteristic. With the inclusion of digital filters this type of scalar network analyser can have a speed performance similar to that obtained using wideband detectors, but with a linearity and selectivity performance similar to that of the vector network analyser.

Fully integrated analysers (Figure 12.3) are now available combining vector network, spectrum, impedance, gain, phase, group delay, distortion, harmonics, spurious

266 *Microwave measurements*

and noise measurements in one instrument. When combined with a test set, these instruments provide reflection measurements, such as return loss, VSWR, voltage reflection coefficient and *S*-parameters in both real and imaginary units that can be displayed as magnitude and phase if desired. These instruments combine tremendous dynamic range (>140 dB is normal) with good linearity and full vector or scalar error-correction creating the ability to perform accurate measurements very quickly. At present these, due their complexity, useful instruments are limited to radio frequencies (RFs).

12.4 Vector network analyser

The vector network analyser consists of a display/processor, source, test set and receivers. Modern vector network analysers are usually encompassed in one compact enclosure (Figure 12.4). They are capable of measuring all of the small signal scattering parameters of a two-port device connected to it in near real time. Because the instrument employs a receiver (often with an adjustable bandwidth) it is able to make reliable measurements over a much wider amplitude range than with the wide band detector based scalar network analyser. The term 'vector' also demonstrates that the analyser is able to measure the quantity in terms of phase and magnitude. By using vector measurements we are able to fully characterise the analyser and then apply corrections when an item is measured. The major part of any errors introduced by the loading effects of the item being measured, or the analyser itself, can be effectively removed by calculation thereby producing very accurate values with reasonable speed.

Modern analysers are able to display the measurements in a variety of formats including phase and magnitude, real and imaginary, impedance co-ordinates, etc. Despite their relatively high-cost vector network analysers are employed to make a variety of measurements where accuracy and speed are important.

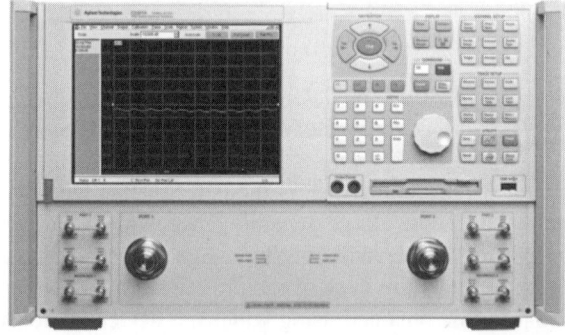

Figure 12.4   *Modern vector network analyser covering the frequency range 10 MHz to 67 GHz*

12.5 Calibration of a scalar network analyser

12.5.1 Transmission measurements

Because scalar network analysers are unable to measure the phase component of any signal the calibration process is much simpler and faster than that necessary with the vector network analyser. Calibration for transmission measurements is simply a process of establishing a reference level to which the measured values will be referred. This is accomplished by connecting the detector to the source, allowing the instrument to sweep through the range of frequencies, and storing the values in the instrument's memory. The device to be measured is then connected between the source and the detector and the instrument swept through the range of frequencies again. The difference between the first set of measurements (stored in memory) and the second set will be due to the device being tested plus any errors within the measurement system. Large potential errors with this type of measurement occur due to the mismatch loss uncertainties where the detector is connected to the source, and where the device being measured is connected to the source and detector. These uncertainties can be reduced by performing measurements through well-matched attenuators or couplers, but it is still likely that the mismatch loss uncertainties will dominate the uncertainty budget. In addition, where attenuators or couplers are used their value has to be chosen very carefully. High-value attenuators often have the best match and provide the best isolation against re-reflections and mismatch effects, but they also allow less of the signal to pass through, therefore reducing the effective dynamic range of the measurement. It is usually not practical to increase the source power as the higher power attenuators required to improve the match at the insertion point are often a poorer match than their lower power counterparts. Another alternative is to use a second detector and a power splitter. The ratio of the power appearing at the output ports of the power splitter is recorded (in the analyser's memory) and the device to be measured is connected between one output port and its detector. The measurements are performed again and the difference between the first and the second measurements will be due to the device being tested. Using this configuration and by connecting an appropriate attenuator between the reference detector and the power splitter, and then, perhaps by using an amplifier increasing the signal generator's amplitude between the first and second measurement it is possible to make measurements using the analyser over a much wider amplitude range than is specified.

Spectrum analyser based instruments will enable a wider variety of attenuators or couplers to be used in the matching process as this type of analyser has a much wider dynamic range which copes with the additional losses much better.

12.5.2 Reflection measurements

Calibrating prior to making reflection measurements follows a similar process of setting a reference and performing measurements relative to it. The input port of the bridge is connected to the generator and a short circuit connected to the bridge's test port. The generator is swept through the range of desired frequencies and the values stored in the scalar network analyser's memory.

The short circuit is then replaced with an open circuit and the source is swept again through the range of desired frequencies and the values stored again in the analyser's memory. The mean of these two sets of measurements is used as a reference and all measured values of reflection referred to it. It is important that the open circuit and short circuit are exactly 180° apart throughout the frequency range or further errors will be present in the measurement. Because an open circuit will always have a capacitance term associated with it and a short circuit effectively shunts any capacitance it is not normally possible to satisfy this requirement over the entire frequency range. The resultant errors are normally included as contributions to the uncertainty budget, having the most effect on the bridge's source match estimate. As with transmission measurements, compromises are often made to ensure that the best quality measurement is performed without compromising speed or cost, etc. For instance, it is good practice to include a power splitter at the input to the bridge and connect a detector to the other output port of the power splitter. The scalar network analyser is then set to measure the ratio of the bridge over the detector's output. The power splitter and detector perform three functions:

(1) They measure and compensate for any variations in the generator's output power which may not have been compensated for with the generator's automatic level control.
(2) When the directional bridge output port is loaded with different impedance devices connected to it (such as the short and open circuits and device being tested) it may cause the generator's output amplitude to change. This phenomenon is almost eliminated by this arrangement.
(3) The mismatch looking into the directional bridge's test port is a contribution to the measurement uncertainties; if it can be improved the uncertainties will reduce. A typical microwave generator has a fairly poor mismatch, whereas power splitters have a fairly good mismatch in comparison. The mismatch of the generator or power splitter is transmitted through the bridge and will have an effect upon the resultant measurement uncertainties. When used in this configuration the effective output match of the power splitter is at its best, therefore transferring the best measurement conditions through the bridge.

Unfortunately, as with transmission measurements, there is a downside. Every power splitter has loss and inserting more loss into the measuring system will reduce the dynamic range thereby increasing the noise floor. Power splitters and detectors also cost money and each item will have a maintenance cost associated with it so including additional items in the measurements will increase costs.

Inserting a good quality attenuator between the directional bridge and the source will also improve the 'effective source match'. To be effective the attenuator will need to have at least 20 dB transmission loss so it will not be suitable for most wideband detector systems. This method could be the most cost-effective for the spectrum analyser based system.

A reasonably high value of attenuator will perform exactly the same function as the power splitter above but at a fraction of the cost.

12.6 Problems associated with scalar network analyser measurements

The scalar network analyser measurement system consists of a microwave generator, detector (or a bridge and detector) and a scalar network analyser. The scalar network analyser is very similar to an oscilloscope in construction and operation. It has an input for the x-scale and several inputs for the detectors which display on the y-axis. The time base or x-axis is usually derived from the sweep output of the signal generator. Modern scalar network analysers also have a digital connection to the signal generator so that the display can be annotated with the start and stop frequencies, enabling easier control of the instruments. In addition, the digital connection is often used to connect to printers, plotters and disk drives to provide a permanent record of the test results. It can also be used to connect a computer so that the entire measurement process, presentation and archiving of results can be automated. The biggest problem with any measurement system employing diode type detectors is that they have different responses depending on the applied power level. At low powers (less than -30 dBm) they typically have a response proportional to the square of the applied power. As the power level increases their response becomes closer to a linear response. The designers of the early scalar network analysers tried to compensate for this effect by having active feedback loops in the conditioning amplifiers in the analyser; more modern instruments compensate for these effects digitally. Another problem is the limited dynamic range when compared to network analyser with a tuned front end. The diode detector often has a very wide frequency response (10 MHz to 26.5 GHz is common and 10–50 MHz is becoming more popular) which results in its ability to detect and add many very small signals across its operating spectrum. Where each of these signals might have a very small amplitude when they are all combined they effectively produce a noise floor of around -70 dBm. At this level the random component in the measurements is usually too large for sensible measurements to be performed so scalar network analyser measurements are often limited to -60 dBm. At the higher powers the detectors might suffer from being over loaded so most diode detectors are limited to a maximum input power of about $+16$ dBm.

12.7 Calibration of a vector network analyser

The vector network analyser as the name suggests also has the capability to measure the relative phase of the signals. The measurement system employs several receivers (usually three or four) to make the measurements as fast as possible without the need for extensive switching of the signals. On modern instruments the 'resolution bandwidth' is switchable allowing the user to make compromises between accuracy and speed. A process known as '*accuracy enhancement*' is usually employed to reduce the errors in measurement due to the network analyser. Expressed simply, accuracy enhancement is the process whereby the network analyser is characterised using known standards so the errors within the measurement are removed mathematically. Each device, which is used for this characterisation, is manufactured to be excellent for only one parameter or purpose (e.g. a short should have 100 per cent reflection or a load should have 100 per cent absorption) so it is a lot easier to manufacture these 'simple' devices than the perfect couplers which might otherwise be required. A potential confusion in terms often occurs, the term *'calibration'* when applied to vector network analysers is usually intended to describe the *'accuracy enhancement'* process. The following paragraphs are taken from the Agilent Technologies 8722ES operating manual [2] and the Hewlett-Packard HP8753A operating manual [3] and describe in some detail the process of *'accuracy enhancement'*.

12.8 Accuracy enhancement

12.8.1 What causes measurement errors?

Network analysis measurement errors can be separated into systematic, random and drift errors. Correctable systematic errors are the repeatable errors that the system can measure. These are errors due to mismatch and leakage in the test setup, isolation between the reference and test signal paths, and system frequency response. The system cannot measure and correct for the non-repeatable random and drift errors. These errors affect both reflection and transmission measurements. Random errors are measurement variations due to noise and connector repeatability. Drift errors include frequency drift, temperature drift, and other physical changes in the test setup between calibration and measurement. The resulting measurement is the vector sum of the test device response plus all error terms. The precise effect of each error term depends on its magnitude and phase relationship to the actual test device response. In most high-frequency measurements the systematic errors are the most significant source of measurement uncertainty. Since each of these errors can be characterised, their effects can be effectively removed to obtain a corrected value for the test device response. For the purpose of vector accuracy enhancement, these uncertainties are quantified as directivity, source match, load match, isolation (crosstalk) and frequency response (tracking). The description of each of these systematic errors follows. Random and drift errors cannot be precisely quantified, so they must be treated as producing a cumulative uncertainty in the measured data.

12.8.2 Directivity

Normally a device that can separate the reverse from the forward travelling waves (a directional bridge or coupler) is used to detect the signal reflected from the test device. Ideally the coupler would completely separate the incident and reflected signals, and only the reflected signal would appear at the coupled output (Figure 12.5).

However, an actual coupler is not perfect. A small amount of the incident signal appears at the coupled output due to leakage as well as reflection from the termination in the coupled arm (Figure 12.6). Also, reflections from the coupler output connector appear at the coupled output, adding uncertainty to the signal reflected from the device.

Calibration of automatic network analysers 271

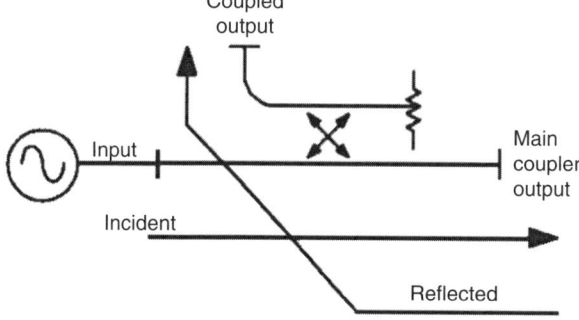

Figure 12.5 Diagrammatic representation of an ideal directional coupler or directional bridge

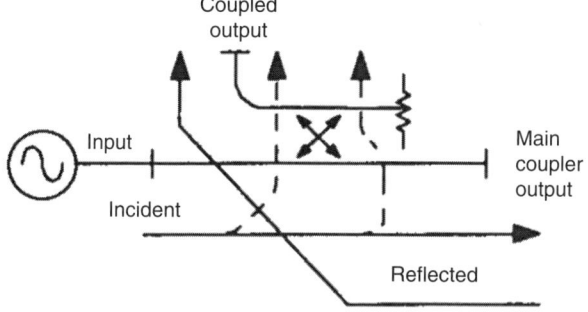

Figure 12.6 Diagrammatic representation of an actual directional coupler or directional bridge showing the various error paths

The figure of merit for how well a coupler separates forward and reverse waves is directivity. The greater the directivity of the device, the better the signal separation. System directivity is the vector sum of all leakage signals appearing at the analyser receiver input. The error contributed by directivity is independent of the characteristics of the test device and it usually produces the major ambiguity in measurements of low reflection devices.

12.8.3 Source match

Source match is defined as the vector sum of signals appearing at the analyser receiver input due to the impedance mismatch at the test device looking back into the source, as well as to adapter and cable mismatches and losses (Figure 12.7). In a reflection measurement, the source match error signal is caused by some of the reflected signal from the test device being reflected from the source back towards the test device and re-reflected from the test device.

In a transmission measurement, the source match error signal is caused by reflection from the test device that is re-reflected from the source.

272 *Microwave measurements*

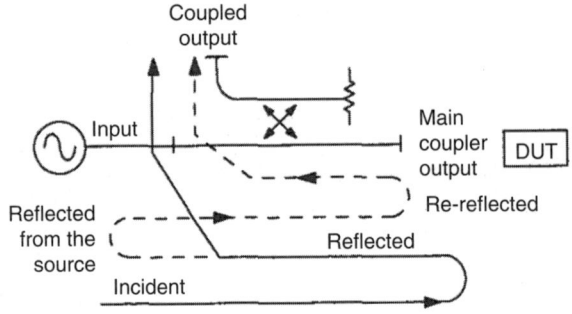

Figure 12.7 Diagrammatic representation of the constituent parts in the formation of source match

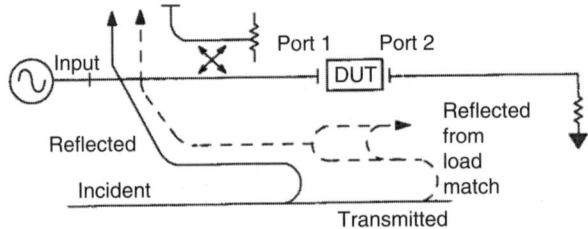

Figure 12.8 Diagrammatic representation of the constituent parts in the formation of load match

The error contributed by source match is dependent on the relationship between the actual input impedance of the test device and the equivalent match of the source. It is a factor in both transmission and reflection measurements. Source match is a particular problem in measurements where there is a large impedance mismatch at the measurement plane (e.g. reflection devices such as filters with stop bands).

12.8.4 Load match

Load match error results from an imperfect match at the output of the test device. It is caused by impedance mismatches between the test device output port and port 2 of the measurement system. Some of the transmitted signal is reflected from port 2 back to the test device. A portion of this wave may be re-reflected to port 2, or part may be transmitted through the device in the reverse direction to appear at port 1. If the test device has low insertion loss (e.g. a filter pass band), the signal reflected from port 2 and re-reflected from the source causes a significant error because the test device does not attenuate the signal significantly on each reflection (Figure 12.8). The error contributed by load match is dependent on the relationship between the actual output impedance of the test device and the effective match of the return port (port 2). It is a factor in all transmission measurements and in reflection measurements of two-port devices.

The interaction between load match and source match is less significant when the test device insertion loss is greater than about 6 dB. However, source match and load match still interact with the input and output matches of the DUT, which contributes to transmission measurement errors (these errors are largest for devices with highly reflective ports).

12.8.5 Isolation (crosstalk)

Leakage of energy between analyser signal paths contributes to error in a transmission measurement, much like directivity does in a reflection measurement. Isolation is the vector sum of signals appearing at the analyser samplers due to crosstalk between the reference and test signal paths. This includes signal leakage within the test set and in both the RF and IF sections of the receiver. The error contributed by isolation depends on the characteristics of the test device. Isolation is a factor in high-loss transmission measurements. However, analyser system isolation is more than sufficient for most measurements, and correction for it may be unnecessary. For measuring devices with high dynamic range, accuracy enhancement can provide improvements in isolation that are limited only by the noise floor. Generally, the isolation falls below the noise floor, therefore, when performing an isolation calibration the performer should use a noise reduction function such as averaging or reducing the IF bandwidth.

12.8.6 Frequency response (tracking)

This is the vector sum of all test setup variations in which magnitude and phase change as a function of frequency. This includes variations contributed by signal π separation devices, test cables, adapters, and variations between the reference and test signal paths. This error is a factor in both transmission and reflection measurements.

12.9 Characterising microwave systematic errors

12.9.1 One-port error model

In a measurement of the reflection coefficient (magnitude and phase) of a test device, the measured data differs from the actual, no matter how carefully the measurement is made. Directivity, source match and reflection signal path frequency response (tracking) are the major sources of error (Figure 12.9).

To characterise the errors, the reflection coefficient is measured by first separating the incident signal (I) from the reflected signal (R), then taking the ratio of the two values. Ideally, (R) consists only of the signal reflected by the test device (S_{11A}, for S_{11} actual) (Figure 12.10).

However, all of the incident signal does not always reach the unknown. Some of (I) may appear at the measurement system input due to leakage through the test set or through a signal separation device. Also, some of (I) may be reflected by imperfect adapters between a signal separation device and the measurement plane. The vector

274 *Microwave measurements*

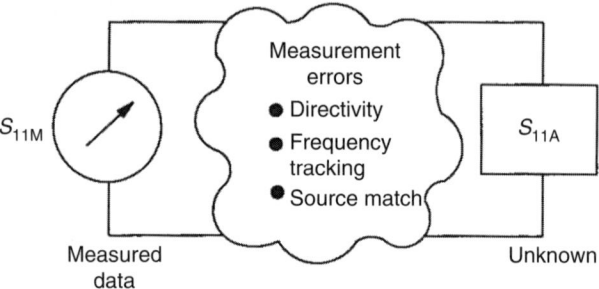

Figure 12.9   Sources of error in reflection measurement

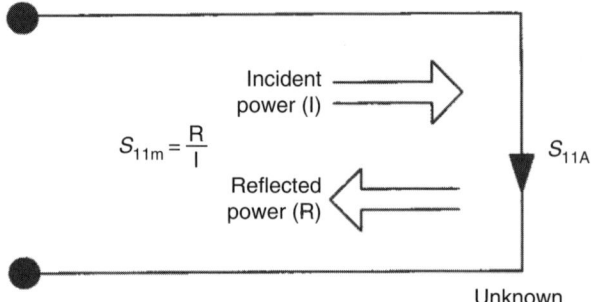

Figure 12.10   Reflection coefficient model

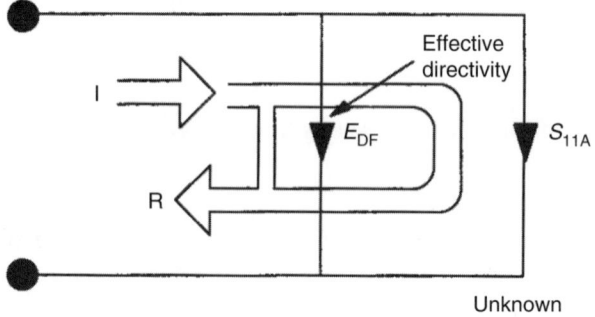

Figure 12.11   Effective directivity (E_{DF}) model

sum of the leakage and the miscellaneous reflections is the effective directivity, E_{DF} (Figure 12.11). Understandably, the measurement is distorted when the directivity signal combines with the actual reflected signal from the unknown, S_{11A}.

Since the measurement system test port is never exactly the characteristic impedance (50 Ω), some of the reflected signal bounces off the test port, or other

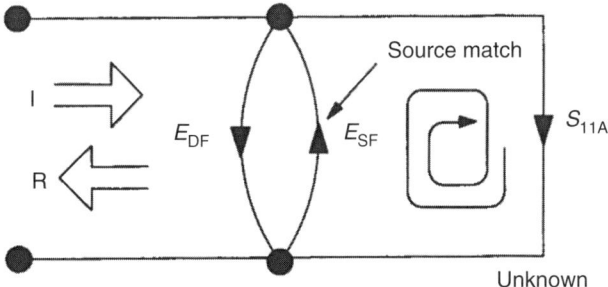

Figure 12.12   Source match (E_{SF}) model

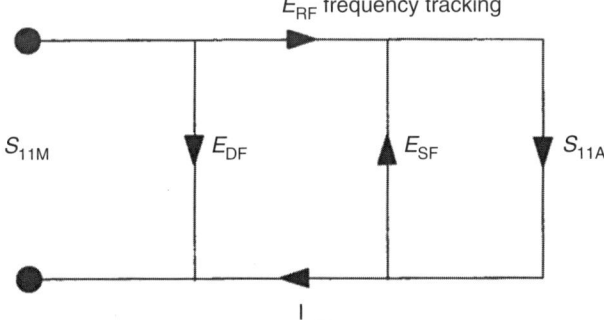

Figure 12.13   Reflection tracking (E_{RF}) model

impedance transitions further down the line, and back to the unknown, adding to the original incident signal (I). This effect causes the magnitude and phase of the incident signal to vary as a function of S_{11A} and frequency. Levelling the source to produce a constant incident signal (I) reduces this error, but since the source cannot be exactly levelled at the test device input, levelling cannot eliminate all power variations. This re-reflection effect and the resultant incident power variation are caused by the source match error, E_{SF} (Figure 12.12).

Frequency response (tracking) error is caused by variations in magnitude and phase flatness versus frequency between the test and reference signal paths. These are mainly due to coupler roll off, imperfectly matched samplers, and differences in length and loss between the incident and test signal paths. The vector sum of these variations is the reflection signal path tracking error, E_{RF} (Figure 12.13).

These three errors are mathematically related to the actual data, S_{11A}, and measured data, S_{11M}, by the following equation:

$$S_{11M} = E_{DF} + \frac{(S_{11A} E_{RF})}{(1 - E_{SF} S_{11A})} \tag{12.1}$$

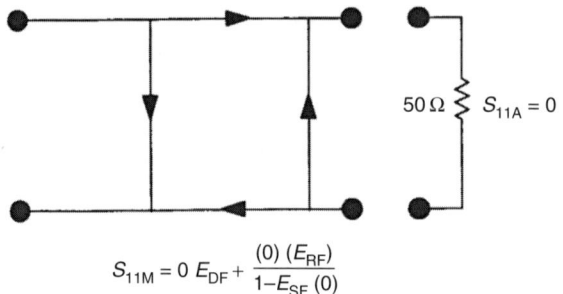

$$S_{11M} = 0 \; E_{DF} + \frac{(0)\,(E_{RF})}{1 - E_{SF}\,(0)}$$

Figure 12.14 'Perfect load' termination model

If the value of these three 'E' errors and the measured test device response were known for each frequency, this equation could be solved for S_{11A} to obtain the actual test device response. Because each of these errors changes with frequency, their values must be known at each test frequency. These values are found by measuring the system at the measurement plane using three independent standards whose S_{11} is known at all frequencies.

The first standard applied is a 'perfect load', which assumes $S_{11} = 0$ and essentially measures directivity (Figure 12.14). 'Perfect load' implies a reflection-less termination at the measurement plane. All incident energy is absorbed. With $S_{11A} = 0$ the equation can be solved for E_{DF}, the directivity term. In practice, of course, the 'perfect load' is difficult to achieve, although very good broadband loads are available in the compatible calibration kits.

Since the measured value for directivity is the vector sum of the actual directivity plus the actual reflection coefficient of the 'perfect' load, any reflection from the termination represents an error (Figures 12.15 and 12.16). System effective directivity becomes the actual reflection coefficient of the near 'perfect load'.

In general, any termination having a return loss value greater than the uncorrected system directivity reduces reflection measurement uncertainty.

Next, a short circuit termination whose response is known to a very high degree is used to establish another condition (Figures 12.17 and 12.18. The open circuit gives the third independent condition (Figures 12.19 and 12.20). In order to accurately model the phase variation with frequency due to fringing capacitance from the open connector, a specially designed shielded open circuit is used for this step (the open circuit capacitance is different for each connector type).

Now the values for E_{DF}, directivity, E_{LF}, source match, and E_{RF}, reflection frequency response, are computed and stored.

This completes the calibration procedure for one-port devices.

12.10 One-port device measurement

The unknown one-port device is measured to obtain values for the measured response, S_{11M}, at each frequency.

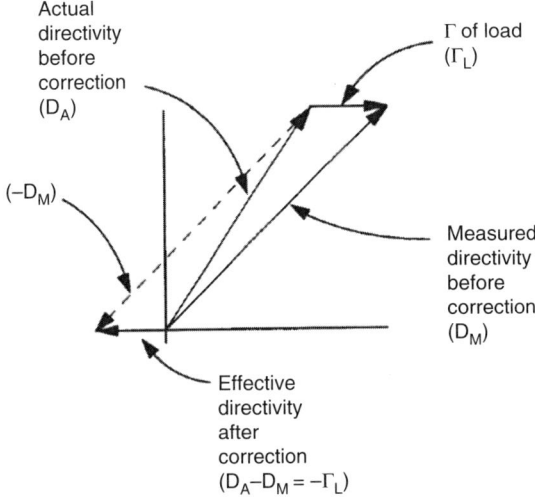

Figure 12.15 Vector diagram showing how effective directivity (E_{DF}) is resolved

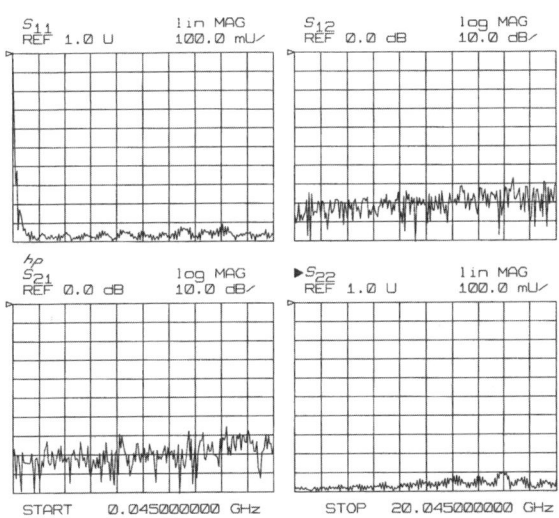

Figure 12.16 Network analyser display with a sliding load on port 1 (S_{11}) and a lowband load connected to port 2 (S_{22})

This is the one-port error model equation solved for S_{11A} (Figure 12.21). Since the three errors and S_{11M} are now known for each test frequency, S_{11A} can be computed using the following equation:

$$S_{11A} = \frac{(S_{11M} - E_{DF})}{E_{SF}(S_{11M} - E_{DF}) + E_{RF}} \tag{12.2}$$

278 *Microwave measurements*

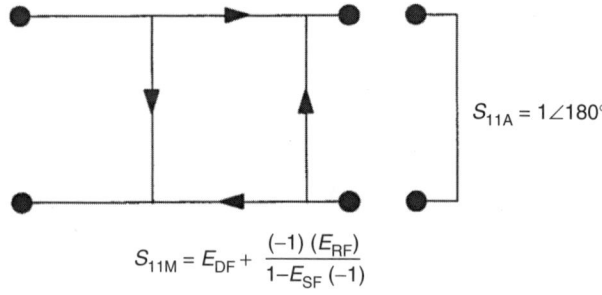

Figure 12.17 Short circuit termination model

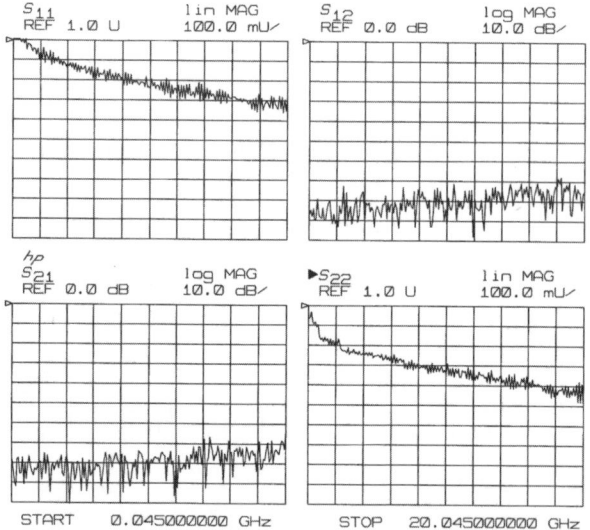

Figure 12.18 Network analyser display with short circuits connected to both ports (S_{11} and S_{22})

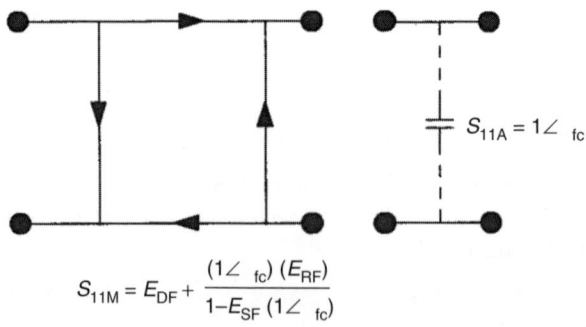

Figure 12.19 Open circuit termination model

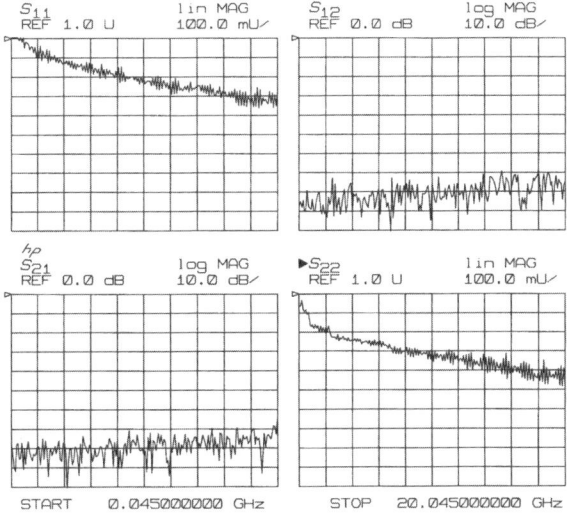

Figure 12.20 Network analyser display with open circuits connected to both ports (S_{11} and S_{22})

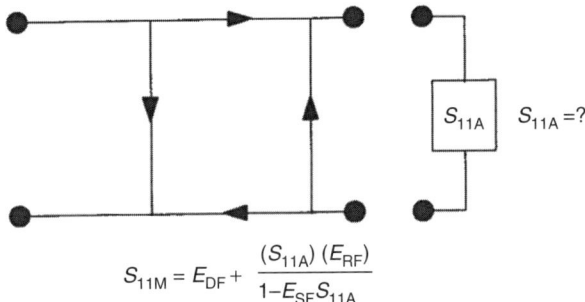

Figure 12.21 Flow diagram representing the individual constituents of an S_{11} reflection measurement

For reflection measurements on two-port devices, the same technique can be applied, but the test device output port must be terminated in the system characteristic impedance. This termination should have as low a reflection coefficient as the load used to determine directivity. The additional reflection error caused by an improper termination at the test device's output port is not usually incorporated into the one-port error model.

12.11 Two-port error model

The error model for measurement of the transmission coefficients (magnitude and phase) of a two-port device is derived in a similar manner. The potential sources of

280 *Microwave measurements*

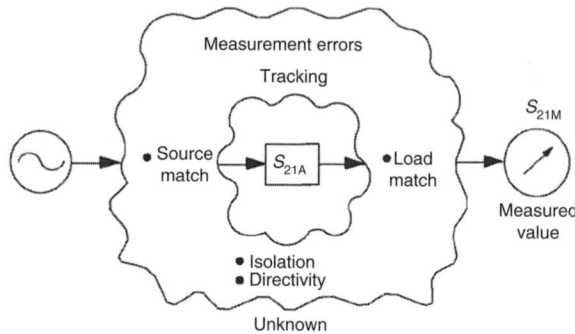

Figure 12.22 *Major sources of error in transmission measurements of a two-port device*

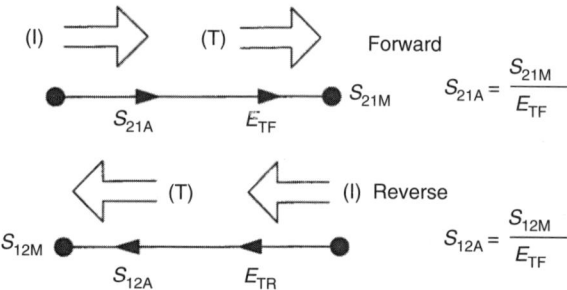

Figure 12.23 *Constituent parts of the transmission coefficient model*

error are frequency response (tracking), source match, load match and isolation as shown in Figure 12.22. On a two-port network analyser these errors are effectively removed using the full two-port error model.

The transmission coefficient is measured by taking the ratio of the incident signal (I) and the transmitted signal (T) (Figure 12.23). Ideally, (I) consists only of power delivered by the source and (T) consists only of power emerging at the test device output.

As in the reflection model, source match can cause the incident signal to vary as a function of test device S_{11A}. Also, since the test setup transmission return port is never exactly the characteristic impedance, some of the transmitted signals are reflected from the test set port 2, and from other mismatches between the test device output and the receiver input, to return to the test device. A portion of this signal may be re-reflected at port 2, thus affecting S_{21M}, or part may be transmitted through the device in the reverse direction to appear at port 1, thus affecting S_{11M}. This error term, which causes the magnitude and phase of the transmitted signal to vary as a function of S_{22A}, is called load match, E_{LF} (Figure 12.24).

The measured value, S_{21M}, consists of signal components that vary as a function of the relationship between E_{SF} and S_{11A} as well as E_{LF} and S_{22A}, so the input and

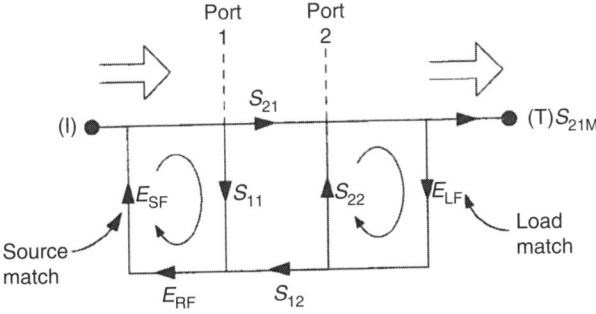

Figure 12.24   Load match error model

output reflection coefficients of the test device must be measured and stored for use in the S_{21A} error-correction computation. Thus, the test setup is calibrated as described for reflection to establish the directivity, E_{DF}, source match, E_{SF}, and reflection frequency response, E_{RF}, terms for reflection measurements on both ports. Now that a calibrated port is available for reflection measurements, the thru is connected and load match, E_{LF}, is determined by measuring the reflection coefficient of the thru connection. Transmission signal path frequency response is then measured with the thru connected. The data are corrected for source and load match effects, then stored as transmission frequency response, E_{TF}.

Note: It is very important that the exact electrical length of the thru be known. Most calibration kits assume a zero length thru. For some connection types such as Type-N, this implies one male and one female port. If the test system requires a non-zero length thru, for example, one with two male test ports, the exact electrical delay of the thru adapter must be used to modify the built-in calibration kit definition of the thru.

Isolation, E_{XF}, represents the part of the incident signal that appears at the receiver without actually passing through the test device (Figures 12.25 and 12.26). Isolation is measured with the test set in the transmission configuration and with terminations installed at the points where the test device will be connected. Since isolation can be lower than the noise floor, it is best to increase averaging by at least a factor of 4 during the isolation portion of the calibration.

Note: If the leakage (isolation) falls below the noise floor, it is best to increase averaging before calibration. If it is not possible to increase the averaging it will be better to omit the isolation measurement.

Thus there are two sets of error terms, forward and reverse, with each set consisting of six error terms, as follows:

- Directivity, E_{DF} (forward) and E_{DR} (reverse)
- Isolation, E_{XF} and E_{XR}
- Source match, E_{SF} and E_{SR}

282 Microwave measurements

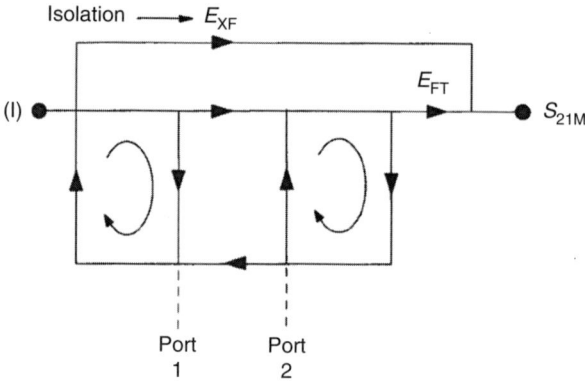

Figure 12.25   *Isolation error model*

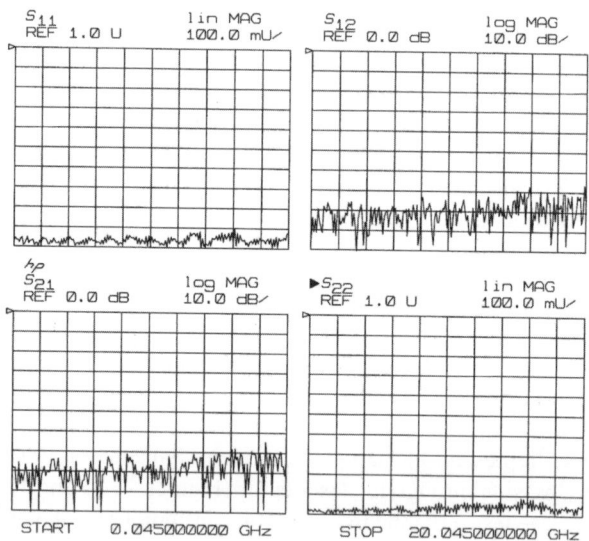

Figure 12.26   *Typical network analyser display during the isolation measurement*

- Load match, E_{LF} and E_{LR}
- Transmission tracking, E_{TF} and E_{TR}
- Reflection tracking, E_{RF} and E_{RR}

Network analysers equipped with S-parameter test sets can measure both the forward and reverse characteristics of the test device without the performer having to manually remove and physically reverse the device.

A full two-port error model is illustrated in Figure 12.28. This illustration depicts how the analyser effectively removes both the forward and reverse error terms for transmission and reflection measurements.

Calibration of automatic network analysers

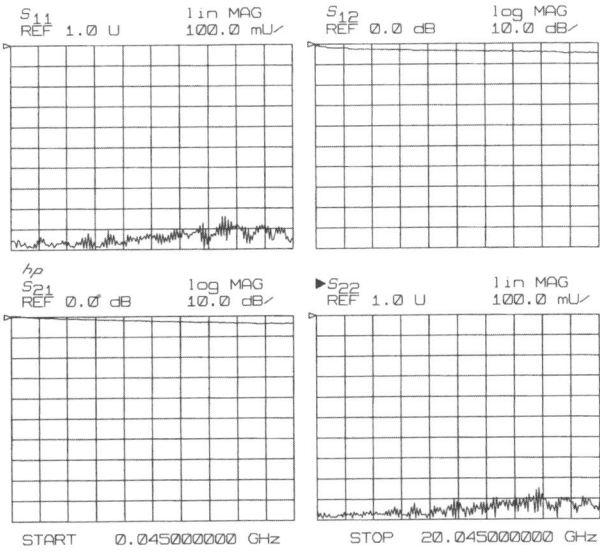

Figure 12.27 Typical network analyser display during the 'through' measurement

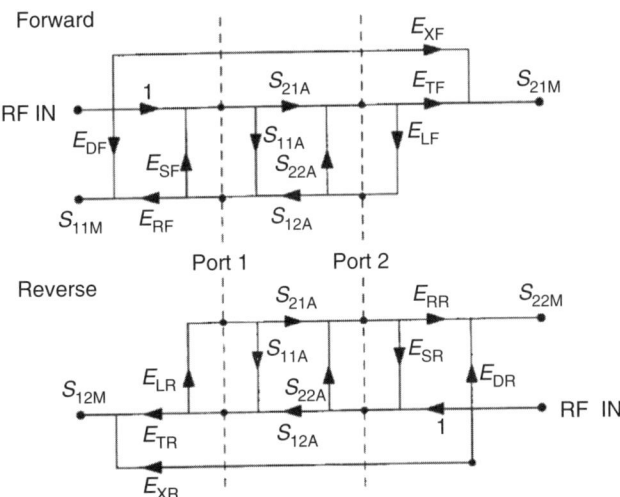

Figure 12.28 Full two-port error model

The equations for all four S-parameters of a two-port device are shown in Figure 12.29. Note that the mathematics for this comprehensive two-port error model use all forward and reverse error terms and measured values. Thus, to perform full error-correction for any one parameter, all four S-parameters must be measured.

$$S_{11A} = \frac{\left[\left(\frac{S_{11M} - E_{DF}}{E_{RF}}\right)\left[1 + \left(\frac{S_{22M} - E_{DR}}{E_{RR}}\right)E_{SR}\right]\right] - \left[\left(\frac{S_{21M} - E_{XF}}{E_{TF}}\right)\left(\frac{S_{12M} - E_{XR}}{E_{TR}}\right)E_{LF}\right]}{\left[1 + \left(\frac{S_{11M} - E_{DF}}{E_{RF}}\right)E_{SF}\right]\left[1 + \left(\frac{S_{22M} - E_{DR}}{E_{RR}}\right)E_{SR}\right] - \left[\left(\frac{S_{21M} - E_{XF}}{E_{TF}}\right)\left(\frac{S_{12M} - E_{XR}}{E_{TR}}\right)E_{LF}E_{LR}\right]}$$

$$S_{21A} = \frac{\left[1 + \left(\frac{S_{22M} - E_{DR}}{E_{RR}}\right)(E_{SR} - E_{LF})\right]\left(\frac{S_{21M} - E_{XF}}{E_{TF}}\right)}{\left[1 + \left(\frac{S_{11M} - E_{DF}}{E_{RF}}\right)E_{SF}\right]\left[1 + \left(\frac{S_{22M} - E_{DR}}{E_{RR}}\right)E_{SR}\right] - \left[\left(\frac{S_{21M} - E_{XF}}{E_{TF}}\right)\left(\frac{S_{12M} - E_{XR}}{E_{TR}}\right)E_{LF}E_{LR}\right]}$$

$$S_{12A} = \frac{\left[1 + \left(\frac{S_{11M} - E_{DF}}{E_{RF}}\right)(E_{SF} - E_{LR})\right]\left(\frac{S_{12M} - E_{XR}}{E_{TR}}\right)}{\left[1 + \left(\frac{S_{11M} - E_{DF}}{E_{RF}}\right)E_{SF}\right]\left[1 + \left(\frac{S_{22M} - E_{DR}}{E_{RR}}\right)E_{SR}\right] - \left[\left(\frac{S_{21M} - E_{XF}}{E_{TF}}\right)\left(\frac{S_{12M} - E_{XR}}{E_{TR}}\right)E_{LF}E_{LR}\right]}$$

$$S_{22A} = \frac{\left[\left(\frac{S_{22M} - E_{DR}}{E_{RR}}\right)\left[1 + \left(\frac{S_{11M} - E_{DF}}{E_{RF}}\right)E_{SF}\right]\right] - \left[\left(\frac{S_{21M} - E_{XF}}{E_{TF}}\right)\left(\frac{S_{12M} - E_{XR}}{E_{TR}}\right)E_{LR}\right]}{\left[1 + \left(\frac{S_{11M} - E_{DF}}{E_{RF}}\right)E_{SF}\right]\left[1 + \left(\frac{S_{22M} - E_{DR}}{E_{RR}}\right)E_{SR}\right] - \left[\left(\frac{S_{21M} - E_{XF}}{E_{TF}}\right)\left(\frac{S_{12M} - E_{XR}}{E_{TR}}\right)E_{LF}E_{LR}\right]}$$

Figure 12.29 Mathematical representation of the full two-port error model algorithms

12.12 TRL calibration

12.12.1 TRL terminology

Notice that the letters TRL, LRL, LRM etc. are often interchanged, depending on the standards used. For example 'LRL' indicates that two lines and a reflect standard are used and LRM indicates that a reflection and match standards are used. All of these refer to the same basic method.

TRL* calibration is a modified form of TRL calibration. It is adapted for a receiver with three samplers instead of four samplers. The TRL* calibration is not as accurate as the TRL calibration because it cannot isolate the source match from the load match, so it assumes that load match and source match are equal.

12.12.1.1 How TRL*/LRL* calibration works

The TRL/LRL calibration used in the network analyser relies on the characteristic impedance of simple transmission lines rather than on a set of discrete impedance

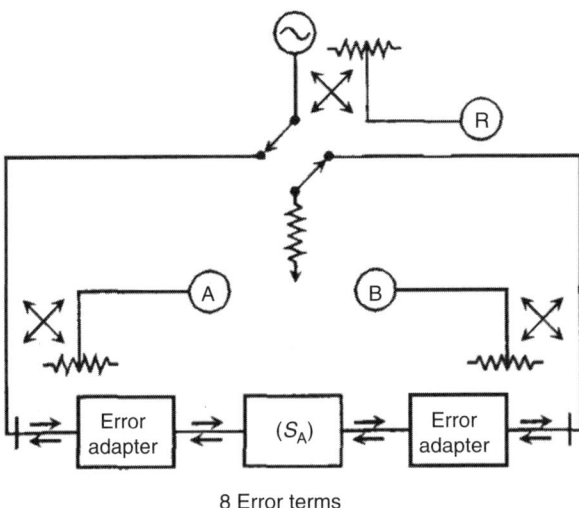

Figure 12.30 Functional block diagram for a two-port error corrected network analyser measurement system employing only three receivers

standards. Since transmission lines are relatively easy to fabricate (e.g. in microstrip or co-axial), the impedance of these lines can be determined from the physical dimensions and substrate's dielectric constant.

For the analyser TRL* two-port calibration, a total of ten measurements are made to quantify eight unknowns (not including the two isolation error terms). Assume the two transmission leakage terms, E_{XF} and E_{XR} are measured using the conventional technique. The eight error terms are represented by the error adapters shown in Figure 12.30. Although this error model is slightly different from the traditional Full two-port 12-term model, the conventional error terms may be derived from it. For example, the forward reflection tracking (E_{RF}) is represented by the product of ε_{10} and ε_{01}. Also notice that the forward source match (E_{SF}) and reverse load match (E_{LR}) are both represented by ε_{11} while both the reverse source match (E_{SR}) and forward load match (E_{LF}) are represented by ε_{22}. In order to solve for these eight unknown TRL error terms, eight linearly independent equations are required.

The first step in the TRL* two-port calibration process is the same as the transmission step for a full two-port calibration. For the thru step, the test ports are connected together directly (zero length thru) or with a short length of transmission line (non-zero length thru) and the transmission frequency response and port match are measured in both directions by measuring all four S-parameters.

For the reflect step, identical high-reflection coefficient standards (typically open or short circuits) are connected to each test port and measured (S_{11} and S_{22}).

For the line step, a short length of transmission line (different in length from the thru) is inserted between port 1 and port 2 and the frequency response and port match are measured in both directions by measuring all four S-parameters.

In total, ten measurements are made, resulting in ten independent equations. However, the TRL error model has only eight error terms to solve for. The characteristic impedance of the line standard becomes the measurement reference and, therefore, has to be assumed ideal (or known) and defined precisely.

At this point the forward and reverse directivity (E_{DF} and E_{DR}), transmission tracking (E_{TF} and E_{TR}) and reflection tracking (E_{RF} and E_{RR}) terms may be derived from the TRL error terms. This leaves the isolation (E_{XF} and E_{XR}), source match (E_{SF} and E_{SR}) and load match (E_{LF} and E_{LR}) terms to discuss.

12.12.1.2 Isolation

Two additional measurements are required to solve for the isolation terms (E_{XF} and E_{XR}). Isolation is characterised in the same manner as the full two-port calibration. Forward and reverse isolation are measured as the leakage (or crosstalk) from port 1 to port 2 with each port terminated. The isolation part of the calibration is generally only necessary when measuring high-loss devices (greater than 70 dB).

12.12.1.3 Source match and load match

A TRL calibration assumes a perfectly balanced test set architecture as shown by the term which represents both the forward source match (E_{SF}) and reverse load match (E_{LR}) and by the (ε_{22}) term which represents both the reverse source match (E_{SR}) and forward load match (E_{LF}). However, in any switching test set, the source and load match terms are not equal because the transfer switch presents a different terminating impedance as it is changed between port 1 and port 2.

In network analysers based on a three-sampler receiver architecture, it is not possible to differentiate the source match from the load match terms. The terminating impedance of the switch is assumed to be the same in either direction. Therefore, the test port mismatch cannot be fully corrected. An assumption is made, such that

Forward source match (E_{SF}) = reverse load match (E_{LR}) = ε_{11}
Reverse source match (E_{SR}) = forward load match (E_{LF}) = ε_{22}

For a fixture, TRL* can eliminate the effects of the fixture's loss and length, but does not completely remove the effects due to the mismatch of the fixture.

Note: Because the technique relies on the characteristic impedance of transmission lines, the mathematically equivalent method (for line-reflect-match) may be substituted for TRL. Since a well matched termination is, in essence, an infinitely long transmission line, it is well suited for low-frequency calibrations. Achieving a long line standard for low frequencies is often physically impossible.

Most of the latest network analysers are equipped with four receiver test-sets. In this configuration they are able to implement the full TRL algorithm.

12.12.2 True TRL/LRL

Implementation of TRL calibration with a network analyser which employs four receivers requires a total of fourteen measurements to quantify ten unknowns as opposed to only a total of twelve measurements for TRL* (both include the two isolation error terms).

Because of the four-sampler/receiver architecture, additional correction of the source match and load match terms is achieved by measuring the ratio of the two 'reference' receivers during the thru and line steps. These measurements characterise the impedance of the switch and associated hardware in both the forward and reverse measurement configurations. They are then used to modify the corresponding source and load match terms (for both forward and reverse).

The four receiver configuration with TRL calibration establishes a higher performance calibration method over TRL*, because all significant error terms are systematically reduced. With TRL*, the source and load match terms are essentially that of the raw, 'uncorrected' performance of the hardware where as with TRL the source and load match terms are reduced in line with the quality of calibration kit components used.

12.12.3 The TRL calibration procedure

When building a set of standards the requirements for each of the standard types specified in Table 12.1 must be satisfied.

Table 12.1   TRL calibration procedure: requirements for each of the standard types

Standard types	Requirements
Thru	No loss Impedance (Z_0) need not be known $S_{21} = S_{12} = 1 \angle 0°$ $S_{11} = S_{22} = 0$
Thru (non-zero length)	Z_0 of the thru must be the same as the line. Attenuation of the thru need not be known. If the thru is used to set the reference plane, the insertion phase or electrical length must be well known and specified
Reflect	Reflection coefficient Γ magnitude is optimally 1.0, but need not be known. Phase of Γ must be known and specified to be within $\pm 1/4$ wavelength or $90°$. Γ must be identical on both ports. If the reflect is used to set the reference plane, the phase response must be well known and specified.
Line/match (line)	Z_0 of the line establishes the impedance of the measurement (i.e. $S_{11} = S_{22} = 0$). Insertion phase of the line must be different from the thru. Difference between thru and line must be $>20°$ and $<160°$. Attenuation need not be known. Insertion should be known
Line/match (match)	Z_0 of the match establishes the reference impedance of the measurement. Γ must be identical on both ports

When calibrating a network analyser, the actual calibration standards must have known physical characteristics. For the reflect standard, these characteristics include the offset in electrical delay (seconds) and the loss (Ω per second of delay). The characteristic impedance, Z_0, is not used in the calculations in that it is determined by the line standard. The reflection coefficient magnitude should optimally be 1.0, but need not be known since the same reflection coefficient magnitude must be applied to both ports.

The thru standard may be a zero ss-length or known length of transmission line. The value of length must be converted to electrical delay, just like that done for the reflect standard. The loss term must also be specified.

The line standard must meet specific frequency-related criteria, in conjunction with the length used by the thru standard. In particular, the insertion phase of the line must not be the same as the thru. The optimal line length is $\frac{1}{4}$ wavelength (90°) relative to a zero length thru at the frequency of interest, and between 20° and 160° of phase difference over the frequency range of interest. (*Note*: these phase values can be $\pm N \times 180°$, where N is an integer.) If two lines are used the difference in electrical length of the two lines should meet these optimal conditions. Measurement uncertainty will increase significantly when the insertion phase nears zero or is an integer multiple of 180°, and this condition is not recommended.

For a transmission medium that exhibits linear phase over the frequency range of interest, the following expression can be used to determine a suitable line length of $\frac{1}{4}$ wavelength at the frequency (which equals the sum of the start frequency and stop frequency divided by 2):

Electrical length (cm) = (Line − Zero length thru)

$$\text{Electrical length (cm)} = \frac{(15{,}000 \times \text{VF})}{f_1(\text{MHz}) + f_2(\text{MHz})} \quad (12.3)$$

where $f_1 = 1000$ MHz, $f_2 = 2000$ MHz and VF = Velocity Factor = 1.

Thus the length to initially check is 5 cm. Next, use the following to verify the insertion phase at f_1 and f_2 (1000 and 2000 MHz):

$$\text{Phase (degrees)} = \frac{(360 \times f \times l)}{v} \quad (12.4)$$

where f is the frequency (MHz), l is the length of line (cm) and $v =$ velocity = speed of light × velocity factor, which can be reduced to the following:

$$\text{Phase (degrees) approximately} = \frac{0.012 \times f\,(\text{MHz}) \times l\,(\text{cm})}{\text{VF}} \quad (12.5)$$

So for an airline (velocity factor is approximately 1) at 1000 MHz, the insertion phase is 60° for a 5 cm line; it is 120° at 2000 MHz. This line would be suitable as a line standard.

Where the standard is fabricated in other media (microstrip for instance) the velocity factor is significant. For example, if the dielectric constant for a substrate is 10, and the corresponding 'effective' dielectric constant for microstrip is 6.5, then the 'effective' velocity factor equals 0.39 $(1 + \sqrt{6.5})$.

Using the above a potential problem using TRL becomes evident. The lengths of airline required at low frequencies become so long that they are difficult to fabricate.

12.13 Data-based calibrations

Traditionally the calibration standards used in any network analyser calibration routine have been defined in terms of the way in which their parameters vary in relation to the measurement frequency; for instance, the open circuit would be defined in terms of capacitance. Three or four frequency terms would be employed, f, f^2, f^3 and sometimes f^4. Open circuits would be defined in a similar manner, in terms of inductance. As correction algorithms progressed some standards were defined in terms of both capacitance and inductance. Loads were usually considered as perfect. These definitions are usually excellent providing that it is possible to define the standards using smooth curves.

As processors and particularly memory have become cheaper another method of defining the calibration standards has become available, the data-based calibration. Each standard is measured across the frequency range of interest using the best equipment and techniques available. These measured values are entered into the network analyser's database and used in the correction algorithms. At frequencies where data are not available the network analyser uses interpolation, thus if measurements are made at more frequencies on the standards, the resulting network analyser measurements will become more accurate. Electronic calibration units, where the standards are in one enclosure and a switch matrix employed to apply them to the network analyser, often use a data-based calibration routine. The accuracy available from the data-based calibration employing the electronic calibration units approaches the best available from TRL calibrations, but without needing the same level of skilled operator.

References

1 J.M. Hawkins (ed.): *The Oxford Reference Dictionary*. (Oxford University Press, Oxford, 1987, reprinted 1989)
2 8719ET/20ET/22ET, 8719ES/20ES/22ES Network Analysers User's Guide, Agilent Technologies, Inc. 2000
3 HP8753A Network Analyser Operating and Programming Reference–08753-90015, Hewlett-Packard Company, 1986. Now Agilent Technologies, Inc.

Chapter 13
Verification of automatic network analysers
Ian Instone

13.1 Introduction

Network analysers are complex instruments that can combine many different instruments within one measurement system. With this in mind it is easy to make apparently similar measurements with a variety of different instrument settings. Each setting may enhance one particular aspect of the measurement, but this is often traded off in another area. For example, to improve repeatability we might increase the averaging or decrease the bandwidth or use a combination of both. The resulting improvement in repeatability will usually be at the expense of the considerably increased measurement time.

This chapter discusses different types of verification which may be applied to network analyser measurements to enable the user to assess or confirm the most appropriate choice of settings on the network analyser for their particular measurement scenario.

13.2 Definition of verification

As with calibration, it is important to understand the interpretation of the word 'verification'. *The Oxford Reference Dictionary* (1989) defines the word 'verify' as 'to establish the truth or correctness of by examination or demonstration; (of an event etc.) to bear out, to fulfil (a prediction or promise)'. This dictionary definition exactly describes the process of verification as applied to automatic network analysers; the quality of measurements which the analyser is capable of making is verified by comparing them with values obtained from another source, whereas calibration characterises the network analyser prior to 'corrected' measurements being performed.

13.3 Types of verification

There are several different methods of verification so the method chosen needs to address the particular requirements of the user. In all cases the method chosen or designed should provide the user with at least acceptable confidence that the measurements being made with the network analyser meet the user's minimum quality requirements. Verification limits are set using a combination of the measurement uncertainties and the acceptable product quality. Uncertainties should be assessed using an accepted method such as that described in EA-10/12, *Guidelines on the Evaluation of Vector Network Analysers*, available free from http://www.euromet.org/docs/calguides/index.html

13.3.1 Verification of error terms

As described in the previous chapter, the corrected network analyser's display is made up of the following elements:

(1) parameters of the device under test (DUT),
(2) errors contributed by the measurement system,
(3) corrections applied to the measurements and
(4) residual errors present after correction.

Verification of the network analyser's residual errors after correction involves measuring and quantifying the residual errors present after the error correction has been applied. This method is perhaps one of the most difficult to perform, is the most time consuming, and requires the highest skill levels, but will enable the user to determine exactly which components may require attention without any additional measurements having to be performed. Typically, this type of verification provides the greatest insight into the characteristics of the network analyser and calibration kit used.

13.3.2 Verification of measurements

This verification scheme involves calibrating the network analyser (usually as part of the normal measurement process) and then measuring a known artefact(s). Appropriate acceptance limits must be set when using this method as it is often possible for one parameter showing poor performance to be masked by other parameters where performance exceeds minimum expectations. Whilst this method provides the best assessment of all the contributors combining in the uncertainty budget, the danger is that one component in the calibration kit or network analyser which is beginning to deteriorate is masked by other parameters that are still exceeding expectations. This method, however, is one of the easiest to implement, easiest to understand and quickest to perform so warrants consideration on these points alone.

On a production line this method might be implemented by periodically taking a 'sample' DUT and re-testing it on a different network analyser or measurement system. If the measurements from both systems are compared and the results found to fall within the user's acceptable quality limits it can be assumed that both systems are making acceptable measurements.

This method is often used by network analyser manufacturers and their service agents when maintaining customer's equipment at the customer's site.

13.4 Calibration scheme

It should be possible to perform verification of the network analyser irrespective of the calibration scheme used. The correction coefficients employed as a result of the calibration may affect the acceptance limits used for the verification but should have little or no influence on the method of verification. Ideally the calibration scheme employed will be identical to that used for measurements, and might even be exactly the same calibration. As the verification verifies the satisfactory operation of the network analyser, test port leads, adapters and calibration kit, it is essential to ensure that all of these items are used in the calibration and verification process.

13.5 Error term verification

For a full two-port measurement seven dominant error terms that could be checked are as follows:

(1) effective directivity,
(2) effective source match,
(3) effective load match,
(4) effective isolation,
(5) effective tracking,
(6) effective linearity and
(7) repeatability.

The term 'effective' as used in the list above refers to the parameter after error correction has been applied. These terms are often referred to as the residual errors, which are also contributors to the uncertainty of measurement. Methods for checking most of these terms are shown in EA-10/12.

13.5.1 Effective directivity

Directivity refers to the ability of a directional device, such as a coupler or directional bridge, to separate the forward and reverse signals. Where the bridge or coupler is embedded in a network analyser the most convenient way to measure this parameter is to first reflect all of the signal using a short or open circuit (the mean between the short and open circuit is considered the most accurate in this simplistic case) and set as a reference. The short or open circuit is then replaced with a fixed termination of the correct characteristic impedance. Where the fixed termination has a good match (negligible voltage reflection coefficient) the network analyser's display will be

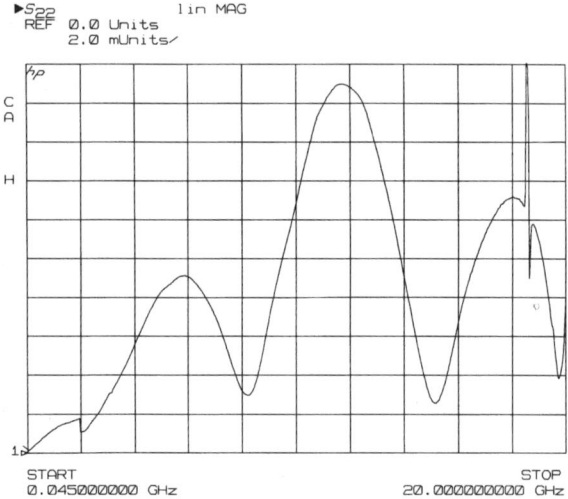

Figure 13.1   Typical network analyser display of the voltage reflection coefficient of a fixed broadband load

predominantly composed of the effective directivity. Since the perfect termination rarely exists, we need some method of separating the network analyser's own errors from those generated by the fixed termination. These errors tend to increase as the measurement frequency increases. Two methods of 'signal separation' are discussed below (Figure 13.1).

13.5.1.1   Sliding load method

A sliding load can be used to separate the directivity from the terminating load. Where possible the network analyser should be set to display the measurements in 'linear mode'. After the reference has been recorded the sliding load is connected in place of the open or short circuits. If the load element is positioned furthest away from the input connector the network analyser will display a curve representing the match of the sliding load's load element with ripple superimposed upon the measurement. The majority of ripple is produced by the directivity either adding 'in phase' or 'anti-phase' with the load element measurement. There will also be a small error produced in this measurement contributed by the effects of imperfect source match and an imperfect sliding load element; however, this error is often so small that it is neglected. The directivity may be assessed by measuring the height of the ripples: directivity will be one-half the ripple amplitude. Sometimes the transitions in match of the sliding load make the measurement of the superimposed ripple difficult or impossible. In these cases it will be necessary to make a continuous waves (CW) measurement. The network analyser's marker is placed at the frequency of interest. The sliding load is adjusted so that a maximum value is observed using the marker and the value noted. The sliding load is now adjusted so that a minimum value is observed using

Verification of automatic network analysers 295

the marker and the value noted. The directivity is one-half of the difference between the two marker values.

The major problem in using a sliding load is that measurements on sliding loads are difficult to perform and traceability for these measurements may not be easy to obtain.

13.5.1.2 Offset load or airline method

This method works in a very similar way to the sliding load method. After the reference has been recorded the airline and fixed termination are connected in place of the open or short circuits. The network analyser will display a curve representing the match of the fixed termination with ripple (from the directivity) superimposed upon the measurement. Half of the amplitude of the ripple is the directivity. This method has the same problem as the sliding load method regarding the effects of source match. Providing the fixed termination has a small reflection coefficient this problem will be kept to a minimum (Figures 13.2 and 13.3).

Where the fixed termination shows a rapid transition between two values of reflection coefficient it may not be possible to make an accurate measurement of directivity. Since this method should be independent of the fixed termination used, it will be perfectly valid to select another fixed termination with a different reflection coefficient profile to provide more reliable directivity measurements at these more difficult frequencies.

The calibration devices used to characterise the effective directivity term are the low-band load (at lower frequencies), and the sliding load or short airline(s) at high frequencies except in broadband load calibrations where the broadband load is used

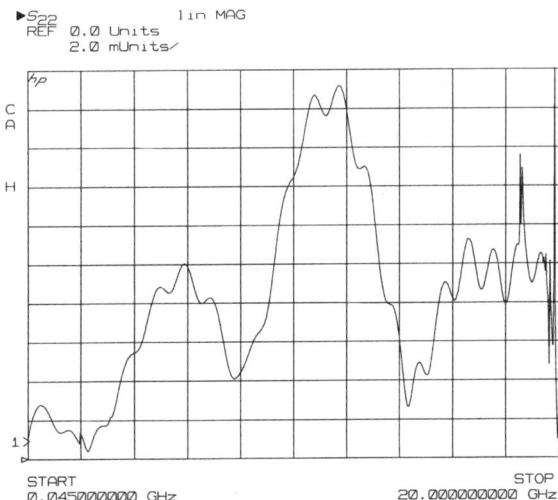

Figure 13.2   Ripple superimposed on the fixed load response caused by the interaction of directivity and the broadband load

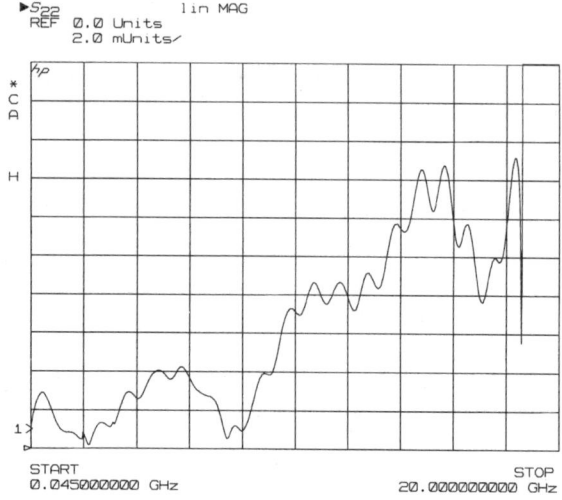

Figure 13.3 Using another broadband load with a different profile can make the ripples easier to determine

exclusively to define the directivity term. The types of measurement most affected by directivity errors are low-reflection measurements; high-reflection measurements will often appear as normal.

13.5.2 Effective source match

This term refers to the impedance of the directional bridge or coupler and associated cables and adapters as they are presented to the DUT. Methods of measurement are very similar to those used to measure effective directivity. However, since we need to measure source match we must feed a reasonable amplitude signal back into the directional bridge or coupler. This task is performed best using either a short or open circuit. The short or open circuit is usually connected to the directional bridge or coupler via an airline, which provides some phase shift enabling the source match to be shown as ripple superimposed on the reflection characteristics of the short or open circuit. One problem in trying to present these data is that the loss of the airline used is often a major part of the displayed measurement. This can make it difficult to determine the ripple amplitude when the source match is fairly small. Shorter airlines will reduce the loss and will also reduce the quantity of ripples observed so a suitable compromise must be achieved. Note in the following plots that there are some ripples of very short period which can be ignored as they are probably generated by other effects within the measurement system (Figures 13.4 and 13.5).

As with directivity, the peak to peak height of the ripple is twice the source match. Note also that this measured source match also contains the directivity, which at any

Verification of automatic network analysers 297

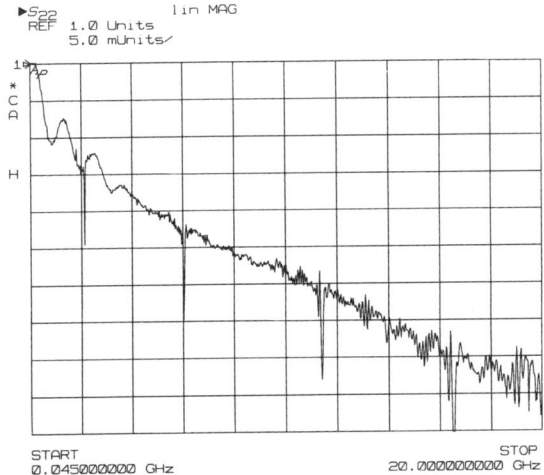

Figure 13.4   Ripple caused by the interaction of the source match and an open circuit

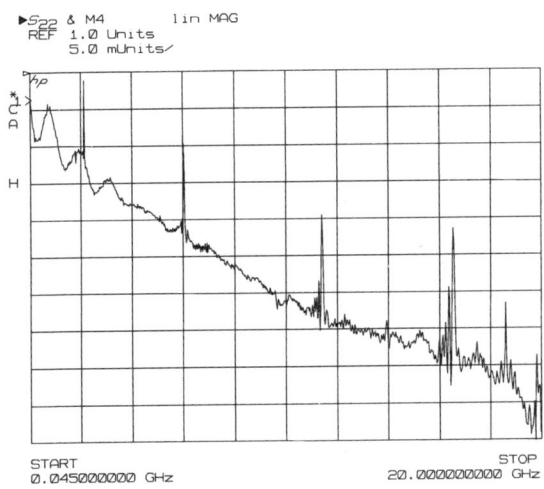

Figure 13.5   Ripple caused by the interaction of the source match and a short circuit

given frequency may either add to or subtract from the source match. Since we have no easy way of separating the source match and directivity, we usually consider directivity as one of the sources of uncertainty when making source match measurements. Directivity is usually much smaller than source match so this assumption causes few problems.

Time-domain gating (explained below) can be used to effectively separate these interacting terms. Unfortunately, it has not been possible to provide traceability for

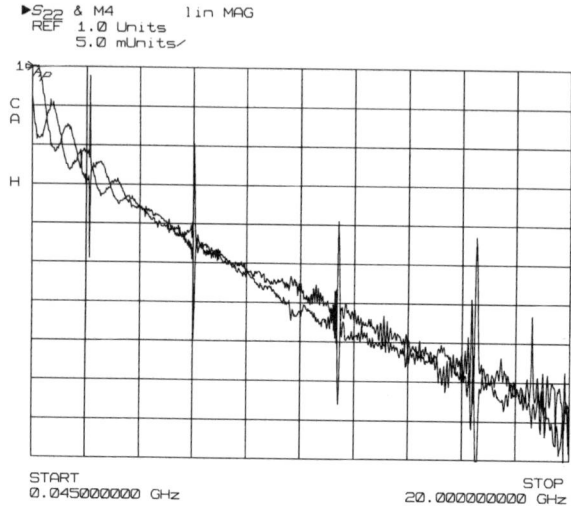

Figure 13.6 Ripple caused by the interaction of the source match and short and open circuits

any measurements in the time-domain so this function is best left to the development laboratories where it provides useful improvements in test development times.

One neat trick that can be employed to provide reliable and easy to read source match measurements is to either store or plot the display with a short circuit connected, then connect the open circuit. Assuming the short and open circuits are approximately 180° apart in reflection phase, the resultant display will be one of two traces where the 'peaks and troughs' occur at approximately the same frequencies (looking similar to the envelope on an Amplitude Modulated signal).

The peaks and troughs can now be read at the same frequency, producing a more accurate value of source match at a particular frequency (Figure 13.6).

It is also possible to use a sliding short circuit to determine source match at any particular frequency, using a similar technique as described for the sliding load in the measurement of directivity. Unfortunately, sliding short circuits fitted with co-axial connectors are now getting harder to obtain. This technique is still useful where rectangular waveguide is employed as the transmission medium because sliding short circuits in rectangular waveguide are still supplied by several manufacturers.

The calibration items used to characterise the effective source match term are the short and open circuits. A poor connection of either of these devices will affect the effective source match. Further, open circuits usually have a centre pin supported with a delicate piece of dielectric; if this dielectric fractures and the centre pin is misplaced the effect on the source match will be massive. The measurements most affected by source match errors are high-reflection measurements and transmission measurements of highly reflective devices. Poor cables can cause both the directivity and source match terms to vary as the cable is flexed. The effect of this variation is that there will be errors in the measured values.

13.5.3 Effective load match

Effective load match is the effective impedance of the load presented to the DUT. For a full two-port measurement the load would be represented by the 'receiving signal port'. As there appear to be no 'classical' methods for measuring load match it is usually assumed that it has a similar value to the source match. Refer to *Network Analyser Uncertainty Computations for Small Signal Model Extractions* by Jens Vidkjær [1] for more detailed information on this subject. The measurements most affected by effective load match are all transmission and reflection magnitude measurements of low insertion loss two-port devices.

13.5.4 Effective isolation

Isolation is a measure of how much signal passes from one channel to the other when both channels are terminated in their characteristic impedance. Although the error correction routines are designed to compensate for some degree of poor isolation it is good practice to maintain as ideal a value as possible. The simplest way to measure isolation is to connect the two test port cables together and set a transmission reference in each direction on the screen. Then connect reasonably well-matched terminations to the DUT ends of the test port cables and repeat the transmission measurement. The screen display will be very noisy and should consist of a combination of the network analyser noise floor and the network analyser's isolation. Poor isolation may be caused by loose connectors within the test set or poor or worn screening throughout the measurement system. In particular, look at the test port extension cables as these are often subjected to plenty of flexing and plenty of wear and tear at the connector. Whilst connectors in poor condition will be obvious to the experienced eye, there will be few visible signs of any deteriorating screening making regular testing desirable. Where isolation is found to be a constant value at any particular frequency corrections are applied. With modern network analysers having very good isolation, often in the same area as the instrument's noise floor, there is often a danger that the values due to the noise floor become entered into the isolation corrections causing further errors rather than correcting them. Poor isolation would affect both reflection and transmission measurements where the test channel signal is at a very low level, that is, reflection measurements and also transmission measurements where the insertion loss of the DUT is large (i.e. greater than a 50 dB attenuator).

13.5.5 Transmission and reflection tracking

This correctable error includes the effects of the insertion loss of the signal separation devices, detectors (or samplers), cables, signal paths and any other items in the signal paths. Residual errors after correction may be analysed by connecting the test port cables together and examining the transmission trace. Any deviation from 0 dB may be due to tracking. Also, there may be an amplitude-dependent tracking error; this would be checked in the same way, but in addition the source power would be varied and the trace deviation from the 0 dB level noted.

The calibration devices used to characterise transmission tracking are the transmission measurements of the 'thru' connection. Large variations in the tracking terms might indicate a problem in the reference or test signal path in the test set or poor connections during the calibration process. All transmission measurements are affected by transmission tracking errors.

The calibration devices used to characterise reflection tracking are the short and open circuits. As with transmission tracking large variations in the tracking term might indicate a problem in the reference or test signal path in the test set or poor connections during the calibration process. All reflection measurements are affected by transmission tracking errors.

13.5.6 Effective linearity

Deviation from linearity may be checked by measuring a previously calibrated step-attenuator. Providing the step-attenuator has been calibrated with a sufficiently low measurement uncertainty, and the step-attenuator has a good match in each direction, it can be assumed that any deviations noted are due to the network analyser's deviation from ideal linearity. Effective linearity is a significant contributor in the uncertainty budget and needs to be assessed with the signal travelling in either direction.

Linearity is not a term characterised using the calibration kit. Some network analysers have corrections for linearity which may be updated when a routine maintenance check is performed. All measurements are affected by linearity.

13.5.6.1 Time-domain and de-embedding

Many of the higher frequency network analysers are capable of performing fast Fourier transforms (FFTs). Where implemented this process allows measurements of components within complex networks to be displayed using a process known as 'time-domain gating'. The component under test or evaluation is mathematically de-embedded from its surrounding network and its response displayed on the screen of the network analyser. This function can be employed to provide values of directivity and source match providing a suitable reference (usually an airline in same characteristic impedance as the coupler or directional bridge) is available. Unfortunately, traceability of measurement has not been developed for this type of time-domain function, so these measurement methods are best left for routine maintenance and diagnostic tasks rather than the task of ensuring traceability of measurement. The concept of time-domain gating refers to mathematically removing a portion of the time-domain response, and then viewing the result in the frequency domain. The intent is to remove the effects of unwanted reflections, say from connectors and transitions leaving just the response of the device being measured. An experienced operator will be able to perform measurements of directivity, source match and load match much faster using time-domain gating rather than using any of the alternative methods described above.

13.6 Verification of measurements

This method of verification is perhaps easier to understand and provides a much easier visualisation of the general health of the network analyser, calibration kit and test port cables. The method involves calibrating the network analyser then measuring an artefact or artefacts. The measurements are then compared either with measurements performed earlier, or if it is desired to obtain traceability this way they would be compared with measurements performed on the same artefacts at a laboratory operating at a higher echelon in the traceability chain. For this method to be effective the artefacts used for the verification need to be stable with both time and temperature. For these reasons 'simple' devices such as fixed attenuators, fixed terminations and certain types of coupler are often chosen. Sometimes an artefact similar to that which it is desired to measure is chosen so that if an error occurs within the measuring system its effect can be seen and assessed immediately.

13.6.1 Customised verification example

To improve throughput on one of the production lines it was decided to use an electronic calibration module with the network analyser testing input impedance. It was also desired to calibrate or check the e-cal module on site as the only alternative was to have it sent overseas to its manufacturer which would cause unacceptable downtime. The specification of the e-cal module is excellent so straightforward testing of it could not be performed to the desired level. It was decided that an artefact which was representative of the manufactured product could be used to access the 'general health' of the complete measuring system. The artefact chosen was a programmable attenuator with a short circuit connected to one port (Figure 13.7). This provides a range of mismatch that can be adjusted using software so maintaining the level of automation.

It was not considered necessary to have all steps of the attenuator measured as this would provide too much information, much of which may never be looked at, hence, the following were chosen:

(1) highest mismatch,
(2) approximate upper specification of DUT,
(3) approximate centre of specification of DUT,
(4) approximate lower specification of DUT and
(5) lowest mismatch.

Figure 13.7   *Artefact chosen for the comparison, an Agilent 84904K programmable step-attenuator with a type N adapter and short circuit fitted*

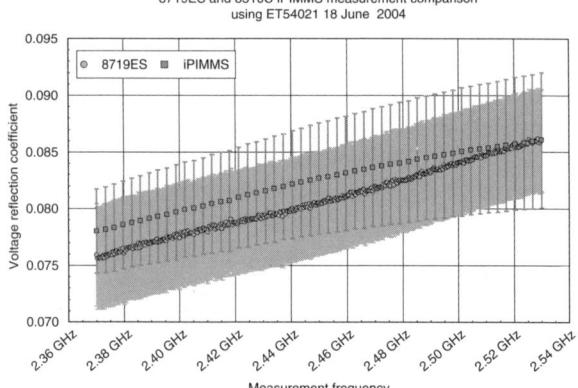

Figure 13.8 Plot produced from the results of a customised verification example showing all of the uncertainty bars overlapping

This list provides plenty of measurements in the range where it is essential for the network analyser to provide the most accurate measurements possible, and some supplementary measurements (highest and lowest mismatch) which could be used to provide some rudimentary diagnosis should the need arise. The attenuator was calibrated using the best and most accurate and traceable equipment possible. The attenuator was then transferred to the production line where it was measured using the network analyser and electronic-calibration system. A graphical representation of the two sets of results obtained is shown in Figure 13.8. The process is fully automated so it can be used each time the network analyser is re-calibrated. Since accurate measurements can take a long time to obtain there were only 51 points measured by the 'accurate' network analyser. This is adequate in this case because the attenuator is a linear resistive device so there is a high probability that linear interpolation can be used between measurement points, if necessary. The production line network analyser, however, is normally measuring active devices so measurements are made at considerably more frequencies, albeit with slightly greater uncertainties in places. In order to make this quantity of measurements within the very short times demanded by production processes they must be made faster, with the trade-off being slightly increased measurement uncertainties.

Note in Figure 13.8 that the reference measurements are performed at considerably fewer frequencies. This is quite normal as 'quality measurements' can be expensive to perform. Sufficient measurements have been performed showing that linear interpolation between measured values is valid.

13.6.2 Manufacturer supplied verification example

Many manufacturers supply verification procedures with their network analysers. The user will normally need to buy a verification kit which is often supplied with a disk containing measurements made on the component parts of the kit. Verification kits and associated procedures are usually designed to provide a quick 'health check' on the network analyser. Testing that the network analyser (and calibration kit) meet their specification will often involve adjusting the settings on the network analyser resulting in the measurements taking far longer. The process begins with the operator performing an appropriate calibration (error correction). Test devices from the verification kit are then measured and the results compared with measurements that were made using a reference measurement system (Figures 13.9 and 13.10). If the comparison reveals that the results fall within prescribed limits the network analyser (and appropriate calibration kit) are said to be verified. This type of verification is intended as a routine 'health check' and is used by some manufacturers as a routine check for equipment installed at a customer's location. To this end the software required to automate this process and therefore improve consistency is often included within the operating firmware of the network analyser.

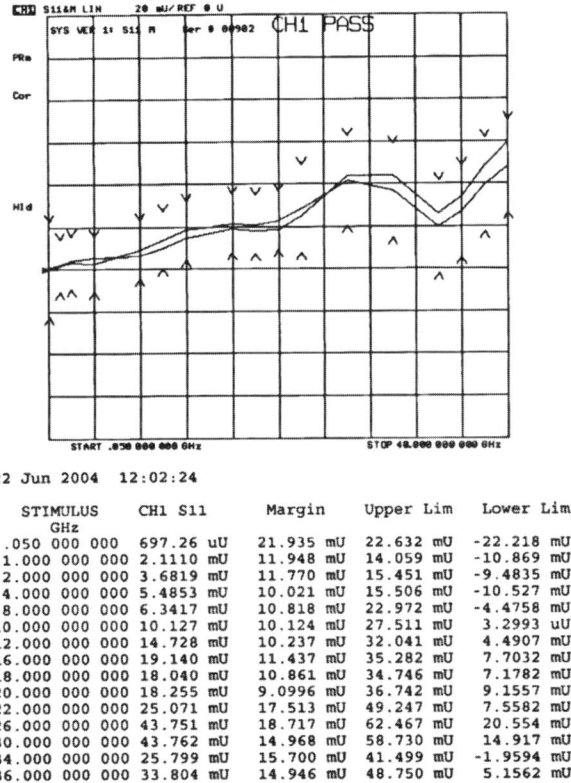

Figure 13.9 Printed output from a typical verification program. A sheet similar to this is produced for both phase and magnitude for each S-parameter of each device tested

304 *Microwave measurements*

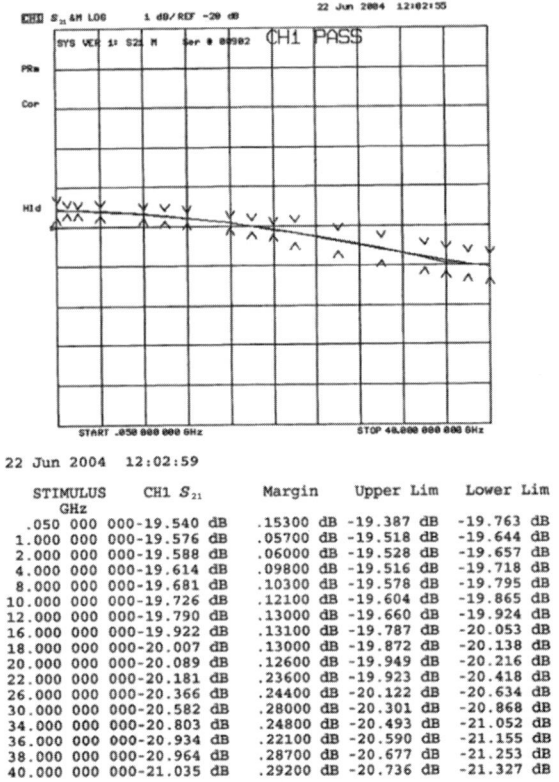

Figure 13.10 *Another example from the same verification routine, this time displaying a transmission parameter*

The major problem with these types of verification (manufacturer supplied and customised) is that all of the 'errors' and measurements are lumped together, the measured values contain both and there is no easy way to separate them. Degraded items can be offset by items still in their prime. This makes it very difficult to identify any one device in the calibration kit or network analyser which may be starting to drift into a problem state, but at least has the advantage of allowing the user to quickly estimate if their system is in a suitable state for measurements.

Presentation of the results can be difficult in certain circumstances, particularly transmission phase where the phase vector often rotates through its full 360° and the test limit can be less than 1°.

References

Vidkjær, J.: *Network Analyser Uncertainty Computations for Small Signal Model Extractions*, Technical University of Denmark, R549, Feb 1994

Chapter 14
Balanced device characterisation
Bernd A. Schincke

14.1 Introduction

For decades high frequency circuits were developed using unbalanced (non-symmetrical) structures. Typical line systems, representing this kind of structure, are coaxial or coplanar line systems. Each unbalanced system consists of a signal line and a ground. The measurable signal is referenced to the ground.

Balanced (or symmetrical) structures are not used that often. A typical balanced structure is a parallel line system (Lecher line), a Low Voltage Differential Signal Line (LVDS-line) or balanced amplifiers and filters. Typically such a structure consists of two lines (simply said, a 'plus' and a 'minus') and a signal can be measured between these two lines. In practice, these structures create some additional phenomena compared to unbalanced systems which must be analysed in detail.

In an unbalanced system only the non-symmetrical TEM mode is present and it can be compared to the so-called common mode, which we will discuss later. In a coaxial system the inner conductor is the signal line and the outer conductor represents the ground. In addition this ground functions as a shield. Unbalanced line systems are normally connected to unbalanced circuits. Under the condition of power matching the measured voltage U_1 against ground is $U_{01}/2$.

We can conclude that such an unbalanced system offers very high noise immunity, it generates less radiation, the integration density is high and the losses are acceptable.

If such a line is connected to, for example, a non-shielded circuit, a signal generated by an interferer (like ground noise or general electromagnetic interference) can be induced on the signal and will be present on the signal at the load. A fundamental disadvantage of an unbalanced structure is its susceptibility against an interferer (Figure 14.1).

By using a balanced system (two-line-system), in an ideal case only one signal between the two lines can be measured. Here, the differential TEM mode is the only

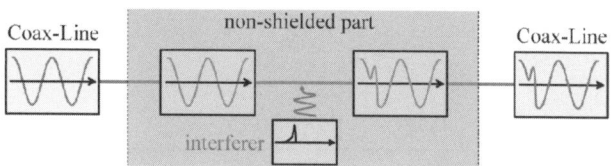

Figure 14.1   Unbalanced system

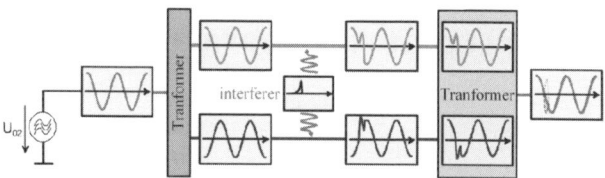

Figure 14.2   Balanced system

one which is present. Analogous to the balanced line system the circuits are performed as balanced structures, too. A disadvantage by using balanced structures is that more components are needed compared with unbalanced structures. Under the condition of power matching a voltage U_2 can be measured between the two lines which can be expressed by $U_{02}/2$.

An important advantage is that the original signal is (theoretically) not influenced by electromagnetic radiation. A balanced system can be performed by using a transformer to transform the signal with 0° phase shift to the 'upper' line and by using a BALUN (BALanced–UNbalanced) to transform the signal with 180° phase shift to the 'lower' line. If an interferer occurs, the signals on both lines are interfered in the same way. Using the same transformer/BALUN structure at the output of the circuit the 0° phase shifted, interfered signal will be superimposed on the 180° phase shifted signal and the interference will be shortened (Figure 14.2).

If a ground is present, again under the condition of power matching, the signal that can be measured between each signal line (upper and lower) and the ground is $U_{02}/4$. Between the two lines $U_2 = U_{02}/2$. The signals have the same amplitude, but they are 180° phase shifted. This means that the needed voltage amplitudes to generate a desired power are half of the needed amplitudes working with an unbalanced structure. The advantage is that components with a lower breakdown voltage can be used.

For narrow band applications especially in the higher GHz range (e.g. low noise converter (LNC)) the band-pass filtering offers sufficient interference suppression. Such a system will in the future also serve as an unbalanced system.

14.1.1   Physical background of differential structures

An essential problem when working with differential structures is that we cannot regard a differential structure as a pure 'two-line-system'. Every balanced system

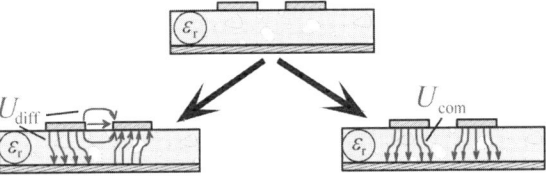

Figure 14.3   Three-line-system

must be regarded together with a ground. When, for example, a twisted pair line is used in an instrument, the instrument wall is normally grounded. A multilayer board needs ground layers in order not to influence each other. Because of the *ground* ('third line') in practice such a differential structure must be regarded as a 'three-line-system' (Figure 14.3).

In a three-line-system two different TEM modes of propagation are possible and must be analysed. On the right-hand side the (wanted) differential energy propagates through the device under test (DUT), on the left-hand side the common mode energy. Especially electromagnetic radiation and ground noise are typical common mode signals. On a board the two modes are quasi-TEM modes with different field distributions in the air and in the dielectric material. This can result in different propagation velocities and the characteristic impedances of the two modes are typically different.

To perform an exact measurement and dimensional design in the RF, an S-parameter description is needed taking both modes into account.

14.1.1.1   Ideal device

An ideal balanced device is characterised by ideal symmetry of, for example, the two lines. This means in detail the same electrical length, same attenuation, same dielectric, etc. In this case only the differential mode signal is transmitted and the common mode signal is suppressed. This is valid for pure differential structures (balanced input/balanced output) and for balanced to single-ended structures (e.g. balanced input/single-ended output) (Figure 14.4).

14.1.1.2   Real device

Caused by asymmetries such an ideal balanced structure is normally not given. Very often it is possible to measure a common mode at the output of a DUT even though the device is powered by a pure differential mode signal. In this case a common mode signal is generated from the differential mode signal. Such a common mode signal can be described as an electromagnetic interferer. This procedure is called 'Differential Mode to Common Mode Conversion'.

If the device is powered only by a common mode signal, it is possible to measure a differential mode signal at the output. Here, from an electromagnetic interferer at the input, a differential mode signal at the output is generated which will be superimposed the original differential signal. Caused by this 'Common Mode to Differential Mode Conversion' the structure becomes susceptible to an EMI (Figure 14.5).

308 *Microwave measurements*

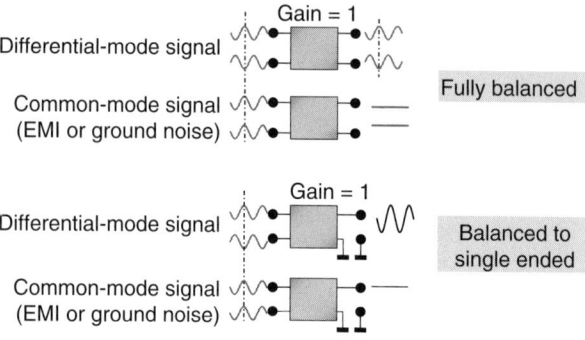

Figure 14.4   Ideal device

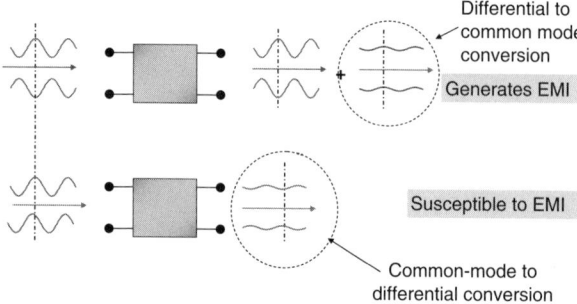

Figure 14.5   Real device, transmission characteristic

These facts are valid for the transmission characteristic of the DUT as well as for the reflection characteristic of the DUT.

We can conclude that real devices are normally non-ideal devices. Non-ideal devices convert differential mode energy into common mode energy and common mode energy into differential mode energy. This conversion can be measured at the input of a DUT (converted, reflected energy) and at the output (converted, transmitted energy).

The complete description of a non-ideal device is shown in the signal-flow diagram in Figure 14.6.

Using this model the balanced device is described by two separate systems, a pure common mode system and a pure differential mode system.

The pure common mode S-parameters are the connection between the common mode stimulus signals and the measured common mode responses. The pure differential mode S-parameters are described by the connection between differential mode stimulus signals and measured differential responses. The conversion parameters, caused by the interaction between the two systems, are also shown in this model.

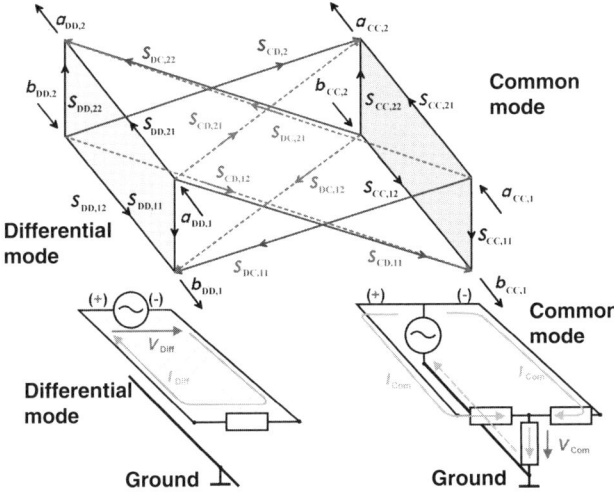

Figure 14.6 Signal-flow diagram

14.2 Characterisation of balanced structures

The typical parameters to be tested are

(1) performance in the pure differential mode,
(2) performance in the pure common mode,
(3) conversion from differential mode to common mode (in both directions) and
(4) conversion from common mode to differential mode (in both directions).

To be able to do all these measurements the unbalanced two-port model must be extended to a balanced two-port model. Such a balanced two-port model has by definition four unbalanced physical ports.

To measure the pure differential mode and the pure common mode behaviour as well as the conversion parameters we must be able to generate differential mode and common mode signals and to measure the desired responses. These measurements must be done in both directions. The connection between the stimulus signals and the measured responses are described by the so-called Mixed Mode S-parameters Matrix.

$$\begin{pmatrix} b_{DD,1} \\ b_{DD,2} \\ b_{CC,1} \\ b_{CC,2} \end{pmatrix} = \begin{pmatrix} S_{DD,11} & S_{DD,12} & S_{DC,11} & S_{DC,12} \\ S_{DD,21} & S_{DD,22} & S_{DC,21} & S_{DC,22} \\ S_{CD,11} & S_{CD,12} & S_{CC,11} & S_{CC,12} \\ S_{CD,21} & S_{CD,22} & S_{CC,21} & S_{CC,22} \end{pmatrix} \cdot \begin{pmatrix} a_{DD,1} \\ a_{DD,2} \\ a_{CC,1} \\ a_{CC,2} \end{pmatrix} \quad (14.1)$$

The differential mode stimulus signals are labelled with a_{DD} at port one and at port two and the common mode stimulus signals with a_{CC}. The differential mode and the common mode response signals are described by b_{DD} and b_{CC} (Figure 14.7).

310 Microwave measurements

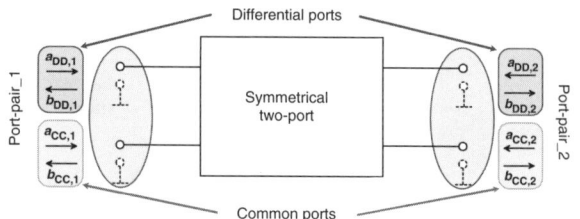

Figure 14.7   Balanced two-port

The S-parameters shown with the indices 'DD' and 'CC' are called self-parameters. These parameters are comparable to the unbalanced S-parameters, because by these the reflection quantity and the transmission quantity for the common mode and for the differential mode operation are described.

The S-parameters shown with the indices 'CD' and 'DC' are the conversion parameters. These parameters describe the reflection behaviour and the transmission behaviour of the DUT under the condition that mode conversion happens. If possible, the conversion parameters must be as low as possible. Ideally the common mode system is completely separated from the differential mode system. Then the conversion parameters are zero. The conversion parameters of differential structures become very low, whenever the lines are symmetric. This means that each line offers the same attenuation, same electrical length, etc.

14.2.1 Balanced device characterisation using network analysis

Network analysers are in general not developed to characterise balanced devices because they are unbalanced and normally only have two ports. They are working with CW (continuous wave) signals and do not generate common mode signals and differential mode signals. In addition, the hardware structure is not designed to measure the common mode and the differential mode response and characterise the common mode and the differential mode behaviour of the DUT. In addition for balanced devices, balanced calibration standards and a normalised reference impedance (Z_0) are not available.

14.2.2 Characterisation using physical transformers

By using physical transformers it is possible to transform a single-ended signal into a balanced common mode signal and using a BALUN it is possible to generate a balanced differential mode signal. Normally line impedances of 50 Ω against ground are used. In this case a common mode impedance of 25 Ω and a differential mode impedance of 100 Ω are generated (Figure 14.8).

By using a four-port network analyser it is possible to connect one transformer at port 1 and one BALUN at port 3 to feed the balanced input of the balanced DUT with a common mode signal and a differential mode signal. Using both, the differential mode and the common mode response can also be measured. At port 2 of the DUT the second transformer and the second BALUN can be connected to measure the transmitted common mode and differential mode signal (Figure 14.9).

Balanced device characterisation   311

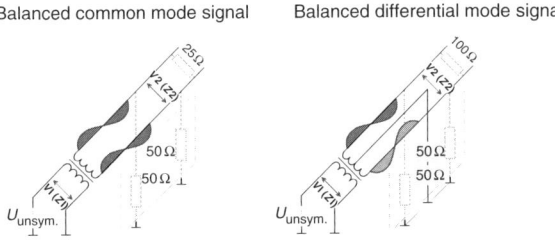

Figure 14.8   Physical transformer and BALUN

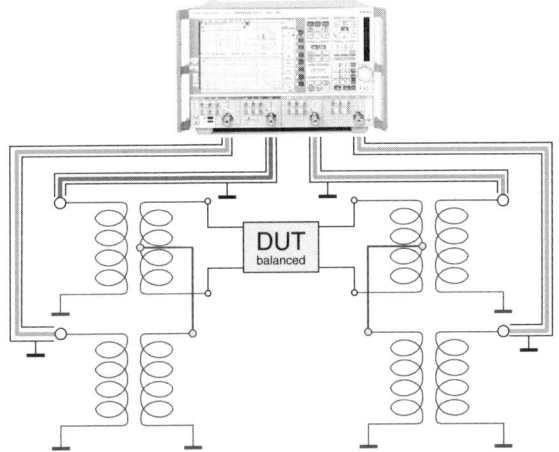

Figure 14.9   BALUN setup

Using such a setup the conversion parameters can be measured as well, because it is possible to generate a differential mode stimulus signal at port 1 and to measure the differential mode and the common mode response at port 1 (reflection characteristics) and at port 2 (transmission characteristics). These measurements can be done bi-directionally.

This simple test setup needs four physical ports and additional external equipment. Caused by this external equipment some disadvantages are known which make it impossible to use this configuration in the range of super high frequencies (SHF range) and partly in the VHF range (very high frequencies).

An important disadvantage is that the calibration plane is different from the measurement plane because of the unavailability of balanced calibration standards. A calibration can be performed at the single-ended ports on the basis of coaxial calibration tools. The measurement results after such a calibration are results of the DUT including the characteristics of the used transformers and BALUNs. Especially, the poor RF performance of a normal BALUN degrades the measurement accuracy. In addition the RF performance is responsible for the limited frequency range.

These problems can be compensated by using the so-called Virtual Ideal Transformers.

312 *Microwave measurements*

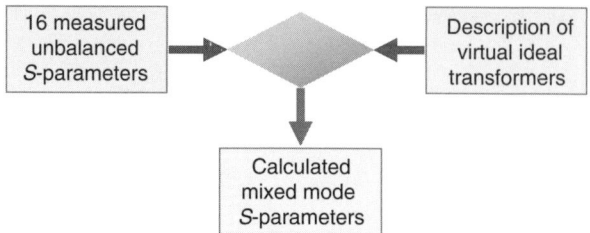

Figure 14.10   *Modal decomposition method, principle*

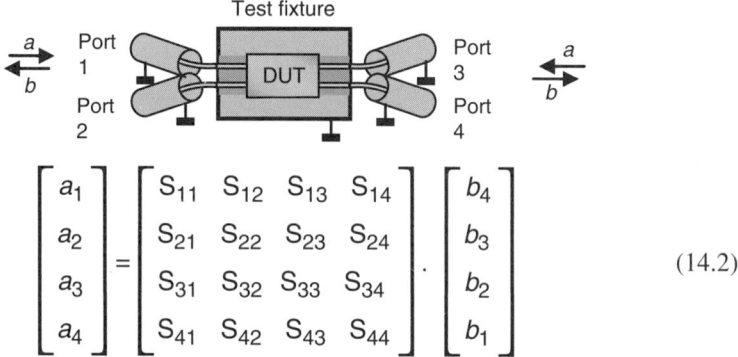

$$\begin{bmatrix} a_1 \\ a_2 \\ a_3 \\ a_4 \end{bmatrix} = \begin{bmatrix} S_{11} & S_{12} & S_{13} & S_{14} \\ S_{21} & S_{22} & S_{23} & S_{24} \\ S_{31} & S_{32} & S_{33} & S_{34} \\ S_{41} & S_{42} & S_{43} & S_{44} \end{bmatrix} \cdot \begin{bmatrix} b_4 \\ b_3 \\ b_2 \\ b_1 \end{bmatrix} \qquad (14.2)$$

Figure 14.11   *Unbalanced measurement*

14.2.3 *Modal decomposition method*

The principle is to measure 16 unbalanced S-parameters and to calculate with the help of (virtual) ideal transformers, the Mixed Mode S-parameters. The whole theory and procedure are given by Bockelman and Eisenstadt [1] (Figure 14.10).

First, measure the 16 unbalanced S-parameters of the balanced two-port model using a four-port analyser (Figure 14.11).

As S-parameters can be converted into all other parameters it is possible to convert the S-parameters into Z-parameters which connect – according to Ohm's law – voltages with currents (Figure 14.12).

$$[V] = [Z] \cdot [I] \qquad (14.3)$$

The next step is to express the unbalanced measured currents and voltages by balanced currents and voltages. This connection can be shown for two coupled lines (14.5) using the Kirschoff laws. The principle is shown only for the balanced port 1. For the balanced port 2 it can be demonstrated in the same way.

The current at the physical port 1 can be expressed by the sum of the differential mode current at the balanced port 1 and half of the common mode current at the balanced port 1. Respectively the current at the physical port 4 can be expressed

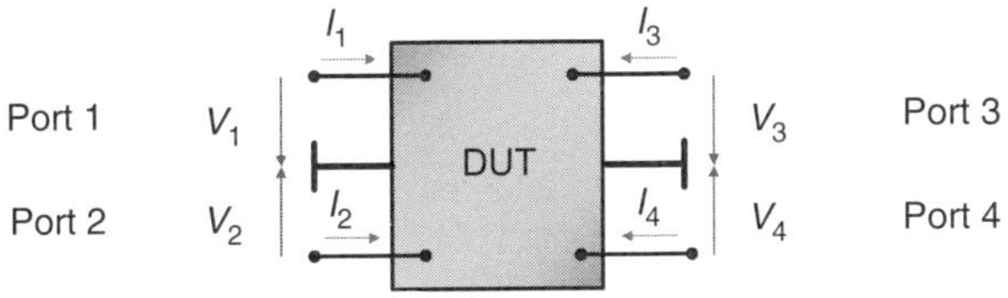

$$\begin{bmatrix} V_1 \\ V_2 \\ V_3 \\ V_4 \end{bmatrix} = \begin{bmatrix} Z_{11} & Z_{12} & Z_{13} & Z_{14} \\ Z_{21} & Z_{22} & Z_{23} & Z_{24} \\ Z_{31} & Z_{32} & Z_{33} & Z_{34} \\ Z_{41} & Z_{42} & Z_{43} & Z_{44} \end{bmatrix} \cdot \begin{bmatrix} I_1 \\ I_2 \\ I_3 \\ I_4 \end{bmatrix} \quad (14.4)$$

Figure 14.12   Conversion S-parameters → Z-parameters

by the sum of the negative differential mode current and half of the common mode current (14.5).

Port 1:

$$\begin{aligned} I_1 &= I_{\text{diff}.1} + \frac{1}{2} I_{\text{com}.1} \\ I_4 &= I_{\text{diff}.1} + \frac{1}{2} I_{\text{com}.1} \end{aligned} \quad (14.5)$$

$$\begin{aligned} I_{\text{diff}.1} &= \frac{1}{2}(I_1 - I_4) \\ I_{\text{com}.1} &= I_1 + I_4 \end{aligned} \quad (14.6)$$

Using this connection the differential mode and the common mode currents at the balanced (or logical) port 1 and the balanced (or logical) port 2 can be expressed by the measured (unbalanced) currents (Figure 14.13).

The voltage at the physical port 1 can be expressed by the sum of the differential mode voltage at the balanced port 1 and the voltage U_4 at the unbalanced port 4. Respectively, twice the common mode voltage at the balanced port 1 can be expressed by the sum of U_1 and U_4 (Figure 14.14).

Port 1:

$$\begin{aligned} U_1 &= U_{\text{diff}.1} + U_4 \\ 2.U_{\text{com}.1} &= U_1 + U_4 \end{aligned} \quad (14.7)$$

$$\begin{aligned} U_{\text{diff}.1} &= U_1 - U_4 \\ U_{\text{com}.1} &= \frac{1}{2}(U_1 + U_4) \end{aligned} \quad (14.8)$$

314 *Microwave measurements*

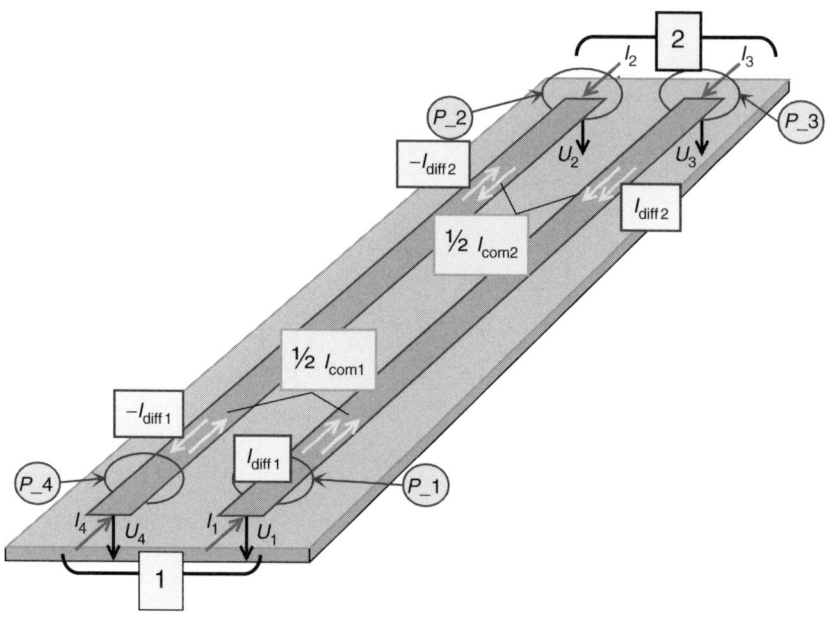

Figure 14.13   Nodal → modal, currents at port 1

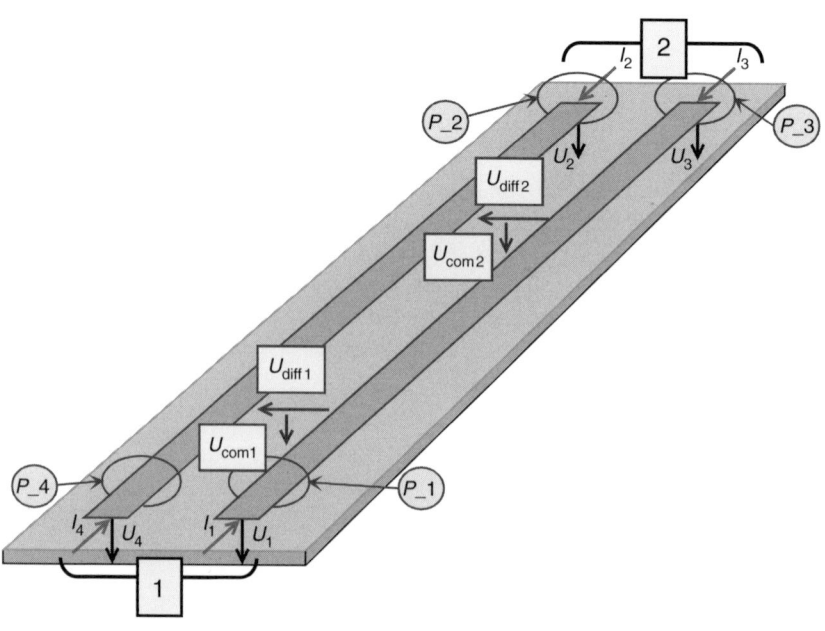

Figure 14.14   Nodal → modal, voltages at port 1

To give a total description according to the (14.5) and (14.7) the following matrixes can be used:

$$\begin{pmatrix} I_1 \\ I_2 \\ I_3 \\ I_4 \end{pmatrix} = \begin{pmatrix} 1 & \frac{1}{2} & 0 & 0 \\ 0 & 0 & -1 & \frac{1}{2} \\ 0 & 0 & 1 & \frac{1}{2} \\ -1 & \frac{1}{2} & 0 & 0 \end{pmatrix} \cdot \begin{pmatrix} I_{\text{diff}.1} \\ I_{\text{com}.1} \\ I_{\text{diff}.2} \\ I_{\text{com}.2} \end{pmatrix} \tag{14.9}$$

$$I = Q \cdot I_m \tag{14.10}$$

$$\begin{pmatrix} U_1 \\ U_2 \\ U_3 \\ U_4 \end{pmatrix} = \begin{pmatrix} \frac{1}{2} & 1 & 0 & 0 \\ 0 & 0 & -\frac{1}{2} & 1 \\ 0 & 0 & \frac{1}{2} & 1 \\ -\frac{1}{2} & 1 & 0 & 0 \end{pmatrix} \cdot \begin{pmatrix} U_{\text{diff}.1} \\ U_{\text{com}.1} \\ U_{\text{diff}.2} \\ U_{\text{com}.2} \end{pmatrix} \tag{14.11}$$

$$U = P \cdot U_m \tag{14.12}$$

At this point the calculation can be done using the following procedure. As the mixed mode voltages and currents are directly connected to the measured unbalanced voltages and currents it is possible to show the measured quantities by the mixed mode quantities virtually linked to 'ideal transformers' (Q, P-matrix).

The quotient U_m/I_m is equal to the mixed mode impedances Z_m. The last step is to convert the mixed mode impedance matrix into the mixed mode S-parameters matrix.

$$V = [P] \cdot V_m$$
$$I = [Q] \cdot I_m$$
$$V = [Z] \cdot I$$
$$V = [Z] \cdot [Q] \cdot I_m$$
$$[P] \cdot V_m = [Z] \cdot [Q] \cdot I_m$$
$$V_m = [P^{-1}] \cdot [Z] \cdot [Q] \cdot I_m$$
$$Z_m = [P^{-1}] \cdot [Z] \cdot [Q] \rightarrow S_m$$

Calculation 1: calculation of S_m using Z-parameter.

Another possibility to explain the calculation of the mixed mode S-parameters is to proceed directly using the wave quantities. Then the differential/common currents and voltages must be expressed according to (14.6) and (14.8)

$$\begin{pmatrix} I_{\text{diff}.1} \\ I_{\text{diff}.2} \\ I_{\text{com}.1} \\ I_{\text{com}.2} \end{pmatrix} = \begin{pmatrix} \frac{1}{2} & 0 & 0 & -\frac{1}{2} \\ 0 & -\frac{1}{2} & \frac{1}{2} & 0 \\ 1 & 0 & 0 & 1 \\ 0 & 1 & 1 & 0 \end{pmatrix} \cdot \begin{pmatrix} I_1 \\ I_2 \\ I_3 \\ I_4 \end{pmatrix} \quad (14.13)$$

$$I_m = M_i \cdot I \quad (14.14)$$

$$\begin{pmatrix} U_{\text{diff}.1} \\ U_{\text{diff}.2} \\ U_{\text{com}.1} \\ U_{\text{com}.2} \end{pmatrix} = \begin{pmatrix} 1 & 0 & 0 & -1 \\ 0 & -1 & 1 & 0 \\ \frac{1}{2} & 0 & 0 & \frac{1}{2} \\ 0 & \frac{1}{2} & \frac{1}{2} & 0 \end{pmatrix} \cdot \begin{pmatrix} U_1 \\ U_2 \\ U_3 \\ U_4 \end{pmatrix} \quad (14.15)$$

$$U_m = M_u \cdot U \quad (14.16)$$

Wave quantities are normalised to the root square of the impedance. By unbalanced system the reference impedance is normally 50 Ω.

$$i_i = I_i \sqrt{Z_0}; \quad u_i = \frac{U_i}{\sqrt{Z_0}} \quad (14.17)$$

Because of the differential mode impedance normally being twice the reference impedance and the common mode impedance normally being half of the reference impedance according to Bockelman and Eisenstadt [1] the normalisation shown in the following equation is common (Figure 14.15).

Form 2 (Bockelmann a.o.)

$$\begin{aligned} i_{\text{diff}.i} &= I_{\text{diff}.i} \cdot \sqrt{2Z_0}; \quad u_{\text{diff}.i} = \frac{U_{\text{diff}.i}}{\sqrt{2Z_0}} \\ &\rightarrow Z_{\text{diff}} = 2Z_0 \\ i_{\text{com}.i} &= I_{\text{com}.i} \cdot \sqrt{\frac{Z_0}{2}}; \quad u_{\text{com}.i} = \frac{U_{\text{com}.i}}{\sqrt{Z_0/2}} \\ &\rightarrow Z_{\text{com}} = \frac{Z_0}{2} \end{aligned} \quad (14.18)$$

The calculation of the mixed mode S-parameters is based on the ratio between the measured wave quantities. Using a normalisation according to (14.17), the measured

Balanced device characterisation 317

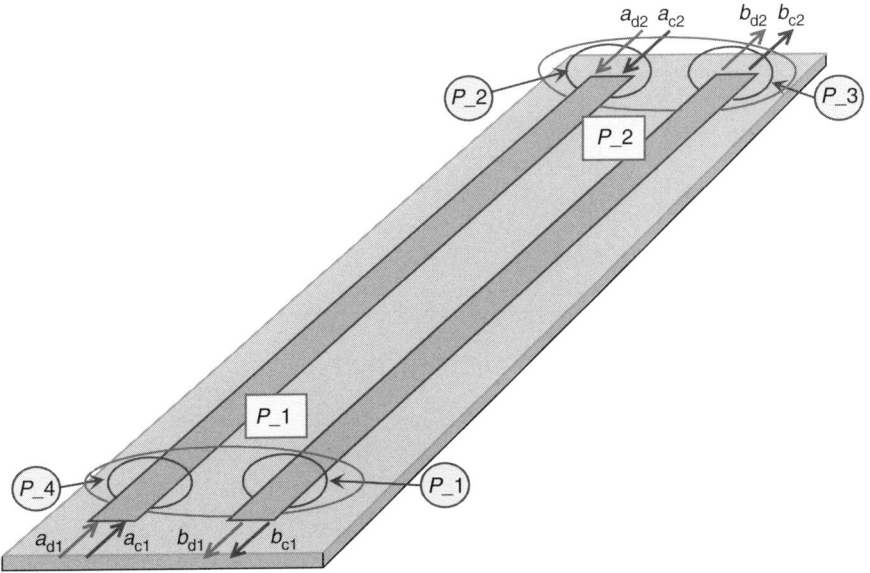

Figure 14.15 Balanced two-port description

mixed mode response can be shown by mixed mode S-parameters multiplied with a mixed mode stimulus signal:

$$b_m = S_m \cdot a_m$$
$$u_i = a_i + b_i; \quad i_i = a_i - b_i$$
$$b_i = \frac{1}{2}(u_i - i_i)$$
$$a_i = \frac{1}{2}(u_i + i_i)$$

Using the M_i and the M_u matrix, it is possible to calculate the mixed mode wave quantities from the single-ended wave quantities and the measured (unbalanced) S-parameters. The ratio between the mixed mode stimulus and response wave quantities describes the mixed mode S-parameters.

Using the normalisation according to (14.18) it can be shown that

$$M_u = M_i$$
$$b_m = \frac{1}{2}(u_m - i_m) = \frac{1}{2}M(u - i) = \frac{1}{2}M \cdot b = \frac{1}{2}M \cdot S \cdot a$$
$$a_m = \frac{1}{2}(u_m + i_m) = \frac{1}{2}M(u + i) = \frac{1}{2}M \cdot a$$
$$M \cdot S \cdot a = S_m \cdot M \cdot a$$

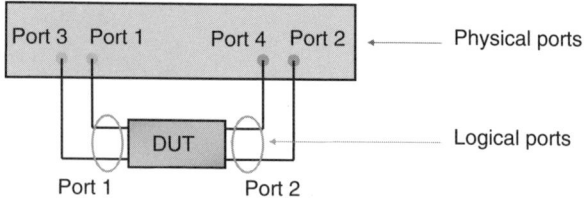

Figure 14.16   Port configuration

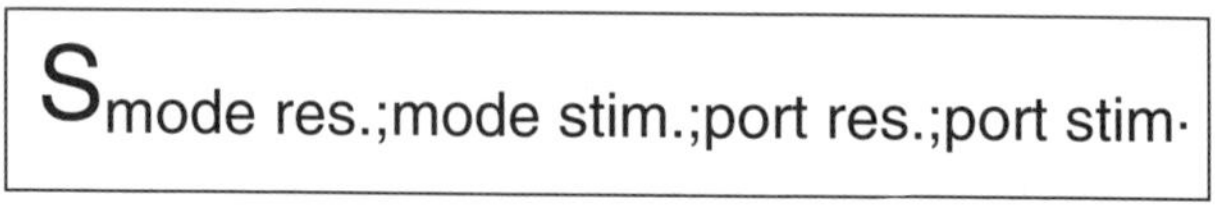

$$S_{\text{mode res.;mode stim.;port res.;port stim.}}$$

Figure 14.17   Naming convention

$$S_m = M \cdot S \cdot M^{-1}$$
$$S = M^{-1} \cdot S_m \cdot M$$

Calculation 2: calculation using form 2.

Important for all calculations is that the port numbering must be known because two single-ended physical ports are combined to a logical (balanced) port 1 and the other two single-ended physical ports are combined to a logical (balanced) port 2 (Figure 14.16).

Especially, the calculations which are based on form 2 are referred to differential mode impedance which is twice the single-ended impedance and a common mode impedance which is half of the single-ended impedance.

14.2.4   Mixed-mode-S-parameter-matrix

The resulting S-parameters matrix, called mixed mode S-parameters matrix, contains S-parameters describing the reflection and the transmission characteristics of a DUT using differential and common mode stimulus signals and measuring differential and common mode responses.

The naming convention is related to the naming convention of the S-parameters. The first letter shows the TEM mode of the measured signal (response) and the second letter the TEM mode of the stimulus signal (Figure 14.17).

The first number names the logical port where the response is measured and the second number names the logical port where the generator is working.

The mixed mode S-parameters matrix describes in the upper left quadrant (or 'DD-quadrant') the fundamental performance of the DUT in pure differential mode operation and the lower right quadrant (or 'CC-quadrant') in pure common mode operation. Each of the four S-parameters represents reflection coefficients at logical port 1 and port 2 and the forward/reverse transmission coefficients between the logical ports (Figure 14.18).

$$\begin{bmatrix} S_{dd11} & S_{dd12} & S_{dc11} & S_{dc12} \\ S_{dd21} & S_{dd22} & S_{dc21} & S_{dc22} \\ S_{cd11} & S_{cd12} & S_{cc11} & S_{cc12} \\ S_{cd21} & S_{cd22} & S_{cc21} & S_{cc22} \end{bmatrix}$$

Figure 14.18 Pure differential and pure common modes

$$\begin{bmatrix} S_{dd11} & S_{dd12} & S_{dc11} & S_{dc12} \\ S_{dd21} & S_{dd22} & S_{dc21} & S_{dc22} \\ S_{cd11} & S_{cd12} & S_{cc11} & S_{cc12} \\ S_{cd21} & S_{cd22} & S_{cc21} & S_{cc22} \end{bmatrix}$$

Figure 14.19 Mode conversion parameter

The upper right and the lower left quadrant provide the information about the conversion parameter. In the upper right quadrant (or 'DC-quadrant') the conversion of a common mode stimulus signal to a differential mode response is described. A conversion of a differential mode stimulus signal to a common mode response is described in the lower left quadrant (or CD-quadrant). It is obvious that these reflection and transmission parameters are equal to zero in the case of ideal symmetry (Figure 14.19).

As the DC-quadrant shows the part of differential mode energy generated from a common mode stimulus signal, this quadrant describes directly the susceptibility to an EMI.

However, the CD-quadrant describes the amount of produced common mode energy from a differential mode stimulus signal. Therefore the information of this quadrant is related to the generation of EMI.

14.2.5 Characterisation of single-ended to balanced devices

Such a typical three-port device is a surface acoustic wave (SAW)-filter. The mixed mode S-parameters of a three-port device can also be calculated from the unbalanced measured S-parameters using the same theory (Figure 14.20).

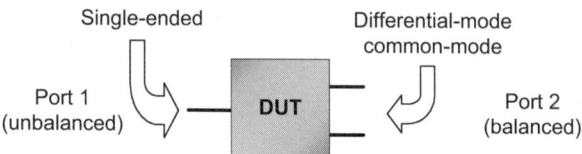

Figure 14.20   Three-port device

$$\begin{bmatrix} S_{ss11} & S_{sd12} & S_{sc12} \\ S_{ds21} & S_{dd22} & S_{dc22} \\ S_{cs21} & S_{cd22} & S_{cc22} \end{bmatrix}$$

Figure 14.21   Nine-parameter mixed mode matrix

The difference to a fully balanced DUT is the single-ended input and the balanced output. A structure containing a balanced input and a single-ended output is also possible.

Here the resulting mixed mode S-parameters matrix is a nine S-parameters matrix. At the single-ended port only the normal reflection coefficient can be measured (labelled with S_{ss}). The mode conversion parameters in the transmission paths describe the mode conversion from the single-ended signal into a common mode signal and into a differential mode signal in the forward and in the reverse direction. At the logical port 2 the reflection coefficients in the pure common mode and the pure differential mode operation can be measured, as well as the mode conversion parameters of the reflected energy at port 2 (Figure 14.21).

14.2.6   Typical measurements

The measurement parameters directly give some information about the differential and common mode insertion loss and the differential and common mode return loss. By using external hardware or network analysers providing more than four ports it is also possible to make Near End Crosstalk (NEXT) and Far End Crosstalk (FEXT) measurements.

Other very popular quality parameters are the amplitude imbalance and the phase imbalance because this information is directly related to the symmetry of the structure and therefore to the mode conversion characteristic of the DUT.

The imbalance parameters can be calculated using (14.19). Using the unbalanced MAG-information it is possible to show the amplitude imbalance and using

Balanced device characterisation 321

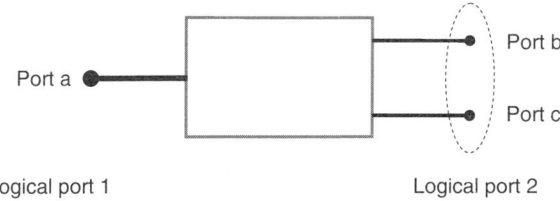

Figure 14.22 Imbalance and CMRR measurement

the unbalanced phase measurements it is possible to show the phase imbalance (Figure 14.22).

$$\text{IMB} = \frac{S_{ba}}{S_{ca}} \qquad (14.19)$$

The common mode rejection ratio (CMRR) is the relation between the S_{ds21} parameter and the S_{cs21} parameter and provides information about the rejected common mode energy.

$$\text{CMRR} = \frac{S_{ds21}}{S_{cs21}} \qquad (14.20)$$

Modern network analysers with powerful Math-functions are able to calculate these results immediately and show the results in an additional trace.

14.3 Measurement examples

14.3.1 Example 1: Differential through connection

This simple example shows the influence of the symmetry on the conversion parameters, and thus on the pure differential and pure common mode parameters.

Two coaxial lines connected at physical port 1 and physical port 2 of the analyser are combined to a logical port 1. The other two coaxial lines are connected at physical port 2 and physical port 4 of the analyser and combined to a logical port 2. The two logical ports are directly connected (through connection). To avoid measurement errors caused by different lengths or different attenuations of the coaxial lines, a coaxial full four-port calibration is recommended (Figure 14.23).

The diagrams in Figure 14.24 show all the 16 mixed mode S-parameters. The traces 1, 2, 5 and 6 show the results under pure differential mode operation. We see that the DUT is well matched at both ports and that the differential mode energy does transmit the DUT with negligible losses.

The traces 3, 4, 7, 8, 9, 10, 13 and 14 display the conversion parameters. It is evident that the mode conversion from differential mode energy to common mode energy and the conversion from common mode energy to differential mode energy are very low.

322 *Microwave measurements*

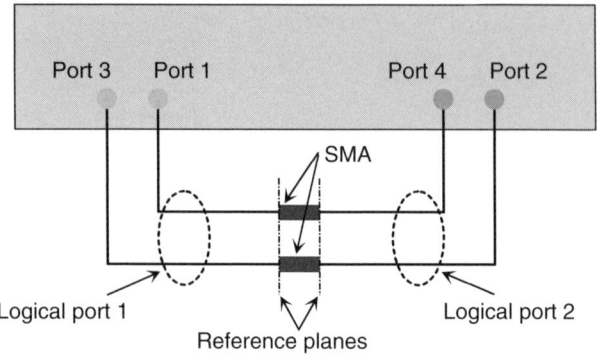

Figure 14.23   Test setup, example 1

The pure common mode behaviour of the DUT is shown in the traces 11, 12, 15 and 16. To show the influence of the symmetry on the structure of a balanced device, we will change the symmetry in two different ways.

(1) Change of the electrical length of one part (line) of the balanced device by implementing an additional small piece of line between ports 2 and 3.
(2) Creation of two different attenuations, by implementing a 3 dB attenuator between ports 2 and 3 and by implementing a 6 dB attenuator between ports 1 and 4. It is essential to use two different attenuators because, otherwise, the electrical length will be changed significantly.

First, change of the electrical length between the ports (Figure 14.25).

As we are using a very small piece of line, the influence in the lower frequency range is lower to the mode conversion than in the higher frequency range. This happens due to the relation between the dimension of the additional line and the wave length (Figure 14.26).

As expected, the mode conversion, especially in the higher frequency range, becomes higher. This is shown in the lower two diagrams. For comparison with the ideal results, these are traced. Another point of interest is that the converted energy (differential to common mode) will no longer be transmitted as differential mode energy. In the higher frequency range the transmitted differential mode energy seems to be attenuated.

The next exercise is to use different attenuations in the two parts of the balanced device (Figure 14.27).

Because of the attenuator being a broadband working device, the influence of the symmetry change in this way will be the same during the whole frequency range.

Compared with the ideal (traced) results the mode conversion becomes higher during the whole frequency range. The transmitted differential mode energy is once again attenuated, because a part of the differential mode energy is converted into common mode energy. In other words, an EMI is produced from the differential mode energy at the input (Figure 14.28).

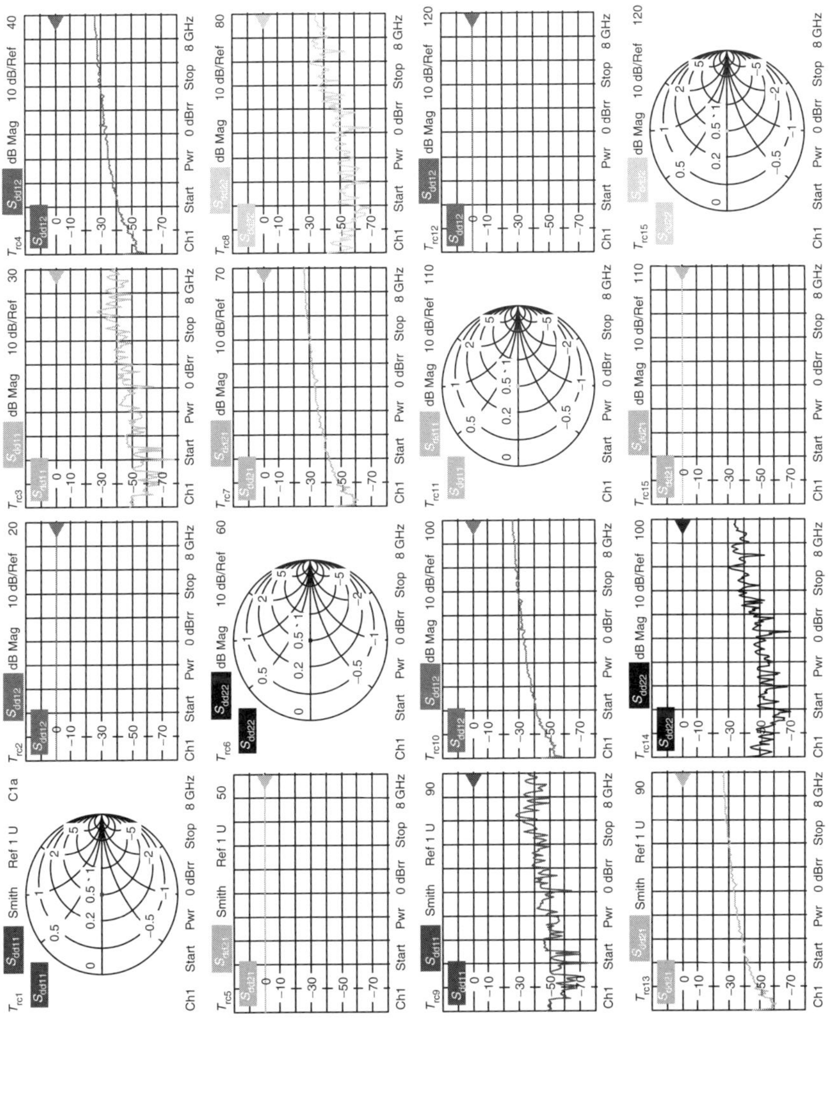

Figure 14.24 Measurement of mixed mode S-parameters acc. to example 1

324 *Microwave measurements*

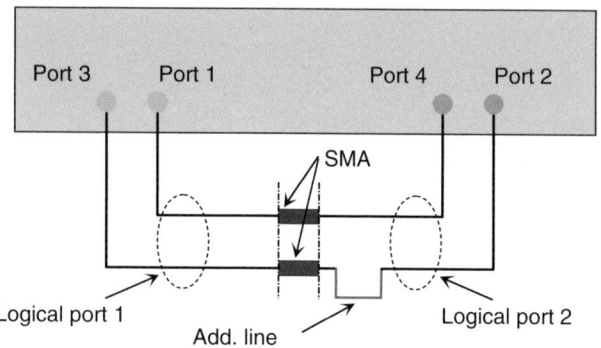

Figure 14.25 Change of electrical length

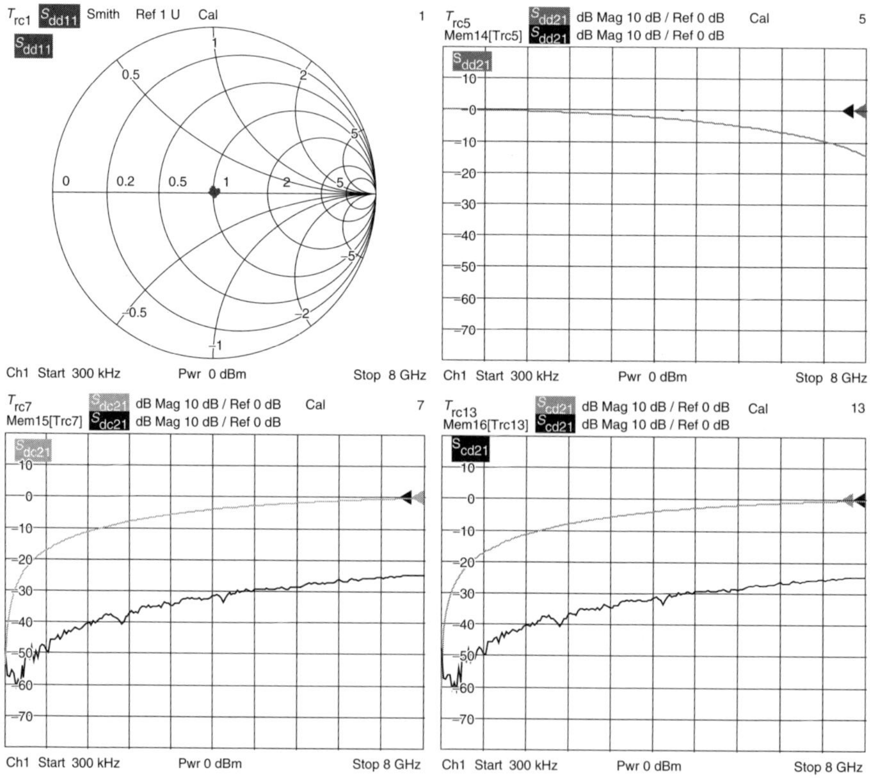

Figure 14.26 Mode conversion with additional line

Balanced device characterisation 325

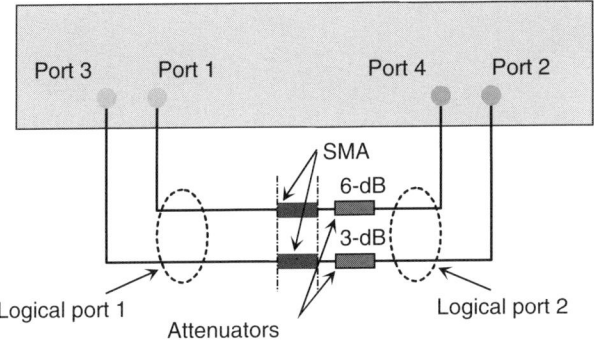

Figure 14.27 Change of the attenuation

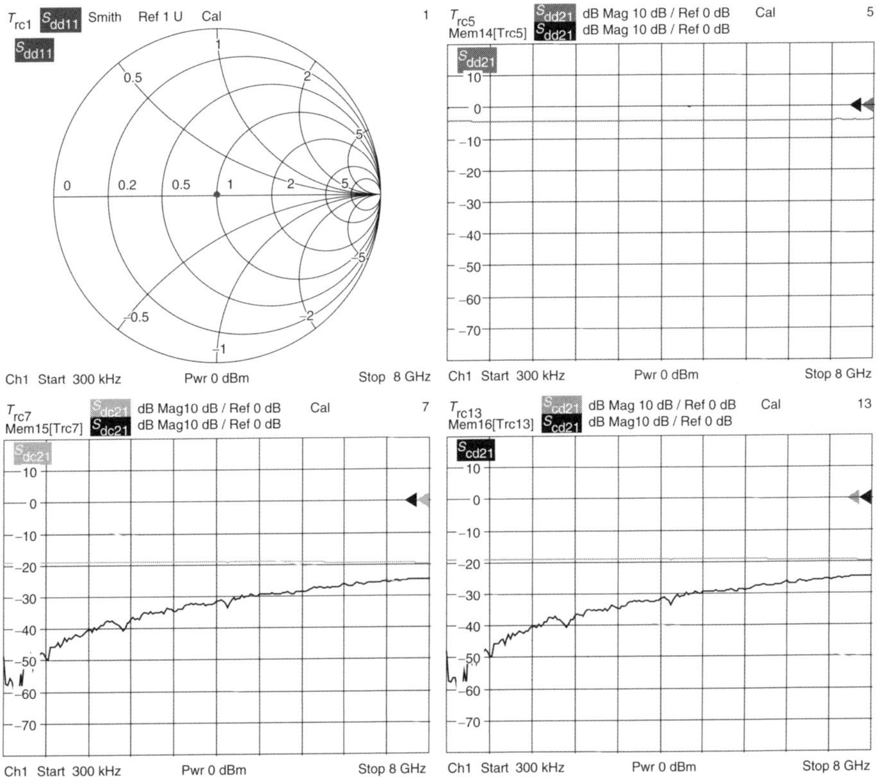

Figure 14.28 Mode conversion by different attenuation

326 *Microwave measurements*

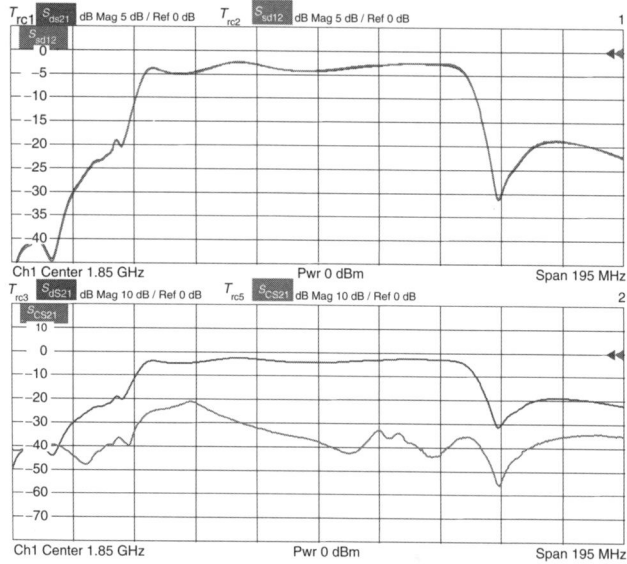

Figure 14.29   *Transmission behaviour of the SAW-filter*

14.3.2 Example 2: SAW-filter measurement

A SAW-filter is a typical three-port device with one single-ended input and a balanced output.

When measuring a SAW-filter it is of importance that most of the single-ended input energy is converted into differential mode energy. It is not intended to receive common mode energy. This type of energy must be rejected because otherwise an interferer is produced from the single-ended signal (Figure 14.29).

To calculate the CMRR we can use (14.20) and using the User Defined Math Editor the result can be shown directly (Figure 14.30).

14.4 (De)Embedding for balanced device characterisation

By working with single-ended 50 Ω systems, the differential mode impedance is 100 Ω and the common mode impedance is 25 Ω. These impedance values are not the normalised reference values. A SAW-filter, for example, provides differential output impedances different from 100 Ω (e.g. 150–240 Ω or other), but the calculation routine works with 25 and 100 Ω.

By working with such a real device a mismatching happens.

This mismatching can be reduced by using physical matching networks. Disadvantages of physical matching networks are the poor reproducibility, their being normally restricted to lower frequencies and the possibility to only use them in the narrow band. Another disadvantage is that using physical networks the user cannot operate as flexibly as possible.

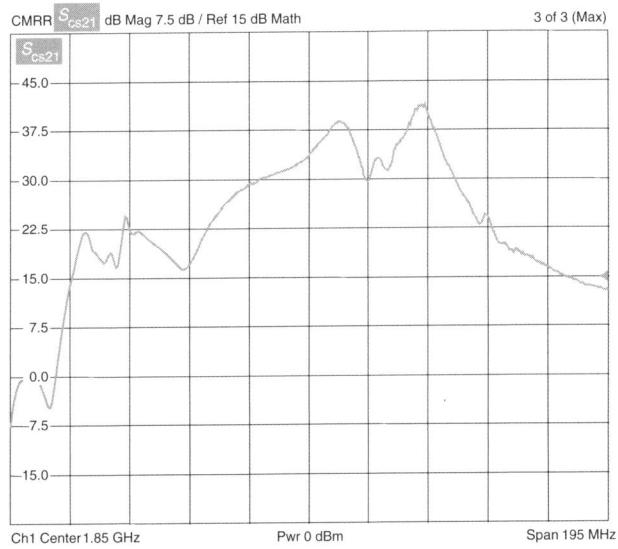

Figure 14.30   CMRR of a SAW-filter

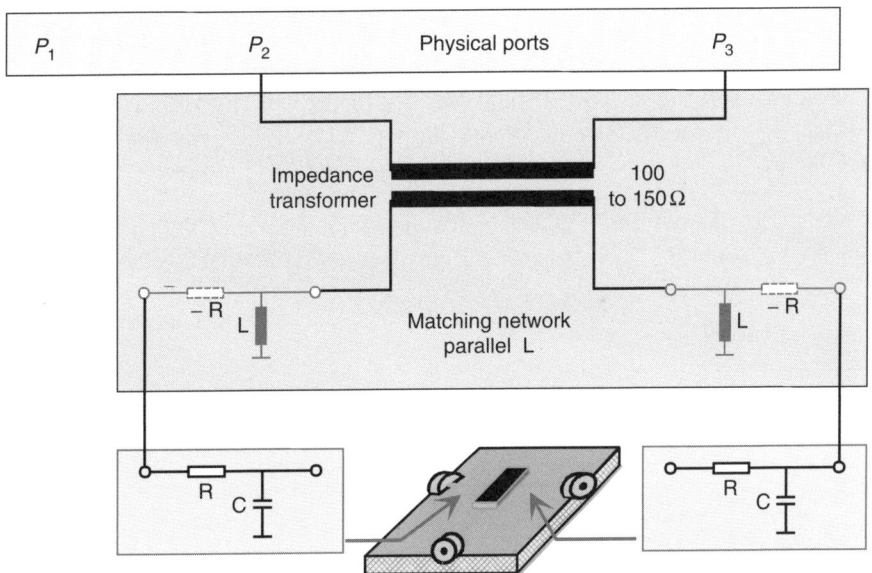

Figure 14.31   Virtual matching

The use of virtual (theoretically) matching networks provides a high range of flexibility with no frequency restriction. Using these virtual networks both embedding and de-embedding are possible.

To provide the impedance transformation from 100 to 150 Ω a virtual transformer must be embedded. In general, 'Embedding' is used to implement virtual additional components and circuits and to show the S-parameters with the influence of these virtual networks.

The de-embedding functionality can be used to remove virtually an influence caused by the hardware, for example, the characteristics of a test fixture when it is possible to give a complete S-parameters description of the fixture.

When, for example, a DUT with 150 Ω output impedance is placed in a test fixture with an RC-characteristic the structure can be matched using a virtual network shown in Figure 14.31.

Virtual embedding and de-embedding are possible for single-ended and for balanced structures at all used physical and logical ports. It is possible to use predefined structures and vary the parameters of the given lumped elements or to import the S-parameters to describe the networks.

Further Reading

1 Bockelman, D. E., and Eisenstadt, W. R.: 'Combined differential and common mode scattering parameters: theory and simulation', *IEEE Transactions on Microwave Theory and Techniques*, 1995;**43**(70): 1530–9
2 Simon, J.: *Measuring balanced components with vector network analyzer ZVB, Rohde and Schwarz Application Note 1EZ53*, September 2004
3 Heuermann H.: 7.10.2003, Grundlagen der Hoch- und Höchstfrequenztechnik, *Umdruckversion 1.1, Fachhochschule Aachen* (script from studies at university)
4 *Concepts in Balanced Device Measurements, Multiport and Balanced Device Measurement Application Note AN1373-2*, Agilent Technologies
5 Martius S.: January 2002, Nodale und Modale Streumatrizen (Dreileiteranordnungen), Lehrstuhl für Höchstfrequenztechnik Universität Erlangen-Nürnberg (script from studies at university)

Chapter 15

RF power measurement

James Miall

15.1 Introduction

This is a brief introduction to guided-wave power measurements in the approximate range of a few MHz to several hundreds of GHz, some devices that can be used to measure RF power and the techniques for calibrating these devices.

15.2 Theory

15.2.1 Basic theory

The instantaneous incident power due to an electromagnetic field can be written as [1]

$$P = \frac{1}{2} \oint \vec{E}_t \times \vec{H}_t \, dS$$

where $\vec{E}_t$ and $\vec{H}_t$ are the electric and magnetic fields at a time t, and S is the surface over which the power is being measured. In terms of voltage (V) and current (I) in a transmission line, power can be written as

$$P_{\text{instantaneous}} = V(t) \times I(t)$$
$$P_{\text{average}} = V_{\text{RMS}} \times I_{\text{RMS}} \times \cos(\phi)$$

where ϕ is the phase angle between the voltage and current waveforms. In many situations $V(t)$ and $I(t)$ are sinusoids and in this case the instantaneous power will vary at twice the frequency of the sinusoid. However, these are not particularly useful definitions at RF and microwave frequencies because the instantaneous voltage, current and field distributions are not easily measured. At RFs and above power becomes the only convenient measure of signal strength.

In practice RF and microwave power is usually measured using substitution techniques based on its heating effect, or by rectification.

The unit of power is Watt (W), where

$$1\,\text{W} = 1\,\text{kg}\,\text{m}^2\,\text{s}^{-3}$$

Power ratios are often more conveniently expressed in decibels where given

$$\text{Power}_{\text{bel}} = \log_{10}(\text{Power Ratio})$$

and

$$1\,\text{decibel} = \frac{1}{10}\,\text{bel}$$

the power ratio in decibels is therefore

$$\text{Power}_{\text{dB}} = 10 \times \log_{10}\left(\frac{\text{Power}_1}{\text{Power}_2}\right)$$

A power in dBm is defined as the ratio with respect to 1 mW, that is

$$\text{Power}_{\text{dBm}} = \log_{10}\left(\frac{\text{Power}}{1\,\text{mW}}\right)$$

That is, 0 dBm corresponds to 1 mW, 20 dBm to 100 mW and −50 dBm to 10 nW.

Often 'power' refers to CW power, that is, the average power produced by a constant sinusoid waveform at a single frequency. It should be assumed that these notes are dealing with this case unless otherwise specified. Discussion of non-CW power occurs in section 15.9.

In general, in power measurements, a source of power will be connected via a transmission line to a load (Figure 15.1). Both the source and the load will reflect some of the incident electromagnetic field and therefore the power delivered into the load will be dependent on the reflection coefficients of both devices.

If we define, at the connection to the load, the forward voltage wave (a), reflected voltage wave (b), reflection coefficient of the load Γ, incident power (P_i), reflected power (P_r) and the power delivered to the load (P_d) then we can write the

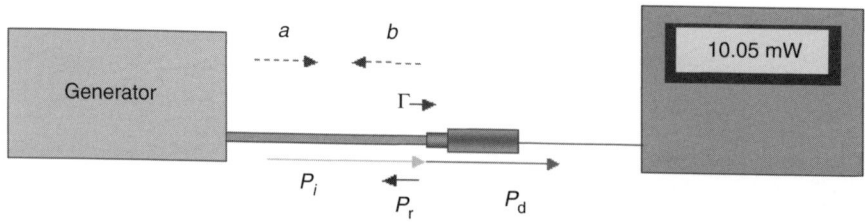

Figure 15.1 Incident, reflected and delivered power from source to load in a general power measurement situation

following relations:

$$P_i = \frac{|a|^2}{Z_0}$$

$$P_r = \frac{|b|^2}{Z_0}$$

$$|\Gamma| = \frac{|b|}{|a|}$$

$$|\Gamma|^2 = \frac{|b|^2}{|a|^2}$$

$$P_r = P_i|\Gamma|^2$$

$$P_d = P_i - P_r$$

$$P_d = P_i - P_i|\Gamma|^2$$

$$P_d = P_i(1 - |\Gamma|^2)$$

The power delivered to the load is never greater than the incident power.

Until now this treatment has ignored the power source. Any generator can be thought of as an idealised source with available power P_a and an internal impedance Z_0 (cf. the DC case) (Figure 15.2). The power dissipated into a load with impedance Z is P_Z. The Z_0-available power, P_{Z_0}, is the power available to a load with impedance Z_0. The maximum power that can be dissipated in a load occurs when the load has the complex conjugate impedance of the source internal impedance.

$$P_d = P_a \frac{(1 - |\Gamma_G|^2)(1 - |\Gamma_L|^2)}{|1 - \Gamma_G \Gamma_L|^2}$$

$$P_{Z_0} = P_a(1 - |\Gamma_G|^2)$$

$$P_{Z_0} = P_d \frac{|1 - \Gamma_G \Gamma_L|^2}{1 - |\Gamma_L|^2}$$

As expected if $\Gamma_L = 0$ the power delivered to the load is the Z_0-available power, P_{Z_0}. In general we can measure P_i or P_d with a power sensor but often we wish to know P_{Z_0}, the power available to a perfectly matched load, which can be found, for example, by using (15.1). A more comprehensive description of the relationship between powers in different points in microwave circuits can be found in Reference 22.

$$P_{Z_0} = P_i|1 - \Gamma_G \Gamma_L|^2 \tag{15.1}$$

No power sensor is a perfect indicator of the delivered power. There will always be losses within the sensor and systematic errors within the measurement process that will mean that the measured power is not the same as the power delivered (Figure 15.2).

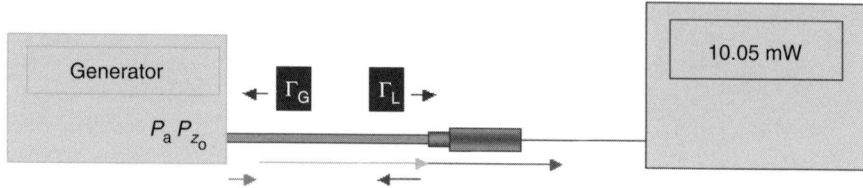

Figure 15.2 Effect of generator match

The Effective Efficiency (η_e) of a power sensor can be defined as the ratio of the measured power to the RF absorbed power

$$\eta_e = \frac{P_{meas}}{P_d}$$

and the calibration factor (K) as the ratio of the measured power to the RF incident power

$$K = \frac{P_{meas}}{P_i} \tag{15.2}$$

so the two definitions are related by

$$K = \eta_e(1 - |\Gamma_L|^2)$$

In a power sensor calibration certificate, K would usually be quoted for each frequency (either relative to absolute power or often relative to the figure at 50 MHz).

15.2.2 Mismatch uncertainty

By combining equations (15.1) and (15.2) we can write

$$P_{Z_o} = \frac{P_{meas}}{K}|1 - \Gamma_G \Gamma_L|^2$$

and if our power meter supplies a reading P_{rdg}, corrected for the sensor calibration factor, as is often the case, then

$$P_{Z_o} = P_{rdg}|1 - \Gamma_G \Gamma_L|^2$$

this has a maximum and minimum given by

$$P_{Z_o} = P_{rdg}(1 \pm |\Gamma_G||\Gamma_L|)^2$$

for small Γ this can be expanded as

$$P_{Z_o} \approx P_{rdg}(1 \pm 2|\Gamma_G||\Gamma_L|)$$

$$P_{Z_o} \approx P_{rdg} \pm P_{rdg} 2|\Gamma_G||\Gamma_L|$$

or in other words the approximate mismatch [3] uncertainty when making a power measurement where only the magnitudes of the source and load reflection coefficients are known is $200|\Gamma_G||\Gamma_L|\%$. The mismatch uncertainty has a U-shaped distribution [4].

15.3 Power sensors

There are a great variety of different techniques for measuring RF and microwave power. All the different sensor types have advantages and disadvantages. Several of the more common sensor types have been covered in earlier chapters and these will only be mentioned briefly along with a short introduction to some more unusual sensor types.

15.3.1 Thermocouples and other thermoelectric sensors

- Coaxial and waveguide sensors easily available on the market:
  - Coaxial: DC to 50 GHz
  - Waveguide: 8–110 GHz (limited supply outside these frequencies)
- Power range: 1 μW to 100 mW (50 dB range)
- Advantages (+) and disadvantages (−):
  + Good long-term stability
  + Reasonably linear
  + Generally lower VRC than thermistor mounts
  + Easily integrated into automatic systems
  − Often require a reference source
  − Only measure average power

15.3.2 Diode sensors

- Coaxial sensors easily available in range: 0.1 MHz to 50 GHz
- Power range: 1 nW to 100 mW (90 dB)
- Advantages (+) and disadvantages (−):
  + Good long-term stability
  + Reasonably linear at low levels
  + Generally lower VRC than thermistor mounts
  + Easily integrated into automatic systems
  + Fast response allowing envelope power to be tracked
  + High dynamic range
  − Require a reference source
  − Poor linearity at higher levels
  − Can be inaccurate for modulated and distorted signals

15.3.3 Thermistors and other bolometers

- Coaxial and Waveguide mounts available on the market:
  - Coaxial: 1 MHz to 18 GHz
  - Waveguide: 2.6–200 GHz
- Operate with DC substitution (closed-loop operation)
- Advantages (+) and Disadvantages (−):
  + Very good long-term stability
  + Fundamentally very linear

- Power range: 10 μW to 10 mW (30 dB range)
- High VRC at high frequencies (especially waveguide)
- Older technology
- Slow response time
- Only measure average power
- Poor dynamic range

Fundamentally, if the requirement is for a fast, high dynamic range sensor then a diode sensor should be used. For slightly higher accuracy over a smaller power range then a thermocouple sensor should be used. Thermistor sensors are generally only used in situations where very high linearity and stability are needed, such as a calibration laboratory.

15.3.4 Calorimeters

Calorimeters measure the heat produced by incident microwave radiation. They are typically constructed from a thermally insulating section of waveguide, a load and a temperature sensor such as a thermopile. They are in most cases the most accurate sensors available and so are used in national standards and some other calibration laboratories. Their main disadvantage is their extremely long time constant (often 20+ minutes) so they are not suitable for use in many measurement situations.

15.3.4.1 Twin load calorimeters

Twin load calorimeters [5] consist of two identical loads at the end of thermally insulating RF line sections within a thermally insulating container (see Figure 15.3). The temperature difference between the two loads is measured with temperature sensors. RF power can be applied to one side of the calorimeter and DC power to the other. When the temperature difference between the two sides is zero the RF and DC powers can be considered equivalent.

15.3.4.2 Microcalorimeters

Microcalorimeters [6] are used to calibrate thermistor type sensors. These sensors operate in a bridge circuit such that the power dissipated in the sensor should be constant whether or not RF power is applied. The microcalorimeter measures the small temperature change caused by the extra losses in the input line of the sensor in the RF case. A microcalorimeter consists of a thin-walled line section connected to the thermistor sensor being calibrated (see Figure 15.4). A thermopile measures the temperature difference between the thermistor and a dummy sensor or temperature reference. By measuring the temperature change due to the RF loss in the sensor and the input line, the efficiency of the sensor can be calculated.

15.3.4.3 Flow calorimeters

Flow calorimeters [7] are suitable for higher power measurements than the other calorimeters mentioned so far. They contain a quartz tube carrying flowing water

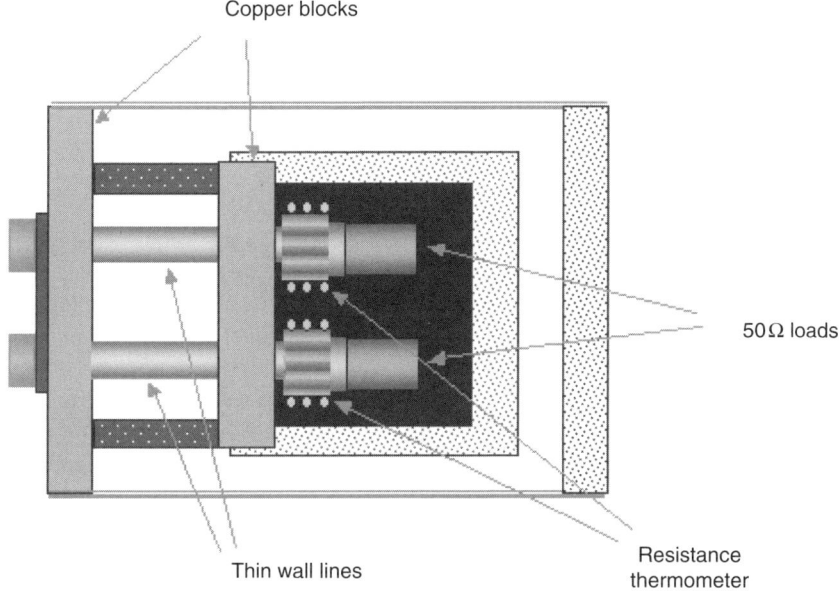

Figure 15.3   A coaxial twin dry-load calorimeter

Figure 15.4   Waveguide calorimeters for WG27 and WG16

which is positioned at an angle across a waveguide. The power into the waveguide can be calculated by measuring the temperature rise of the water and the flow rate. The RF heating is compared to DC heating by means of heating wires within the quartz tube that provide an identical temperature distribution.

15.3.5 Force and field based sensors

There are several unusual types of power sensor such as the torque vane [8] and electron beam sensors [9] that have existed previously but are not found in a commercially available form today. With improved fabrication techniques and component quality some of these designs may form the basis of future generations of power sensors.

In the torque vane sensor a conductor vane is hung in a waveguide. In the presence of electromagnetic fields a torque will be produced on the vane and if this torque can be measured then the power level can be determined. A commercial version of this type of sensor has been produced in the past. The electron beam power sensor operates by measuring the field strength in a cavity of known geometry necessary to just stop a beam of electrons of known energy. From this the RF power can be calculated. Other novel power sensors include atomic fountain based power sensors, which have been the subject of recent research [10,11] and Hall Effect sensors [12] which measure the voltage across the faces of a piece of semiconductor in the presence of an RF magnetic field.

15.3.5.1 MEMS

MEMS is an abbreviation of Micro Electro-Mechanical Systems and typically refers to a moving system fabricated on a silicon wafer with measurements made using electrical methods. MEMS interest has greatly increased over recent years as the cost of wafer production, a spin-off from the semiconductor industry, has decreased. MEMS-based power sensors offer the possibility of easily measuring the very small forces produced by the strength of electromagnetic field associated with a power level in the mW range or lower. Some recent MEMS power sensor designs [13,14] have been based on variations on the theme of capacitively measuring the deflection of a thin bridge caused by power passing along a coplanar waveguide structure beneath it.

15.3.6 Acoustic meter

This type of quasi-optic sensor [15] is designed for use from mm wave up to optical frequencies. Pulse-modulated incident power is absorbed in a thin metal film supported by a mylar substrate within a closed cell. This generates a sound wave within the gas of the cell at the modulation frequency, which is picked up by a microphone. A DC current pulse of opposite phase is then applied until there is no microphone response, at which point the microwave and DC power can be said to be equal.

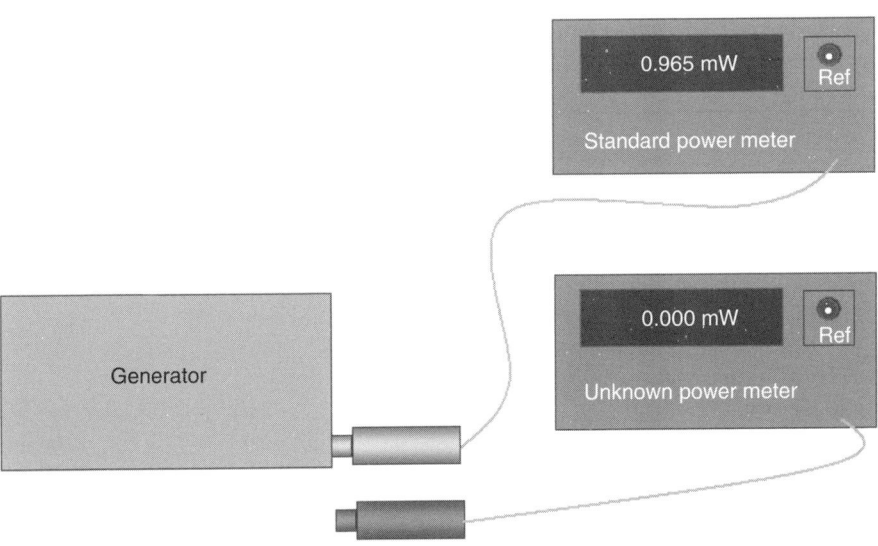

Figure 15.5   A 'simple' power measurement made by exchanging power sensors

15.4   Power measurements and calibration

15.4.1   Direct power measurement

In a direct power measurement such as the one shown in Figure 15.5 the calibration factor of the device under test (CF_{DUT}) in terms of the calibration factor of the standard sensor is given by

$$CF_{DUT} = CF_{std} \frac{|1 - \Gamma_{Src}\Gamma_{Std}|^2}{|1 - \Gamma_{Src}\Gamma_{DUT}|^2} \frac{P_{DUT}}{P_{Std}} \tag{15.3}$$

where P_x is the power measured by device x, Γ_x is the reflection coefficient of device x and Src refers to the generator or source.

15.4.2   Uncertainty budgets

Table 15.1 is an example uncertainty budget for a power measurement made by connecting a calibrated power sensor (such as a thermocouple) on to a badly matched source, where neither reflection coefficient is known. The numbers are for illustration only but are typical of real uncertainties in certain situations. There are several ways to improve this measurement and lower the uncertainties such as:

(1) Measuring the complex voltage reflection coefficient (VRC) of the source and load and performing a full mismatch correction.
(2) Evaluating the connector repeatability by doing several repeat connections using torque spanners.

Table 15.1 Uncertainty budget for basic power measurement without mismatch correction

Source of uncertainty	Divisor	Uncertainty contribution	Standard uncertainty
Calibration factor	2	1.0	0.500
Drift in calibration factor	$\sqrt{3}$	0.2	0.116
Reference source (including mismatch)	2	0.7	0.350
Mismatch	$\sqrt{2}$	1.2	0.857
VRC magnitude of sensor: 0.03			
VRC magnitude of source: 0.20			
Power meter	2	0.2	0.100
Repeatability	1	0.1	0.100
Combined standard uncertainty			1.07
Expanded uncertainty ($k = 2$)			2.14

(3) Referencing the power sensor to a better characterised (with known output port match, for example) 50 MHz reference source than the one on the power meter.

15.5 Calibration and transfer standards

Calibration of a power sensor involves comparing it against another power sensor of known calibration factor. Rather than just connecting the two sensors in turn to a source this is generally done using a transfer standard. Usually this would take the form of a power splitter (or coupler) with a power sensor permanently attached to one arm of the splitter. The use of a transfer standard has many advantages: it allows the ratio of the instantaneous powers to be taken; the transfer standard can be measured against the standard, which may be a slow device, and then the device under test can be measured more rapidly against the transfer standard; the full S-parameters of the coupler (or splitter) do not need to be known; and the repeatability of the transfer standard can be evaluated over time.

15.5.1 Ratio measurements

Many power meters have an internal reference source with an RF connector on the front panel for calibrating or checking the operation of the power sensor. These sources produce a known power level (generally 1 mW) at a single frequency (generally 50 MHz or DC). The sensor should be referenced to this known power level when it is first turned on and periodically after that. Often when calibrating this type of sensor a calibration of the response of the sensor at the calibration frequency compared to the reference frequency is required, such as the following definition:

$$\text{Cal Factor} = \text{Reference Cal Factor} \times \frac{\text{Incident Power at 50 MHz}}{\text{Incident Power at Cal Freq}}$$

When calibrating this type of sensor on a splitter based transfer standard against a calibrated standard sensor the calibration factor of the DUT (CF_{DUT}) is given by

$$CF_{DUT} = CF_{Std} \frac{R_{Std50}}{R_{StdRF}} \frac{R_{DUTRF}}{R_{DUT50}} \frac{M_{StdRF}}{M_{Std50}} \frac{M_{DUT50}}{M_{DUTRF}}$$

where CF_{Std} is the calibration factor of the standard sensor, R_{Std50} is the ratio of the power indicated on the standard sensor to the power indicated on the transfer standard at 50 MHz, M_{DUTRF} is the mismatch factor for the DUT at the RF calibration frequency.

$$M_{DUTRF} = |1 - S\Gamma_{DUT}|^2$$

where S, the equivalent output port match (here at port 2 of the splitter), is given by $S = S_{22} - S_{21}S_{32}/S_{31}$. The other definitions follow the same logic. Note the similarity of this equation to (5.3).

15.6 Power splitters

Couplers and splitters can both be used to make the power ratio measurements necessary to calibrate a power sensor (Figure 15.6) [11].

Two resistor power splitters (not to be confused with three resistor power dividers) are well-matched devices that are extremely useful for coaxial calibrations

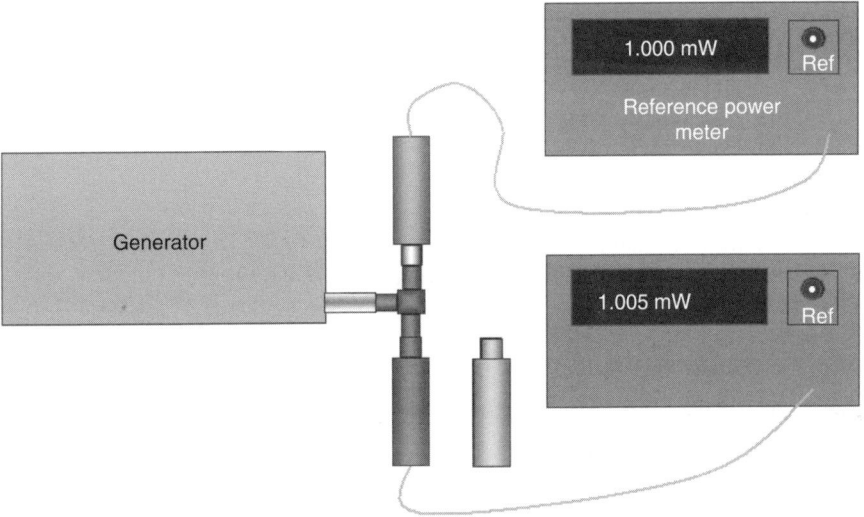

Figure 15.6   Calibration using a power splitter

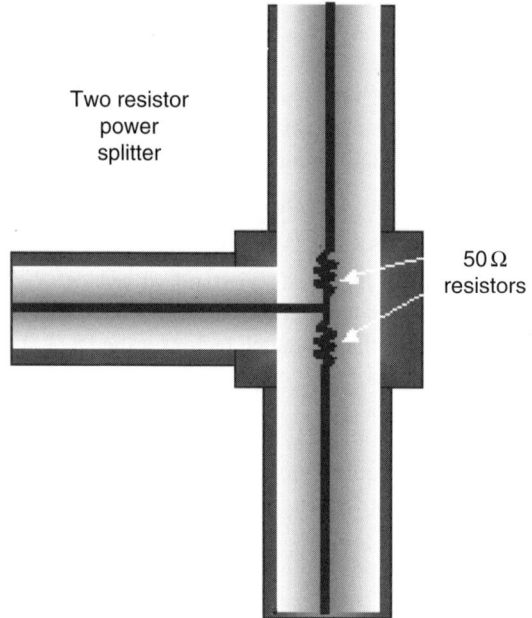

Figure 15.7   A two resistor splitter

(Figure 15.7 and Table 15.2). When used in a leveling loop the voltage at the centre of the T is held constant and a device on the other arm of the splitter sees an ideal source with a 50 Ω characteristic impedance.

15.6.1   Typical power splitter properties

- Wide frequency range of operation
- 7 mm DC to 18 GHz
- 3.5 mm DC to 26.5 or 33 GHz
- 2.4 mm DC to 50 GHz
- Reasonably good output match
- Good long-term stability
- Limited power capability (6 dB loss: Max. input ≈ 0.5 W)

15.6.2   Measurement of splitter output match

Knowledge of the splitter output port source match is necessary in order to perform a full mismatch correction: there are several ways to measure this. The easiest is probably to measure the S-parameters of a reasonably high value attenuator and attach this on to one arm of the splitter. The match of the attenuator and splitter together will be approximately that of the attenuator on its own. This method has the disadvantage of significantly reducing the output power which is often a particular problem at higher

Table 15.2 Example uncertainty budget for measurement of the calibration factor of a power sensor wrt 50 MHz using a power splitter based transfer standard

Source of uncertainty	Divisor	Uncertainty contribution	Standard uncertainty
Calibration factor	2	0.60	0.300
Drift in calibration factor	$\sqrt{3}$	0.20	0.116
Ratio 1	$\sqrt{3}$	0.10	0.058
Ratio 2	$\sqrt{3}$	0.10	0.058
Ratio 3	$\sqrt{3}$	0.10	0.058
Ratio 4	$\sqrt{3}$	0.10	0.058
Drift in transfer standard	$\sqrt{3}$	0.20	0.116
Mismatch Std at 50 MHz	1	0.04	0.029
Mismatch DUT at 50 MHz	1	0.08	0.057
Mismatch Std at 18 GHz	1	0.15	0.150
Mismatch DUT at 18 GHz	1	0.20	0.200
Repeatability of standard	1	0.10	0.100
Repeatability of DUT	1	0.10	0.100
Combined standard uncertainty			0.50
Expanded uncertainty ($k=2$)			1.00

frequencies where broadband, higher power sources are not so readily available. Another method is to measure all the S-parameters of the splitter with a load (or other known impedance) attached to the other ports in turn. The equivalent output port match can then be calculated from these values [16]. A third method, known as the 'direct method' [17,18], involves performing a one-port calibration on one arm of the splitter using a network analyser connected to the other two ports (Figure 15.8).

15.6.3 The direct method of measuring splitter output

If for a setup similar to Figure 15.8, the uncalibrated S-parameter data ($S_{11,\mathrm{raw}}$ and $S_{21,\mathrm{raw}}$) from the ANA is extracted, for any item connected to the splitter port, and the ratio x taken of the two S-parameters

$$x = \frac{S_{11,\mathrm{raw}}}{S_{21,\mathrm{raw}}}$$

then following the procedure below, the splitter output match can be calculated.

Three devices of known VRC such as Short (Γ_{SC}), Open (Γ_{OC}) and Load (Γ_{L}) can be connected to the splitter port in turn. The three ratios for the Load, Open and Short can be defined as being A, B and C, respectively, and the three one-port error terms defined as Directivity E_{DF}, Source Match E_{SF} and Reflection Tracking E_{RF}.

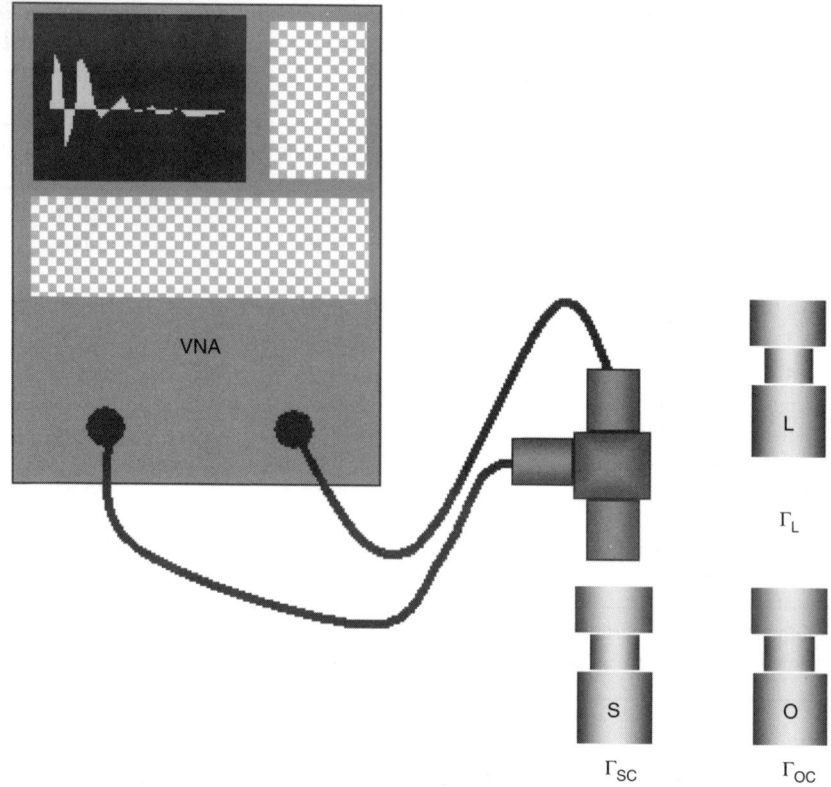

Figure 15.8 Measurement of splitter output match using the direct method

The equations relating all these can be written in matrix form as

$$\begin{bmatrix} 1 & A\Gamma_L & \Gamma_L \\ 1 & B\Gamma_{OC} & \Gamma_{OC} \\ 1 & C\Gamma_{SC} & \Gamma_{SC} \end{bmatrix} \begin{bmatrix} E_{DF} \\ E_{SF} \\ E \end{bmatrix} = \begin{bmatrix} A \\ B \\ C \end{bmatrix}$$

where $E = E_{RF} - E_{DF}E_{SF}$.

These can be solved by finding the inverse of the square matrix (or by any other matrix equation solving technique). Writing the solutions out in full gives

$$E_{SF} = \frac{A(\Gamma_{SC} - \Gamma_{OC}) + B(\Gamma_L - \Gamma_{SC}) + C(\Gamma_{OC} - \Gamma_L)}{A\Gamma_L(\Gamma_{SC} - \Gamma_{OC}) + B\Gamma_{OC}(\Gamma_L - \Gamma_{SC}) + C\Gamma_{SC}(\Gamma_{OC} - \Gamma_L)}$$

$$E_{DF} = \frac{A(C-B)\Gamma_{OC}\Gamma_{SC} + B(A-C)\Gamma_L - \Gamma_{SC} + C(B-A)\Gamma_{OC}\Gamma_L}{A\Gamma_L(\Gamma_{SC} - \Gamma_{OC}) + B\Gamma_{OC}(\Gamma_L - \Gamma_{SC}) + C\Gamma_{SC}(\Gamma_{OC} - \Gamma_L)}$$

$$E = \frac{A(B\Gamma_{OC} - C\Gamma_{SC}) + B(C\Gamma_{SC} - A\Gamma_L) + C(A\Gamma_L - B\Gamma_{OC})}{A\Gamma_L(\Gamma_{SC} - \Gamma_{OC}) + B\Gamma_{OC}(\Gamma_L - \Gamma_{SC}) + C\Gamma_{SC}(\Gamma_{OC} - \Gamma_L)}$$

Once E_{SF}, E_{RF} and E_{DF} are calculated the values can be checked against a reference device of known VRC by performing a 'normal' one-port VRC measurement using (15.4).

$$S_{11} = \frac{x - E_{DF}}{E_{RF} + E_{SF}(x - E_{DF})}$$
$$= \frac{x - E_{DF}}{E + E_{SF}x} \tag{15.4}$$

where x is, as above, the ratio of the two uncalibrated S-parameters for the known device. E_{SF}, the splitter output match, has now been established.

15.7 Couplers and reflectometers

Calibration of waveguide sensors and higher power calibrations in coaxial line are often done using couplers. If the DUT and coupler S-parameters are already known then a calibration can occur in a manner similar to using a power splitter. If the coupler or DUT S-parameters are not known then two couplers and power sensors can be combined (Figure 15.9) to form a basic reflectometer that gives an indication of the forward and reverse powers and hence the DUT VRC. The directivity of the couplers will limit the accuracy of any calibrations made using this method.

Calibrations at higher power levels than those at which calibrated standards exist can be performed using multiple, well-matched, high coupling factor couplers with power sensors on the sidearm and by varying the power level at each stage of the calibration process over the linear range of the sidearm power sensors.

15.7.1 Reflectometers

If the transfer instrument is a Reflectometer such as a VNA, six-port or multi-state reflectometer, the DUT reflection coefficient can be determined at the same

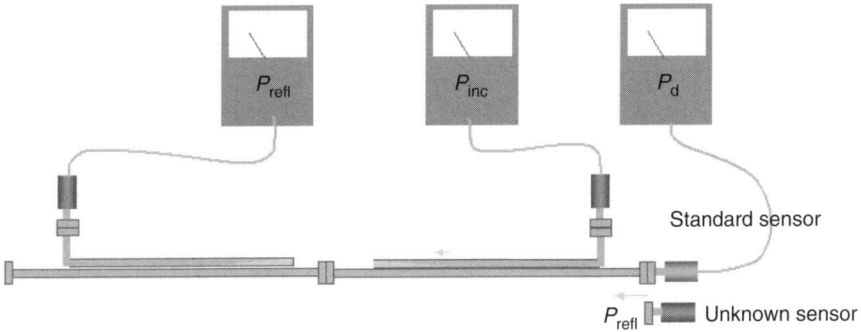

Figure 15.9   Calibration of a waveguide power sensor using two couplers

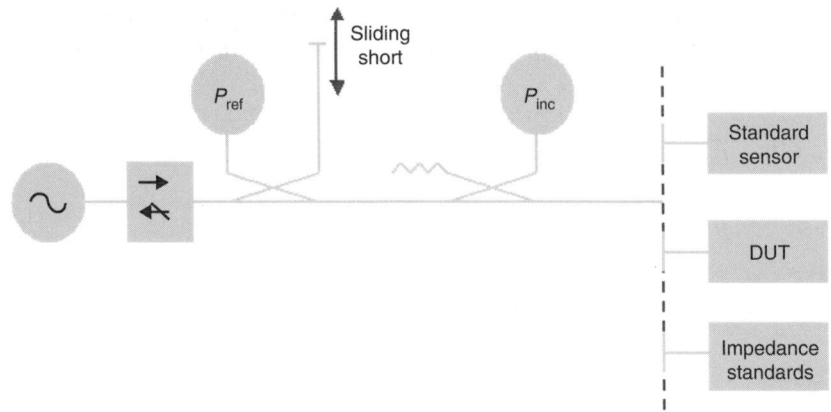

Figure 15.10   Multistate reflectometer

time as taking the necessary power ratios to calibrate the device allowing mismatch corrections to be made.

The multistate reflectometer [19] (Figure 15.10) consists of two couplers with power sensors to measure the forward and reverse powers and a sliding short circuit which alters the 'state' of the system. By measuring the power ratio between the forward and reverse coupler arms in several states for several known impedances the system properties (such as coupler directivity) can be found.

For each state k the ratio of the powers on the forward and reverse arms of the couplers is given by

$$\frac{P_{\text{ref}}}{P_{\text{inc}}} = \left| \frac{d_k \Gamma + e_k}{c_k \Gamma + 1} \right|^2$$

where c_k, d_k and e_k are state-dependent complex constants and Γ is the reflection coefficient of any device attached to the output port.

At NPL multistate reflectometers using three states and four known impedances are used for waveguide calibrations between 8.2 and 110 GHz.

15.8   Pulsed power

The topic of non-CW power measurements is extremely large and cannot be covered adequately in the space available here. The notes here present a brief introduction.

In a CW signal, such as the one illustrated in Figure 15.11 the instantaneous power will cycle at twice the frequency of the voltage or current. The power reported by a sensor with much longer time constant than the RF frequency will remain constant at the average power.

In a case such as the one shown in Figure 15.12 a slowly varying signal is then modulated onto a much higher RF frequency. Here, there are three obvious definitions of power: the instantaneous power, the average power and the envelope power

RF power measurement 345

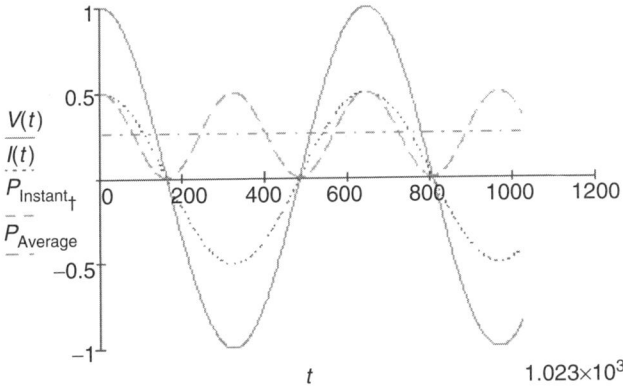

Figure 15.11 Voltage, current and instantaneous and average power

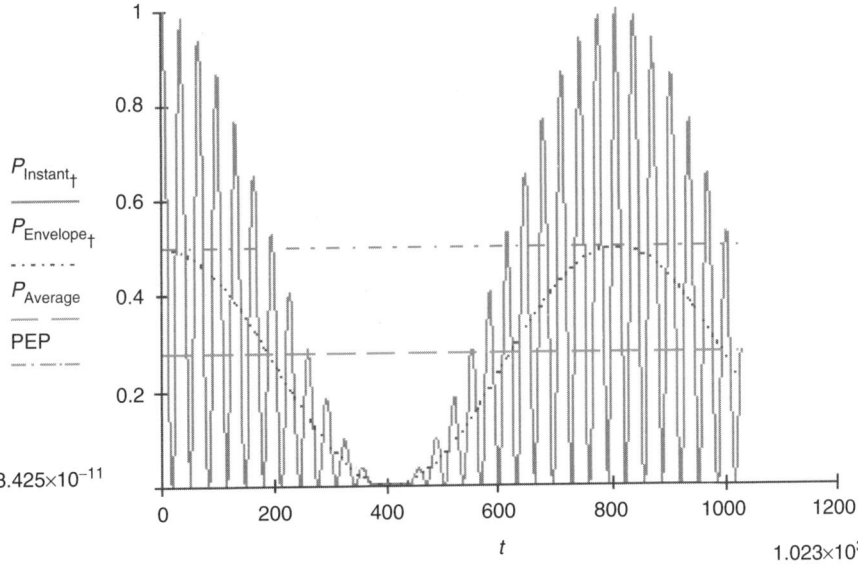

Figure 15.12 Modulated RF – instantaneous, average and envelope power

averaged over each RF cycle. A slow-response power sensor will still measure the average power but a faster diode-based power sensor may be able to follow the envelope and provide details about the power–time template. A sensor with a very fast response, such as an oscilloscope, may be able to trace the RF frequency, and the envelope can then be extracted by various methods providing even greater details of the pulse shape.

346 *Microwave measurements*

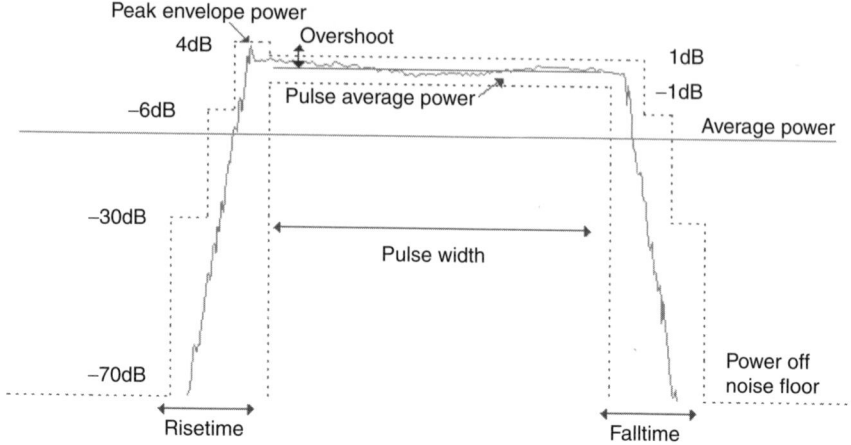

Figure 15.13   GSM pulse specifications

Perhaps the simplest pulsed measurement involves measuring the average power of a repetitive pulsed signal and multiplying by the duty cycle to arrive at a 'pulse' power. However pulses are never exactly square, or of constant power and so this method tells us relatively little about the pulse.

The GSM Pulse specification for envelope power shown in Figure 15.13 is typical of the usual pulsed measurements that are required – peak envelope power, pulse average power, average power and pulse risetime and falltime. The calibration factor of a diode power sensor capable of performing measurements on fast pulses will not, in general, be the same as its CW calibration factor. These sensors do not measure true RMS power and factors such as sensor impulse response time, recovery time or nonlinearity will lead to errors.

15.9   Conclusion

These notes have briefly covered the techniques and instruments needed to make a variety of common power measurements and power sensor calibrations. A more detailed guide to power measurements across a wide range of topics is the book by Alan Fantom [20] and this is recommended as a good starting point for those who wish to learn more about this area.

15.10   Acknowledgements

The author would like to thank Geoff Orford and Alan Wallace who both previously worked in the Power Measurement area at NPL and RSRE and contributed greatly to previous versions of these notes and viewfoils.

References

1 Ramo, S., Whinnery, J. F., and van Duzer, T.: *Fields and Waves in Communication Electronics* (Wiley, New York, 1965).
2 Kerns, D. M., and Beatty, R. W.: *Basic theory of waveguide junctions and introductory microwave network analysis* (Pergamon Press, London, 1967)
3 Warner, F. L.: *Microwave Attenuation Measurements*, (Peter Peregrinus, London, 1977)
4 'The expression of uncertainty and confidence measurements', United Kingdom, Accreditation Service (UKAS) document M3003, London 1997
5 Fantom, A.: 'Improved coaxial calorimetric RF power meter for use a primary standard', *Proc. Inst. Electr. Eng.*, 1979;**126**(9):849–54
6 Macpherson, A. C. and Kerns, D. M.: 'A microwave microcalorimeter', *Review of Scientific Instruments*, 1955; **26**(1):27–33
7 Abbott, N. P., Reeves, C. J., and Orford, G. R.: 'A new waveguide flow calorimeter for levels of 1–20 w', *IEEE Trans. Instrum. Meas. Inst. Electr. Eng.*, 1974; **IM-23**(4):414–20
8 Cullen, A. L. and Stephenson, L. M. A.: Torque operated wattmeter for 3 cm microwaves, *Proc. Inst. Electr. Eng.*, 1952;**99**(4):112–20
9 Oldfield, L. C., and Ide, J. P.: A fundamental microwave power standard, *IEEE Trans. Instrum. Meas.*, 1987;**IM-36**(2):443–9
10 Paulesse, D., Rowell, N., and Michaud, A.: Realization of an atomic microwave power standard, *Digest of Conference on Precision Electromagnetic Measurements*, Ottawa, Canada, 2002; pp. 194–5
11 Donley, E. A., Crowley, T. P., Heavens, T. P., and Riddle, B. F.: 'A quantum-based microwave power measurement performed with a miniature atomic fountain', *Proceedings of the 2003 IEEE International Frequency Control Symposium*, Tampabay, FL, 2003; pp. 135–7
12 Barlow, H., and Katoaka, S.: The Hall Effect and its application to power measurement at 10 G c/s, *Proc. Inst. Electr. Eng.*, Part B, 1958;**105**:53–60
13 Fernandez, L. J., Visser, E., Sese, J., *et al.*: 'Development of a capacitive MEMS RF power sensor without dissipative losses: towards a new philosophy of RF power sensing', *Digest of Conference on Precision Electromagnetic Measurements*, London, UK, 2004; pp. 117–18
14 Alastalo, A.Kyynanainen, J., Sepa, H. *et al.*: 'Wideband microwave power sensor based on MEMS technology', *Digest of Conference on Precision Electromagnetic Measurements*, London, UK, 2004; pp. 115–16
15 *NPL News*, Spring 1990, no. 369, p. 12
16 Tippett, J. C., and Speciale, R. A.: 'A rigorous technique for measuring the scattering matrix of a multiport device with a 2-port network analyser', *IEEE Transaction on Microwave Theory and Techniques*, 1982;**MTT-30**: 661–6
17 Juroshek, J.: 'A direct calibration method for measuring equivalent source mismatch', *Microwave Journal*, 1997;**40**:106–18

18 Rodriguez, M.: 'A semi-automated approach to the direct calibration method for measurement of equivalent source match', *ARMMS Conference*, Bracknell, UK, April 1999, pp. 35–42.
19 Oldfield, L. C., Ide, J. P., and Griffin, E. J.: A multistate reflectometer, *IEEE Trans. Instrum. Meas*, 1985;**IM-34**(2):198–201
20 Fantom, A.: '*Radio frequency and microwave power measurement*', *Electrical Measurement Series* no. 7', (Peter Peregrinus, London, 1990)
21 Johnson, R. A.: Understanding microwave power splitters, *Microwave Journal*, Dec 1975, pp. 49–51, 56
22 Engen, G. F.: Power equation: a new concept in the description and evaluation of microwave systems, *IEEE Transactions on Instrumentation and Measurement*, 1971;**IM-20**(1):49–57

Chapter 16
Spectrum analyser measurements and applications
Doug Skinner

Spectrum analyser is a measuring instrument, which is used to display many different kinds of signal. This chapter is an introduction to the spectrum analyser and covers the most important parts of the analyser performance that need to be understood. This overview of spectrum analysers is split into four parts.

Part 1: Introduction. Describes the basics of signal analysis and compares the oscilloscope time-domain display with the spectrum analyser frequency-domain display and some basic spectrum analyser measurements are also described.

Part 2: How the spectrum analyser works. Provides an explanation of how a basic spectrum analyser works. It includes a description of the importance and significance of the main operator controls and how they are used to ensure a clear understanding of the display and to reduce or prevent mistakes.

Part 3: The important specification points of a spectrum analyser. Describes the important specification points that need to be known and understood in order to select the correct instrument for a particular measurement. Some sources of errors and measurement uncertainties are also covered in this section.

Part 4: Spectrum analyser measurements. Discusses some of the common measurements that are made using a spectrum analyser and the measurements reviewed include harmonic and intermodulation measurements as well as the measurement of modulated and pulsed signals.

16.1 Part 1: Introduction

16.1.1 Signal analysis using a spectrum analyser

Before making any measurements using a spectrum analyser the user should prepare the spectrum analyser for use by carrying out any pre-calibration procedure (Auto Cal)

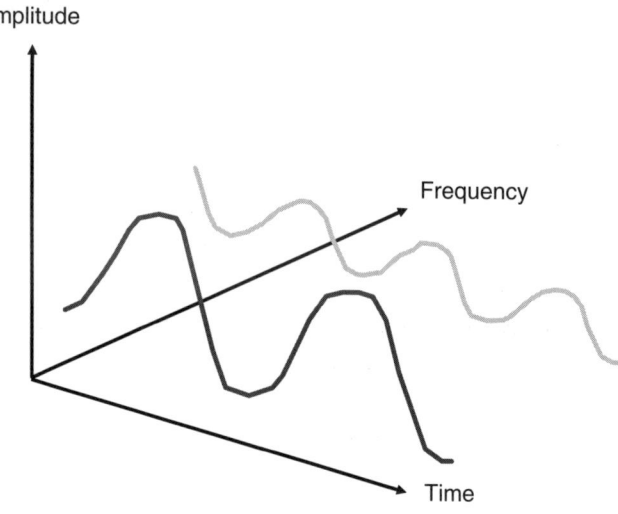

Figure 16.1   Three-dimensional graph

recommended by the manufacturer. Some spectrum analysers include a RESET button to return the analyser to the initial set of conditions if the user has problems in interpreting the display.

The next step is to consider the type of input signal and power level that are to be applied to the spectrum analyser to avoid overloading or damaging the input circuitry. The final step is to interpret and understand the displayed results.

16.1.2   Measurement domains

Suppose that there is a requirement to analyse a signal that consists of a sine wave with a second harmonic component.

Consider the three-dimensional graph shown in Figure 16.1.

It can be seen that the graph has three mutually perpendicular axes that are calibrated in terms of time, amplitude and frequency. The objective of the signal analysis is to display the components of such a signal and there is a choice to view the signal in terms of Amplitude against Time or as Amplitude against Frequency.

16.1.3   The oscilloscope display

The display when viewed as an Amplitude against Frequency display is shown in Figure 16.2 and is recognisable as a typical oscilloscope display and it is known as a time-domain display. In this situation only a single combined waveform is shown on the display. This is the waveform in the Figure shown as a solid line, but there are in

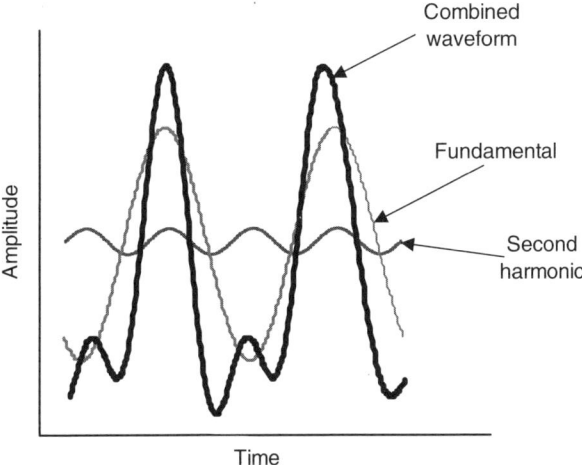

Figure 16.2   Oscilloscope amplitude against time display

fact at least two sinusoids present as shown by the two thin lines. The oscilloscope time-domain display does not separate out the individual frequency components and its shape changes depending on the relative amplitudes and phase of the sinusoids present.

16.1.4   The spectrum analyser display

The spectrum analyser display of Amplitude against Frequency is shown in Figure 16.3 and is known as a frequency-domain display. In this case, it reveals the two separate frequency components of the applied signal, the fundamental and the harmonic. The fundamental frequency is represented on the display by the first single vertical line. The shorter vertical line that can be clearly seen to the right of the fundamental represents the second harmonic. How the Amplitude against Frequency display is achieved using the spectrum analyser is explained later in this chapter.

16.1.5   Analysing an amplitude-modulated signal

16.1.5.1   Amplitude modulation – Oscilloscope

The first analysis example is to look at the relatively simple amplitude-modulated signal as displayed on an oscilloscope. Figure 16.4 shows the familiar oscilloscope display of an amplitude-modulated signal.

It can be seen that the high-frequency carrier has a low-frequency signal superimposed upon it. The modulation envelope can also be seen on the display. It is possible to measure the modulation frequency (f mod) and modulation depth from

352 *Microwave measurements*

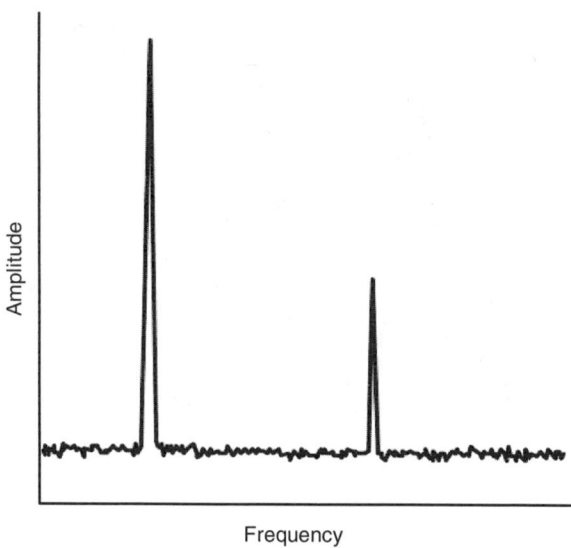

Figure 16.3   Spectrum analyser amplitude against frequency display

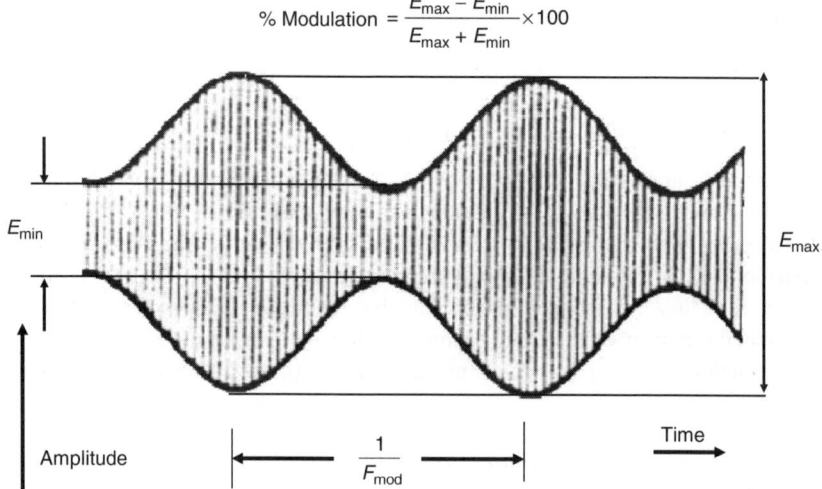

Figure 16.4   The oscilloscope display

the display but it is difficult to obtain any further information over and above modulation depth and modulation frequency. Consequently, the oscilloscope is not widely used to analyse radio frequency (RF) and microwave signals because of the limitations described.

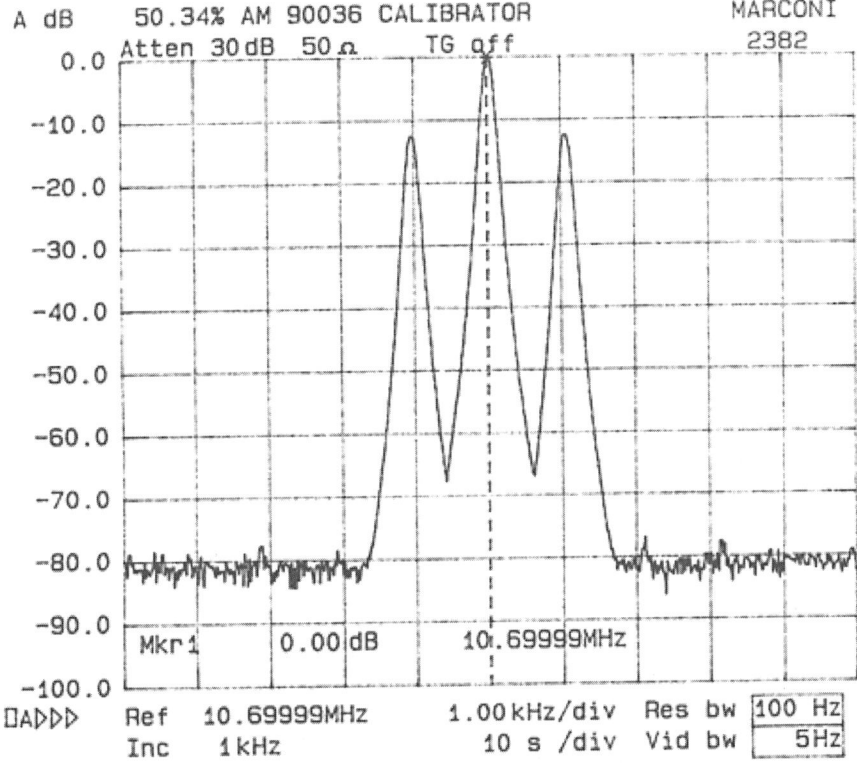

Figure 16.5 The spectrum analyser display

16.1.5.2 Amplitude modulation – spectrum analyser

Figure 16.5 shows an amplitude-modulated signal as displayed by a spectrum analyser. The carrier, upper and lower side frequencies and noise can all be clearly seen.

Note that the analysis of the amplitude-modulated waveform clearly demonstrates the superior analytical powers of the spectrum analyser.

Spectrum analyser display of Amplitude against Frequency is more useful because the harmonics, spurious signals, sidebands and noise can be observed. One further advantage of a spectrum analyser is its high sensitivity, which means that it can measure very low-level signals down to less than 0.1 μV because it is selective rather than broadband. It can also display low-level signals at the same time as high-level signals because logarithmic amplitude scales are used. An oscilloscope, which generally has a linear vertical scale, does not have this capability.

Many other measurements can also be made on many different and complex signals using a spectrum analyser as will be described later in Part 4. It should be emphasised at this stage that the interpretation of some spectrum analyser displays of complex waveforms requires careful study.

16.2 Part 2: How the spectrum analyser works

16.2.1 Basic spectrum analyser block diagram

A greatly simplified block diagram of a basic swept-tuned heterodyne spectrum analyser is shown in Figure 16.6. In practice, the implementation is considerably more complex as there are many more frequency conversion stages.

The input signal is applied to the input mixer through an input attenuator, which adjusts the sensitivity and optimises the signal level at the mixer to prevent overload or distortion. An input low-pass filter is also included at this stage to avoid intermediate frequency (IF) feed-through and to reject the upper image frequency.

The mixer converts the input signal to a fixed IF, at which point a range of Gaussian band-pass filters or digital filters are switched in to change the selectivity or resolution. To give a vertical scale, calibrated in dB, the signal at the IF stage is passed through a logarithmic amplifier. The signal is then applied to a detector and passes through selected video filters before being applied to the vertical scale of the display.

The horizontal input of the spectrum analyser display (frequency) is achieved by using a variable amplitude ramp generator, or saw-tooth generator, which is also applied to a voltage-controlled oscillator that feeds the mixer. As the ramp voltage is increased, the receiver tunes to a progressively higher frequency and the trace on the display moves from left to right. Using this technique, an Amplitude against Frequency display is shown on the spectrum analyser.

16.2.2 Microwave spectrum analyser with harmonic mixer

The basic block diagram of Figure 16.6 is generally only used for spectrum analysers covering up to around 4 GHz. For a 4 GHz instrument the first local oscillator would have to cover from approximately 5 to 9 GHz but the local oscillator for a 26.5 GHz spectrum analyser would have to cover approximately 30–56.5 GHz. This is a major engineering challenge especially as the oscillator needs to be at a high level and have good voltage frequency linearity, low-noise, low-level spurious signals and an output

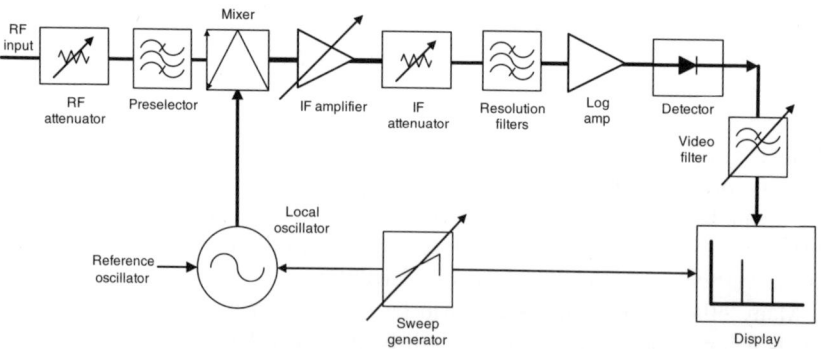

Figure 16.6 Block diagram of a basic spectrum analyser

level that is adequately independent of frequency. Furthermore, the design has to be implemented at an economical price.

An alternative more practical approach, used in most microwave spectrum analysers, is to use a harmonic mixer. This concept is shown in Figure 16.7. The fundamental frequency of the local oscillator is used for the lower frequencies and higher harmonics are used to cover the higher frequencies. A separate harmonic multiplier is not actually used in practice; the mixer is designed to mix with harmonics of the local oscillator.

16.2.3 The problem of multiple responses

The system described in Figure 16.7 will operate to high microwave frequencies but there is a major limitation. The type of analyser shown in the previous diagram has a fundamental flaw: one signal at the input generates multiple responses such that one signal has many other signals associated with it, as shown in Figure 16.8 which is obviously incorrect.

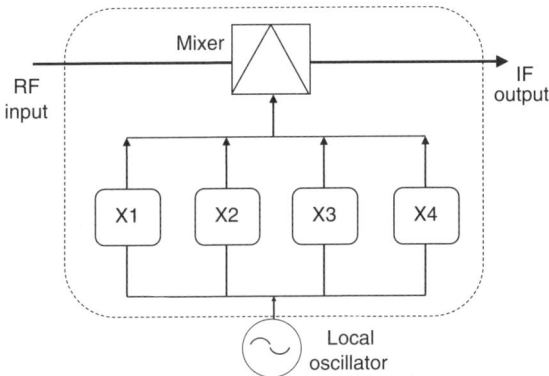

Figure 16.7 Microwave spectrum analyser with harmonic mixer

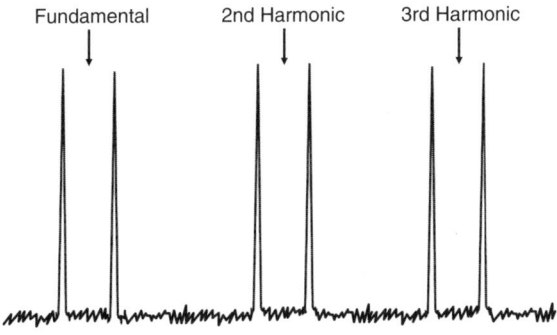

Figure 16.8 Multiple responses for a single input frequency

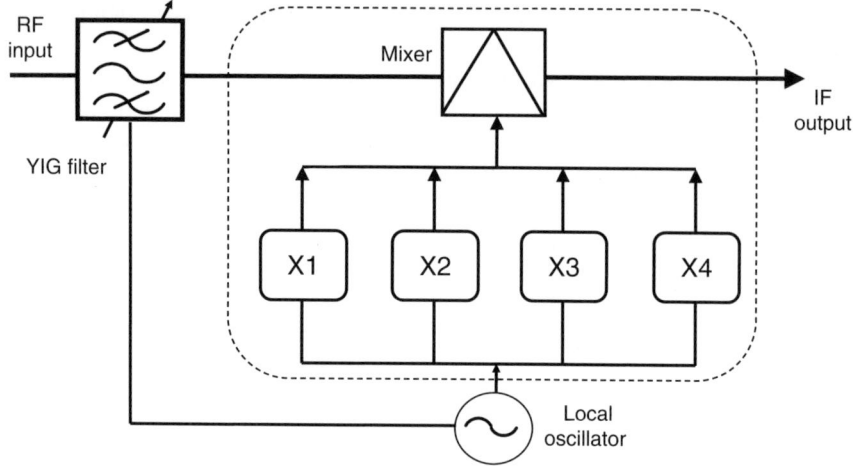

Figure 16.9   Microwave spectrum analyser with a tracking preselector

Not only does this one signal mix with each of the harmonics of the local oscillator to produce multiple responses but additional responses are also generated at the image frequencies. Some of the earlier microwave spectrum analysers used this technique but the limitations are so severe that it is very rarely, if ever, used today.

16.2.4   Microwave spectrum analyser with a tracking preselector

The diagram in Figure 16.9 shows how adding a band-pass filter at the input of the spectrum analyser can refine the harmonic mixer technique.

This is known as a tracking preselector and the microwave spectrum analyser uses a YIG (Yttrium Iron Garnet) swept band-pass filter for the tracking filter and is usually referred to as a preselector.

16.2.5   Effect of the preselector

The effect of using a preselector is shown in Figure 16.10.

The swept band-pass filter selects only the wanted signal so that all the unwanted signals are rejected to make the measurement valid. A quality instrument has a preselector with high out of band rejection and the ability to track closely the input tuned frequency.

Certain earlier spectrum analysers required the preselector to be 'peaked' before a measurement was made to ensure that the preselector is tuned correctly but this is not necessary with the latest and more complex instruments.

16.2.6   Microwave spectrum analyser block diagram

In practice, modern microwave spectrum analysers are usually a combination of a fundamental frequency analyser and a harmonic analyser. The fundamental frequency

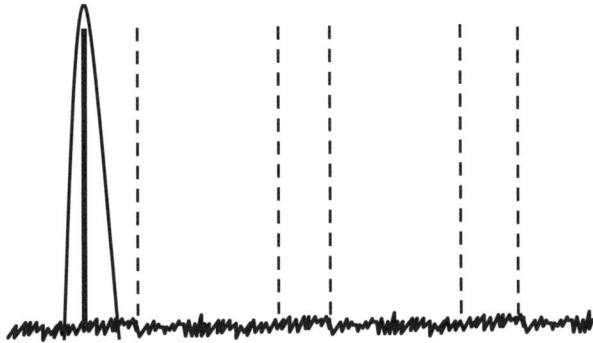

Figure 16.10   Effect of the preselector

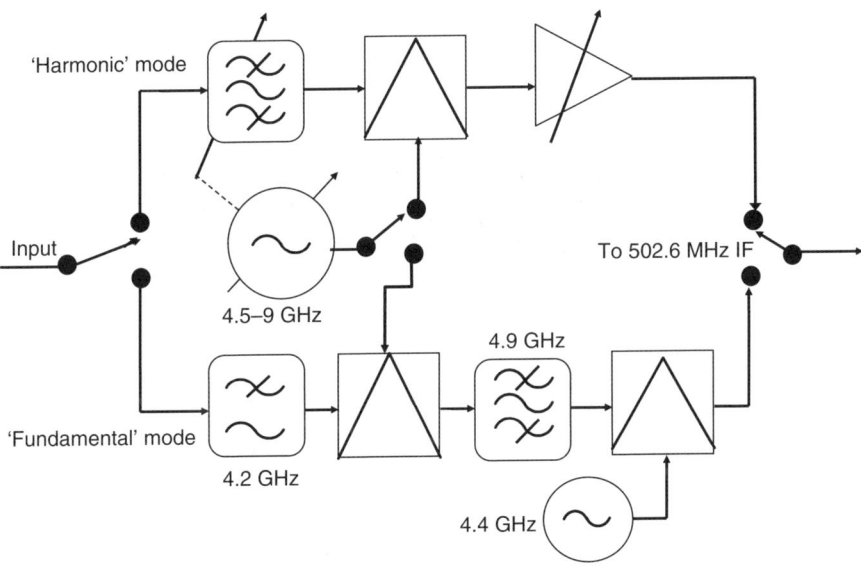

Figure 16.11   100 Hz to 4.2 GHz spectrum analyser

method of operation is used at the lower frequencies but at the higher frequencies the multiplication technique, with a preselector, is used. Figure 16.11 shows the architecture of a typical 100 Hz to 4.2 GHz spectrum analyser.

In the fundamental mode the input signal is mixed with a local oscillator covering from 4.5 to 9 GHz. The IF is then downconverted to a 502.6 MHz signal by a fixed 4.4 GHz local oscillator.

To cover the higher frequencies the change over switch operates to bring the swept harmonic mixer into play and the 4.5–9 GHz local oscillator is used to downconvert the signal to the IF of 502.6 MHz.

Microwave spectrum analysers that use a harmonic mixer have a characteristic 'stepped' noise floor as illustrated in the display in Figure 16.12. The rise in the noise

358 *Microwave measurements*

occurs at the frequency break points where the higher harmonics of the local oscillator are used. From Figure 16.12 it can be seen that the instrument is approximately 10 dB less sensitive at 22 GHz compared with the sensitivity at 2 GHz.

16.2.7 Spectrum analyser with tracking generator

Spectrum analysers are made even more useful by the addition of a tracking generator. A tracking generator is a swept signal whose instantaneous frequency is always the same as the frequency to which the spectrum analyser is tuned. Many spectrum analysers incorporate tracking generators to increase the applications of the instrument to include wide dynamic range swept frequency response measurements. The use of

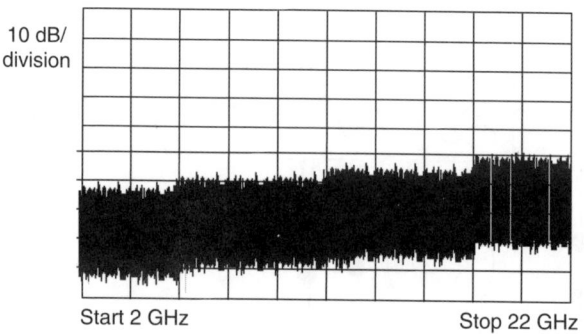

Figure 16.12   Noise floor display

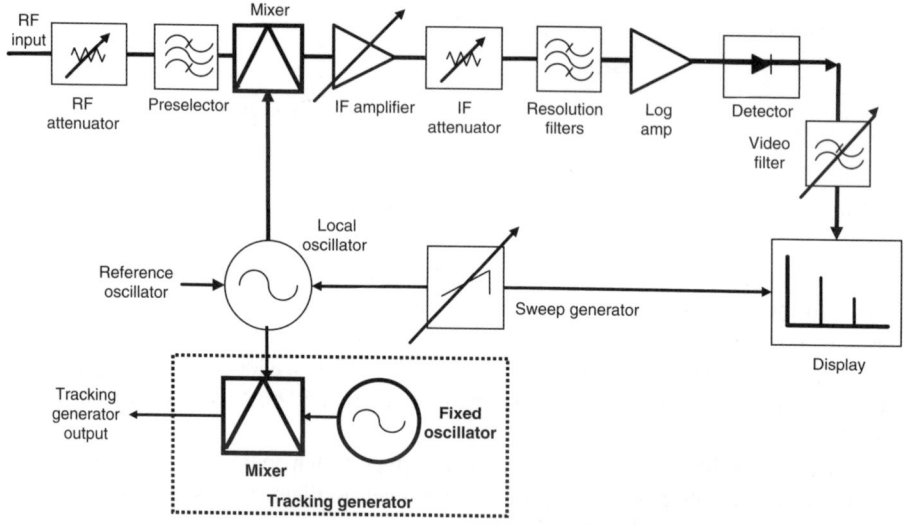

Figure 16.13   A spectrum analyser with tracking generator

a tracking generator means that it is not always necessary to have an external signal source when making some measurements.

Figure 16.13 shows how a tracking generator facility can be added to a spectrum analyser. The output signal synchronously tracks the input tuned frequency of the instrument with the advantage that the dynamic range is better than would be obtained if a broadband detector was used. A dynamic range of over 110 dB can be achieved with a spectrum analyser using a tracking generator.

16.3 Part 3: Spectrum analyser important specification points

Spectrum analysers are complex items of test equipment and they can easily be misused. At worst, a wrong result can be obtained; at best, the operator may not be getting the best performance from the instrument.

The latest spectrum analysers have many automatic functions, but incorrect results are still possible. When using a spectrum analyser it is important that the operator understands the function of the basic controls of the instrument in order to be able to use it effectively and to avoid incorrect results.

The spectrum analyser block diagram (Figure 16.14) is repeated here to show how the controls change the instrument functions. There are four main controls on a spectrum analyser and they are

(1) RF Attenuator and IF gain,
(2) sweep speed,
(3) resolution bandwidth and
(4) video bandwidth.

The reason for highlighting the four controls listed above is that they are probably the most commonly misunderstood and abused. Incorrect settings of these controls can cause serious measurement errors, so it is important to realise their significance. The frequency and amplitude are also important controls, but they are more easily understood and less likely to cause problems.

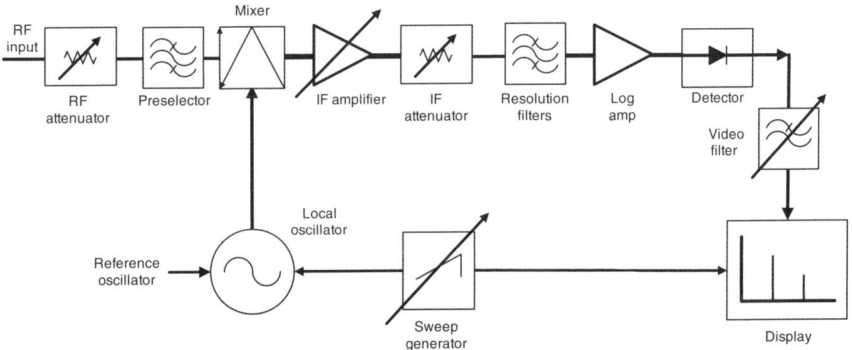

Figure 16.14   Spectrum analyser controls

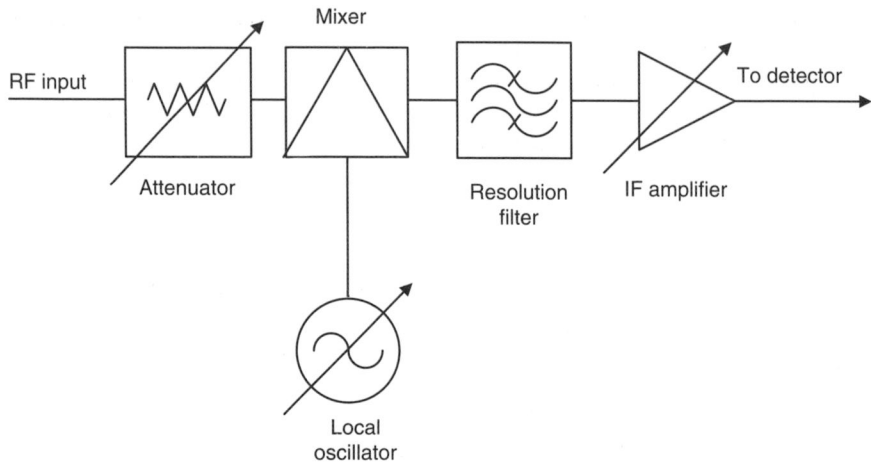

Figure 16.15   Input attenuator and IF gain controls

16.3.1   The input attenuator and IF gain controls

The block diagram Figure 16.15 shows how the sensitivity of a spectrum analyser can be changed.

To increase the sensitivity of the spectrum analyser the operator has two options, either the input attenuation can be reduced or the IF gain can be increased, but if the wrong option is chosen then the measurement may become invalid. It is essential to arrange the correct signal power input level to the mixer to ensure correct operation.

If the input attenuation is reduced too much then the input mixer could be overloaded with the result that unwanted distortion products are generated within the spectrum analyser. If the IF gain is increased then the risk of overloading the input mixer is removed but the noise level could rise to an unacceptable level with the result that some signals of interest could be masked in the noise. A further problem that could arise is the introduction of distortion or intermodulation in the IF stages.

Many spectrum analysers automatically select the optimum RF attenuation and IF gain settings once the reference level at the top of display has been selected. Under certain circumstances, however, it may be an advantage to override the automatic selection to select a mode of operation with either lower noise or lower intermodulation.

16.3.2   Sweep speed control

The spectrum analyser sweep speed must be swept sufficiently slowly to allow the signal level in the narrow resolution filters to settle to a stable value. Figure 16.16 shows two different analyser responses to the same signal and the effects produced when sweeping too fast are clearly shown. First, the amplitude of the displayed signal is reduced because the filter does not have sufficient time to respond to the signal

Spectrum analyser measurements and applications   361

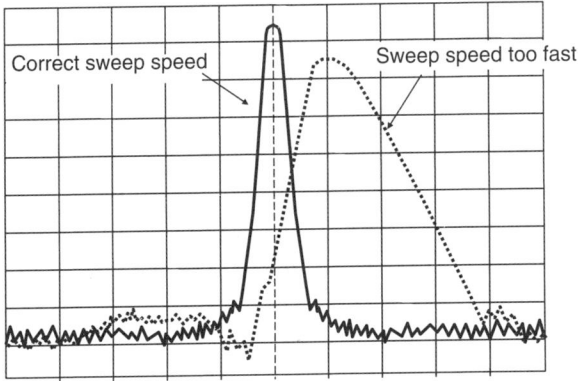

Figure 16.16   *Shows the effect of sweeping too fast*

and second the maximum is moved to the right due to the delay in the response. This effect is sometimes referred to as 'ringing'.

The Sweep bandwidth factor is given by the following relationship:

$$\text{Sweep} \propto \frac{\text{Span}}{\text{Resolution bandwidth}^2} \qquad (16.1)$$

We can see that for a given Span (total frequency scale across the screen) if the resolution bandwidth is changed then the sweep speed will change. For most spectrum analysers this is carried out automatically and modern instruments incorporate software control to ensure that the correct sweep speed is achieved. But under certain conditions, where high resolution is required, the sweep speed may need to be as slow as 100 seconds and then some form of digital storage is essential to ensure that a visible display is achieved. Manual adjustment of the sweep speed is sometimes provided on some instruments to override the automatic selection.

Sweeping faster than the optimum value can be useful to carry out a rapid uncalibrated search for spurious signals or to study the effects of rapidly changing transient signals. However, the operator must be aware of the display errors that can be caused. Sweeping slower than the optimum sweep can be used, for example, when sweeping a filter with very steep skirts by using the Tracking Generator.

16.3.3   Resolution bandwidth

Resolution filters are a very important part of the spectrum analyser operation and they need to be carefully used. Resolution bandwidth is the bandwidth of the IF filter that determines the selectivity of a spectrum analyser. It is basically the ability of the analyser to separate closely spaced signals. A wide resolution bandwidth is required for wide sweeps whilst a narrow filter is used for narrow sweeps. Figure 16.17 shows three displays of an amplitude-modulated signal, they illustrate why it is necessary to be able to change resolution bandwidth.

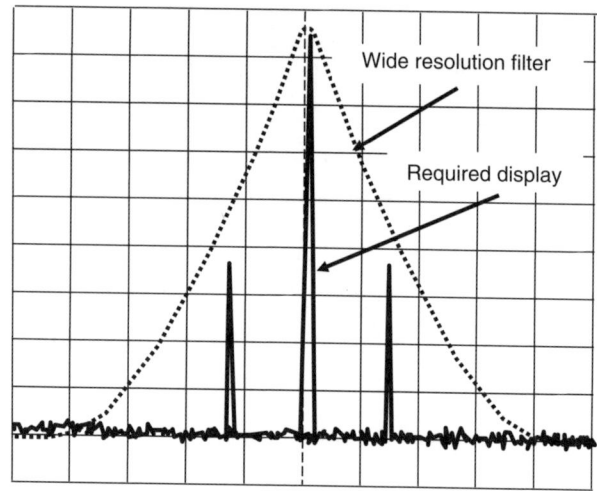

Figure 16.17 Using a wide resolution bandwidth

The wide resolution bandwidth is effectively a plot of the response of the resolution filter of the spectrum analyser. As the resolution filter is swept across the frequency scale of the spectrum analyser, any signal that is within the pass-band of the filter will result in a response on the display. Figure 16.17 shows that if the signal that is being measured is a carrier with two side frequencies when the resolution bandwidth (shown dotted) is too wide it is not possible to display the signal correctly. We can see frequency response of the instrument's filter is swept by the local oscillator and the side frequencies are not seen in this situation.

The detail of the response on the display is clearly dependent upon the bandwidth of the resolution filter and speed that it is moved across the display.

However, by using progressively narrower resolution filter bandwidths as shown in Figure 16.18, the display can resolve the side frequencies. However, the penalty for high resolution is that a slower sweep speed needs to be used.

Most spectrum analysers have a number of resolution bandwidth filters. The wide resolution bandwidth filters are only normally used when the display needs to be updated rapidly.

16.3.4 Shape factor of the resolution filter

Figure 16.19 shows two types of filter in use as resolution filters in spectrum analysers and they have defined filter shapes.

$$\text{Shape Factor} = \frac{60 \text{ dB Bandwidth}}{3 \text{ dB Bandwidth}}$$

The shape factor is defined as the ratio of the 60 dB bandwidth to the 3 dB bandwidth. The first type of filter is the Gaussian filter and it has a shape factor of 11:1 for high quality to 15:1 for a lower quality filter. The second type of resolution filter is a digital

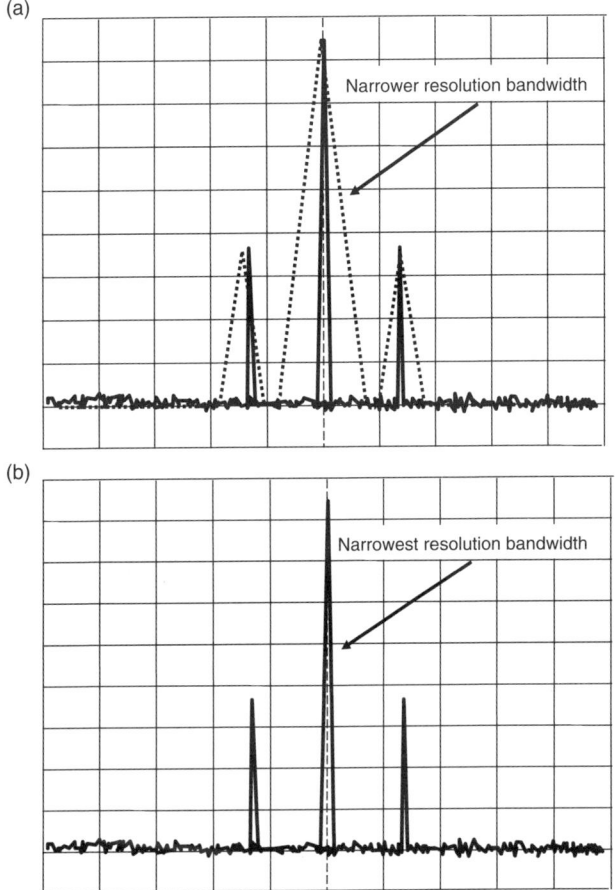

Figure 16.18 (a) Using a narrower resolution and (b) the narrowest resolution

filter that has a shape factor of 5:1. The digital filter is particularly useful where a narrow resolution filter is needed, say from 1 Hz to 30 Hz.

This minimum resolution bandwidth of a spectrum analyser is a key measure of the ability to measure low-level signals adjacent to high-level signals. Many spectrum analysers have a combination of Gaussian and digital filters included in their design.

A measurement that illustrates the importance of minimum resolution bandwidth is the determination of low-level signal such as a 50 Hz side frequency (hum sidebands) close to a large signal.

For example in Figure 16.20, the upper trace is achieved by using a 10 Hz resolution bandwidth and only one signal is discernible. The lower trace, which uses a 3 Hz resolution bandwidth, clearly shows the low-level signals.

For example, if the sidebands are 70 dB down then a 10 Hz resolution bandwidth filter with a shape factor of 11:1 could not resolve the side frequencies because if the

364 *Microwave measurements*

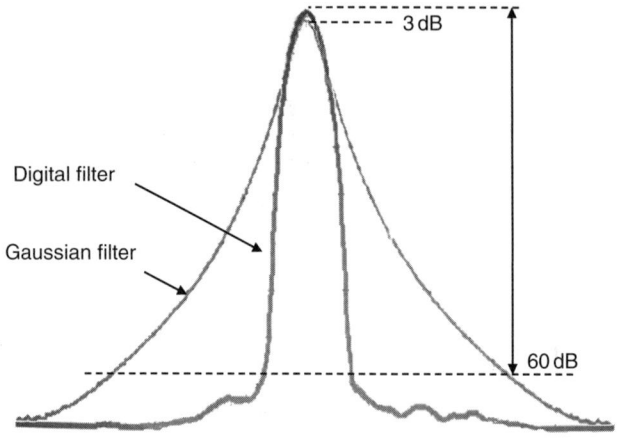

Figure 16.19   Resolution bandwidth filter shape factor

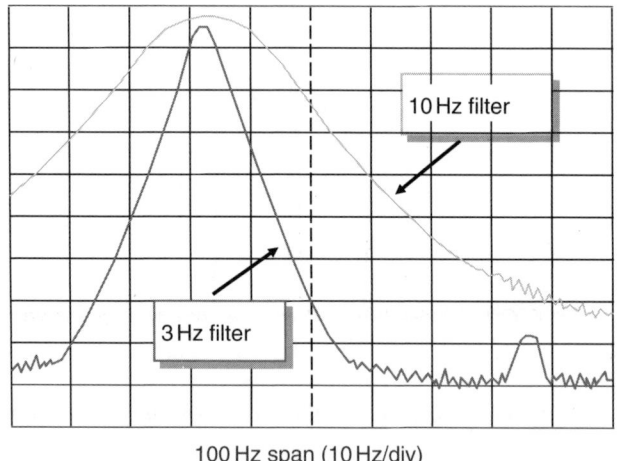

Figure 16.20   Resolution bandwidth change

3 dB bandwidth is 10 Hz then the 60 dB bandwidth is 110 Hz. A signal 60 dB down and 55 Hz away could just be discerned but a signal 70 dB down and 50 Hz away would not be resolved.

By using a 3 Hz filter with a shape factor of 11:1 a signal 16.5 Hz away can be resolved if it is less than 60 dB down; it follows that a signal 70 dB down and 50 Hz away can be easily measured.

Digital filters are now common in spectrum analyser and they have a shape factor of 5:1 enabling close-in signals to be resolved and measured.

Spectrum analyser measurements and applications 365

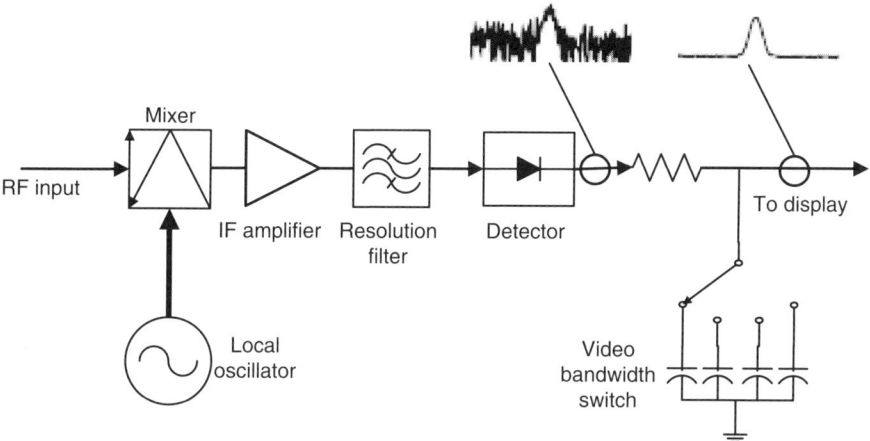

Figure 16.21   Video bandwidths

16.3.5 Video bandwidth controls

The previous section explained that spectrum analysers are often used to measure very low-level signals that may be almost indiscernible from the system noise. Using a narrower resolution bandwidth filter will reduce the average displayed value of the noise. However, to make the signals even easier to view it is often necessary to smooth out the random fluctuation of noise so that a coherent signal can be more clearly viewed. The traditional way to smooth the noise is to use a low-pass video filter after the detector as shown in Figure 16.21

In order to achieve the noise smoothing it is necessary to sweep more slowly because the time constant of the filter is reduced as the bandwidth of the filter is reduced. Modern instruments couple the video bandwidth controls to the sweep speed control so that the instrument automatically selects a slower sweep speed if the video bandwidth is reduced. Conversely, a lower frequency video bandwidth is automatically selected if the sweep speed is increased.

A useful general rule is to set the video bandwidth to be one-tenth of the resolution bandwidth being used.

16.3.5.1   Video averaging

An alternative method of noise averaging that has become increasingly popular on software-controlled instruments is to use multiple sweep video averaging. Successive sweeps are averaged so that the amplitudes of coherent signals are unchanged whilst the levels of varying noisy signals are averaged out.

The effect of using video averaging is to see the noise level slowly fall. Any low-level coherent signals that have been obscured by noise may become visible.

Clearly, it is most important that an operator is aware of the difference between the video bandwidth controls and the resolution bandwidth controls and not to confuse their different functions. Additional critical aspects of the performance of a spectrum analyser are noise, dynamic range, accuracy and local oscillator phase noise.

16.3.6 Measuring low-level signals – noise

The problem when measuring low-level signals is that even a component such as passive resistor generates noise due to thermal effects. The noise voltage generated is given by the equation:

$$V^2 = 4KTBR$$

where K is the Boltzmann's constant (1.374×10^{-23} J°K^{-1}), T is the temperature in K (absolute temperature), B is the bandwidth of the system (Hz) and R is the resistor value (generally 50 Ω for most measurements).

Using the figures given above results in a value for V^2 of 8.927×10^{-10} V EMF and converting this to dBm gives a value of -174 dBm.

If a spectrum analyser has a typical noise figure of 20 dB then with a 1 Hz resolution bandwidth, the lowest level signal that could be discerned would be 20 dB higher in amplitude than the noise of -174 dBm of a passive termination. This means that with a 1 Hz filter, a spectrum analyser with a 20 dB noise floor could theoretically measure $-174 + 20 = -154$ dBm.

An analyser with the same noise Figure but with a minimum resolution bandwidth of 3 Hz could discern a signal at -149 dBm and with a 1 kHz resolution bandwidth could only measure down to -119 dBm, which is 30 dB worse (Figure 16.22).

The use of a pre-amplifier at the input of a spectrum analyser can assist to measure lower amplitude signals.

16.3.7 Dynamic range

A useful definition of the dynamic range is that it is the ratio of the largest to the smallest signal simultaneously present at the input of the spectrum analyser that

Resolution bandwidth	Noise floor
10 kHz	−110 dBm
1 kHz	−120 dBm
100 Hz	−130 dBm
10 Hz	−140 dBm
3 Hz	−145 dBm
1 Hz	−150 dBm

Figure 16.22 Shows how the noise floor drops as the resolution bandwidth is reduced

permits the measurement of the smaller signal taking into account the uncertainty of the measurement. The dynamic range is usually quoted in dB. Note that uncertainty of measurement is included in the definition so we need to consider, how the internally generated distortion and noise affect the measurement that we make. For a constant local oscillator level the mixer output is linearly related to the input signal level and for all practical purposes this is true provided that the input signal is more than 20 dB below the local oscillator drive level. The input signal at the mixer determines the dynamic range. The level of signal we need for a particular measurement can be calculated using data from the manufacturer's specification for the analyser and in some cases the manufacturer's data sheets include graphs showing the information.

16.3.7.1 Intermodulation and distortion

A spectrum analyser can introduce intermodulation and cause distortion on a measurement; certain measurements cannot be made if the instrument itself generates excessive distortion. The distortion is normally described by its order and is noted by its relationship to the signal frequency, therefore second harmonic distortion is known as second order and the third harmonic distortion is known as third-order.

Let us consider the second-order distortion first. Suppose that the information from the manufacturer's specification gives the following data that the second harmonic distortion is 75 dB down on the fundamental for a signal level of −40 dBm at the mixer input. We can plot the data on the graph in Figure 16.23.

This means we can measure distortion down to 75 dB. The value can be plotted on a graph of Distortion (dBc) against the mixer input level. Now if the mixer level is changed to −50 dBm we know that distortion changes by 10 dB to −85 dBm. Now if

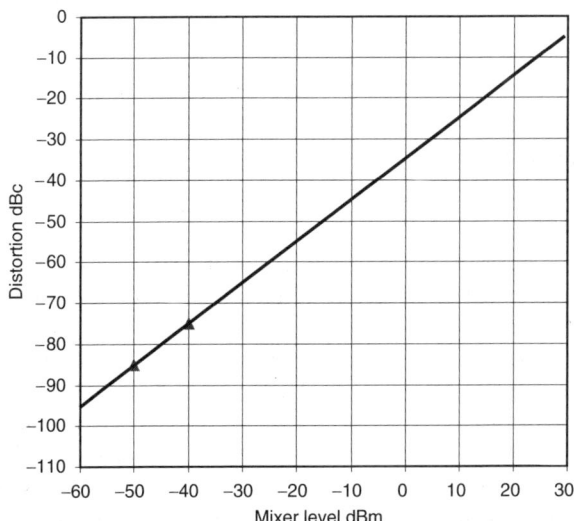

Figure 16.23   *Second-order distortion*

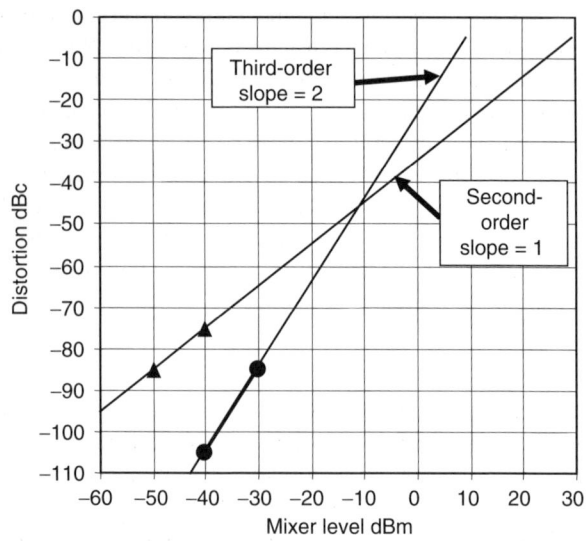

Figure 16.24   Third-order distortion added

the signal level at the mixer changes to −50 dBm then the internal distortion and the measurement range changes from −75 dBc to −85 dBc. From mathematical analysis of the mixer, it is known that for the second-order distortion the two points are on a line whose slope is 1 so we can draw a line on the graph giving the second-order performance for any level at the input to the mixer. Similarly, we can now construct a line for the third-order distortion. The manufacturer's data sheet gives −85 dBc for a level of −30 dBm at the mixer input and this value is plotted on the graph in Figure 16.24. If the difference between the two values changes by 20 dB the internal distortion is changed to −105 dBc. Again from mathematical analysis of the mixer these two points are on a line of slope 2 giving the third-order performance for any level at the input to the mixer.

16.3.7.2 Noise

There is a further effect on the dynamic range and that is the noise floor of the spectrum analyser. Remember that the definition of the dynamic range is the ratio of the largest to the smallest signal that can be measured on the display. So the noise level places a limit on the smaller signal. The dynamic range is relative to the noise and becomes the signal-to-noise ratio where the signal is the fundamental we require to measure.

To plot the noise on a dynamic range chart we take the data from the manufacturer's data sheet, which gives −110 dBm for a 10 kHz resolution bandwidth. If our signal level at the mixer is −40 dBm it is 70 dB above the average noise. Now for every dB we lose at the mixer input we lose 1 dB of signal-to-noise ratio so the noise curve is a straight line having a slope of −1 and this can be drawn on the graph as shown in Figure 16.25.

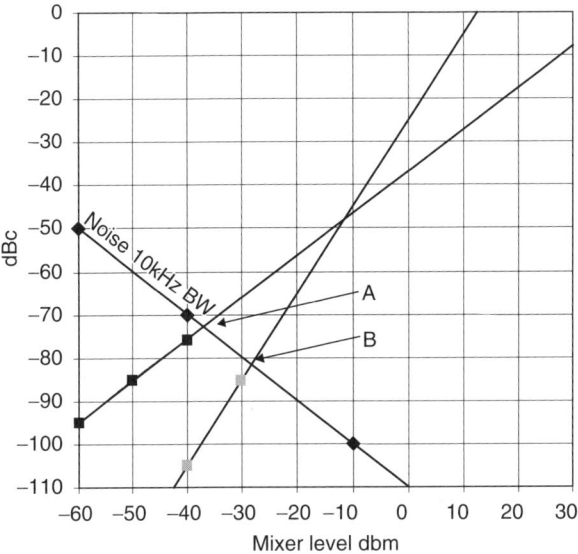

Figure 16.25   Dynamic range versus distortion and noise

Figure 16.25 shows two intercepts marked A and B. A is the second-order maximum dynamic range and B is the third-order maximum range.

Therefore, the best dynamic range for the second-order distortion is therefore $A = 72.5$ dB and for the third-order distortion it is $B = 81.7$ dB. Practically, the intersection of the noise and distortion graph is not sharply defined because the noise adds to the continuous wave (CW) like distortion and reduces the dynamic range by a further 2 dB.

The plot for other resolution bandwidths can be added to the graph as required and shows that by reducing the resolution bandwidth the dynamic range can be improved.

The two points A and B in Figure 16.26 show the second and third dynamic range improvement by changing the resolution bandwidth from 10 kHz to 1 kHz.

Unfortunately, there is no one to one change between the lowered noise floor and the improvement in the dynamic range. And for the second order the change is one-half of the change in the noise floor and for the third-order distortion two-thirds of the change in the noise floor.

16.3.7.3   Spectrum analyser local oscillator phase noise

The final item affecting the dynamic range is the local oscillator phase noise on the spectrum analyser and this affects only the third-order distortion measurements.

For example, if a two-tone third-order distortion measurement was being made on an amplifier and the test tones were separated by 10 kHz, the third-order distortion components are also separated by 10 kHz. Now, suppose we choose the resolution bandwidth of the spectrum analyser to be 1 kHz allowing for a 10 dB decrease in

370 *Microwave measurements*

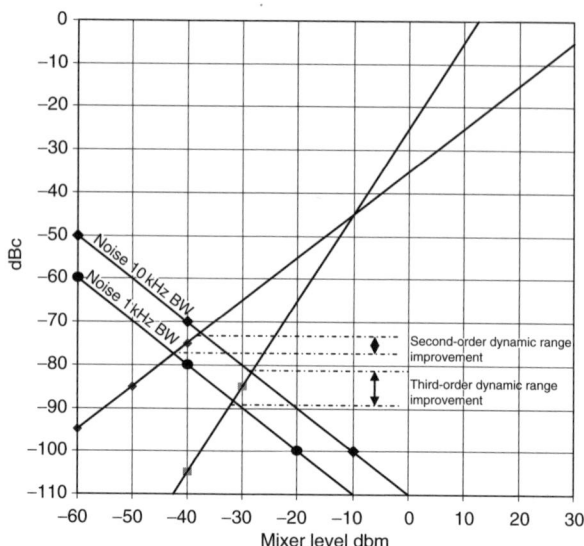

Figure 16.26 *Reducing resolution bandwidth improves dynamic range*

the noise curve then the maximum dynamic range is approximately 88 dB. But if the phase noise at a 10 kHz offset is only −80 dBc then this value becomes the limit of the dynamic range.

16.3.7.4 Selecting the optimum conditions

Figure 16.27 combines the graphs given in the two previous illustrations. From this combined graph the optimum dynamic range can be determined. The signal-to-noise ratio improves as the input mixer level is increased.

An example illustrates the use of the graph.

To determine the optimum dynamic range available to measure third-order intermodulation products the '1 kHz bandwidth (BW)' line is followed; at −34 dBm mixer level the signal-to-noise ratio is almost 90 dB.

No further improvement is possible because as the mixer level is increased further the level of the third-order intermodulation products increases. At a mixer level of −30 dBm, the dynamic range is reduced to 80 dB.

In addition to the three key aspects highlighted above, other points are also covered in this section, such as sideband noise, residual responses, residual FM and input overload, where experience shows that these areas are also frequently misunderstood.

16.3.7.5 Sideband noise

Three specification points affect the ability of a spectrum analyser to measure low-level signals close to high-level signals. Two of the points have already been described; they are minimum resolution bandwidth and resolution filter shape factor.

The third point is the sideband noise of the local oscillators in the instrument.

Spectrum analyser measurements and applications 371

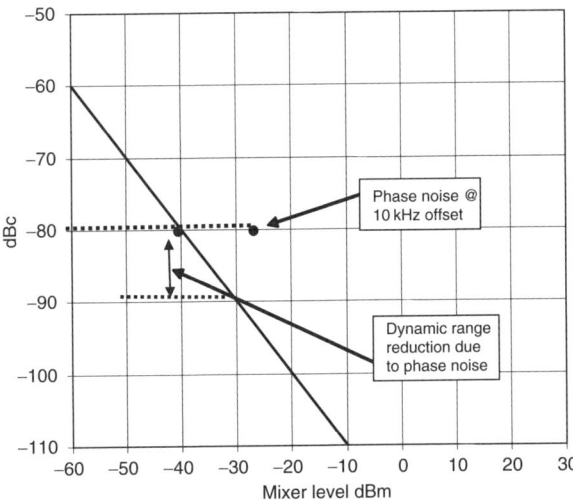

Figure 16.27 Phase noise limit

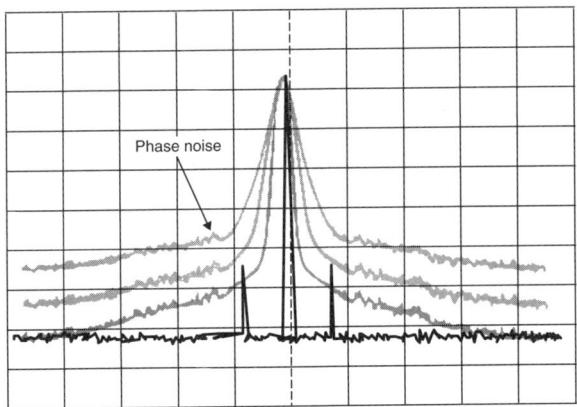

Figure 16.28 Local oscillator noise sidebands

Figure 16.28 shows the sideband noise of the instrument's local oscillator superimposed on the resolution bandwidth response. Measurement of low-level signals close to a carrier can be impaired if sideband noise is too high. When developing spectrum analysers designers endeavour to keep the local oscillator phase noise as low as possible.

16.3.7.6 Checking for internal distortion

Some spectrum analysers have an 'Intermodulation Identify' key (Figure 16.29) to automate and simplify the self-test procedure. In the latest spectrum analysers the

372 *Microwave measurements*

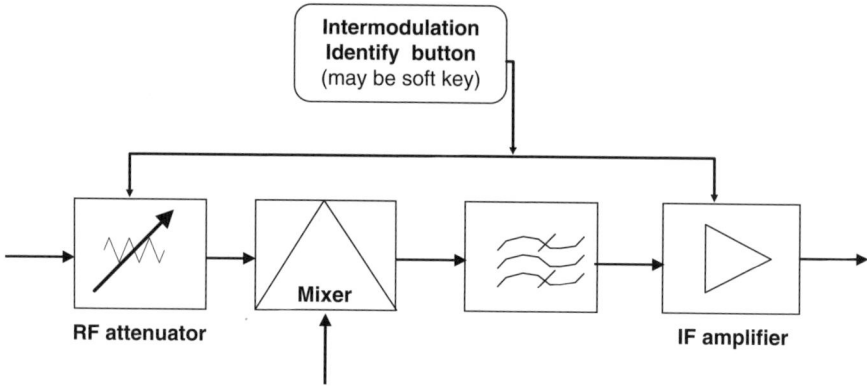

Figure 16.29   Intermodulation distortion identification button

intermodulation key may be a 'soft key' and is included as a part of the software functions that appear on the display. However, when the key is pressed additional input attenuation is introduced and the IF amplification is simultaneously increased by an equal amount. If signal levels seen on the display do not move then the measurement is valid. This is a useful, quick and effective way to check for a possible mixer overload situation.

If this feature is not available then a useful way to check for any internal overload is to introduce temporarily additional RF attenuation. If a further 10 dB of attenuation is introduced, then all the signals on the screen should drop by 10 dB. If the level changes by a different amount then this indicates that the spectrum analyser is being overloaded and distortion is present.

16.3.8   Amplitude accuracy

A good amplitude accuracy specification is essential for accurate and repeatable measurements, but there can be considerable measurement uncertainty if the input match is poor.

16.3.9   Effect of input VSWR

The input match, generally expressed as VSWR, reflection coefficient or Return Loss, is a measure of the proportion of the signal incident at the input that is reflected back. Amplitude measurement uncertainty deteriorates; as the match becomes worse, the effect is aggravated more if the source match is poor.

The graph of Figure 16.30 shows a convenient plot to give an estimate of the uncertainty limits for a variety of source and load values. The uncertainties rise considerably as the matches become worse. For example, Figure 16.30 shows the mismatch uncertainty for a source VSWR of 2.0:1 and the spectrum analyser input VSWR of 1.5:1 gives a mismatch uncertainty of 1.2 dB.

Spectrum analyser measurements and applications 373

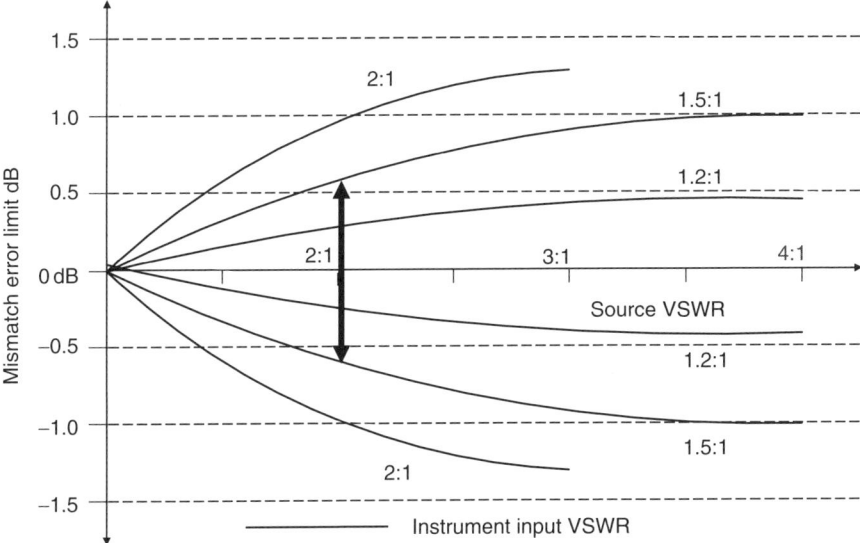

Figure 16.30   Input mismatch uncertainty

16.3.10 Sideband noise characteristics

Figure 16.31 shows the typical sideband noise performance of a quality spectrum analyser. The Figure shows how the sideband noise can reduce close-in resolution as well as reducing dynamic range even for measurements 200 kHz away from the carrier.

16.3.11 Residual responses

In an earlier section, the problem of spurious responses was highlighted. A spectrum analyser can display a signal on the screen although no signal is present at the input. Instrument designers endeavour to eliminate this undesirable phenomenon but these residual responses are known to be present in all instruments to a greater or lesser extent. Residual responses occur because within a spectrum analyser there are a number of local oscillator frequencies and their harmonics which can mix with each other to produce signals which can fall within the IF bandwidth and will appear as false signals.

Active RF and microwave systems frequently generate non-harmonically related signals that need to be identified and measured. Tracking down and then reducing the level of unwanted spurious signals is a very common application of a spectrum analyser.

Inexperienced spectrum analyser users can have problems with such a measurement if they are unaware of the limitations of the instrument. The problem of internally generated harmonically related distortion products has been described but a spectrum

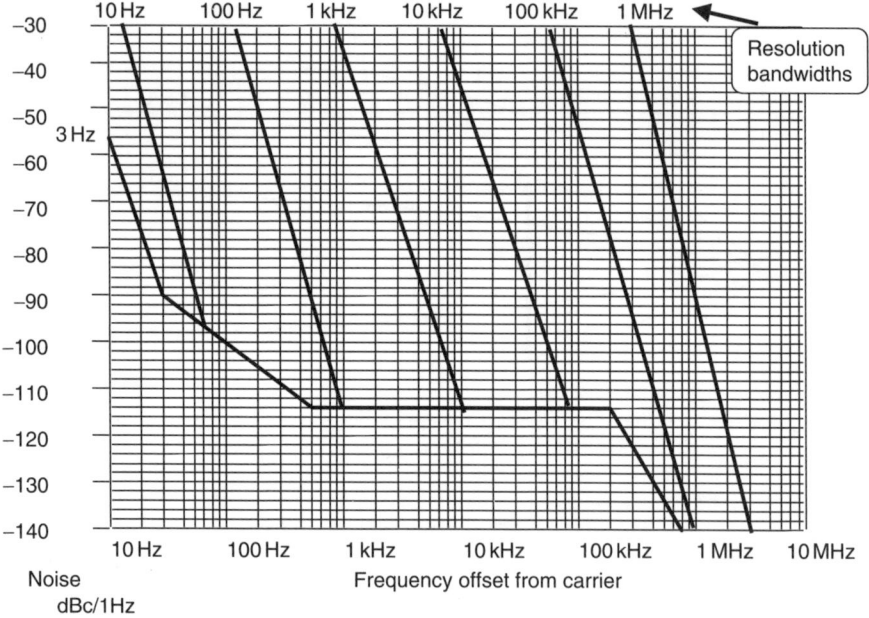

Figure 16.31   Sideband noise graph

analyser itself can have spurious responses. It is essential to ensure that a signal visible on the screen is not generated within the spectrum analyser. The internally spurious signals generated can either be caused by residual responses that are an inherent limitation of the design or caused inadvertently by the operator if the instrument is overloaded. Image responses and multiple responses are also encountered in microwave spectrum analysers if a preselector is not used.

Residual responses (see Figure 16.32) can create significant measurement problems so it is important to purchase an instrument with a very good specification. Residual responses of a quality instrument are typically less than -120 dBm to -110 dBm. Some instruments can have inferior specifications or in some cases, the residual responses are not even quoted at all.

To be absolutely certain that a signal is not being internally generated it may sometimes be necessary to replace the signal being analysed with a known pure signal and to investigate the difference.

16.3.12   Residual FM

An important specification point is residual FM. If the local oscillator in the spectrum analyser has appreciable FM on it then close to carrier measurements cannot be made. Residual FM on a quality instrument will vary from around 1 Hz to 10 Hz depending on frequency range. Figure 16.33 shows how poor residual FM can invalidate close-in measurements.

Spectrum analyser measurements and applications 375

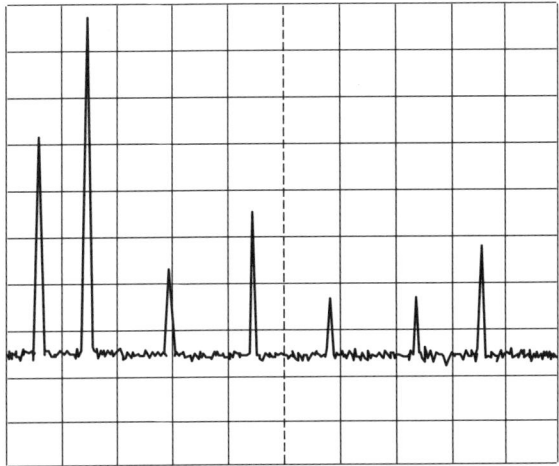

Figure 16.32   Residual responses

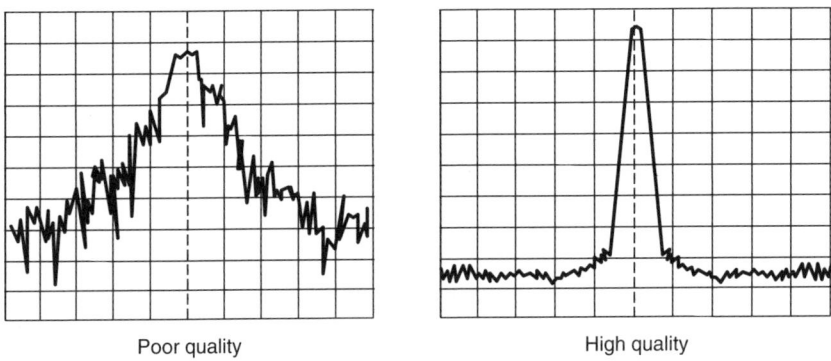

Poor quality   High quality

Figure 16.33   Residual FM

16.3.13   Uncertainty contributions

A spectrum analyser is a very complex device with many elements, which can change with frequency, temperature and time. Each element contributes towards the inaccuracy or uncertainty of a measurement. Figure 16.34 shows a simplified block diagram of a typical instrument with uncertainty contributions added. These figures are taken from the specification of an instrument in present widespread use. For a given measurement, all the uncertainties may not necessary apply, but the accuracy of such an instrument is poor. The problem can be worse when it is realised that with many instruments it is necessary to adjust front panel presets to obtain such accuracy. This relies on the diligence and skill of the operator and is therefore not reliable.

Some spectrum analysers use an automatic self-calibration process and at the touch of a button on the front panel or a soft key the instrument runs through a self-calibration

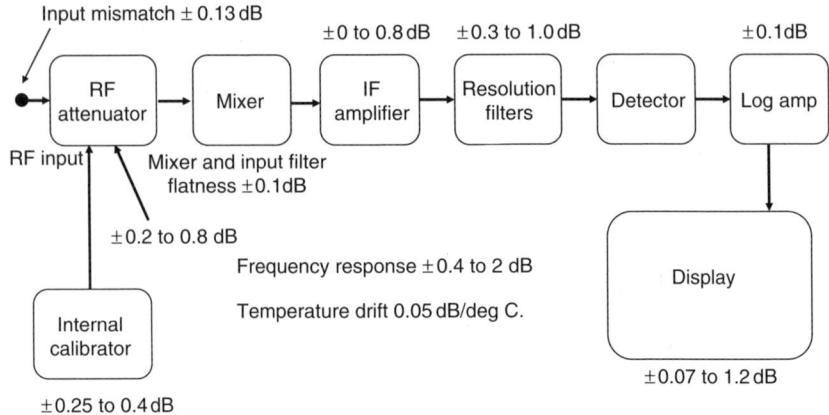

Figure 16.34 Uncertainty contributions

routine. A typical self-calibration routine includes setting up the amplitude and frequency of each of the resolution filters, measuring and correcting for the attenuation of each of the input attenuator steps.

Instruments that have a built-in tracking generator can also correct for the frequency response of the system by sweeping through the entire frequency range whilst routing the amplitude levelled tracking generator into the input. The advantage of automatic self-calibration is that total level accuracy is improved dramatically and the specification is valid for all levels and frequencies and for any span or resolution bandwidth. For engineers who need to produce uncertainty budgets a useful approach is to list all the contributions to the uncertainty of measurement and then to include only those that affect a particular measurement in a final budget as shown in Figure 16.35.

16.3.14 Display detection mode

Modern spectrum analysers use digital methods for acquiring and manipulating the data to display. The input data at the input of the spectrum analyser is placed in to segments sometimes called bins and the bins are digitally sampled for further processing and then displayed. The point in the bin where the data are sampled will clearly affect the displayed information. Spectrum analysers may have a number of selectable detector modes and the mode of detector chosen will determine how the input signal is displayed. Table 16.1 shows the advantages and disadvantages of the various detector modes.

16.4 Spectrum analyser applications

Spectrum analysers are used to make a very wide range of measurements. It is not possible to cover all the possible applications but the more common measurements are included in this section.

Spectrum analyser measurements and applications 377

	CW signal	Harmonic distortion	3rd order inter modulation products	3rd order intercept	Channel power	Adjacent channel power ratio	Power versus time for TDMA signals	Phase noise far from carrier	Phase noise close to carrier
Absolute level	✓				✓	✓			
Frequency response	✓	✓			✓	✓			
RF attenuation	✓				✓	✓		✓	
IF gain	✓				✓	✓		✓	
Linearity error	✓	✓	✓	✓	✓	✓	✓	✓	✓
Bandwidth switching	✓				✓	✓			
Resolution Bandwidth	✓				✓	✓		✓	✓
Sampling					✓	✓			
Mismatch	✓	✓			✓	✓			

Figure 16.35 Typical uncertainty contributions for some spectrum analyser measurements

Table 16.1 Detector modes

Detector mode	Method	Advantages	Disadvantages
Peak	Detects the highest point in the bin	Good for analysing sinusoidal waveforms	Over responds to noise
Sample	Detects the last point in the bin whatever the power	Good for noise measurement	Not good for CW signals with narrow bandwidths and will miss signals that do not appear at the same point in the bin
Negative peak	Detects the lowest power level in the bin	Good for AM/FM demodulation and can distinguish between random and impulse noise	Does not improve the analyser sensitivity although the noise floor will appear to fall
Rosenfell	Dynamically classifies the data as either noise or signal	Gives an improved display of random noise compared with peak detection and avoids the missed signal problem of sample detection	Only used in the high performance spectrum analysers

378 Microwave measurements

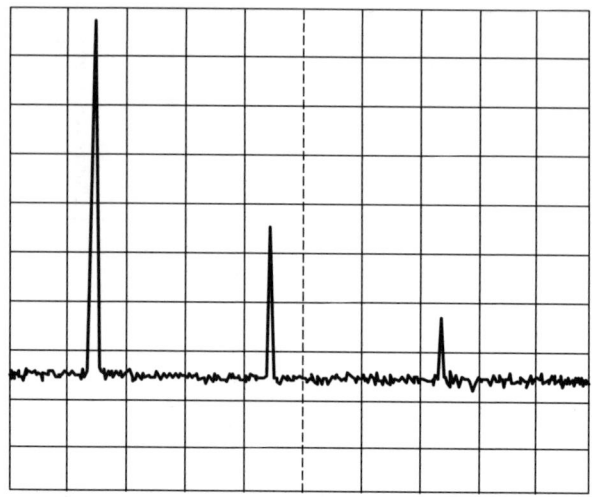

Figure 16.36 Harmonic distortion

16.4.1 Measurement of harmonic distortion

A spectrum analyser can be used to measure the amplitudes of the fundamental and even very low-level harmonics. Sometimes, however, it is necessary to quote not only the level of the harmonic distortion products but also to give the total harmonic distortion.

The total harmonic distortion as shown in Figure 16.36 can be calculated by measuring the amplitudes of all the harmonics and then take the square root of the sum of the squares.

16.4.2 Example of a tracking generator measurement

The display shown in Figure 16.37 is a typical tracking generator measurement, the analysis of a 10.7 MHz band-pass filter over a wide dynamic range. The display shows two different traces simultaneously.

The upper trace shows the overall response of the filter over a dynamic range in excess of 80 dB. The other trace shows the ripple on the pass-band of the filter displayed with a resolution of 0.5 dB per division.

16.4.3 Zero span

The principal function of a spectrum analyser is to sweep through a selected part of the frequency spectrum. In certain circumstances, however, it may be necessary to analyse the characteristics of just one fixed portion of the spectrum. The zero span mode is used for such applications. In this mode, the local oscillator of the instrument is no longer swept; the oscillator is held at a fixed frequency so that the signal of interest can be studied. If sweeping ceases one would expect to merely see a dot or

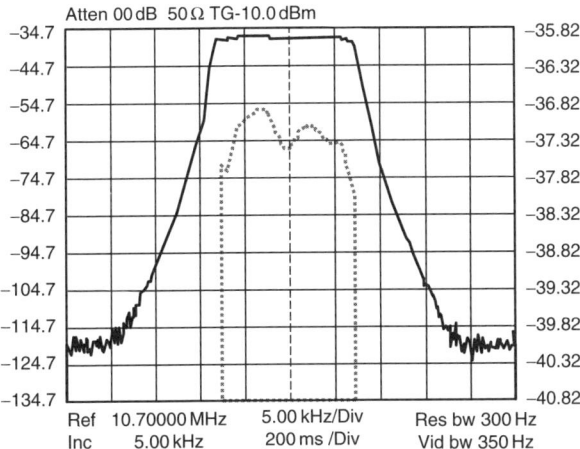

Figure 16.37   Measurement of a 10.7 MHz band-pass filter

line on the display, which moves up and down according to the change in amplitude of the signal to which the instrument is tuned. This would provide a certain amount of information, but much more information is obtained if a time base sweeps the spot horizontally in a manner similar to the technique used in oscilloscopes.

By sweeping the spot horizontally the display will show amplitude versus time variations of the signal to which the instrument is tuned.

16.4.4   The use of zero span

There are many applications of zero span mode but one of the most obvious is to demodulate an amplitude-modulated carrier as shown in Figure 16.38. Another common use is to measure response times, one example is the measurement of transmitter decay time at switch off; this can be a critical measurement since it may determine how quickly an adjacent sensitive receiver can be enabled. Synthesiser switching times and overshoots can also be evaluated using the zero span mode.

The time base of modern sophisticated instruments is derived from the reference oscillator. This ensures the very best accuracy when timing measurements are made. Some instruments only have an inaccurate time base, so it is a wise precaution to check the specification of the instrument before making a measurement.

16.4.5   Meter Mode

In addition to the zero span mode some instruments incorporate a 'Meter Mode'. This is used for applications where a spectrum display needs to be retained whilst still monitoring the changing amplitude of a part of the spectrum.

A typical application of 'Meter Mode' is shown in Figure 16.39. The amplitude of the FM carrier is continuously updated in real time whilst the rest of the display is saved. Any part of the display, selected by the movable marker, can be updated

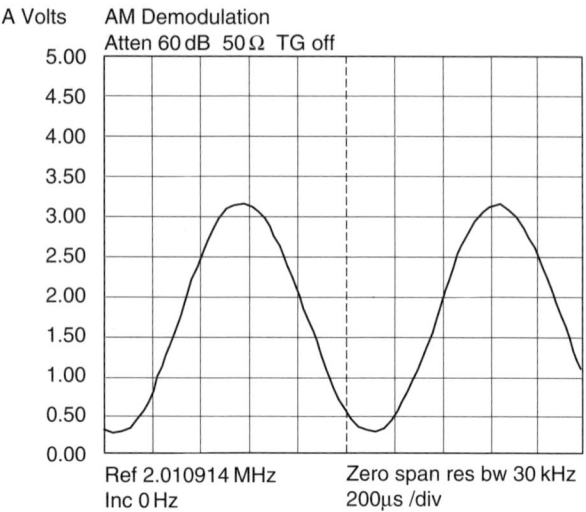

Figure 16.38   AM demodulation

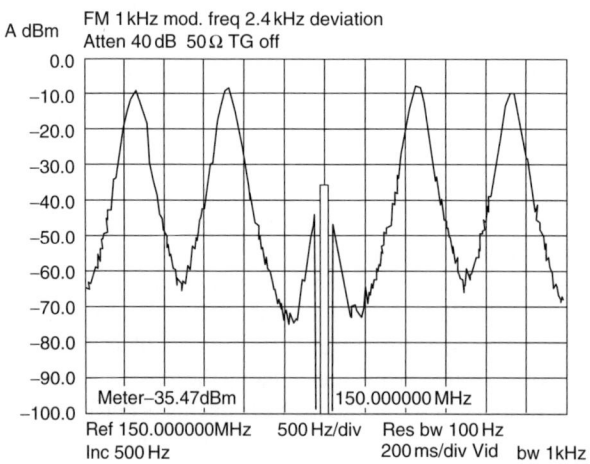

Figure 16.39   Meter Mode

and monitored. This method is very useful when measuring carrier deviation by the Bessel Disappearing Carrier Technique.

16.4.6   Intermodulation measurement

Measuring the harmonic distortion caused by a device is not a very discriminating measurement. A more searching method is to use two or more test signals and to

Spectrum analyser measurements and applications 381

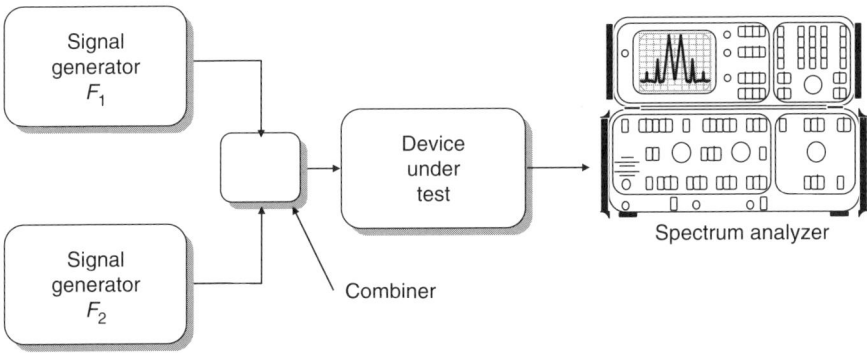

Figure 16.40   Two-tone test set up

measure the intermodulation products that are generated at the output of the device under test. By using more than one test signal the device receives signals that are closer to the more complex signals that are generally encountered in practical systems. Two separate signal generators and a combiner are needed as shown in Figure 16.40. There are also special signal sources developed that contain two or more sources in order to provide the best possible signals for this test.

Another problem is that any non-linearity in the output amplifiers of the signal generators can produce intermodulation. Further problems can arise if the automatic level control (ALC) detector at the output of one signal generator also detects the signal from the other signal generator. It is for these two reasons that it is good practice to insert an attenuator between the signal generator output and the combiner. This may not be practical in some circumstances, because the signal level may be too low. For higher frequency measurements, an isolator is recommended to improve the measurement integrity.

16.4.7   Intermodulation analysis

A typical spectrum analyser display of a two-tone intermodulation test is shown in Figure 16.41. Annotation has been added to explain the origin of the intermodulation products. Signal generator 1 has a fundamental frequency of F_1 and signal generator 2 has a fundamental frequency of F_2. Non-linearity in the device under test will cause harmonic distortion products of frequency $2F_1$, $2F_2$, $3F_1$, $3F_2$, etc. to be generated.

Spectrum analyser will record these harmonic distortion products but the significance of the intermodulation test is that the non-linearity causes the harmonic products to mix together to generate additional signals. Numerous intermodulation products can be generated but the two most commonly encountered ones are known as the third-order and fifth-order products.

Third-order products have frequencies of $2F_1 - F_2$ and $2F_2 - F_1$

Fifth-order products have frequencies of $3F_1 - 2F_2$ and $3F_2 - 2F_1$

382 *Microwave measurements*

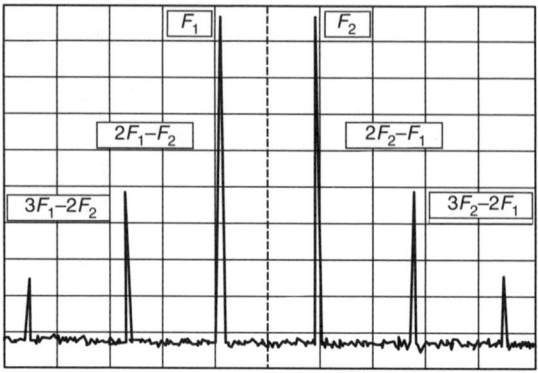

Figure 16.41   *Intermodulation display*

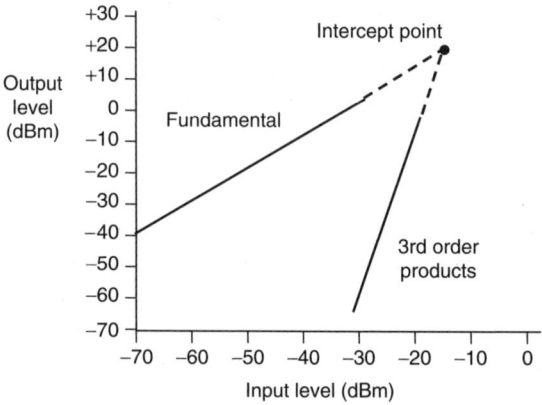

Figure 16.42   *Intermodulation intercept*

Even order products such as $F_1 + F_2$ and $F_2 - F_1$ are also seen but are less significant since the intermodulation products are widely separated from the two frequencies (F_1 and F_2) and they can generally be readily rejected.

High-performance spectrum analysers have an intermodulation distortion of typically −95 dBc or better with a signal level of −30 to −40 dBm at the input mixer to allow for the measurement of low levels of distortion.

16.4.8   *Intermodulation intercept point*

The amplitudes of intermodulation products change according to the amplitudes of the test signals applied; therefore, it is necessary to specify the level of the test signals. It can be difficult to compare the performance of different devices if they were measured at different levels. The solution is to use the concept of an intermodulation intercept point. An intercept point is the theoretical point at which the amplitudes of the intermodulation products equal the amplitudes of the test signals, the illustration shows the concept. There are two lines on the graph in Figure 16.42.

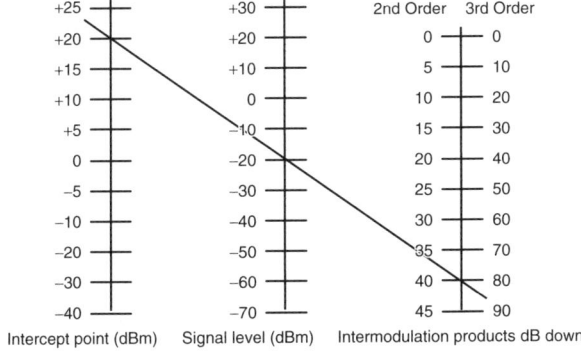

Figure 16.43   Nomograph to determine intercept

The fundamental line shows a linear relationship between the input and output signals, the line has been extrapolated beyond the output level of +5 dBm since at such levels the response becomes non-linear. Input and output signal levels have also been plotted for the third-order products and the line is extrapolated. The two lines meet at the intermodulation intercept point.

The slope of the intermodulation product lines is equal to their order, that is, the second-order lines have a slope of 2:1, the third-order lines have a slope of 3:1. Practically, this means that if the level of the test signal is reduced by 10 dB then the third-order product will theoretically drop by 30 dB, provided that the device is operating in a linear mode.

16.4.9 Nomograph to determine intermodulation products using intercept point method

The nomograph in Figure 16.43 gives a rapid but not very accurate means of determining the intercept point. A straight edge is used to join the two known values so that the unknown can be determined.

16.4.10 Amplitude modulation

Figure 16.44 shows an idealised spectrum analyser display of an amplitude-modulated signal. The carrier frequency is F_c; the frequency of the modulating signal is F_m.

Three separate frequency components are seen

> The carrier frequency F_c
> Lower side frequency $F_c - F_m$
> Upper side frequency $F_c + F_m$

The modulation depth in per cent is given by the following formula:

$$\text{Per cent modulation} = 2 \times \frac{\text{side frequency amplitude}}{\text{carrier amplitude}}$$

$$\times 100 \text{ (measured on a linear scale)}$$

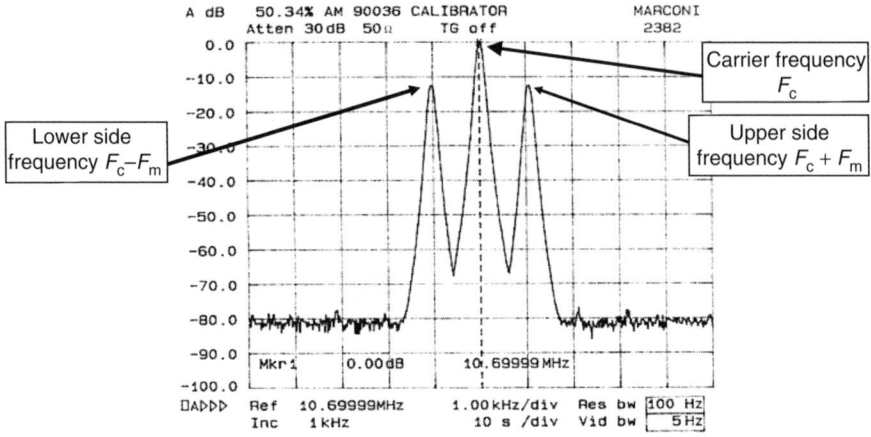

Figure 16.44 Amplitude modulation measurement

The amplitude of the carrier always remains constant as the modulation depth changes but the sideband amplitudes will change in proportion to the modulation depth. The frequency separation between the carrier and either sideband changes as the modulation frequency changes.

When the modulation depth is 100 per cent half of the power is in the sidebands and each sideband frequency amplitude will be 6 dB less than that of the carrier. For lower modulation depths, the sideband amplitude is proportionately less. To measure modulation depth it is thus necessary to measure the amplitude difference between the carrier and the sidebands.

16.4.11 AM spectrum with modulation distortion

In practice, there will be harmonics of the modulation frequency also present at $F_c \pm nF_m$. Figure 16.45 shows distortion produced at $F_c \pm 2F_m$.

16.4.12 Frequency modulation

An FM spectrum theoretically has an infinite number of sidebands, which are symmetrical about the carrier and separated by the modulation frequency.

The FM spectrum display shown in Figure 16.46 is thus considerably more complex than an AM spectrum display. Sideband and carrier amplitudes are determined by the unmodulated carrier amplitude and the modulation index (β) which is expressed as

$$\text{Modulation index, } \beta = \frac{\text{Frequency deviation}}{\text{Modulation frequency}}$$

In practice, although there are an infinite number of sidebands the amplitudes of the higher frequency ones rapidly reduce to near zero and can be neglected.

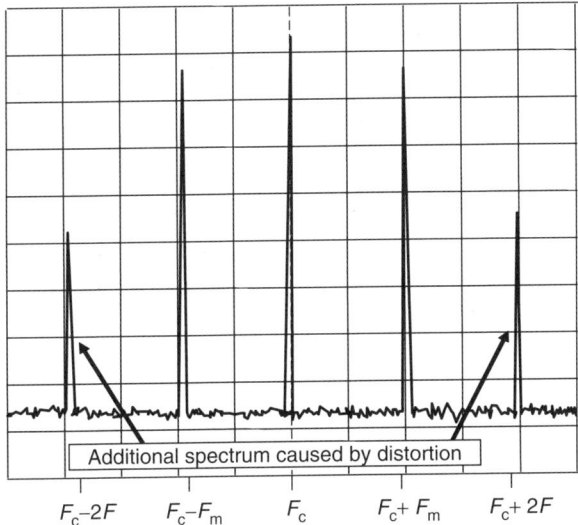

Figure 16.45   Modulation distortion

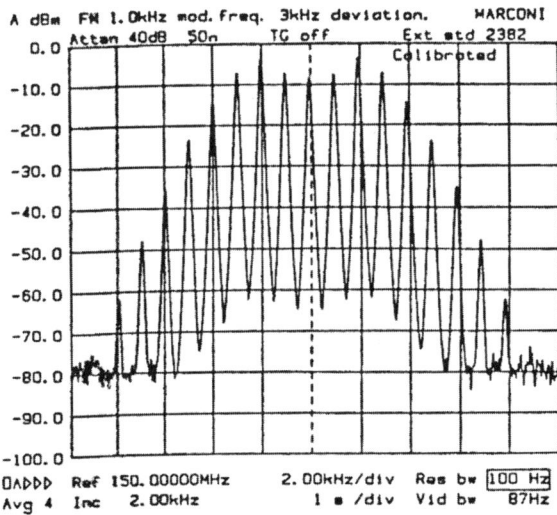

Figure 16.46   Frequency modulation spectrum

16.4.13   FM measurement using the Bessel zero method

With frequency modulation the carrier amplitude is not constant; it varies according to the modulation index and will become zero at times. The sideband amplitudes also become zero at specific values of modulation index. Modulation indices at which the carrier or sidebands have zero amplitude can be calculated. Tables are available

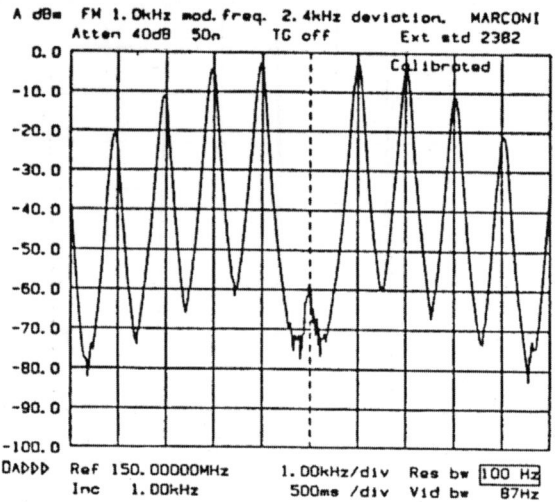

Figure 16.47   Bessel null

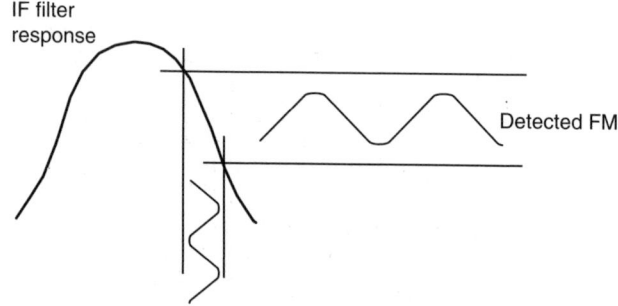

Figure 16.48   FM demodulation

listing the zeros, or Bessel nulls as they are more commonly called. Bessel zeros (see Figure 16.47) are used for accurate calibration of signal generators and modulation meters.

16.4.14   FM demodulation

If zero span mode is used on a spectrum analyser no information should be seen if frequency modulation is applied since zero span shows amplitude variation with time. However, if the spectrum analyser is de-tuned by a small amount then the demodulated signal will be seen. This occurs because the slope of the resolution filter acts as a slope detector as shown diagrammatically in Figure 16.48.

Accurate measurements are not possible but this does provide a convenient method to view a demodulated signal. It should be noted that the technique might be invalid if significant spurious AM is present in addition to the FM. Some spectrum analysers can measure FM directly.

The demodulated FM signal is displayed on a graticule that is vertically calibrated in FM deviation; the horizontal scale is calibrated in time as for the zero span. The illustration shows the technique used.

16.4.15 FM demodulation display

Some spectrum analysers incorporate a function that demodulates the FM signal and displays deviation vertically against time horizontally.

A typical FM demodulation screen display from a spectrum analyser is shown in Figure 16.49. The peak-to-peak FM deviation can be readily measured from the vertical scale.

16.4.16 Modulation asymmetry – combined AM and FM

Simultaneous amplitude and frequency modulation is usually an undesired effect rather than a deliberate form of modulation. It usually results when amplitude modulation is being generated. What happens is that the carrier oscillator frequency is pulled by the modulating signal and hence introduces a small amount of FM together with the desired AM. The result is that AM together with narrowband FM is present at the same modulating frequency producing a combined spectrum.

The AM spectrum consists of the carrier and two sidebands but the FM spectrum will consist of a carrier and an infinite number of sidebands but it must be remembered that the amplitude of the FM sidebands falls off very quickly outside of the peak deviation $\pm \Delta F$.

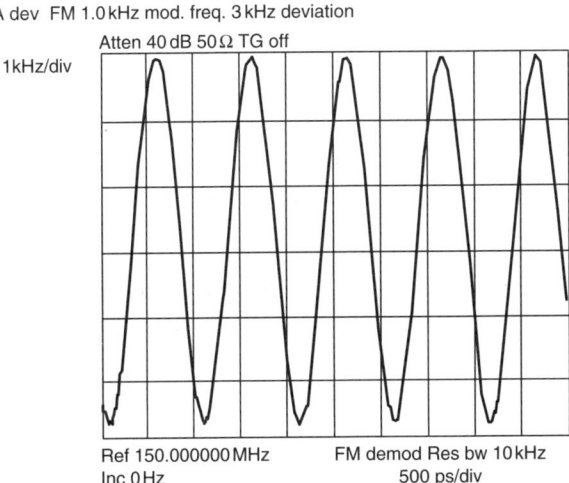

Figure 16.49 Peak-to-peak FM deviation display

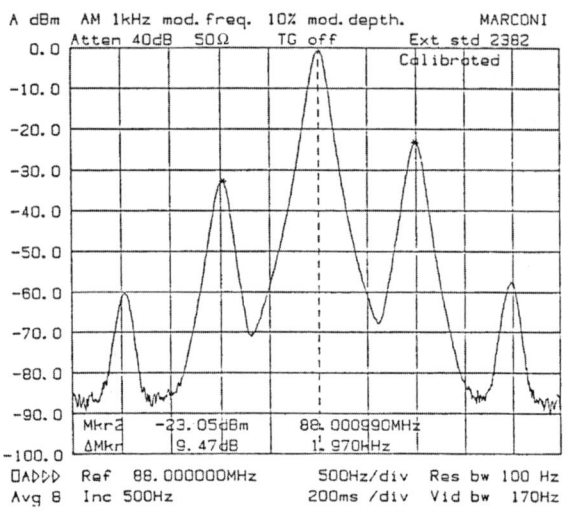

Figure 16.50   AM and FM asymmetry

For narrowband FM where ΔF is considerably less than the modulating frequency f, the higher-order sidebands fall off so rapidly that only the first sideband need be considered. The narrowband FM spectrum differs from AM in that one side band is 180° out of phase with respect to the other sideband. So the resulting spectrum when seen on the spectrum analyser as in Figure 16.50 will be a spectrum where one sideband is larger than the other and it clearly shows the presence of FM on AM.

Where the difference in the amplitudes is less than 20 per cent the modulation depth can be calculated by taking the mean of the two-sideband amplitudes to represent the amplitudes of the sidebands due to the AM alone.

16.4.17   Spectrum of a square wave

Pulsed RF waveforms are most commonly encountered in radar systems both at IF and at microwave frequencies. To understand the analysis of pulsed RF it is first necessary to study the spectrum of a square wave. Figure 16.51 shows the idealised oscilloscope display of a train of rectangular pulses of pulse repetition frequency, F, and pulse width, t.

The corresponding spectrum analyser display in the illustration shows that the individual spectral lines are spaced by the pulse repetition frequency $1/t$. The spectrum analyser display also shows that the amplitudes of the individual spectral lines rise and fall in a regular way; the pulse envelope of the spectral lines follows a curve of the form represented by the expression $y = \sin x/x$. The first zero of the sin x/x envelope occurs at a frequency equal to $1/t$. Subsequent zeros occur at multiples of $1/t$. Each of the rising and falling patterns is referred to as a lobe. In theory, the lobes continue to infinity but in practice the amplitudes of the lobes soon become negligible as the frequency rises.

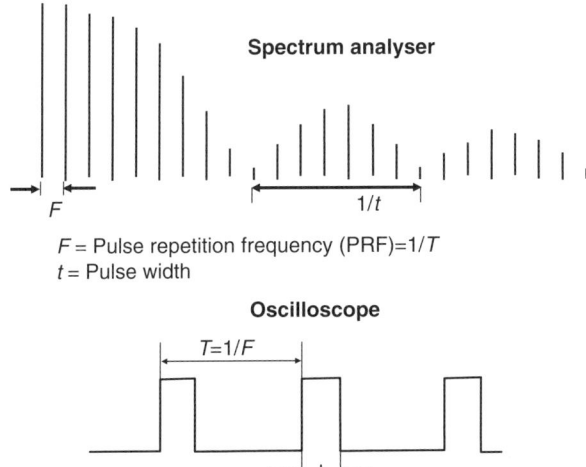

Figure 16.51   Spectrum of a square wave

16.4.18   Pulse modulation

Figure 16.52 shows a typical spectrum analyser display of a pulse-modulated carrier. The spectral line, which can be seen at the centre of the display, is the RF carrier.

The individual spectral lines, which are symmetrical about the carrier, are separated by a frequency equal to $1/T$ as for the basic pulse train; refer back to Figure 16.45 for clarification. The $\sin x/x$ zeros again occur at multiples of $1/t$. The display is only theoretically symmetrical about the carrier since in some practical radar systems, where there are imperfections, the display may be asymmetrical.

16.4.19   Varying the pulse modulation conditions

Pulsed RF can be confusing since the spectrum analyser display depends on both the pulse repetition frequency and the period of the modulating signal. The illustration helps to clarify the situation by showing how the characteristics of a pulse-modulated spectrum change according to the changes in the pulse width and pulse repetition frequency. The upper portion of each of the four displays shows the oscilloscope representation of the modulating waveform; the lower portion of each of the four displays is the spectrum analyser representation of the pulsed RF signal.

The Display 1 (top left) of Figure 16.53 is an arbitrary starting point.

In Display 2 (top right) of Figure 16.53 the pulse width of the modulating signal is increased whilst the pulse repetition frequency is the same. Increasing the pulse width reduces the value of $1/t$ so the first zero is at a lower frequency; the lobes are thus narrower.

In Display 3 (bottom left) of Figure 16.53 the pulse width is the same as for Display 1 but this time the pulse repetition frequency is lower. Spectral lines are spaced

390 *Microwave measurements*

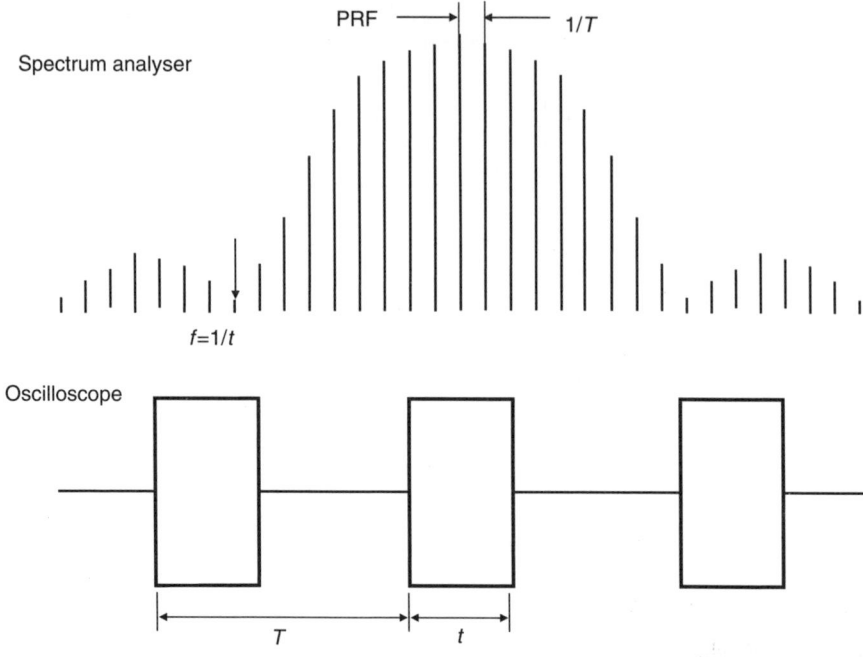

Figure 16.52   Pulse modulation

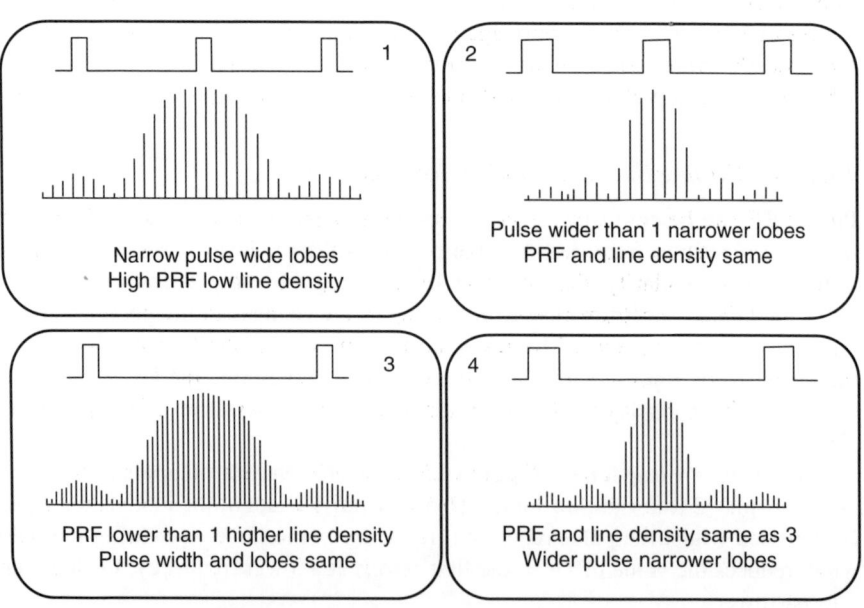

Figure 16.53   Varying pulse modulation

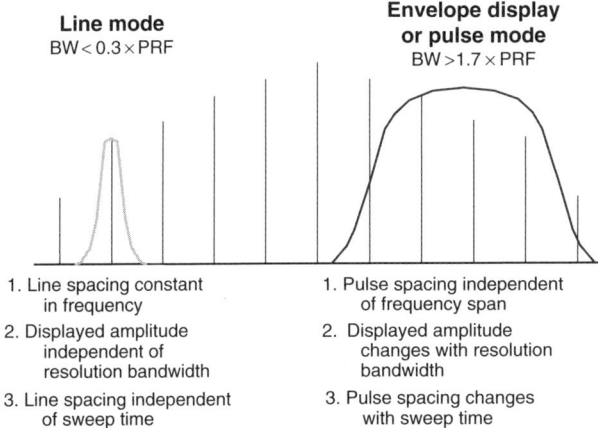

Figure 16.54   Line and pulse mode

according to $1/T$ so the line density is increased as the pulse repetition frequency is decreased.

The Display 4 (bottom right) of Figure 16.53 again shows that a wider pulse causes narrower lobes.

16.4.20   'Line' and 'Pulse' modes

Pulsed RF spectrum analysis is complicated because the display changes according to the resolution bandwidth selected; if it is significantly higher than the pulse repetition frequency then individual spectral lines will not be resolved. Figure 16.54 shows the frequency response of the resolution filter superimposed over a train of pulses. On the left the resolution bandwidth is shown to be less than the pulse repetition frequency so individual spectral lines are resolved; this is known as 'line mode'. On the right the resolution bandwidth is greater than the pulse repetition frequency so individual spectral lines are not resolved; this is known as 'pulse mode'.

In the pulse mode, the display seen is not a true frequency-domain display; it is a combination of a time and frequency display. The lines are displayed when a pulse occurs irrespective of the instantaneous tuned frequency of the instrument. The display is in fact a time-domain display of the spectrum envelope.

One can rapidly determine that a pulse mode display is occurring by changing the scan time or sweep time; the pulse line spacing will change. The line spacing will not change when the span is changed, as one would expect for a normal spectrum analyser display. A further characteristic of a pulse display is that the displayed amplitude increases as the bandwidth increases. A 'rule of thumb' to apply for line mode is to use a resolution bandwidth of less than $0.3 \times$ pulse repetition frequency. For pulse mode, the resolution bandwidth should be greater than $1.7 \times$ pulse repetition frequency.

392 Microwave measurements

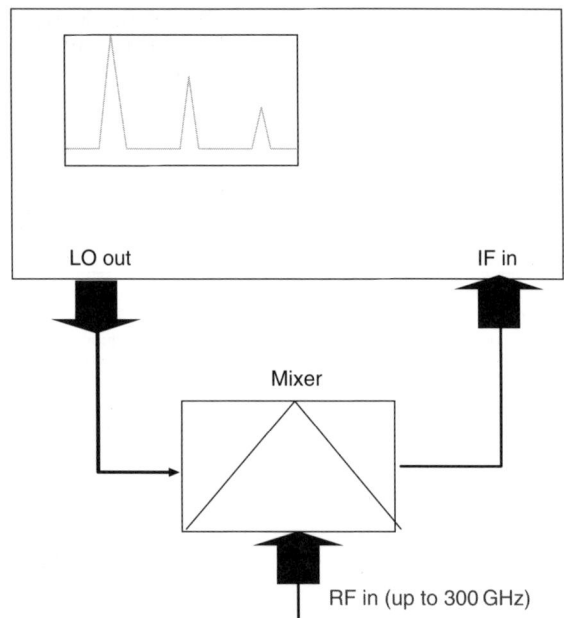

Figure 16.55 Extending the frequency range

16.4.21 Extending the range of microwave spectrum analysers

Most RF spectrum analysers have 'Local Oscillator Output' and 'IF Input' connectors on the front panel to allow the frequency range to be extended higher with the use of external millimetric mixers.

Although this can be useful for a 'quick look see' the measurements can be misleading due to poor amplitude accuracy and multiple responses. This feature shown in Figure 16.55 should be used with extreme caution!

16.4.22 EMC measurements

The spectrum analyser can be used to make EMC measurements and in some cases additional types of detector are included such as Peak, Average, RMS and Quasi Peak detectors. The spectrum analyser is a very useful instrument to use as a diagnostic tool for EMC measurements.

Figure 16.56 shows a plot from a spectrum analyser used for a conducted measurement measured in a semi-lined screened room. The test was made to comply with the European EMC Standards EN 55011 and EN55022 limits.

16.4.23 Overloading a spectrum analyser

Applying AC or DC signals of greater than approximately 0.5 W (+27 dBm) can permanently damage the input of a spectrum analyser. The input attenuator and mixer can be destroyed resulting in costly repair and loss of use of the spectrum analyser.

Peak levels plotted. Class A and B Quasi-peak and Average limits shown reference EN55011 and EN55022.

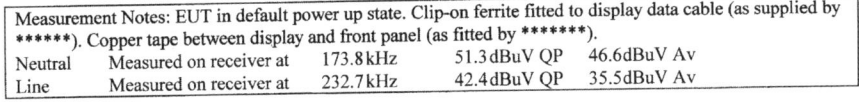

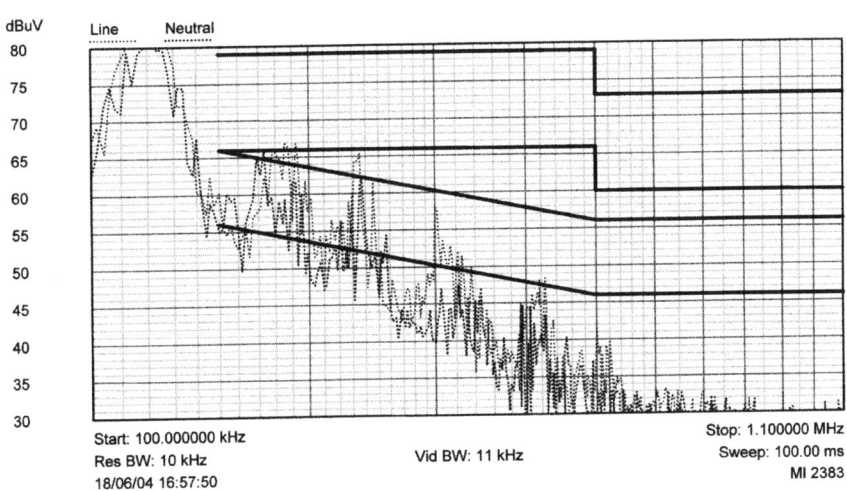

Figure 16.56   Spectrum analyser EMC display

Two techniques to protect against overload are incorporated in modern spectrum analysers. In VHF and UHF instruments, a coaxial relay is generally incorporated in the input. Protection to 50 W is possible on VHF instruments. Microwave instruments can be switched to AC input so that DC voltages up to 50 V can be safely applied.

16.5 Conclusion

The above paragraphs show how useful the spectrum analyser is in design and calibration and there are many more fields of measurements and applications where it can be used. The purpose of this chapter is to act as an introduction to the types and applications of spectrum analysers. Finally, remember what you see is not necessarily what you have got! There is a trend to develop multifunction RF and microwave instruments and they often include a spectrum analyser.

Further reading

1 Hewlett Packard: *Spectrum Analyser Basics*, Application note 150, Publication number 5952-0292, November 1989
2 Hewlett Packard: *8 Hints to Better Spectrum Analyser Measurements*, Publication number 5965-6854-E, December 1996

3 Hewlett Packard: *Amplitude and Frequency Modulation*, Application note 150-1, Publication number 5954-9130, January 1989
4 Witte, R. A.: *Spectrum and Network Measurements* (Prentice Hall Inc., Englewood Cliffs, NJ, 1993)
5 Rauscher, C.: *Fundamentals of Spectrum Analysis*, 2nd edn (Rhode and Schwarz GMBH, Munich, 2002)

Chapter 17

Measurement of frequency stability and phase noise

David Owen

An ideal frequency source generates only one output signal with no instability in its output frequency. In reality, however, all signal sources exhibit some uncertainty in their instantaneous output frequency. The uncertainty can be expressed in a number of different ways. The method of expressing the uncertainty is likely to depend upon the intended application as well as the performance of the signal source, and in many cases a source may be characterised in more than one way.

High-accuracy frequency sources, such as crystal oscillators and rubidium or caesium frequency standards, are principally measured in terms of their long- and short-term frequency stability by directly measuring the source with a frequency counter. Aging rate is used to express the long-term change of the frequency of the source period of many hours (or more) over a while the short-term stability is a measure of the random fluctuation of the source over a period of the order of seconds. Provided a frequency counter with enough frequency resolution and a frequency standard with adequate performance are used as a reference, the measurement of stability presents no serious problems.

Adjustable sources tend to have their frequency stability measured in other ways. For communication systems the most common method of expressing the frequency uncertainty is either as residual phase or frequency modulation or as phase noise. Residual modulation is typically measured by demodulating the carrier and filtering the base band signal through a band-pass filter and measuring the signal in terms of peak, average or RMS radians or Hertz deviation.

Phase noise is the most generic method of expressing frequency instability. The carrier frequency instability is expressed by deriving the average carrier frequency and then measuring the power at various offsets from the carrier frequency in a defined bandwidth. The result is then expressed as a logarithmic ratio compared with the total carrier power. The power ratio is usually normalised to be the equivalent signal power present in a measurement bandwidth of 1 Hz. For some applications (e.g. specifying adjacent channel power on a transmitter) it can be expressed in other bandwidths (in the case of adjacent channel power the receiver bandwidth).

The various ways of expressing frequency stability are all measurements of the same physical characteristics but are specified in terms of a critical characteristic of their application. The most useful general measurement, however, is the phase noise characteristics since other measurements can be derived from a phase noise plot.

For communication systems the most important offset frequencies are those around 1 kHz, since this strongly influences the residual FM and therefore the ultimate signal–to-noise ratio, offsets between 10 and 25 kHz. At the higher frequency offsets phase noise affects transmitted adjacent channel power and adjacent channel selectivity measurements on narrow band receivers.

Phase noise characteristics are important for digital as well as analogue communication systems. The 1 kHz phase noise characteristics of oscillators in transmitters using time-domain multiple access (TDMA) or time-domain duplex (TDD) techniques often determine the residual phase or frequency jitter within a single burst of the carrier frequency. As wider bandwidth systems are adopted phase noise at larger offsets will become increasingly specified, but in general the toughest target is likely to remain the 1 kHz offset performance.

The sensitivity to the noise in the 1 kHz offset region on digital modulation systems arises because the signal is split into blocks of information, typically with a duration of 1–20 ms, for the purpose of encoding speech or adding error correction. The details of this are beyond the scope of this chapter. The blocks of information usually have within them a sequence of digital bits that are used to extrapolate the phase and frequency reference of the transmitted signal over the entire block. Having obtained this phase reference the digital data can be derived. This phase or frequency estimation process means that phase noise at low carrier frequency offsets is removed whereas noise at frequencies corresponding to the data block length can directly lead to an increase in measured modulation error. The longer the length of the data block used the more susceptible the system is to lower frequency noise.

The measurement of phase noise is likely to continue to be an important activity in the design of communication systems.

17.1 Measuring phase noise

The performance of frequency sources varies considerably and consequently making measurements can be complex. Different methods can be used according to the expected performance and the controls available to set up a measurement (Figure 17.1).

The phase noise performance of oscillators can vary greatly according to the type of oscillator and the complexity (and hence cost) of the design.

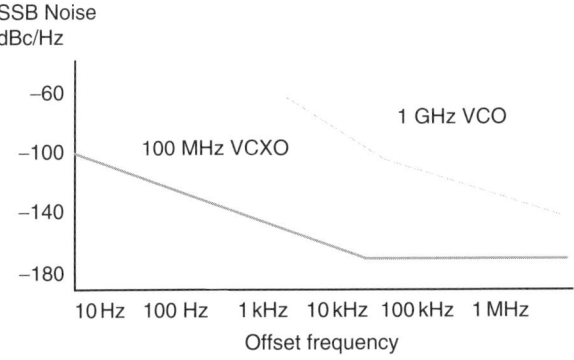

Figure 17.1   Phase noise of a voltage controlled crystal oscillator compared to a typical 1 GHz VCO

A crystal oscillator can exhibit a phase noise of -170 dBc Hz^{-1} at a 20 kHz offset from a carrier of 100 MHz.

A well-designed voltage controlled oscillator covering a frequency range of one-quarter of an octave at 1 GHz will produce a phase noise of -115 dBc Hz^{-1} at a 20 kHz offset frequency.

A typical microwave yttrium iron gasnet (YIG) oscillator could have a significantly worse performance and have the added complication of including large amounts of low-frequency uncertainty caused by disturbance of its magnetic tuning field from mains transformers, switch mode power supplies and display drive circuits.

Measuring such widely divergent oscillators causes considerable measurement problems and it is not surprising that none of the techniques solves all the problems.

Four basic measurement techniques are described based on spectrum analysers, delay line discriminators, quadrature technique and FM discriminators. All of these methods can be used to successfully measure the characteristics of a signal source and each has their advantages and disadvantages.

The methods of all measurements rely on a similar basic principle (Figure 17.2). The signal to be measured is frequency converted to a baseband or IF and then passed through a device which extracts either phase or frequency information from the carrier. A frequency selective measuring device is then used to measure the noise as a function of offset frequency. A calibration system is used to scale the results into meaningful units.

17.2 Spectrum analysers

Since spectrum analysers measure the RF signal power in a specific bandwidth they can clearly be used to measure phase noise. Most modern analysers include software functions which will convert a measured signal level from its measured value (in the

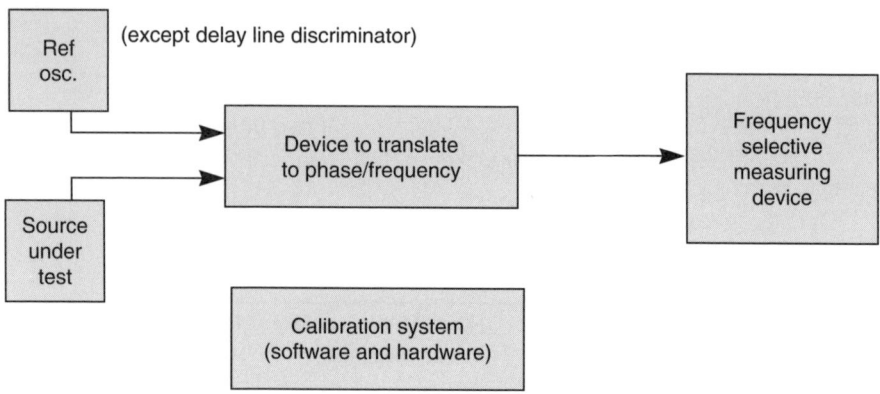

Figure 17.2 Principle of all phase noise measurements

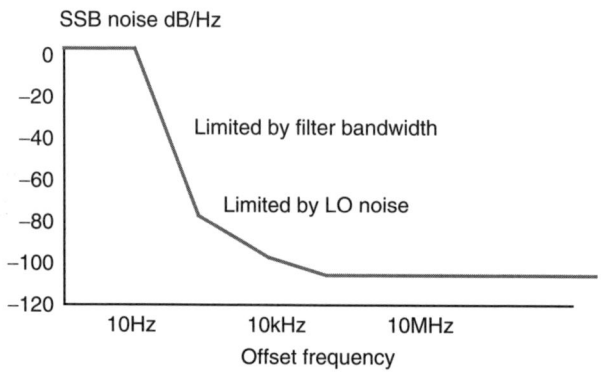

Figure 17.3 Limitations of phase noise measurements with a low/medium cost spectrum analyser

spectrum analyser filter bandwidth) to the equivalent noise signal in a 1 Hz bandwidth provided the noise can be treated as Gaussian. By measuring the total carrier power (on a wide filter setting) and then measuring the noise signal normalised to a 1 Hz bandwidth, a phase noise measurement can be derived.

In practice, the performance of simple spectrum analyser measurements is very limited (Figure 17.3). Typical spectrum analyser noise performance is not adequate to measure noise at offset frequencies much beyond 1 kHz and the minimum filter resolution bandwidth of 3 or 10 Hz limits measurements to offset frequencies above 50 Hz. Those spectrum analysers which perform narrow band measurements using digital techniques can generally perform better close to carrier measurements than analogue versions and, with the right software, provide more reliable conversion of measurement results to phase noise.

The performance at larger offsets is limited by the performance of the synthesisers used to convert the input frequency signal to the spectrum analyser measuring frequency and the relatively poor noise figure of a spectrum analyser front end converter. The noise figure arises because the spectrum analyser usually has to be optimised to obtain the best linear operating range to its maintained intermodulation and spurious specification.

17.3 Use of preselecting filter with spectrum analysers

For some applications, the noise floor and synthesiser noise limitations of spectrum analysers can be partially overcome by the use of band-pass filters. A typical measurement will require the use of second reference RF or microwave source and a mixer to convert the signal to an intermediate frequency (IF). The signal from the mixer is then passed through a band-pass filter and amplifier before being measured by the spectrum analyser. Typically, the band-pass filter is a commercial inductor/capacitor, crystal or ceramic IF filter commonly used in radio receivers. Alternatively the filter can be a band stop filter to reject the IF, but such filters are not as commonly available (Figure 17.4).

Some care needs to be taken in making measurements in this way. Suitable filters with narrow bandwidths are rarely designed for 50 Ω systems and often have severe changes of impedance with frequency. The mixer has to be buffered from this impedance variation to avoid errors due to reflected signals re-mixing. The filters can also exhibit non-linear behaviour at both low levels (particularly crystals) and high levels (as crystal or ceramic devices exceed their linear power ratings). These problems, combined with frequency response unflatness in the pass band, can make the measurement accuracy unreliable unless precautions are taken. It also has to be

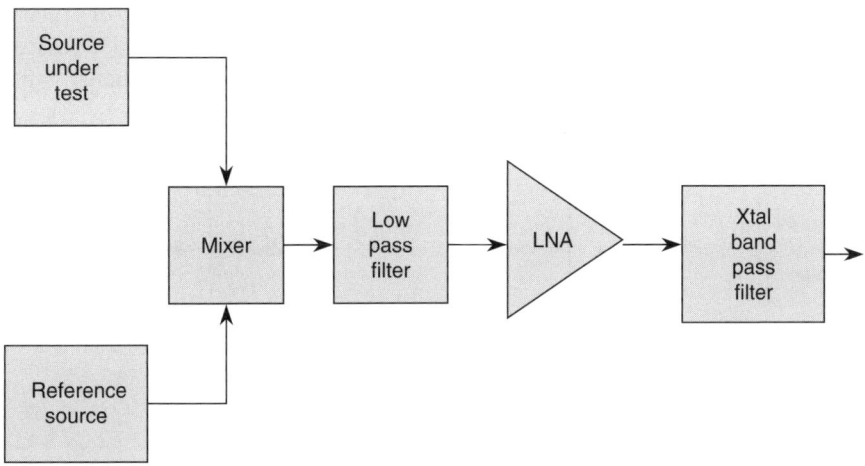

Figure 17.4 Using a spectrum analyser with preselection

remembered that both the phase and the amplitude components of noise are being measured.

The technique is also restricted to measurements at offsets of typically greater than (typically) 10 kHz since it relies on the filter having to reject a significant proportion of the carrier signal at the IF.

The improvement in performance using this method occurs because the signal from the local oscillators inside the spectrum analyser no longer mixes with the carrier frequency of the signal being measured because it is rejected by the band-pass or band stop filter. The spectrum analyser local oscillator noise no longer dominates the measurement and the ratio of the signal to be measured to the total power at the input of the spectrum analyser is much lower, which considerably lowers the dynamic range required of the spectrum analyser.

17.4 Delay line discriminator

A broadband FM discriminator can be constructed by taking the RF signal to be measured and splitting it into two paths (Figure 17.5). One path is fed directly into a mixer and the second path is passed through a delay line and the output is mixed with the non-delayed signal. The delay line includes a variable phase shifter or a mechanically adjustable transmission line so that the phase of the two signals applied to the mixer can be set for phase quadrature.

The bandwidth of the discriminator is a classic $\sin x/x$ response with the first null at a frequency equivalent to the time delay between the two RF paths. The conversion sensitivity of the discriminator is dependent on the RF level applied, the conversion loss of the mixer and the time delay of the delay line. The longer the delay line the greater the sensitivity of the measurement but the more restricted is its measurement bandwidth.

The great advantage of this technique is that it does not require the use of a second RF source to convert the frequency of the source to be measured to a fixed IF (or base band signal). This removes one potential source of error, that is, an additional source

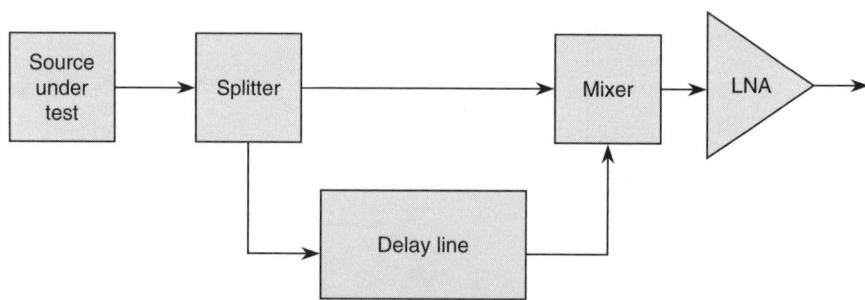

Figure 17.5   Delay line discriminator

of noise. Also, since the method is based on the use of a frequency discriminator it is not too prone to being overloaded by low-frequency sources of phase noise (e.g. power supply related signals).

It does have the practical disadvantage of not being easily automated. It needs to be calibrated which can be troublesome. The sensitivity of the system is dependent on the applied RF signal levels. The normal method of calibration is to adjust the time delay to find the peak positive and negative voltage that can be obtained from the mixer. From this the sensitivity can be deduced if it is assumed that the mixer is behaving in a linear fashion. At microwave frequencies the insertion loss of the delay line can result in the sensitivity of the measurement being limited.

Commercial products are available based on the use of this measurement method.In general, this method is not capable of measuring high-performance oscillators over a significant bandwidth but is capable of measuring typical free running YIG and RF voltage–controlled oscillator (VCO) sources.

17.5 Quadrature technique

In the quadrature system, two oscillators at identical frequencies are used (Figure 17.6). Typically, one of the oscillators will be the source being tested and the other will be a reference source whose performance is known to be better than the source under test. The oscillator outputs are combined in a mixer and the resulting output signal is filtered and amplified by a low noise amplifier (LNA). A fast Fourier transform (FFT) analyser or a spectrum analyser typically measures the output from the mixer.

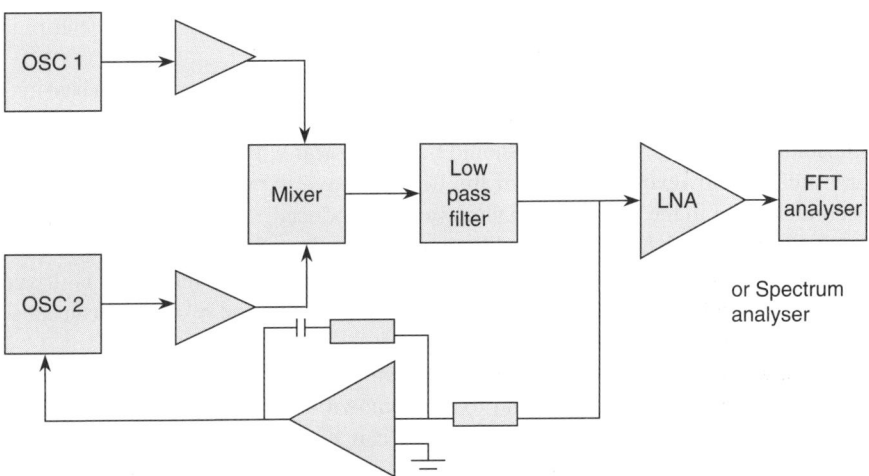

Figure 17.6 *Quadrature method shown with a feedback loop to maintain phase quadrature at the mixer*

In order to provide a valid measurement the phase of the two oscillators has to be set so that they are in phase quadrature at the mixer input. The mixer output will then be close to 0 V and the mixer will behave as a phase detector.

Setting the sources to be in phase quadrature is not always very easy. If both frequency sources are synthesisers with good long-term stability, then there is usually not a great problem – the phase adjustment controls can be used to set the signals in quadrature. However, in the more typical applications where measurements are undertaken under less than ideal conditions, a feedback system has to be used to maintain phase quadrature. The feedback system forms a phase locked loop which drives one of the oscillators to correct for departures from quadrature.

The use of phase locked loop to maintain phase quadrature does imply some knowledge of the tuning characteristics of one of the oscillators and the mixer drive levels since the bandwidth of the phase locked loop is affected by both of these parameters. In practice for many sources the availability of a low noise signal generator with a high-performance DC coupled FM capability, such as the IFR 2040 series, can considerably simplify the measurement system.

If the peak phase excursion of the noise exceeds 0.1 radians, the mixer phase detector response becomes non-linear and degrades the measurement accuracy. Since the peak phase excursion is caused primarily by low-frequency noise then under these conditions, the phase locked loop bandwidth has to be widened in order to restrict the peak phase excursion.

In order to carry out a measurement, the quadrature system has to be calibrated since the sensitivity of the measurement is dependent on the insertion loss and drive level used for the mixer.

If the local oscillator (LO) input level required for the mixer is substantially greater than the RF port drive, a calibration assessment can be obtained by offsetting the frequency of one of the sources by a small amount. A low-frequency sine wave is produced at the output of the mixer whose amplitude can be measured to determine the sensitivity of the mixer.

If both ports of the mixer are driven at a high level to give the maximum sensitivity then the waveform from the mixer will be more like a triangle waveform than a sinusoid and the mixer sensitivity should be more linear with errors in phase quadrature present. However, the slope of the triangle wave is more difficult to measure accurately than in the case where a sine wave is produced.

A further complication in the calibration process can arise if the drive signals are not well matched to the source impedance. Whichever port of the mixer is driven hard, the mixer tends to convert the signal to a square wave and reflections can cause re-mixing and slope perturbations in the output.

An alternative, and often more reliable method of calibration, is to use a signal generator as one of the sources and to set a known amount of phase or frequency modulation. Measuring the resulting output can provide the required calibration information. The phase modulation applied has to have a modulation index of less than 0.1 radians to avoid mixer overload, and a modulation frequency significantly in excess of the phase locked loop bandwidth used for setting up phase quadrature.

This in itself can be another source of error in the measurement since the phase locked loop is used to set up phase quadrature. The loop tends to remove low-frequency phase noise that is present on the source under test.

The errors introduced by the phase-locked loop must either be set so that they are below the frequency offset of interest or they have to be corrected for by measuring the loop characteristics and then mathematically correcting the measurement result.

There is a further practical problem that needs to be assessed. If there is a lack of isolation between the two RF sources then, as their frequencies are brought close together, there will be a tendency for them to become injection locked. If one of the oscillators is a VCO then this is certain to happen and will need to be characterised. Under these conditions it is advisable to ensure that the deliberate phase locked loop bandwidth exceeds the injection locked bandwidth. Even then the loop must be characterised if accurate measurements are to be made on the source (Figure 17.7).

The phase locked loop response can be measured by injecting a calibration signal into the loop. The calibration signal can be swept signal (e.g. the tracking generator output of a spectrum analyser or the modulation oscillator of a signal generator) or a noise source (often available on an FFT analyser). Outside the loop bandwidth the analyser measures the amplitude of the calibration signal but inside the loop bandwidth the phase-locked loop (PLL) reduces the level of calibration signal measured. From the frequency response plotted on the analyser, a correction plot can be deduced and applied to correct the phase noise measurement results.

Care needs to be taken when interpreting results that include high correction factors: the software may display the answers to a high degree of precision not reflected in the real accuracy of the numbers.

Commercial systems are available from a number of vendors based on the use of the Quadrature Technique.

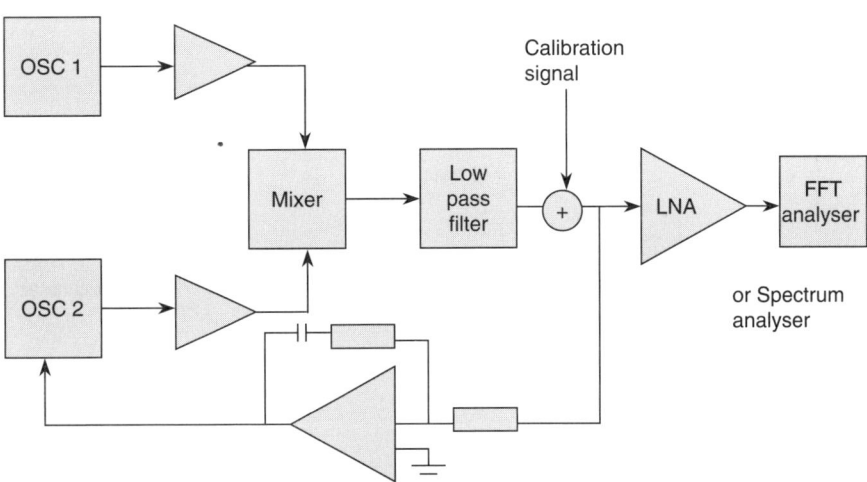

Figure 17.7 Method for calibrating the effects of phase locked loop

17.6 FM discriminator method

This method uses a mixer and a reference source to convert the signal to an IF where it is demodulated by an FM discriminator (Figure 17.8). In principle any FM discriminator, including a discriminator of the type found in a modulation analyser, can be used. However, the noise performance of the discriminator is likely to have a critical effect on the ability to make a phase noise measurement. A proprietary FM discriminator phase noise measuring system is used at IFR Ltd (now part of AeroFlex) to measure high-performance signal generators based on a high performance. A 1.5 MHz discriminator is shown in Figure 17.8. This system is used to aid the design of the oscillator systems deployed in the company's signal generators.

The discriminator is based on the use of a splitter, a band-pass filter and a mixer acting as a phase detector. The band-pass filter uses a coupled resonator design that ensures that at the centre frequency of operation, the phase shift through the filter is 90° so the inputs to the phase detector are in quadrature. In the practical implementation two band-pass filters are available, one allowing a measurement bandwidth of up to 20 kHz and the other allowing measurements to 100 kHz offset.

In principle, the system behaves in a way similar to the delay line discriminator method but it does have some substantial advantages. In particular, since the discriminator operates at an IF, a limiter can be used to control the amplitude of the signal into the discriminator and hence the conversion gain of the discriminator is independent of RF input level. Operation at an IF also allows the FM discriminator to be implemented using a different type of phase detector operating at much higher signal levels. The design used in the IFR version uses two transformer coupled full wave rectifiers operating at very high signal levels to increase the signal-to-noise ratio.

As with the Delay Line Discriminator, it is important to remember that FM noise is being measured rather than phase noise. Conversion between the two measurements

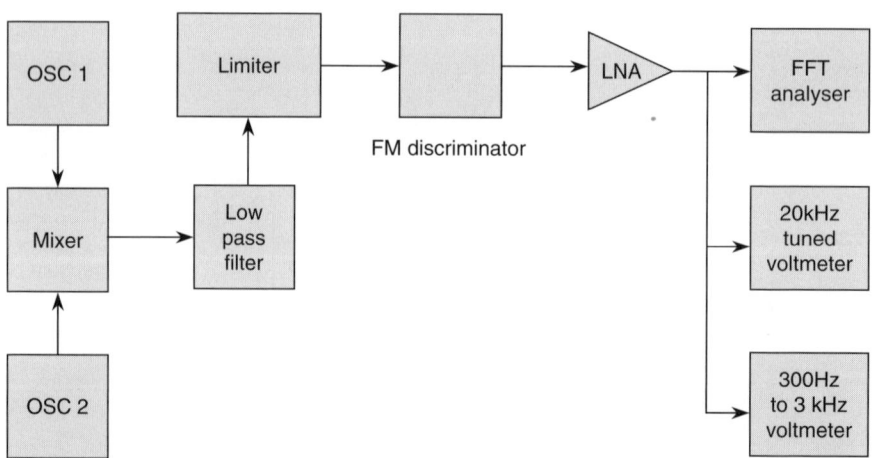

Figure 17.8 Method of measurement using an FM discriminator

is relatively straightforward (by differentiation of the spectrum measurements) and some FFT analysers are available which can mathematically convert the measurement result automatically.

Calibration of the system is very straightforward since the system sensitivity is independent of the input drive level to the frequency conversion mixer. Once a system has been constructed, the calibration factors are constants that can be allowed for by periodic (six monthly) calibration checks. Calibration is typically performed by making one of the sources a signal generator with calibrated amounts of FM or phase modulation.

The system used at IFR Ltd includes a meter to measure the residual FM of a signal source in a 300 Hz to 3 kHz bandwidth (a common signal generator specification parameter) and a tuned voltmeter to measure phase noise at 20 kHz offset to give a fast measurement of these two signal generator parameters.

The performance of an FM discriminator system is limited by the noise figure of the amplifiers and limiters which recover the signal from the output of the mixer and by the performance of the discriminator itself. In the case of the system described previously the discriminator consists largely of passive components, which exhibit very good noise characteristics, and very high signal levels can be used to maximise the $Hz\,V^{-1}$ at the output of the discriminator. Performance tends to be controlled by the slope of the discriminator and it is for this reason that two band-pass filters are used to allow a compromise between sensitivity and measurement bandwidth.

A high-performance FM discriminator is capable of measuring very low levels of phase noise. The above system is capable of measuring residual phase noise of $-170\,dBc\,Hz^{-1}$ at 20 kHz offset and residual FM of 0.003 Hz in a 300 Hz to 3 kHz bandwidth.

The FM discriminator system also has some disadvantages. The need to have sources at different frequencies can be inconvenient. The mixing process can also generate intermodulation products that can give a false indication of there being spurious signals present. With a 1.5 MHz IF, however, this is unlikely to be a problem for carrier frequencies above 50 MHz. A less obvious problem is that if a source exhibits a flat noise profile from the offset frequency being measured to the image frequency (approximately 3 MHz offset for a 1.5 MHz IF) then the image frequency noise will be added to the noise at the required offset. Most signal sources, however, tend to exhibit better noise performance at the image frequency than at the closer offset frequencies. The substitution of a single-sideband (SSB) mixer for the double balanced mixer can eliminate this problem.

17.7 Measurement uncertainty issues

The measurement methods for measuring phase noise are often subject to large error bands. There are a number of basic problems that lead to difficulties. For this reason it is difficult to obtain truly traceable measurement results, and indeed improving the traceability of phase noise is a subject being actively pursued by the National Physics Laboratory and other standards organisations.

Some of the issues are as follows:

- Removal of the carrier signal means the reference value has been removed and needs to be reinstated by the calibration system
  - inherent to quadrature and delay line methods (but important to its performance)
  - implies software correction (hard to prove under all conditions, hard to assign unique values, susceptible to changing levels, increases test time)
- Reliance on phase coherence at a mixer
  - more calibration and software problems for PLL effects
  - restricts the type of on device that can be tested
- The spectrum or FFT analysers are subject to significant errors
  - filter BW correction numbers
  - absolute level errors
  - inherent frequency response
  - absolute level errors
  - scale shape errors
  - detector response to noise like signals (noise is a power measurement)
  - display formatting algorithms (especially FFT analysers)
  - input attenuator accuracy and VSWR
  - reference level accuracy
  - IF switched gain errors
  - amplitude display non-linearities (not to be confused with display distortion)
- LO residual phase noise
- Injection locking defects

These errors make it difficult to assign a traceability figure to phase noise measurements. Often measurements of the same device will lead to differing answers, even when measured on the same system. It is not uncommon for instance to see 'stitching' errors in the results where the test system changes settings to measure noise at different offset frequencies.

17.8 Future method of measurements

There are, however, other techniques that may be developed in the future which could offer other solutions. A promising area is the use of direct Analogue to Digital conversion of IF from a mixer. Current levels of performance are limited by A to D linearity, quantisation error and aperture dither, the resulting noise tending to restrict the usefulness of this measurement technique to offset frequencies of a few kHz. However, improvements in converters are being steadily made.

17.9 Summary

From the above discussion it can be seen that no measurement scheme for phase noise can be said to offer a complete solution in all applications. Of the methods discussed

Table 17.1   Advantages and disadvantages of the various methods described

Method	Advantages	Disadvantages
Spectrum analyser	Simple to use Simple to get a result	Poor dynamic range Cannot measure close to carrier noise Difficult to measure sources with frequency drift
Delay line discriminator	Requires no additional RF source Can measure drifting RF sources	Requires frequent calibration Restricted dynamic range Restricted bandwidth Difficult to automate Requires direct manipulation of microwave sources
Quadrature technique	Reference oscillator is at the same frequency Large dynamic range Measurements can be made at small and large offsets	Requires calibration of every measurement Requires the use of PLL and prior knowledge of the source Takes a long time to make accurate measurements Easy to make errors
FM discriminator	Large dynamic range Does not require frequent calibration Very accurate and hard to make errors Can measure drifting sources easily Tolerates LF noise	Requires a frequency offset source Limited frequency offset range Frequency conversion can make the results pessimistic due to image signals

the most reliable technique is the FM Discriminator Method, since it is the most 'fail safe' technique. The quadrature technique is the most widely used since it offers better overall capability (and the greatest number of commercial solutions) but requires much more care (and time) to undertake a measurement. The spectrum analyser based techniques are often the most convenient to undertake since the equipment is likely to be readily available in most laboratories. Table 17.1 shows the advantages and disadvantages of the various methods described in a summary.

Chapter 18

Measurement of the dielectric properties of materials at RF and microwave frequencies

Bob Clarke

18.1 Introduction

In all RF and microwave (RF and MW) applications electromagnetic (EM) fields interact with materials, that is, with solids, liquids or gases. Viewed on a macroscopic scale, all materials will allow EM-fields to pass into them to some extent: even good conductors and superconductors. Therefore, in one sense, all materials may be said to be *dielectrics*: the word 'dielectric' is a contraction of '*dia*-electric', which means that electric fields can (to some extent) pass through them. In designing RF and MW components for applications, therefore, we need to understand just how electric and magnetic fields propagate into and through materials. We also need to know how they behave (reflect, transmit, scatter, etc.) when they meet interfaces between different materials, for example, between air and an insulator or between a crystal and a metal. To achieve this understanding we need to be able to measure the dielectric and magnetic properties of the materials.

The characterisation of the EM properties of materials is a very broad topic and we can only touch on the most general points in this overview. A wide variety of important *functional* or *active* materials are used in RF and MW applications. They include semiconductors, superconductors, chemically active media and non-linear media in general. Studies of such materials require special detailed consideration and they are generally regarded as specialist topics in their own right. We will not be considering such materials here, we will rather be concerned with materials that respond *linearly* to low-field strength electric fields and it is this subset of materials (which excludes semiconductors, superconductors and so on in their active functional roles) to which the term '*dielectric*' is usually applied on a day-to-day basis. Good conductors such as metals are usually excluded from this class too, though virtually everything we will be saying about dielectrics at RF and MW frequencies applies to metals too – they can be regarded as dielectrics with very high conductivity. Even with this restricted definition '*dielectrics*' nevertheless make up a very wide and important class of materials for RF and MW applications. They are variously used to transmit, absorb, reflect, focus, scatter or contain EM-fields and waves. Bear in mind also that in many applications – radar, RF and MW processing (e.g. in microwave ovens), also in biomedical studies and treatments – the materials that are being detected, studied or treated are themselves dielectrics, so we need to understand their EM properties too.

At RF and MW frequencies, with wavelengths in the centimetre to millimetre range, we are normally dealing with macroscopic components – that is, to say that the linear scale of the components that we are concerned with is much greater than the molecular scale – this is true even in modern 'micro'-devices such as RF MEMs (microelectromechanical devices). It is for this reason that we are normally concerned with parameters that characterise the materials on a macroscopic scale in our practical studies and uses of dielectrics. Of course, on a microscopic scale there are complex EM interactions between atoms, electrons, holes and molecules – and for full understanding of the theory of dielectrics we must also study dielectrics on this scale. However, in the practical realm of RF and MW activities we wish to deal with parameters that capture the macroscopic consequences of all of the microscopic phenomena in the material. The most important of these parameters are the complex permittivity, ε^*, and the complex magnetic permeability, μ^*.

This chapter is mainly concerned with the characterisation of dielectric materials at RF and MW frequencies. There are a number of existing publications that deal with this topic in depth that are well worth consulting. The most recent is the Good Practice Guide from NPL [1] (on which this chapter is based), which deals with the topic comprehensively. The long-standing 'bible' of dielectric metrology is the book by von Hippel [2]. It has deservedly stood the test of time and remains one of the most useful treatises on RF and MW dielectric measurement ever written. It covers background theory in greater depth than Reference [1], though the measurement technology described in it is now rather dated. A number of reviews (e.g. [3] and [4]) and book with details on measurement (e.g. [5]), and conference proceedings (e.g. [6]) also provide a useful background to this discipline. However, before we take up the topic of measurement, we must turn first to the dielectric theory behind the parameters, ε^* and μ^* – the parameters that we wish to measure.

18.2 Dielectrics – basic parameters

Introductory theoretical treatments on dielectrics can be found in most standard textbooks on electromagnetism and in books on dielectrics (e.g. see [1,5,7–12]).

The quantity with which we are most concerned here is the *relative permittivity*, ε^*. This can be converted to the *absolute* permittivity if we multiply it by the *permittivity of free-space*, $\varepsilon_0 = 8.8542 \times 10^{-12}$ F m^{-1}. Note that *absolute* permittivities in the SI system have units of *farads per metre*, whereas *relative* permittivities are dimensionless quantities. In the practical world we are usually concerned with relative permittivities. As it is a complex quantity, ε^* has two components:

$$\varepsilon^* = \varepsilon' - j\varepsilon'' \qquad (18.1)$$

where $j = \sqrt{-1}$. If we assume that our dielectric material is placed between plane-parallel electrodes to form a capacitor, as in Figure 18.1, ε', the *real part of the permittivity*, characterises the *capacitative* part of the admittance, Y, of the capacitor and ε'' characterises the *conductive* or *lossy* part of the admittance. In all materials, ε' and ε'' depend on ambient parameters such as the temperature, relative humidity, as well as the frequency, so neither ε^* nor ε' should be given the name 'dielectric constant' – they are not constant (the only true dielectric constant is ε_0).

The properties of the capacitor in Figure 18.1 can be captured in simple *equivalent circuits*, such as those shown in Figure 18.2a and b, in which the resistive and conductive components R and G represent the dielectric loss of the specimen. It is usually better to use the parallel equivalent circuit of Figure 2b for dielectric materials because C and G are proportional to ε' and ε'', respectively.

In some circumstances it is better or more conventional to quote the *loss tangent*, $\tan \delta$, to quantify the conductive or *lossy* part of the complex permittivity, rather than ε'':

$$\tan \delta = \frac{\varepsilon''}{\varepsilon'} \qquad (18.2)$$

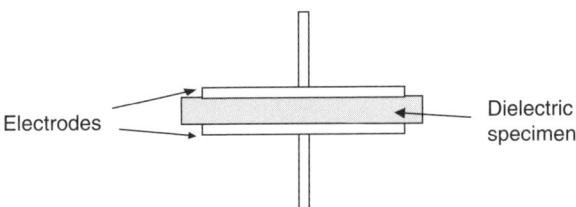

Figure 18.1   A dielectric specimen in a plane-parallel-electrode admittance cell

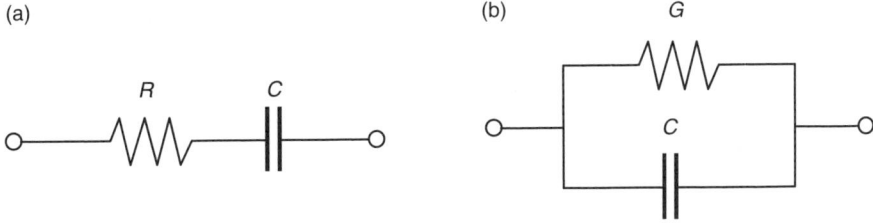

Figure 18.2   Simple series (a) and parallel (b) equivalent circuits of the dielectric specimen in the admittance cell of Figure 18.1. R and G take account of the dielectric loss in the specimen. R is the equivalent series resistance and G is the equivalent parallel conductance of the dielectric

For small losses it is often convenient instead to refer to the *loss angle*, δ, measured in *radians*, which is the arctangent of tan δ. As $\delta \approx$ tan δ for small angles, they can be numerically equivalent for most practical measurements. The advantage of using loss angle is that one can use the convenient unit '*microradian*' (μrad) to quantify small losses [similarly, '*milliradian*' (mrad) can be used for medium losses].

As implied in Section 18.1, we may encounter many different types of dielectric material in our work and they will display a wide range of properties. ε', for example, can vary from 1.0 for air, 2–3 for typical low to medium loss polymers, 5–100 for typical electroceramics, 80 for water, 5 to >1000 for biological tissues and up to >2000 for ferroelectrics. The loss tangent, tan δ, may be as low as 3×10^{-5} for crystals such as quartz and sapphire at room temperature (or as low as 10^{-7} at cryogenic temperatures). Typical low-loss polymers have tan δ in the range $10^{-4} - 10^{-3}$, while absorbing materials like radiation absorbing materials (RAM) and bodily tissues can have tan $\delta > 0.1$ or even >1. With such a wide range of dielectric properties and an equally wide range of physical properties (as dielectrics can be solids, liquids, powders, malleable, hard, etc.) it is not surprising that many different dielectric measurement techniques have been developed over the years. It is good practice in dielectric measurement, if we wish to reduce our measurement uncertainties to *match* the technique to the material. In this overview, for convenience, we refer to materials with tan δ less than 3×10^{-4} as low-loss, above 3×10^{-2} as 'high-loss' and those with tan δ in between these two values as 'medium loss'.

The *magnetic* parameter that corresponds to the *dielectric* parameter ε^* is the *complex relative magnetic permeability*, μ^*, which may be defined and treated by analogy with ε^*

$$\mu^* = \mu' - j\mu'' \tag{18.3}$$

In many materials *for all practical purposes* $\mu^* \approx (1.0 - j0.0)$ and we can refer to these as 'non-magnetic materials' or just simply as '*Dielectrics*'. If μ^* differs significantly from $(1.0 - j0.0)$ we have a *magnetic material* and we need to take account of its magnetic properties in our measurements. There is an important difference between dielectric and magnetic responses in most materials. The dielectric response to small signals is typically linear while the magnetic response can be non-linear even for small signals, so that μ^* is a function of signal strength – one well-known manifestation of this non-linearity is magnetic hysteresis. At sufficiently high field-strengths all dielectric materials will also respond non-linearly to applied fields (e.g. see [2,13]). However, we will restrict our discussion here to the normal, low signal and linear regime.

As we move up in frequency through to the millimetre-wave region of the spectrum and up through THz frequencies, it becomes more convenient to characterise dielectrics in terms of *quasi-optical* or *optical* parameters such as the *complex refractive index*: $n^* = n - jk$ [14]. This quantity is related to ε^* and μ^* by

$$n^* = n - jk = \sqrt{\varepsilon^* \mu^*} \tag{18.4}$$

For non-magnetic materials,

$$\varepsilon^* = n^{*2} = (n-jk)^2, \quad \text{so } \varepsilon' = n^2 - k^2 \text{ and } \varepsilon'' = 2nk \quad (18.5)$$

Rather than using k to quantify the loss, it is more conventional in optical and quasi-optical systems to employ the *power absorption coefficient*, α_p,

$$\alpha_p = \frac{4\pi f k}{c} \quad (18.6)$$

where f is the frequency and c is the speed of light and α_p is the *power* absorption coefficient per unit length of signals transmitted through the medium. α_p is conventionally measured in the units *nepers per metre* (Np m^{-1}). As 1 $Np = 8.69$ dB, the power attenuation of a signal passing through the dielectric may also be expressed in decibels as $8.69 \times \alpha_p$ dB m^{-1}.

We can express the loss tangent of the material in terms of the (quasi-)optical parameters n and α_p as follows:

$$\tan \delta = \frac{8\pi f n c \alpha_p}{((4\pi f n)^2 - c\alpha_p)^2} \quad (18.7)$$

which for low-loss materials having $\alpha_p \ll 4\pi fn/c$ reduces to $\tan \delta \cong c\alpha_p/2\pi fn$.

18.3 Basic dielectric measurement theory

As most dielectric measurements make use of *cells* to contain the dielectric, and as we can regard the cell as an RF and MW component, many dielectric measurement techniques, viewed at the instrumental level, are similar to other S-parameter measurement techniques that are covered in-depth in other chapters and also in textbooks on RF and MW measurements (e.g. [15–17]). A more comprehensive version of the treatment given here can be found in the Good Practice Guide [1].

Dielectric measurement methods and measurement cells largely fall into two broad classes:

(1) Those in which the dielectric properties are measured as an impedance, Z, as in Figure 18.2a, or more commonly, as an admittance, Y, as in Figure 18.2b. These may collectively be called *lumped-impedance methods* and are generally used at low frequencies (LF) and in the RF region of the spectrum up to 1 GHz.
(2) Those in which the dielectric is considered to be interacting with *travelling and standing electromagnetic waves* – these may collectively be called 'Wave Methods'.

Both lumped-impedance and wave techniques can be used in *resonators*. Resonators are measurement cells with *resonating* EM-fields inside them that are used to obtain high sensitivity for measuring the loss of low-loss dielectrics (see below).

18.3.1 Lumped-impedance methods

In these methods we use an impedance/admittance analyser or bridge to perform the measurements on the cell. If we carry out a measurement in the cell of Figure 18.1, we usually measure the *cell admittance* $Y = G + jB$, where G is the equivalent *parallel conductance* and B is the equivalent *parallel susceptance*, both measured in units of siemens. The equivalent circuit for a lossy dielectric is shown in Figure 18.2b, where C is related to B by $B = \omega C$, where $\omega = 2\pi f$ and f is the frequency in hertz. The cell shown in Figure 18.1 is commonly called an *admittance cell* because one determines ε^* by measuring its admittance. In lumped-impedance analysis one therefore treats continuous dielectric media, and specimens thereof, in terms of *lumped equivalent-circuits*, that is, circuits containing discrete components: inductors, capacitors and resistors. This is perfectly acceptable as long as their physical dimensions are very small compared with the wavelength, λ, of the radiation. It is this requirement that limits the usefulness of lumped-impedance methods as one moves up through the spectrum.

18.3.2 Wave methods

Measurements at MW frequencies usually differ from electrical measurements at lower frequencies because they are *conceived of* differently: they deal with *waves* rather than with *impedances* and *admittances*. Wave methods may be *travelling-wave* or *standing-wave* (resonant) methods and they may employ a *guided-wave* or a *free-field* propagation medium. Coaxial, metal and dielectric waveguide, microstrip, co-planar waveguide and optical-fibre transmission lines are examples of *guided-wave* media while propagation between antennas in air uses a *free-field* medium.

In *guided-wave travelling-wave* methods the properties of the measurement cell are measured in terms of *Scattering Parameters* or '*S*-parameters' (e.g. see [16] and Figure 18.3). It is good practice in such measurements to ensure that as much of the measurement system as possible is matched to the transmission-line *characteristic impedance*, Z_0, because mismatches produce reflections and *multiple* reflections that can seriously reduce the accuracy of measurement. The reflection coefficient, Γ, from an impedance Z that terminates a transmission-line of characteristic impedance Z_0 [18] is given by

$$\Gamma = \frac{Z - Z_0}{Z + Z_0} \tag{18.8}$$

so reflections are only absent if $Z = Z_0$. In fact, the propagation of EM waves through any uniform isotropic medium, or any uniform transmission-line containing such a medium, is governed by two parameters: the *characteristic impedance*, Z_0, and the *complex propagation constant*, γ, of the medium [10]. Both are functions of the complex permittivity, ε^*, and permeability, μ^*, of the medium. The propagation constant governs both the wavelength and attenuation of waves moving through the medium:

$$\gamma = \alpha + j\beta = 2\pi f \sqrt{\varepsilon_0 \varepsilon^* \mu_0 \mu^*}$$

or, because

$$c = \frac{1}{\sqrt{\varepsilon_0 \mu_0}}$$

$$\gamma = \frac{2\pi f \sqrt{\varepsilon^* \mu^*}}{c} = \frac{2\pi \sqrt{\varepsilon^* \mu^*}}{\lambda_0} \tag{18.9}$$

where c is the speed of light in free-space and λ_0 is the free-space wavelength, α is called the *attenuation constant* and β is the *phase constant*. For low-loss dielectrics in which it is assumed that $\alpha \approx 0$, β is sometimes itself called the propagation constant. In non-magnetic materials, $\mu^* = (1.0 - j0.0)$ and

$$\gamma = \frac{2\pi f}{c}\sqrt{\varepsilon^*} = \frac{2\pi}{\lambda_0}\sqrt{\varepsilon^*} \tag{18.10}$$

In a low-loss dielectric α is small ($\alpha << \beta$) and we find

$$\lambda \approx \frac{\lambda_0}{\sqrt{\varepsilon'}}, \beta \approx \frac{2\pi\sqrt{\varepsilon'}}{\lambda_0} \text{ and } \alpha \approx \frac{\pi\sqrt{\varepsilon'}\tan\delta}{\lambda_0} \text{ or } \alpha \approx \frac{\pi f n \tan\delta}{c} \tag{18.11}$$

so $\tan\delta \approx c\alpha/\pi fn$ and referring back to (18.18), we find $\alpha = \alpha_p/2$. This relationship actually follows more fundamentally from the definitions of α and α_p. First is the exponential coefficient for decay of *field-strength* as the wave propagates through the medium, the second describes the decay of *power*, so the relationship $\alpha = \alpha_p/2$ is valid even if α is not small.

The measurement of *propagation* or *transmission* parameters, such as β, γ and λ, provides a means for deriving dielectric parameters. Many dielectric measurement methods are based on this principle. The same is true of the measurement of *reflection* parameters, Γ, S_{11}, S_{22}, etc. Reflections at plane interfaces between two uniform media are governed by the *intrinsic wave impedances* of the media, η_1 and η_2, respectively, where $\eta_1 = \sqrt{\mu_1^*/\varepsilon_1^*}$, and $\eta_2 = \sqrt{\mu_2^*/\varepsilon_2^*}$, where μ_1^*, μ_2^*, ε_1^* and ε_2^* are the *complex relative permeabilities* and *permittivities* of the two media, respectively. By analogy with (18.8), the reflection coefficient in Medium 1 of a wave *at normal incidence* (i.e. at 0° angle of incidence) on Medium 2 is

$$\Gamma = \frac{\eta_2 - \eta_1}{\eta_2 + \eta_1} \tag{18.12}$$

Such reflections are dependent on the *ratio* of μ^* and ε^*, whereas the transmission parameters defined in (18.9) are governed by the *product* of μ^* and ε^*. In order to separate μ^* and ε^* we must normally, therefore, measure the specimen *both* in transmission and reflection (though other techniques are possible). This is readily achieved by measurements in transmission cells, such as the one shown in Figure 18.3, in which all four S-parameters are easily determined. If we are sure that we have a non-magnetic material, however, we have $\mu^* = (1.0 - j0.0)$ and we can measure ε^* *either* by transmission *or* by reflection methods. The choice should normally be made on the basis of which is the most accurate. For non-magnetic materials, the reflection

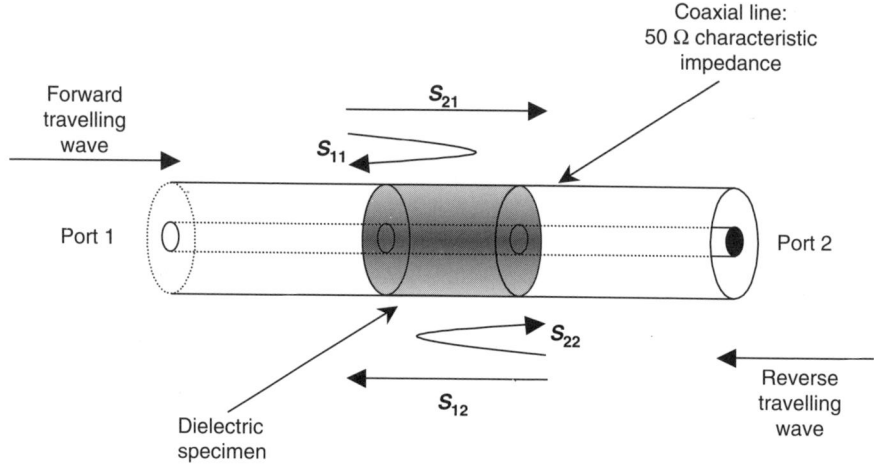

Figure 18.3 Electromagnetic travelling waves in a coaxial measurement cell that contains a dielectric specimen. The cell is placed in a measurement system in which there are waves travelling in both directions, designated forward and reverse waves. The figure shows the reflection coefficients of the specimen, S_{11} and S_{22}, for the forward and reverse wave, respectively, and the corresponding transmission coefficients, S_{21} and S_{12}. The S-parameter subscripts refer to the measurement ports (that is the two ends) of the coaxial cell

from a dielectric interface between two media with permittivities ε_1^* and ε_2^* travelling at normal incidence from Medium 1 to 2 is

$$\Gamma = \frac{\sqrt{\varepsilon_1^*} - \sqrt{\varepsilon_2^*}}{\sqrt{\varepsilon_1^*} + \sqrt{\varepsilon_2^*}} = \frac{n_1^* - n_2^*}{n_1^* + n_2^*} \tag{18.13}$$

where $n_1{}^*$ and $n_2{}^*$ are the complex refractive indices of the two media. For free-field measurements in which the angle of incidence is not necessarily normal these equations may be generalised for any angle of incidence by using Fresnel's Equations [10,19]. In the free-field it is also important to take account of whether the electric field polarisation of the incoming radiation is *parallel* to the dielectric surface or in a plane *perpendicular* to the surface. There are two sets of Fresnel's equations – one each for the parallel and perpendicular cases.

18.3.3 Resonators, cavities and standing-wave methods

Resonance methods are generally used for measuring the loss of low-loss dielectrics. The cell in such measurements is commonly referred to as a *cavity* or *resonator*. In such cells the real permittivity, ε', is typically determined by measuring the change in resonant frequency when the specimen is inserted into the resonator or else it can be measured by changing the length (or some other key dimension) of the

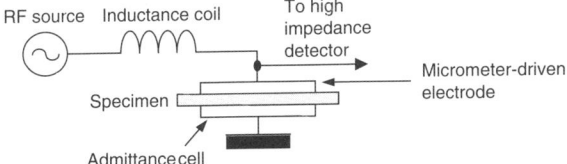

Figure 18.4   The basic principles of the Hartshorn and Ward method for low-loss specimens

resonator to return it to resonance at the same frequency. These contrasting methods are known, respectively, as the *frequency-change* and *length-change* methods for determining ε'. It is possible to measure ε' by the length-change method because the wavelength in the dielectric medium, λ, differs from that in free-space (18.11). The determination of dielectric loss, ε'' or tan δ, in such resonators usually proceeds via the measurement of *Quality-Factor*, otherwise known as *Q-factor* or *Q* [20], (see Section 18.7).

Both lumped-impedance and wave techniques can be employed in resonators. In the former case an admittance cell can be resonated with an external inductor; (Section 18.9.2 and Figure 18.4). This RF '*dielectric-test-set*' method was first used for measuring low-loss dielectrics in the 1930s. At higher frequencies where wave methods are more appropriate, the resonator is often modelled using *standing-wave* equations rather than a *travelling-wave* analysis. Note that wave resonance methods are also often referred to as '*multi-pass techniques*' because a travelling-wave in the cell, in bouncing backwards and forwards between the two ends of the resonator, will pass through the dielectric specimen many times before it is absorbed. This concept helps to explain why resonant methods are more sensitive to low-losses than single (transmission) and double-pass (reflection) techniques.

18.3.4   The frequency coverage of measurement techniques

It is worthwhile drawing attention to the frequency coverage of the various techniques used in different parts of the spectrum. At MW frequencies we generally need to adapt equipment dimensions to the radiation wavelength (this is almost a definition of what 'microwave' means in practice). We therefore tend to need many differently sized measurement cells in this region of the spectrum. In contrast, at LF and from THz frequencies upwards (above approximately 300 GHz), a single instrument/cell combination may measure over many decades of frequency. For example, a bridge/admittance cell combination can cover five decades of frequency at LF, while a single Fourier transform spectrometer [21] can cover five decades in the sub-millimetre to infrared regions of the spectrum. Note also that many resonance techniques operate at one spot-frequency only! We need to bear these limitations in mind when we set out to measure dielectrics – as explained at the end of Section 18.4, measurements at a few spot frequencies may suffice to characterise a *low-loss* material over a wide frequency band, but much more detailed frequency coverage is needed for high-loss materials.

18.4 Loss processes: conduction, dielectric relaxation, resonances

It is important, before we set out to measure our dielectric materials, that we know something about *the physics of dielectric response behaviour*. The most relevant physical processes here are those that give rise to power loss in bulk dielectrics. In the RF and MW region of the spectrum such power losses arise largely from four different physical processes – see [22–24] or [5] for details and [12] for an introduction. These four physical processes are associated with mechanisms for generating loss – that is, conversion of EM energy into other forms of energy – and so are commonly called 'loss processes'. They are (1) Electrical Conduction, (2) Dielectric Relaxation, (3) Dielectric Resonance and (4) Loss from Non-Linear Processes.

(1) Electrical Conduction in which charge carriers (electrons, ions or holes) in a material medium are relatively free to move physically through the medium under the influence of an electric field. The ease of conduction is quantified by the *conductivity* of the medium, σ, measured in siemens per metre (S m^{-1}). In general, σ will depend on frequency, temperature, concentration of carriers, etc., though in many materials σ will change only slowly with frequency in the RF and MW range. In metals the effective conductivity depends on the physical properties of the metal surface (e.g. scratches, machining or grinding marks or debris) and so σ will depend on frequency because the skin-depth [18] is a function of frequency.

(2) Dielectric Relaxation refers to the response of the *electric dipoles* in a material medium to the applied alternating EM-fields. Many different types of dipole can be present in dielectric media. Some materials have *permanent* molecular dipoles inside them and they are called *polar materials*. Their molecules are referred to as *polar molecules*. Materials in which the dipoles are induced only by the application of the electric field itself are called *non-polar materials*. Polar molecules typically exhibit a number of different relaxation processes. If they are in a liquid phase they can rotate bodily to try to align themselves with the field, giving rise to *rotational polarisation*. Otherwise portions of large molecules can move with respect to each other, giving rise to one or more *distortional polarisation* processes, each with its own relaxation behaviour. Any interfaces in a material that prevent or inhibit the passage of charge carriers will have dipolar layers set up across them when the electric field is applied and they give rise to *interfacial polarisation*, a phenomenon, essentially similar to the Maxwell–Wagner effect [5,22], which exhibits its own relaxation behaviour. The membranes in the cells of biological tissues typically exhibit this behaviour [25]. In a complex medium, many or all of these relaxation processes may be present, giving rise to very complex relaxation behaviour. Each process will have its own *strength*, which is a parameter that measures the extent to which it contributes to the total magnitude and behaviour of ε' and ε''. *All* relaxations give rise to *very slow* changes in ε' and ε'' with frequency. The behaviour shown in Figure 18.5 for water is typical.

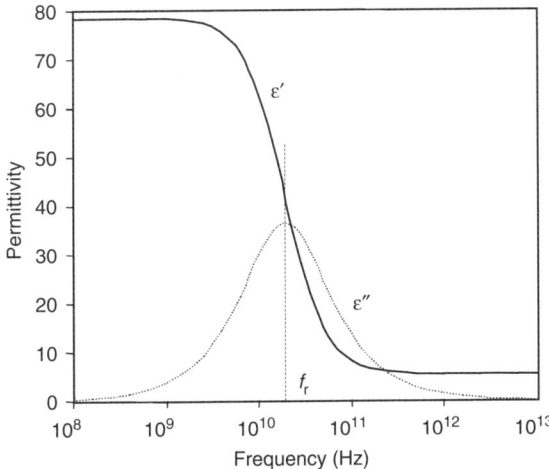

Figure 18.5 A typical Debye Relaxation response, see (18.14). This plot is for deionised water. The diagram omits the effects of all other loss processes in the water, for example, conduction. The use of a logarithmic scale demonstrates just how slowly dielectric properties change with frequency when governed by a relaxation process

(3) Dielectric Resonance. Dielectric polarisation *resonance* should not be confused with *dielectric relaxation*. The physics of relaxations and resonances is completely different and the two should not be confused (see Figures 18.6 and 18.7). A resonance may appear either as a sharp or broad feature in the frequency domain, depending on its Q-factor (see Section 18.7), whereas relaxations *always* exhibit *very broad* spectral features. Furthermore, a resonance gives rise to a frequency dependence for ε' in which ε' can either *fall* or *rise* with frequency, whereas in relaxation behaviour ε' (or μ') can only *fall* with frequency. The physics of linear, homogeneous, non-composite solid and liquid dielectrics does not normally allow a resonance to occur at RF and MW frequencies. The molecules of such materials, because of their close proximity to one another, interact to such an extent that all potential resonances are damped effectively into non-existence. (This is *not* the case with gases in which spectral lines, that is, resonances in the EM response *can* readily be distinguished at MW frequencies). Therefore, it is a useful rule of thumb that for *linear, homogeneous, non-composite solid* and *liquid* dielectric materials any sharp features that we *appear* to measure in the spectrum of ε' and ε'' at RF and MW frequencies, and any *apparent* increases in ε' with frequency are invariably caused by *non-intrinsic* effects, usually imperfections in our measurement cells! There *are*, however, genuine dielectric resonances in the *infrared* region in the spectra of solids and liquids. These processes give rise to *small but measurable* effects at RF and MW, usually a slowly increasing

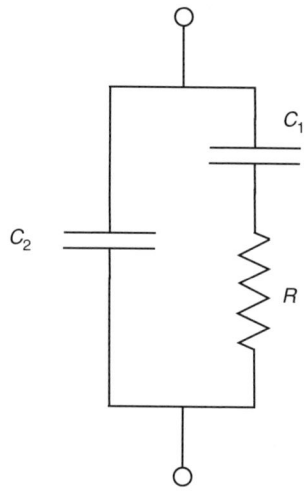

Figure 18.6 An equivalent circuit of a Debye Relaxation

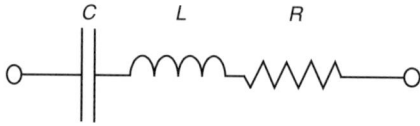

Figure 18.7 An equivalent circuit of a dielectric resonance

value of tan δ with frequency caused by the LF tail of the resonance – this is only visible in low-loss materials. Special notice should also be taken of RF and MW resonances that can occur in *composite materials*. Resonances can occur within the meso- or macroscopic *particles* and *structures* that make up such materials. This is particularly prominent if one component of the material is a high permittivity low-loss component, such as a sintered ceramic. Bear in mind that if the free-space wavelength is λ_0, the wavelength in the material will be $\lambda_0/\sqrt{\varepsilon'}$. So, if ε' for a particle is approximately 1000, resonances can occur in the millimetre-wave region of the spectrum even if the typical particle width, referred to as its *structure-length*, is as small as 0.1 mm. It is usually this phenomenon that limits the useful upper frequency of dielectric composites.

(4) Loss from non-linear processes. It is well known that *hysteresis* in magnetic materials leads to loss [26]. Ferroelectrics [1,27] can exhibit much the same phenomenon in their electrical properties, leading to an independent source of electrical loss. The relative magnitude of these non-linear effects usually rises with the amplitude of the applied fields.

Having listed the four likely sources of loss in dielectrics we should return to *dielectric relaxation* (2) because it is the most characteristic of dielectric loss processes in the RF and MW region of the spectrum.

Relaxation Models. The *Debye Relaxation* is our simplest relaxation model [5,13,22]. It corresponds to the equivalent circuit in Figure 18.6. Three *real-number* parameters (corresponding to the three lumped components in the equivalent circuit) are used to characterise the frequency response of the relaxation: (1) the 'static' *permittivity*, ε_s, is the value of ε' at very LFs; (2) the *relaxation frequency*, f_r, giving the typical speed of the relaxation; and (3) the *high-frequency permittivity limit*, ε_∞, is the value of ε' at high frequencies, well above f_r. All three parameters are functions of temperature. The *Single-Debye* dielectric response is then described by the following equation:

$$\varepsilon^* = \varepsilon_\infty + \frac{\varepsilon_s - \varepsilon_\infty}{1 + jf/f_r} \tag{18.14}$$

A typical response is shown in Figure 18.5. This behaviour is exactly analogous to the behaviour of the equivalent circuit shown in Figure 18.6 with $f_r = 1/2\pi\ RC_1$

Debye relaxations are exhibited by a number of liquids including water at RF and MW frequencies, but many other materials have more structured responses with frequency, for example, they exhibit *Double-Debye or Multiple Debye* responses [28,29], or else they relax *even more slowly* with frequency, as typified by the *Cole-Cole* [30], *Cole-Davidson* [31] or Havriliak–Negami relaxation formulae [32]. In most complex structured materials many different relaxation processes contribute to the dielectric response, which can therefore be quite difficult to describe in terms of any one simple model. In scientific studies, variable temperature measurements are always carried out because the features of relaxations (e.g. loss peaks) typically shift in frequency by different rates as temperature is varied, so temperature variation can be used to differentiate between the various processes.

In all of these relaxation models the contribution from dielectric relaxations to the real permittivity, ε', *always falls* with frequency (or else, in the limit, at very low and high frequencies, it may asymptotically be approximately constant; see Figure 18.5). Given (1) that free particle conductivity, σ, has no effect upon ε'; (2) that intrinsic dielectric resonance does not occur in homogeneous non-composite dielectric solid and liquid materials in the RF and MW region; and (3) that dielectric behaviour in this region is normally dominated by relaxations, we can make the following general statement:

> In the RF and MW region of the spectrum ε' always falls with frequency (or, in the limit, remains stationary) in homogeneous solids and liquids that exhibit a linear response to the applied fields.

Note though (as discussed above) that this statement does not necessarily apply to (1) gases, (2) composite materials with a structure-length close to the EM wavelength in any one of the components in the material and (3) non-linear materials.

There are other important features of relaxations that we should know about. For example, the Kramers–Kronig relations [23,33] provide a formula which relates $\varepsilon'(f)$ to $\varepsilon''(f)$ over a broadband of frequencies. In 1971 Arnold Lynch derived and published [34] a simplified equation for relating changes in ε' with frequency to tan δ when the electrical response of the material is determined by dielectric relaxation behaviour. It is good practice to use such formulae to check our measurements; see also [1]. Another important practical rule of thumb follows from consideration of the Kramers–Kronig relations and from inspecting relaxation models such as the Debye relaxation of (18.14):

> In the RF and MW region of the spectrum for *low-loss* materials both ε' and ε'' change *very slowly* with frequency, so we do not need to measure them at all frequencies.

Thus, the fact that we normally use resonators to measure low-loss materials and that the resonators may be restricted to one frequency only is mitigated to some extent: we need to use only a few such resonators to cover a broad bandwidth. In fact, it is usually a waste of time and effort to measure *low-loss* materials at frequencies more closely spaced than once every octave (i.e. more closely than f, $2f$, $4f$, etc.). This statement does *not* apply to high-loss materials for which continuous frequency coverage is usually advisable.

18.5 International standard measurement methods for dielectrics

Internationally agreed standard methods of measurement allow us to share good practice and ensure the compatibility of measurements that are made in different laboratories. These methods are of particular importance when specifying a procedure for the determination of properties of a material for a well-defined end-use. That said, however, methods outlined in such standards often lag behind the state-of-the-art in metrology. This is inevitable because test houses and product-control laboratories cannot be expected to be as up-to-date or as well equipped as calibration laboratories or those that have the task of developing new measurement methods for new types of material. Standard methods can be read as a guide to the art of measurement and they clearly must be followed, where possible, in many commercially oriented measurements. But there are many areas of dielectric metrology where effective written standards do not exist or else where significant practical detail is lacking from them. In these circumstances one often has to fall back on the scientific literature for a more detailed account of the methods. Dielectric standards may be found in the *American Society for Testing and Materials* (ASTM) catalogs and in the *International Electrotechnical Commission* (IEC) and *Comité Européen de Normalisation* (CEN) systems. *British Standards* (BS) are nowadays subsumed into the IEC and CEN ranges. Note that CEN *electrical* standards are normally referred to as 'CENELEC' standards. A short overview of RF and MW dielectric standards is given in Reference [1].

18.6 Preliminary considerations for practical dielectric measurements

Please see the Good Practice Guide [1] for more details on the points that follow.

18.6.1 Do we need to measure our dielectric materials at all?

Perhaps we can consult manufacturers' data sheets or consult the scientific literature. Either option can be adequate for the early *design* stages of the system development process, but for end-use applications it is usually advisable to measure (or have measured) the materials that are being used. There are a number of reasons for this. For example, the dielectric properties of all ceramics depend critically on how they are prepared and sintered, so there is no such thing as a 'standard' or 'representative' batch of sintered alumina. Likewise, a polymeric material like 'PVC' is far from being a uniquely defined substance – polyvinyl chloride is prepared in many different ways, having different amounts of plasticiser and other additives admixed with it. Both the additives and the processing of the polymer give rise to a whole range of different 'PVC' polymers with varying dielectric properties, so the uncritical adoption of data on 'PVC' from a database will not do. Usually, if we have a 'one-off' job, unless we already have a suitable measurement system available, it will be most cost-effective to ask a test house or calibration lab to perform measurements for us. If we have a long-term interest in measuring or testing dielectric materials we may well decide to set up a measurement system ourselves, but even in this case it may be more cost-effective to seek advice from an experienced laboratory first.

18.6.2 Matching the measurement method to the dielectric material

Dielectric materials come in many different forms, physical phases, shapes and sizes. For example, high-loss solids (e.g. RAM), low-loss solids (e.g. quartz and other pure crystals, and ceramic dielectric resonator materials), hard or malleable solids, liquids with very different viscosities, toxic materials (e.g. many liquid organics and some solid inorganics like beryllia), magnetic solids, thin films, dielectric resonators, substrates, materials available only in small quantities (e.g. expensive crystals, trial samples of ceramic), materials available in copious quantities (e.g. radome materials, some substrates), composite materials, anisotropic materials, significantly inhomogeneous materials (e.g. many composites, foodstuffs, human tissues, etc.), moist materials and powders and so on. It is not too much of an exaggeration to say that each of these different physical classes of dielectric requires a different measurement technique for optimum measurement in the most accurate or the fastest way. Cost often forces us to carry out measurements on non-ideal equipment but it is good to bear in mind that we should ideally think of adapting the methods we use to the material rather than the other way around. Section 18.9 provides a short survey of methods and tells us which material types they are best used for. However, the following rule generally holds.

18.6.2.1   For low-loss materials
It is invariably best to use a resonance method to obtain better sensitivity for loss. As noted above such methods may either be lumped-impedance methods (typically at RF) or standing wave methods (typically at MW frequencies and above).

18.6.2.2   For medium to high-loss materials
It is usually better to use non-resonant methods. At RF, below approximately 1 GHz, the use of admittance bridges and admittance cells is effective. At MW frequencies, where it is appropriate to think in terms of travelling waves, the use of *single-pass* (transmission) or *double-pass* (reflection) techniques is usually more appropriate.

18.6.2.3 Cleanliness, specimen decomposition and contamination

Dielectric specimens can be easily contaminated – the net result in many cases is that one measures a spuriously high-loss for the specimen. *Especially* in the case of low-loss materials, it is important to keep them clean and never touch them directly with the human hand – use clean tweezers or gloves instead – sweat from human fingers can enter the matrix of sintered materials and increase their loss considerably. In the case of hygroscopic liquids (e.g. organic liquids such as short-chain alcohols) it is important to store them in a well-sealed container as they will readily pick up moisture from the atmosphere. Some materials decompose in bright light (typically ultraviolet or sunlight), so store materials in the dark; other materials oxidise in the atmosphere so store them in sealed containers.

18.6.2.4 Specimen dimensions and preparation

It is important to know that, with few exceptions, the uncertainty with which a specimen can be measured depends critically on one or more of its dimensions, typically either its thickness or diameter. Therefore one should use *micrometers* or *optical techniques* to measure critical specimen dimensions, and not less accurate methods, for example, callipers. For all but the most rudimentary or preliminary measurements, it is usually worthwhile ensuring that critical specimen dimensions are machined carefully to a uniform and known size. Grinding is often more accurate (though more expensive) than turning and low tool speeds are recommended for some polymers (to prevent melting the polymer). Cutting fluids must be avoided if they will contaminate the specimen. Specimen thickness, t_s, should ideally be measured at a representative number of points across the area to be tested in order to check its uniformity. Remember that the thickness of hard specimens measured by micrometer may be too high, as it will only measure the high points, while for soft specimens it may measure too low if the specimen is compressed. Care should be taken to detect and correct for these effects in accurate measurements. For most techniques the actual specimen size, and not just its uniformity, will itself also have a major effect on the measurement accuracy. It is therefore good practice to model the measurement method numerically beforehand to determine the dimensions that the specimen should have to optimise measurement accuracy. Of course, in many cases we will not be able to choose the specimen dimensions ourselves and we will then have to do the best we can, but it is often worthwhile asking our specimen suppliers whether they can produce specimens of a more optimum size or shape. In some techniques specimen *diameter* is a critical parameter, for example, in the coaxial line method of Figure 18.3 and Section 18.9.6. It has been found that the use of air-gauging [35] is cost-effective and particularly convenient for the measurement of inner and outer diameters of *both* specimens and the coaxial cells in which they are contained.

18.6.2.5 Anisotropic and magnetic materials

It is important to know whether the material one is measuring is an anisotropic or a magnetic material. Some measurement methods assume isotropy of the specimen's dielectric properties or else they assume a non-magnetic material – if these conditions do not hold they can lead to significant errors in the dielectric measurements. Some of the methods described in Section 18.9 can measure anisotropy or magnetic properties and it will sometimes be necessary to use such methods (perhaps in a preliminary study) if one does not know the EM properties of one's materials at all.

18.6.2.6 Inhomogeneous, composite and structured materials

As explained in Section 18.4, when discussing dielectric resonances it is always important to consider the structure-length of the inhomogeneities in the material (i.e. the typical linear dimension of the particles or components in the material) in relation to the wavelength inside the components of the material. If the structure-length and wavelength are comparable in size the component parts may resonate and the material may scatter radiation rather than reflect or transmit it. Except in ('meta-')materials which are specifically designed to do this – for example, photonic band gap (PBG) or frequency selective (FS) materials – this behaviour is usually undesirable, so that most composites will have an upper frequency limit of usefulness. But there is another reason for being aware of the structure-length in inhomogeneous dielectrics: one wants the specimen that one is measuring to be sufficiently large to be statistically representative of the material as a whole and this should influence one's choice of method: for example, one would choose a large waveguide cell to measure foodstuffs rather than a small coaxial cell. The alternative may be to measure a large number of small specimens and take a statistical average, but this can be time consuming. Significantly, inhomogeneous materials can never be measured as accurately as homogeneous specimens and often the inhomogeneity will be the greatest source of uncertainty in the measurement.

18.6.2.7   Ferroelectrics and high-permittivity dielectrics, $\varepsilon' > 100$

The RF and MW properties of ferroelectrics [11,12,27] are most easily measured by techniques that can measure thin film specimens, for example, the split-post resonator, see Section 18.9.4. Very high permittivity ferroelectric films may only be measurable if they are of sub-micron thickness, so special thickness measurement techniques must be used. Thicker specimens may be measured in admittance cells at RF frequencies if their permittivity is not excessively high (e.g. if $\varepsilon' < 3000$, depending on frequency and specimen size) by applying metal electrodes to both surfaces.

18.7   Some common themes in dielectric measurement

Section 18.9 presents a survey of a number of techniques used for RF and MW measurements upon dielectric materials. However, there are many practical features that these techniques have in common and we consider some of them together briefly here. The details are covered in much greater detail in the Good Practice Guide [1].

18.7.1   Electronic instrumentation: sources and detectors

At MW frequencies it is common practice to use *automatic network analysers* (ANAs) connected to dielectric measurement cells as source/detector combinations. ANAs

[15,16,36,37] are the 'workhorses' of microwave laboratories and are described in detail in other chapters. ANAs can be used both for measuring S-parameters of reflection/transmission cells (as in Figure 18.3) and for measuring frequency changes and Q-factors of resonance cells or cavities. Some ANA models have a synthesised time domain facility as one of their features [38]. This can be useful in a number of techniques for gating or de-embedding the S-parameters of a specimen from any mismatched components and imperfect transmission lines that lie between it and the ANA. ANAs can also be used at RF, but at lower frequencies, say below 100 MHz, and especially where the dielectric measurement cells are significantly mismatched to 50 Ω, it is often better to use Impedance Analysers, 'Materials Analysers', Admittance or Impedance Bridges or Frequency Response Analysers (FRAs) [39,40] as they can be more accurate and sensitive. Some FRAs can work down to frequencies as low as a few microhertz (μ Hz) while others have an upper limit at about 100 MHz. FRA manufacturers provide accessories and cells that are specifically designed for dielectric measurements as a function of temperature, time, DC and AC voltage bias.

18.7.2 Measurement cells

General advice: keep cells clean so that they do not contaminate specimens. It is important to be aware of the EM-field configuration in one's cell in order to choose the optimum size and shape for specimens. Note, in particular, that electric fields usually pass through, that is, from face to face of, laminar specimens in 'lumped-impedance' methods (see Sections 18.9.1 and 18.9.2) but, in contrast, they usually lie in the plane of the lamina in travelling- and standing-wave cells (see Sections 18.9.3, 18.9.4, 18.9.6, 18.9.9, 18.9.11 and 18.9.12). This fact can be important if the material to be characterised is anisotropic (i.e. if ε^* is a function of electrical field direction). With the exceptions of (1) free-field measurements and (2) some resonance measurements, it is usually best to keep the measurement volume of cells as small as possible, consistent with their achieving the required resolution for ε' and tan δ. Also keep connections as small and simple as possible to reduce impedance residuals and mismatches. If a cell is to be actively temperature controlled, small cell size may also be an advantage because power requirements can be lower and temperature-settling times faster. However, in some cases a contrary policy may be advantageous. Whenever a high resolution for a measured quantity such as capacitance is required, it can be an advantage to have a cell with a long thermal time-constant (a 'massive' cell), so that the constant temperature cycling of a temperature-control system does not compromise resolution. Cavities and resonators form a special class of measurement cells; see [8–10,41] for background theory. In general, we use resonators for measuring specimens of low-loss, and so we need to know how to reduce loss in the resonator itself and in its coupling mechanisms – this will increase the resonator Q-factor and increase its resolution for specimen loss. The benefits of resonators do not arise only in the measurement of dielectric loss – the increased resolution offered by resonator techniques also applies to real permittivity ε': some of the most accurate methods for measuring ε' at MW and millimetre-wave frequencies are resonator methods.

In resonators, we are also concerned with the measurement of Q-factor, which we consider next.

18.7.3 Q-factor and its measurement

Q-factor is measured in resonance techniques to determine the loss of dielectric specimens. This is a topic that requires a book in its own right to cover – see the book by Kajfez [42] and the Good Practice Guide [1]. The term Q-factor is actually a contraction of 'Quality Factor' and its symbol is usually Q. It is defined in the frequency domain for resonating systems as follows:

$$Q = 2\pi \times \frac{\text{Energy stored in the resonance}}{\text{Energy lost per cycle}} \qquad (18.15)$$

In the resonators with which we are concerned here the 'energy stored' is the EM energy stored in the fields in the resonator, the 'energy lost' is the EM energy lost by whatever means from such storage and 'per cycle' refers to a cycle of the sinusoidal resonating EM signal at the frequency that is present in the resonator. For a typical resonator we can write

$$\frac{1}{Q} = \frac{1}{Q_{\text{specimen}}} + \frac{1}{Q_{\text{resonator walls}}} + \frac{1}{Q_{\text{coupling}}} + \cdots \qquad (18.16a)$$

where Q_{specimen} accounts for the dielectric loss in the specimen – this is the parameter we are trying to measure, $Q_{\text{resonator walls}}$ accounts for the power lost in the metal walls of the resonator due to conduction losses, Q_{coupling} accounts for power lost through all coupling mechanisms into the resonators: note that power is lost both through output and input coupling ports. There are usually many other sources of loss in resonators, as illustrated in Figure 18.8.

In a typical resonance measurement we must distinguish between the influence of the loss of the dielectric material on the Q-factor and the loss from all other causes. Our measurement technique must be designed either (1) to allow us to estimate these other sources of loss, or else (2) it must cancel them out, for example, by use of a substitution method. For our purposes (18.16a) can further be refined as follows [1]:

$$\frac{1}{Q_l} = F_f \tan \delta + \frac{1}{Q_{\text{resonator walls}}} + \frac{1}{Q_{\text{coupling}}} + \cdots \qquad (18.16b)$$

where $\tan \delta$ is the loss tangent of the specimen and F_f is the 'filling-factor' of the specimen in the resonator defined as follows:

$$F_f = \frac{\text{Average EM energy contained in the speciman}(W_s)}{\text{Average EM energy contained in the whole resonator}(W_r)} \qquad (18.17)$$

F_f can be seen to be another important factor which depends on the cavity and specimen dimensions and on the specimen positioning. We need to know the value of F_f in order to measure the dielectric properties of the specimen. The value of F_f can vary enormously from one technique to another. For example, $F_f = 1$ if the specimen completely fills the resonator but F_f can be less than 0.01 in perturbation techniques

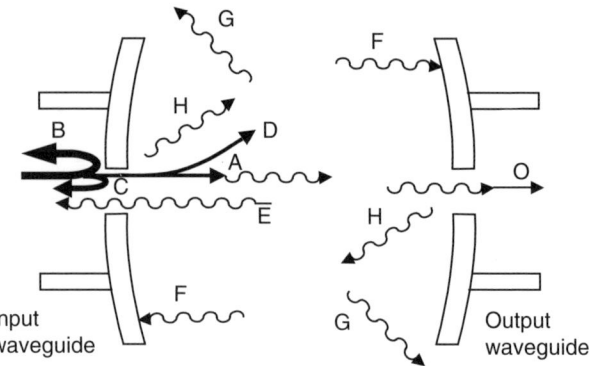

Figure 18.8 *Power transfer and loss processes in a resonator. An open-resonator (see Section 18.9.11) is used as an example. The resonator has two small coupling apertures for input and output from waveguides (shown overscale for clarity). (A) Power coupled from the input into the resonant mode, (B) input power reflected by mismatch, (C) attenuation of input signal by the 'cut-off' section of the input coupling aperture, (D) power coupled through the input aperture that does not enter the resonant mode, (E) power lost from the resonant mode via the input coupling aperture, (F) dissipation in the metal reflectors (conversion to heat), (G) diffraction at reflector edges, (H) scattering (diffraction) from the coupling apertures and (O) power transmitted to the output waveguide. Losses (E) to (O) load (i.e. reduce) the Q-factor of the resonant mode (After Clarke and Rosenberg [102].)*

(see Section 18.9.10). F_f also affects the extent to which the resonant frequency, f_r, is shifted when a specimen is placed in a resonator, and so it is also an important parameter in the measurement of real permittivity, ε' as well as loss. Computation of F_f requires an analytical model of the resonator, many of which can be found in the scientific literature.

Q-factor is traditionally measured by the resonance-width technique, (Figure 18.9).

With fully automatic detection systems such as ANAs it is often better to use curve-fitting techniques, which can take account of signal leakage in the detection system and can be more immune to noise. Signal leakage can give rise to apparently distorted resonances and so to significant measurement errors for Q-factor – the traditional technique is prone to these errors. If the ANA employed is a Vector ANA, that is, one that measures the S-parameters of the resonator as complex parameters, one should carry out a full complex-parameter fitting of the S-parameters to gain maximum benefit from the measurement technique [1,42]. One should always bear in mind that many resonators exhibit a number of resonances that may overlap in the frequency domain – they may in fact be completely degenerate (i.e. coincide in frequency). This is a condition that should be tested for and avoided or compensated for in any serious measurement.

Measurement of the dielectric properties of materials 429

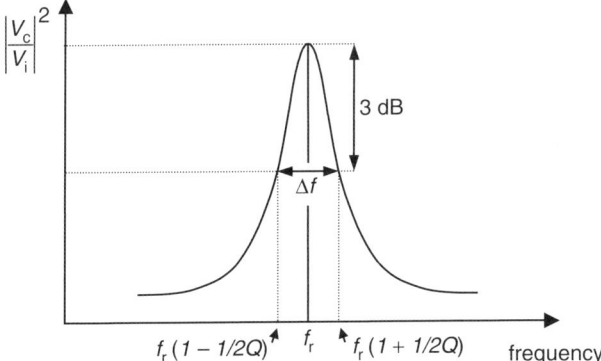

Figure 18.9 A typical resonator resonance curve, showing how the width of the curve, Δf, at the 'half power' or '3dB' points can be used to measure the Q-factor by the resonance-width method: thus $Q \approx f_r / \Delta f$, where f_r is the resonant frequency. The vertical axis gives the square of the voltage excitation of the resonance, that is, it is proportional to the power in the resonance

18.8 Good practices in RF and MW dielectric measurements

It is not always possible to follow these practices in every measurement, but one can try to abide by their spirit.

(1) Calibrate all significant equipment traceably to international standards.
(2) Once the measurement system has been calibrated and *prior* to the measurements upon the specimens under test, *check* the calibration, either by a measurement upon a *check specimen* that has known dielectric properties or else by measuring a *Traceable Dielectric Reference Material* [29,43]. This can save considerable time and embarrassing problems if the calibration is faulty!
(3) Measure a number of specimens of the same material but of different thickness (or length, diameter, and so on as appropriate for the technique) to check for systematic errors. *However*, when the aim of the measurements is to determine the *differences* in the intrinsic dielectric properties of a number of similar specimens, they should all have the *same* size and shape, so as to remove differences caused by systematic errors.
(4) Whenever possible measure specimens across a broad frequency range to check for consistency of properties across that range.
(5) All dielectric specimens should be individually identified. To prevent similar specimens from getting mixed up only one specimen at a time should be out of its container. Record the provenance of specimens.
(6) Dielectric properties usually change rapidly with temperature. The temperature of dielectric measurements should always be recorded. In the case

of ambient temperature measurements, an uncertainty of ±0.2 °C or better should ideally be achieved, though this may not be possible at elevated temperatures. A relative humidity of less than 50 per cent is recommended for measurements, unless otherwise required by one's experiment. In some cases, especially with low-loss specimens, it will be necessary to record relative humidity at the time of measurement and possibly also during the prior storage of the specimen.

(7) Record all relevant information on measurements in a lab book or computer file. If doubts arise about a measurement at a later date, this may enable you to trace the cause of the problem. On the other hand, if there is nothing wrong with the measurement, the record will help you to demonstrate this fact at a later date and so provide confidence in the measurement.

(8) It is good practice to generate a measurement uncertainty budget following the practices of 'The Guide' (GUM) [44] or of the UKAS document M3003 [45].

(9) Be Safety Conscious: always be aware that many materials are toxic or flammable. Follow COSHH guidelines [46] for handling and disposing of materials, especially liquids. *Never* accept a material for measurement unless you know what it is and how to handle it safely.

18.9 A survey of measurement methods

In this survey we attempt only to illustrate the wide range of methods that have been developed to meet the challenge of the wide range of material types discussed in Section 18.6. Some indication is given of the main purpose of the methods and the types of material they are best suited to, but the reader is referred to the Good Practice Guide [1] and the other references provided for more details.

18.9.1 Admittance methods in general and two- and three-terminal admittance cells

If the admittance, Y, is measured between any two electrodes with and without a dielectric material of complex permittivity ε^* present, we have

$$\varepsilon^* = \frac{(Y \text{ when the space around the electrodes is } \textit{completely} \text{ filled with the dielectric})}{(Y \text{ when it is completely filled by a vacuum})}$$

(18.18)

This equation forms the basis for a wide range of dielectric measurement techniques that are in use almost from DC up to 1 GHz – the higher the frequency the smaller is the cell that must be used. The most common methods are those based on adjustable parallel-plate electrode systems (see Figure 18.10). Liquid dielectrics will readily fill the full volume of a measurement cell, whereas solid specimens cannot do so and this gives rise to measurement errors for solids caused by electrode/specimen air-gaps and by fringing-fields. By immersing a solid specimen in a liquid of similar permittivity, these disadvantages can be overcome. This principle is used in *Liquid Immersion Techniques* for solids – see the standards BS 7663:1993 and ASTM D1531-01 and the Guide [1]. A more common way of avoiding errors which arise with two electrodes is to use a guard electrode around the low voltage electrode – this ensures that the electric field lines through the specimen and between the measurement electrodes are parallel, which greatly simplifies the computation of the permittivity, as we have

$$Y = j\omega \frac{\varepsilon_0 \varepsilon^* A}{d} \qquad (18.19)$$

where ω is the angular frequency $= 2\pi f$, A is the area of either of the two measurement electrodes and d is their separation, assuming that the dielectric completely fills the gap between the electrodes. Corrections to this formula can account for the small gap I in Figure 18.10, but they are usually very small. Such a cell is referred to as a three-terminal cell. Cells as shown in Figure 18.10 with only two electrodes are referred to as two-terminal cells.

There are many ways of using such two- and three-terminal cells for dielectric measurements and the best for any given specimen must depend on its properties. In some cases it is better to leave a deliberate air-gap above the specimen and measure its

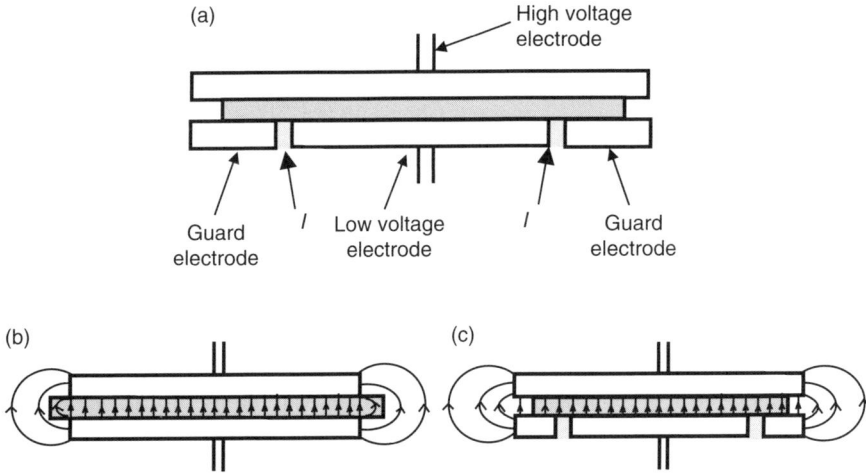

Figure 18.10 (a) A three-terminal admittance cell. 'I' is a thin low-loss insulating ring (it is exaggerated in thickness for clarity in the diagram), which separates the low voltage measurement electrode from the annular guard electrode. (b) In a two-terminal cell the E-field lines fringe out around the edge of the electrodes, giving rise to a fringing-field that can pass partly through the specimen, thereby generating measurement errors. (c) In a three-terminal cell the field lines between the high and low voltage measurement electrodes are straight and the field is uniform so measurement errors caused by fringing-fields are removed

432 *Microwave measurements*

dielectric properties by adjusting the gap d with and without the specimen to produce the same capacitance across the cell in both cases. For some materials (especially those of high permittivity) it is better to metallise the specimen. Please see [1] for details.

These cells are usually used at LF and RF frequencies in conjunction with admittance or impedance analysers or FRAs, (see Section 18.7). They can be employed over many decades of frequency for dielectric measurements.

Two-terminal admittance cells come into their own at higher RF where it is difficult to keep the guard ring in the three-terminal technique at ground potential. Traditionally their uncertainty of measurement has been much lower than that of three-terminal cells (often worse than ±5 per cent for ε') because of the presence of fringing fields around the electrodes, as in Figure 18.10, the effect of which was difficult to calculate. However with the advent of EM-field modelling packages such as finite difference time domain (FDTD) or finite integration (FI), modern software can correct for the fringing-field errors, if one so desires.

18.9.2 *Resonant admittance cells and their derivatives*

Two-terminal techniques came into use many years ago when interest in measuring dielectrics in the 100 kHz to 100 MHz range first arose. A resonant two-terminal cell method was very widely used for many years from the 1930s onwards [47], and was named the Hartshorn and Ward (H & W) Technique, after the two NPL scientists who developed the method. It was used both for medium and low-loss specimens until the advent of sensitive impedance analysers in the 1970s, which were found to be more convenient and accurate for medium-loss specimens. However, the method may still be used to advantage for low-loss specimens. The principle of the method is to resonate a micrometer-driven admittance cell with an inductor, as shown in Figure 18.4. One commonly measures the permittivity by adjusting the gap between the electrodes with and without the specimen so as to resonate the system at the same frequency in both cases – this approach is often called an 'equivalent-thickness method'. Loss is computed from the change of Q-factor. A given specimen can be measured at a range of frequencies by using a number of inductance coils.

The H & W cell is the lowest frequency member of a family of resonance cells that can be employed all the way up through the spectrum into the microwave region (see Figure 18.11). They have in common the feature that the E-field passes directly through the specimen from bottom to top, as shown in the figure, so for cylindrical specimens, they are sensitive to ε^* in a direction parallel to the axis of the cylinder. In such a resonator $f_r \approx 1/\sqrt{LC}$ so the resonant frequency of measurement cells in this family can be increased either by reducing L or by reducing C, or both. One can reduce L by abandoning the coil inductors of the H & W method (Figure 18.11a) and by effectively 'wrapping' the inductance L around the capacitor, as shown in Figure 18.11b. This configuration gives one a re-entrant cavity or hybrid cavity. Such cavities can be used conveniently in the frequency range 100 MHz to 1 GHz. To resonate at even higher frequencies one can effectively reduce C by using a thicker and narrower specimen. One ends up with the TM_{010}–mode (or TM_{020}–mode)

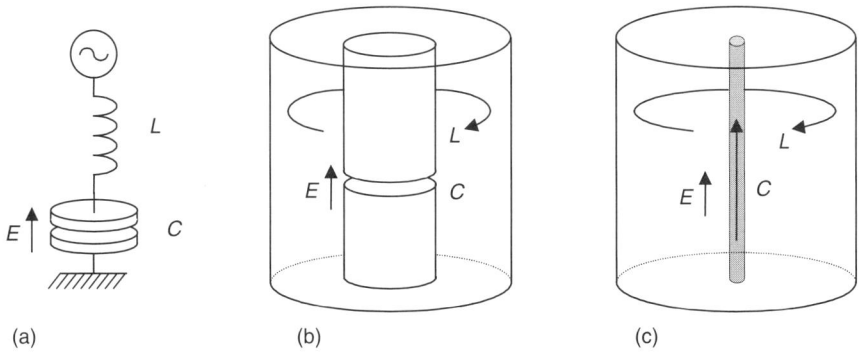

Figure 18.11 A family of dielectric measurement resonators. (a) The principle of the Hartshorn and Ward method: the admittance cell is resonated with an inductor. (b) The re-entrant cavity is a coaxial cavity with a capacitance gap in its inner-conductor where the dielectric specimen is placed. The inductance L is that of the magnetic field in the region between the inner- and outer-conductors. (c) The TM_{010}-mode cavity can, in principle, be seen as a re-entrant cavity that has its inner-conductor fully retracted so that only a cylindrical cavity remains. This reduces the capacitance, C, of the specimen, which in the TM_{010} cavity typically takes the form of a rod, as shown

cylindrical cavity shown in Figure 18.11c. Such cavities typically operate in the range 1–10 GHz.

Re-entrant cavities should always be considered as an option for low-loss materials in the range 100 MHz to 1 GHz as other forms of cavity (see below) are physically large in this frequency range. Details of measurement methods are described in International Standard IEC 60377-2, 'Part 2: Resonance Methods' and in [48].

The TM_{010}–mode cavity (Figure 18.11c) is the next in our family of resonators (see also Figure 18.12a), typically used from 1 to 10 GHz. The last subscript '0' in TM_{010} indicates that the strength of the E- and H-fields does not change as one moves from top to bottom along the axis of the cavity. The axial E-field is at a maximum on the axis of the cavity and falls away to zero amplitude on the cylindrical sidewall, as required by the electrical boundary conditions there. As in the re-entrant cavity, the magnetic or H-fields loop around the axis of the resonator and so the coupling arrangements are similar. Specimens normally take the form either of rods [49] or else tubes, which can be used for containing liquid dielectrics [50,51]. Specimens commonly reach from top to bottom of the cavity, but this is not necessary [5].

Traditionally [52,53] this method was used as a *perturbation method*; (see Section 18.9.11.) If perturbation computations are used, the diameter of the specimen must be small compared with that of the resonator. However, more recently, advantage has been taken of the relatively simple internal geometry of the TM_{010}-mode cavity to employ modal analysis to derive exact equations for the cavity [54]. With such an analysis the diameter of the specimen need not be restricted. Similar advantages

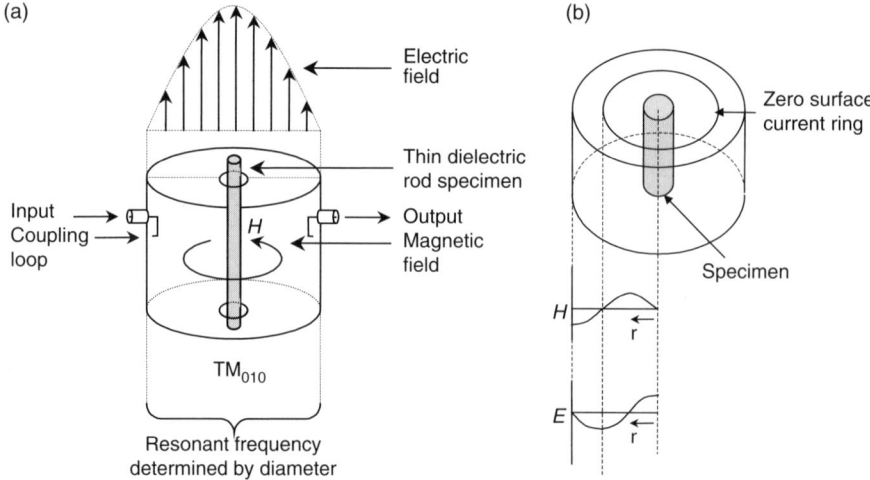

Figure 18.12 (a) A TM_{010}-mode cavity resonator containing a dielectric rod specimen, which in this case is introduced though two small holes (not shown to scale) on the axis of the resonator. (b) A TM_{020}-mode cavity resonator, showing the position of the circle on the top of the resonator where there are no surface currents, this is the best place to make a break in the surface if a lid is needed (After [5].)

accrue if discretised FDTD or FI modelling packages are used for the modelling – such packages are available commercially.

One disadvantage of the TM_{010}-mode cavity arises if its lid has to be removed to insert a specimen. Currents in the top of the cavity have to cross the gap between the lid and the rest of the cavity. Thus, whenever a lid is removed in order to introduce/remove the specimen, contact impedances can change, introducing measurement errors for tan δ. With the TM_{010}-mode it is impossible to find any points or lines in the walls or the top or bottom of the cavity where currents do not flow (except the two points on the axis of the cavity). Fortunately this problem can be avoided by using the TM_{020}-mode rather than the TM_{010}-mode, as explained in Figure 18.12b. Both cavities are relatively easy to use, especially if employed with the perturbation method [5,53].

18.9.3 TE_{01}-mode cavities

The TE_{01}-mode cavity [55,56] (see Figure 18.13), is used for measurements on laminar low-loss specimens, typically in the 8–40 GHz range. These resonators have a high resolution for low-losses and they are typically used for measuring cylindrical disc specimens that notionally have the same diameter as the resonator. The electric fields in the resonator are circularly or ('azimuthally') polarised with respect to the resonator (z-) axis, as in dielectric resonators (see Section 18.9.8). Tuning can be achieved easily, as shown in the figure, by changing the resonance length of the cavity with a micrometer-driven metal piston. Typically, the cylindrical dielectric specimen

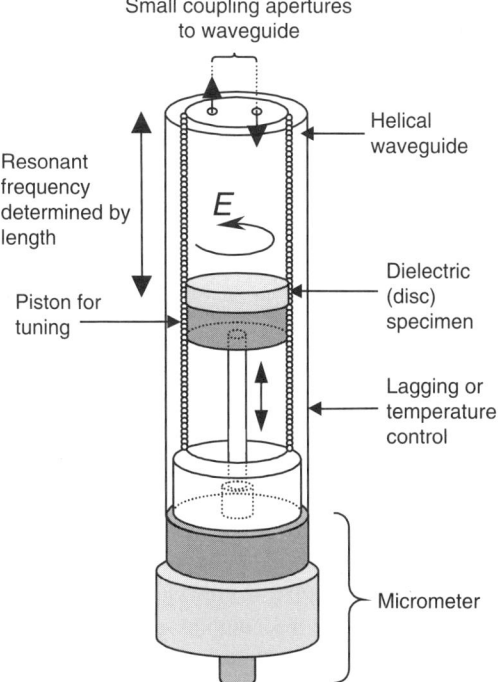

Figure 18.13 *A micrometer-tuned TE_{01}-mode cavity for dielectric measurements*

sits on the piston during measurements. As the electric field must fall to zero on the outer wall of the resonator, the specimen itself need not be an exact fit to the diameter of the cavity. Typically the specimen can be up to 0.5 mm smaller in diameter in a 50 mm diameter cavity without having a significant effect on the measurement accuracy. The technique can be used across a range of temperatures [57].

One potential problem with the TE_{01}-mode in a circular cylinder is that it is always degenerated with (i.e. it always resonates at the same frequency as) the TM_{11}-mode, so a method must be found for filtering out the latter. One of the most effective methods was implemented at NPL in the 1970s [55,58], when the cylinder of the cavity was manufactured from helical waveguide to remove axial currents in the walls, which are necessary for the TM_{11}-mode to resonate.

Coupling into the cavity is typically from waveguide via small coupling holes, as shown in Figure 18.13, delivering a high insertion loss of the cavity on resonance. As in many resonant techniques ε' can be measured by a length-change/equivalent-thickness method. Q-factors may be as high as 60,000 at 10 GHz for the mode with nine half-wavelengths along the cavity in a 50 mm diameter cell. As in many standing wave techniques (e.g. also in open resonators, Section 18.9.11) specimens should ideally be an integral number of half guide-wavelengths thick, both because this renders the measurements insensitive to surface contamination on the specimen (both surfaces are at a field node) and because it can be shown that this is the condition that gives best uncertainties. Experience over many years with such a cavity operating at 10 GHz shows that resolutions for loss angles may be as low as 10μ rad, while uncertainties for $\varepsilon' < 5$ can be as low as ± 0.5 per cent for 'ideal' specimens (i.e. for flat homogeneous low-loss specimens that are an integral number of half-wavelengths thick).

18.9.4 Split-post dielectric resonators

The split-post dielectric resonator (SPDR) (Figure 18.14) was developed by Krupka and his collaborators [59] and is one of the easiest and most convenient techniques to use for measuring microwave dielectric properties. It uses a fixed-geometry resonant measurement cell for characterising low- or medium-loss laminar specimens (such as substrates and thin films) in the frequency range 1–36 GHz. The main drawback with this technique is that each SPDR can only operate at a single fixed frequency, but it is practicable and cost-effective to measure the same specimen in several SPDRs operating at different frequencies. The SPDR uses two identical dielectric resonators of the usual cylindrical disc or 'puck' shape. They are placed coaxially along the resonator axis leaving a small laminar gap between them into which the specimen is placed for measurement. Once the SPDR is fully characterised, only three parameters need to be measured to determine the complex permittivity of the specimen: its thickness and the changes in resonant frequency, Δf, and in the Q-factor, ΔQ, obtained when it is placed in the resonator. Specimens are typically a millimetre or so in thickness, but the method is also suitable for thin films. The resolution for thin specimens can be improved by stacking a number of them in the gap.

The two dielectric pucks resonate together in a coupled resonance in the $TE_{01\delta}$-mode, and so a circularly polarised evanescent E-field exists in the gap region between

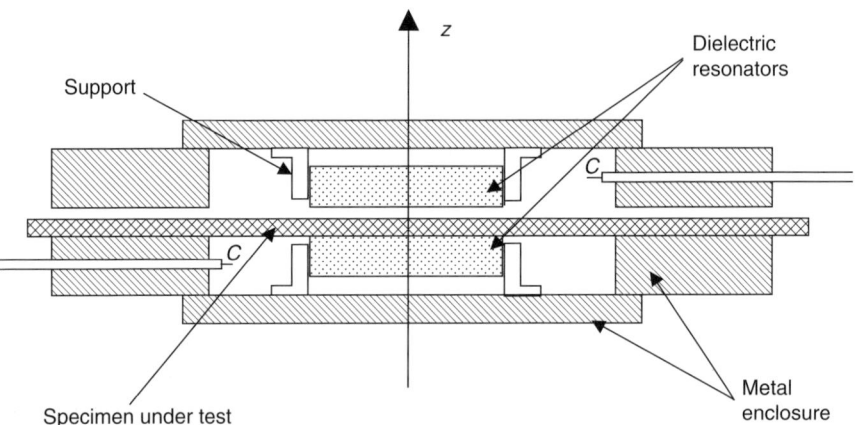

Figure 18.14   *A Split-Post Dielectric Resonator (SPDR) cell for dielectric measurements. Transmission coupling is via both dielectric resonators, 'C' marks the coupling loops*

them. The geometry of the system is such that simple analytical models cannot be used to relate ε^* to Δf and ΔQ, so numerical solutions must be employed. The geometry of the SPDR lends itself to modelling by mode-matching or Finite Difference (FD) techniques.

The main limitation of the SPDR technique is that its resolution for very low-losses is reduced by the Q-factors of the two dielectric resonators it employs, but it can typically be used to measure loss angles down to 100 μrad with reasonable accuracy. Typical uncertainties can be ±1 per cent for ε' up to 10 and ±5 per cent for tan δ. A conference paper [60] gives details of studies that show that measurements on reference specimens in SPDRs agree well with those carried out by other well-established methods.

18.9.5 Substrate methods, including ring resonators

Substrate and printed-circuit manufacturers need to know the dielectric properties of their substrates. Special techniques have been developed to enable them to measure these properties with the facilities commonly at their disposal. Most notable among these are facilities for depositing metal resonant structures lithographically onto substrate surfaces. Such structures include ring [61,62], T-shaped [63] and strip-line [64] resonators, all of which may be used to measure dielectric properties at a number of harmonically related frequencies across a wide band. The main advantage of this approach is that it gives designers of integrated circuits precisely the information that they need: an effective value for permittivity, ε'_e, that is appropriate for their applications. This is not necessarily the absolutely 'true' value of ε' for the substrate material because, like all measurements, these measurements are subject to systematic errors. However, if one is using a strip-line technique to determine a parameter, ε'_e, subsequently to be used for designing similar strip-line circuits, some degree of beneficial compensation of errors must occur. Problems can arise here if one is interested in measuring loss. High-temperature processes used in manufacture can cause deposited metal to in-diffuse into a substrate, so measured properties may differ from those of pure bulk material. The loss tangent of low- and medium-loss substrate materials is generally not measurable by these techniques because radiative and conductivity losses (surface and bulk) in the deposited metal structure tend to dominate: in practice the measured Q-factors can be below 100 even for low-loss materials.

18.9.6 Coaxial and waveguide transmission lines

This method makes use of *annular* specimens, which should fit closely between the inner- and outer-conductors of the coaxial conductors of precision coaxial air-line, as shown in Figure 18.3. The air-line should ideally be fitted with precision connectors that allow for a well-matched coaxial connection to an ANA. The complex permittivity and permeability of specimens are computed from the S-parameters of the specimen, as measured by the ANA. Similar techniques apply with other types of uniform air-line – the use of rectangular waveguide for such measurements is very common.

438 Microwave measurements

The first of such methods was the Roberts and Von Hippel method [65] – in which the specimen is placed hard up against a short-circuit and its reflection coefficient is measured. Such one-port techniques may still be recommended in many instances (e.g. for high-temperature measurements) but there are two advantages to be gained from measuring both reflection *from* and transmission *through* specimens in a matched transmission-line. (1) For purely dielectric specimens up to 10 mm in length, one often finds that reflection coefficient measurements tend to be more accurate at lower frequencies (<500 MHz), while transmission measurements tend to be more accurate at higher MW frequencies. The combination therefore allows broadband measurements to be performed on just one specimen from, say, 100 MHz to 18 GHz in a 7 mm diameter air-line, taking advantage of the best uncertainties from both methods. (2) As explained in Section 18.3, the measurement of both transmission and reflection coefficients allows one to determine the magnetic properties of the specimen as well as its dielectric properties. For magnetic materials, both transmission and reflection data are normally required, though reflection-only data can be used if the specimen is moved axially in the line.

Inevitably, for solid specimens metrological problems arise from the presence of air-gaps between the specimen and the inner- and outer-conductors of the line. They dilute the apparent permittivity obtained from the measurement but, more seriously for accurate measurements, they also help to launch higher-order modes. Both effects give rise to significant measurement errors. Air-gap problems do not usually arise for liquid measurements, but a well-designed liquid cell is required instead. The liquid must be contained between solid dielectric windows, as shown in Figure 18.15, so a multi-layer theory based on cascaded two-ports is necessary for the S-parameter analysis of the specimen/cell combination.

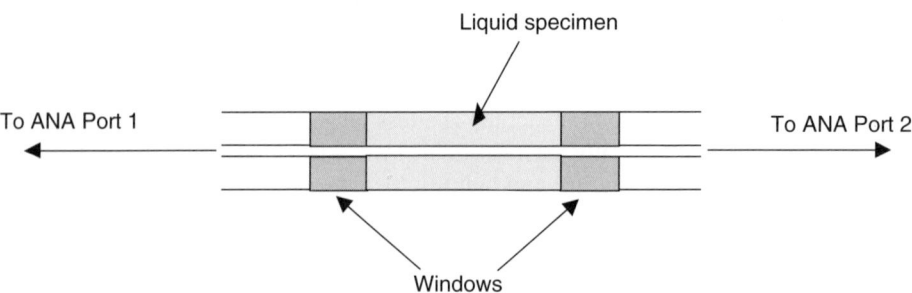

Figure 18.15   *A coaxial line cell for measuring liquid dielectrics*

These transmission-line methods are often the most cost-effective choice for (1) broadband measurements, (2) magnetic materials, (3) medium- to high-loss materials and (4) materials that are only available in small volumes. Uncertainties for ε' can be as low as ± 1 per cent for low-permittivity materials if a correction for air-gaps is made but it may be higher than ± 5 per cent for high permittivity materials, so use of other methods is advisable if more accurate measurements are needed.

NIST Technical Note 1341 [66] provides an excellent guide to the theory of this method as does the NIST follow-up work by Baker-Jarvis and his colleagues [67,68]. The paper by Jenkins *et al.* [43] explains how uncertainties can be computed in these measurements. It is concerned with dielectric liquid measurements but with the exception of the uncertainties caused by air-gaps, the analysis can be extended to solids. The NIST Technical Note provides the formulae that correct for air-gaps, based on a simple capacitative model. There are two published standard methods for the transmission-line technique: ASTM D5568-01 and a UTE (*Union Technique de L'Electricité*) standard from France [69]. The paper by Vanzura *et al.* [68] illustrates just how dominant the resonances of higher-order modes can be at higher frequencies. It shows that it is not advisable to use this method at frequencies above these resonances.

Most of the considerations discussed for coaxial transmission lines apply also to waveguide measurements. The main advantage of using a waveguide is that one does not have to machine axial holes through the specimen to tight tolerances to allow it to be fitted onto a coaxial inner-conductor. The absence of the inner-conductor also makes waveguide cells more suitable for temperature control. The main disadvantage of waveguide is that one is normally restricted in frequency coverage to a single waveguide band (less than an octave in frequency coverage). Please see [1] and [66] for more details of measurements in both types of line.

18.9.7 Coaxial probes, waveguide and other dielectric probes

Coaxial probes [43,70–76] (see Figure 18.16) are extremely popular measurement tools. The principle of operation of the conventional flat-faced probe is illustrated in Figure 18.16. A TEM travelling-wave propagates in the coaxial line up to its end where it launches fringing EM-fields from the open end of the probe into the dielectric specimen. Their magnitude and geometry depends on the complex permittivity, ε^*, of the dielectric, so the reflection coefficient of the TEM-wave from the end of the probe will depend on the value of ε^*. One can relate the measured reflection coefficient, Γ, to ε^* by using (1) a modal analysis of the fields in the coaxial line and (2) an analysis of the fields in the dielectric under test (DUT) that treats the probe as an antenna. Simpler analyses have been used, especially those based on capacitative models for the fringing-fields, but they have their limitations [70]: at sufficiently high frequencies the probe must be treated as an antenna as it actually radiates into the dielectric specimen so that $|\Gamma| < 1.0$ even if the dielectric is lossless [43].

Flat-faced coaxial probes, such as the one shown in Figure 18.16, represent just one member of a whole family of reflectometric and non-invasive probe designs that can be used for dielectric measurements. Another flat-faced probe option is the open-ended waveguide probe [77–79]. This type of probe has the capacity to measure anisotropic materials [80]. Coaxial probes are widely used for characterising lossy solids like biological tissues because of their ability to perform measurements by contacting just one face of the specimen, rather than having to machine the specimen to fit into a measurement cell. This makes them very convenient to use. They are also ideal for measuring lossy liquids and are widely used for SAR liquid characterisation;

440 *Microwave measurements*

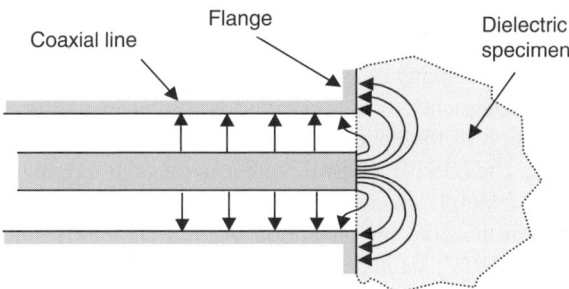

Figure 18.16   A coaxial probe, showing the fringing-fields that emerge from its ends. The field lines shown are those of the electric field, which is seen to fringe out into the dielectric specimen at the end of the probe

see British Standard BS EN 50361:2001, and IEEE Standard P1528 (D1.2). But with liquids one does not need to use a flat-faced probe and there are often advantages to be gained by using other geometries (see below). A single flat-faced coaxial probe can typically operate effectively over a frequency range of about 30:1 (e.g. 100 MHz to 3 GHz for a 15 mm diameter probe) whilst retaining a reasonable uncertainty performance of the order of ±4 per cent for ε' for suitable materials. The actual frequency range depends on the diameter of the coaxial aperture of the probe and the permittivity of the dielectric specimen.

The coaxial probe method has its limitations. Measurement uncertainties are usually of the order of ±3 per cent at best for ε' and ε'' and a number of other techniques described in this survey can be more accurate (e.g. coaxial cells for liquids; Section 18.9.6). There are many types of measurement for which the probe would be the wrong choice. Thus, the assumption is commonly made in the theory that is used to relate ε^* to Γ, that there are reflections of waves neither from the extremities of the specimen nor from permittivity steps or gradients within inhomogeneous specimens. In small, layered or low-loss specimens this is usually not the case. Furthermore coaxial probes are much better suited to measuring malleable materials (or liquids) that accommodate themselves to the shape of the probe than to hard specimens because they invariably leave uneven air-gaps between the specimen face and the probe face. The conventional theory assumes that there are no such air-gaps, and as the probe is particularly sensitive to the permittivity of the material closest to its face, even a small gap can give rise to a large error of measurement [81]. Layered structures and air-gaps can be modelled [73,74], however, see Figure 18.17, if the thickness of the layers is known, and the utility of the probe extended thereby.

The measurement geometry of coaxial and waveguide probes, even for complex geometries similar to those shown in Figure 18.17, can be analytically calculable. An alternative approach that widens the range of probe designs and their range of application is to reduce one's dependence on full calculability and to rely more on probe calibration with a set of known reference liquids to supply the accuracy that the theory lacks. Thus, one may prefer to use a probe without a flange (e.g. [75]) – it

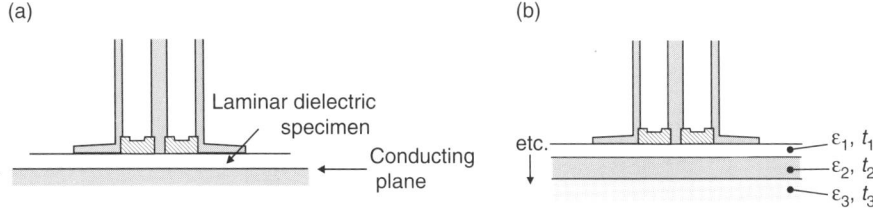

Figure 18.17 Use of a coaxial probe (a) for measuring a dielectric lamina backed by a metal conducting plane, (b) for measuring a multi-layered specimen. Each layer, i, has thickness t_i and permittivity ε_i

is smaller and more convenient than a calculable probe with a (supposedly infinite) flange. One can regard probes having other non-calculable shapes as 'black boxes': whenever they have the property of being reasonably stable and of presenting one with a one-to-one and smoothly varying relationship between Γ and ε^* they can be effective. By measuring a suitably large number of reference liquids having known ε^* values, one can interpolate measured values of Γ to calculate ε^* for other dielectrics. Probes that are used in this way may be called non-calculable probes (e.g. [82]). Some of these probe designs can have much better field penetration into the specimen than coaxial probes [83,84].

Dielectric probes are normally used in conjunction with ANAs, so suitable calibration schemes must be developed for the ANA/probe combination. Typically for a coaxial probe one calibrates with (1) an open-circuit into air, (2) a short-circuit and (3) a measurement of a known reference liquid (e.g. [29]). Great care must be taken, particularly with the short-circuit, where one has to make a good electrical contact to the inner-conductor, and also with the reference liquid, which can easily change temperature by evaporation or absorb contaminants from the atmosphere, for example, water vapour. Both temperature change and contamination can easily lead to the dielectric properties of the liquid not being close to those assumed by the calibration software, and so they can result in significant calibration errors. Given such calibration difficulties it is always good practice to measure a second dielectric reference liquid immediately after a calibration to check calibration validity. It is not uncommon for calibrations to have to be repeated a number of times until a good calibration is obtained. This can be very time consuming. One approach that has been developed to overcome this problem is to use a least-squares calibration technique [1]. Least-squares calibration has another advantage as well: it enables some of the errors and uncertainties of the measurement to be estimated statistically.

For liquid specimens it is often better to extend the outer-conductor of the probe to form a cell – see Figure 18.18 [43]. Extension of the inner-conductor as well, as in Figure 18.18b increases the capacitance of the cell and so makes it more sensitive at lower frequencies. The liquid can be poured in until its meniscus is sufficiently far away from the end of the inner-conductor for no change in reflection coefficient Γ to be detected if more liquid is poured in.

442   Microwave measurements

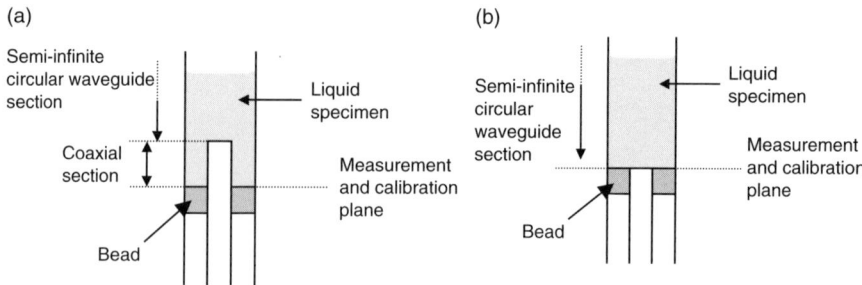

Figure 18.18   *Liquid Cells. (a) A modification of the coaxial probe shown in Figure 18.16 to allow easy measurement of liquids. (b) The Discontinuous Inner-Conductor Cell: a further modification that can be used to increase the measured capacitance of the dielectric liquid. For a given liquid, geometry (b) will be more sensitive at lower frequencies than (a). The inner-conductor extension is of any appropriate length. Both cells are fully calculable*

Open-ended rectangular waveguide probes [77–79] are used less often than coaxial probes, partly because, like all waveguide-based systems, they are limited in frequency range and they may be physically quite large. However, the required probe size for a given frequency range can be reduced if the probe waveguide is itself filled with a 'loading' dielectric material. Reference [80] describes one such probe, based on a WG16 (normally 8.2–12.4 GHz) waveguide adaptor, which was loaded with glass dielectric ($\varepsilon' \approx 6$) to allow it to operate in the range 3.5–5 GHz. Waveguide probes offer two potential advantages [80] over coaxial probes for specific applications. One is the fact that such probes are better matched for measuring lower permittivity than coaxial probes of similar size and at a similar frequency. The other perhaps more important advantage is that waveguide probes, being linearly polarised, can measure anisotropy in dielectrics [80].

18.9.8   Dielectric resonators

Dielectric resonators (DRs) are widely used in electronics and telecommunications applications as high-Q components for narrow-band filters. The theory of their resonances is well developed [85]. The resonators typically take the form of 'puck'-shaped cylinders of dielectric material. They can retain the EM-fields that are resonating inside them because the fields are totally internally reflected from the interior of the dielectric surfaces. A formal analysis of the fields reveals that there are also EM evanescent fields in the air (or other dielectric medium) that surrounds the resonator and that these fields decay exponentially in magnitude as one moves away from the resonator. It is the presence of these fields that allows one to couple RF and MW power into the DR, typically via coupling loops at the end of coaxial line feeds. However, these fields also interact with other objects in the vicinity of the resonator (e.g. its support or its container) and if these nearby objects are lossy the resonance

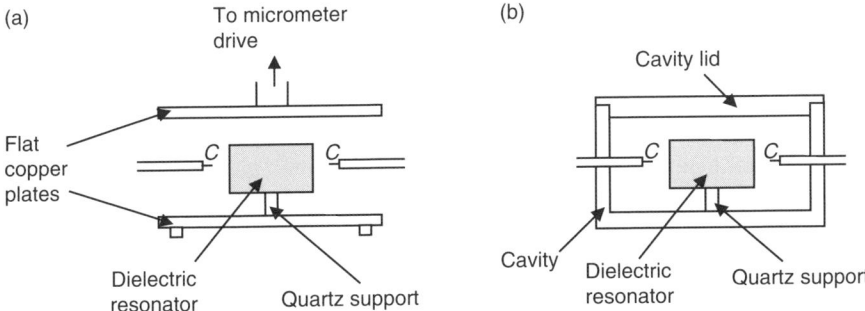

Figure 18.19 Dielectric resonator cells. (a) A Hakki-Coleman Cell, otherwise known as the Courtney Holder or the Parallel Plate Cell, and (b) a Cavity Cell. In both diagrams 'C' marks the coupling loops.

becomes loaded and the Q-factor falls, giving rise to errors if one is measuring the loss of the resonator. Figure 18.19 shows the coupling geometry for two configurations commonly used in dielectric metrology. Typical resonator sizes range from tens of centimetres on a side in 900 MHz cell-phone base-station applications, down to a centimetre on a side or less at around 10 GHz. The frequency depends on the size and the permittivity of the resonator.

Dielectric measurements on 'puck'-shaped specimens offer one of the most accurate and sensitive methods available to us for measuring the permittivity and loss of low-loss dielectric materials. They have a major advantage over other resonant techniques for characterising low-loss materials: the attainable filling-factors, F_f (see Section 18.7), in this technique are normally close to unity because most of the energy in the resonance is contained in the dielectric itself. Measurements are usually performed using the $TE_{01\delta}$-mode in which the E-field is azimuthally circularly polarised, but higher-order 'whispering-gallery modes' have also been used to obtain dielectric data at much higher frequencies [86,87]. When viewed on an ANA or spectrum analyser, one can see that DRs resonate in many different modes. One of the main preliminary steps to be undertaken before measurement, therefore, is to identify the $TE_{01\delta}$ mode. $TE_{01\delta}$-mode resonators must be used in a container or cell, otherwise their Q-factors are loaded by radiative losses. The two types of cell most commonly employed in dielectric metrology are shown in Figure 18.19 and they are designed to prevent this. In the Courtney Holder or Hakki–Colemen Cell of Figure 18.19a the distance between the top and bottom plates must be less than $\lambda/2$ in air if this radiation loss is to be avoided. (NB. radiation loss is not so important for whispering-gallery modes). The Cavity Cell of Figure 18.19b prevents radiation escaping by completely enclosing the specimen with metal walls. These walls should be well separated, ideally by at least one specimen diameter, from the specimen to minimise losses from currents flowing in them. For the same reason, the specimen is normally placed on a small post or tube made from low-loss dielectric (e.g. quartz) to displace it away from the base of the cavity. Cells are often made from copper because its high conductivity reduces the metal losses in the walls. See Reference [1] for the relative advantages and disadvantages of the two types of cell in Figure 18.19.

There may be two reasons why we may wish to perform DR dielectric measurements. First, we may wish to know the intrinsic dielectric properties of the DR specimen material: its real permittivity, ε', and loss tangent, tan δ, or we may ultimately wish to evaluate it as a dielectric resonator, for example, as a component of a filter in an electronic circuit. In the latter case we will want to measure its extrinsic parameters: its resonant frequency, Q-factor and its temperature coefficient of resonant frequency (TCRF). Similar measurement configurations can be used for either type of measurement, but extra analysis is required for intrinsic measurements. The TCRF is one of the most important factors in practical applications: one normally wishes it to be as close to zero as possible.

Examples of the recent use of DRs for dielectric measurements can be found in the literature [88] and they demonstrate the versatility of the technique. Use of dielectric resonators for dielectric measurements is also described in some International Standards, for example, IEC 61338, Sections 18.1–18.3 and IEC 60377-2, Part 2.

18.9.9 Free-field methods

Figure 18.20 shows three typical geometries employed for free-field measurements on dielectrics. These methods are best suited for materials that are intended for end-uses in the free-field, as they are likely to be the only materials available in large enough cross sections to allow free-field methods to be effective. Typical materials are RAM – high-loss materials used for absorbing free-field waves – and Radome materials – typically low-loss materials used for protecting antennas from the elements (rain, wind, snow, etc.).

Free-field methods may be categorised by three contrasting pairings of practical approaches to measurement: (1) Transmission or Reflection methods; (2) Intrinsic or Extrinsic methods. Intrinsic measurements determine the intrinsic dielectric and magnetic properties of the specimen, that is, ε^* and μ^* while Extrinsic measurements measure extrinsic parameters like reflectivity or scattering from materials and transmission through materials; (3) Focused (Quasi-Optical)-Beam or Unfocused-Beam Methods.

Free microwave fields are typically launched as diverging beams from antennas, as in the unfocused measurement systems of Figure 18.20. The inevitable diffraction around the edges of antennas and specimens in these methods limits their accuracy. In focused-beam or quasi-optical methods, see Figure 18.21, lenses or concave mirrors are used to prevent the divergence of the beam. An attempt may also be made to ensure that it is fully calculable in its geometry. Such beams can be launched from *corrugated-horn antennas* [89,90] as *Gaussian Beams* (GBs) [3,91,92]. They are potentially fully calculable all along their length, they decay exponentially to insignificant amplitudes as one moves away from their axis of propagation and they can be focused by concave mirrors or bloomed lenses [92]. Thus, diffraction problems may potentially be made negligible.

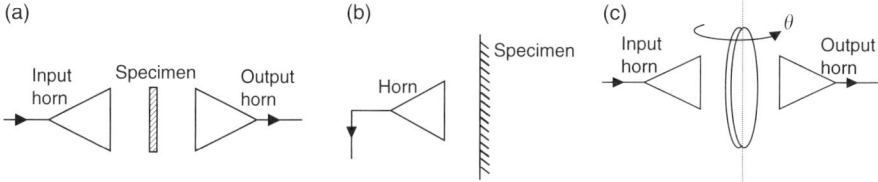

Figure 18.20   *Free-field methods: (a) normal transmission through a specimen between two horn antennas, (b) normal reflection from a specimen and (c) measurement of transmission as a function of the angle of incidence, see text*

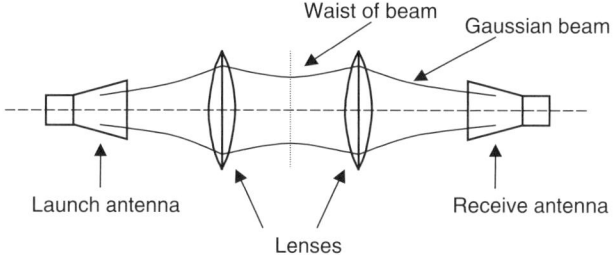

Figure 18.21   *A focused free-field measurement system. The specimen is placed at the waist of the beam. If corrugated-horn antennas are used, then the beam can be launched as a calculable Gaussian Beam*

End-users of free-field materials such as RAM and radome laminates are often more interested in their extrinsic parameters, such as reflection or transmission coefficient, than the intrinsic parameters ε^* and μ^*. Intrinsic measurements are required, however, for design and optimisation purposes. The actual measurement set-ups needed to implement these two approaches may be very similar, but extrinsic measurements ought ideally to be performed in a geometry that approximates to that of the end use of the material. For example, if the reflectivity of RAM at a 45° angle of incidence is required, then extrinsic measurements of reflectivity should be performed at this angle, whereas intrinsic measurements can be performed by any suitable method and the reflectivity at 45° incidence can subsequently be computed from Fresnel's equations [10,19].

Free-field travelling-wave methods have much in common with guided travelling-wave methods in that ε' is generally determined from the phase change of the transmission coefficient and dielectric loss from the attenuation of the beam on insertion of the specimen. The method of time domain gating [38] is particularly useful for improving the accuracy of free-field measurements (especially unfocused measurements) as it allows the wanted signals from the dielectric specimen to be separated from spurious reflections originating from elsewhere in the measurement system and from elsewhere in one's laboratory!

There are many approaches to unfocused free-field measurement, they range from the The Arch Method for RAM at arbitrary angles [93,94], to normal incidence methods [95,96], to Brewster angle methods [97] and to measurement of transmission as a function of the angle of incidence [98]. Focused or quasi-optical methods usually use travelling GB waves [91]. Complete measurement systems for laminar specimens are constructed either with lenses [99] or mirrors [94] for focusing – the latter generally give better performance.

18.9.10 The resonator perturbation technique

High-Q cavity resonators are very sensitive measurement devices and this makes it possible to measure very small dielectric specimens inside them. If a sufficiently small specimen is inserted into a resonator and if no other changes are made to the measurement geometry, the resonant frequency f_r and Q-factor, Q, of the resonator will both change by a small amount: Δf_r and ΔQ, respectively. If both $\Delta f_r/f_r$ and $\Delta Q/Q$ are small (typically less than 5 per cent) and the volume of the specimen is small compared with the volume of the resonator, first-order perturbation theory may be used to calculate the permittivity, ε', and the loss tangent, $\tan \delta$, of the specimen. Such measurement techniques are referred to as Resonator (or Cavity) Perturbation Techniques [41]. One example of this approach is the application of perturbation theory to measurements in TM_{010} or TM_{020}-mode cavities when they are used with narrow rod specimens as described in Section 18.9.2. A number of advantages can accrue from using such a perturbation technique: (1) The measurement theory is much simpler: ε' is usually proportional to Δf_r and $\tan \delta$ to ΔQ. (2) Even high-loss materials may be measured, if the specimens are sufficiently small, and if $\Delta Q/Q$ is also small. In fact, perturbation methods are normally the only resonator methods commonly used for measuring high-loss materials. (3) By placing small specimens in a cavity at points where the direction of the E- and/or H-fields are well defined, the *anisotropy* of permittivity, ε^*, and of permeability, μ^*, can be measured. The perturbation method has been in use at least since the 1940s [52,53], but Waldron later extended and popularised it in the 1960s. His book [41] still provides one of the best guides to its application, both for permittivity and permeability measurement. There are a number of more recent applications described in the scientific literature, however, for example [100,101].

18.9.11 Open-resonators

Open-resonators are millimetre-wave Fabry–Perot interferometers [102,103]. They provide one of the most accurate methods for low-loss dielectrics at millimetre-wave frequencies. Achievable uncertainties can be as low as ± 0.2 per cent for ε' below 3 and better than ± 1 per cent for $\varepsilon' \approx 50$. Uncertainties better than ± 10 per cent for $\tan \delta$ are also possible for specimens with loss angle above 200 μrad, while resolutions for loss down to 20 μrad or less are possible if the unloaded Q-factor of the resonator is greater than 150,000. A typical open-resonator configuration for dielectric measurement is shown in Figure 18.22. Typical sources of loss in an open-resonator are shown in Figure 18.8, Section 18.7. An advantage of open resonators

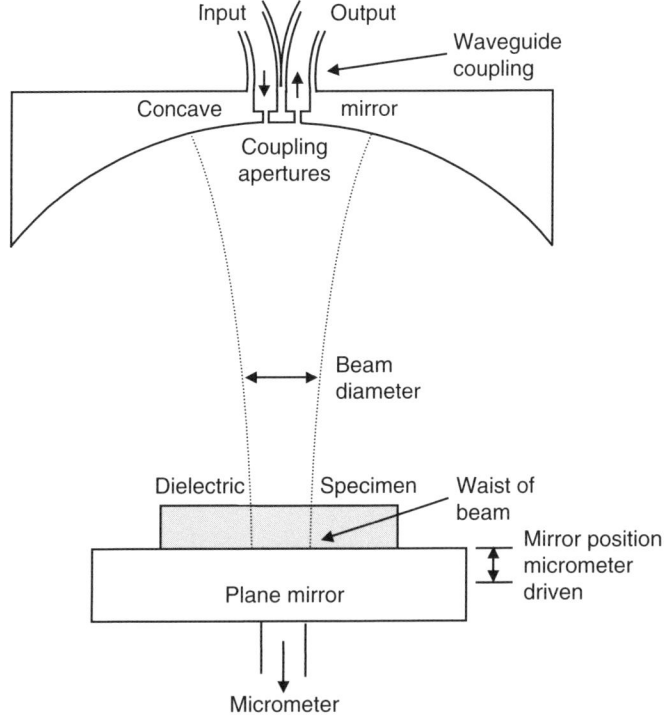

Figure 18.22 An open-resonator geometry that can be used for characterising dielectrics at millimetre-wave frequencies

over most microwave cavities (e.g. TE_{01}, TM_{01n} and $TE_{01\delta}$) is that the resonant mode employed in these measurements is the fundamental transverse electromagnetic (TEM) mode which is linearly polarised – thus enabling specimen anisotropy to be measured [104].

The resonant mode used in the open-resonator is the fundamental TEM_{00n} GB mode [91] and the high Q-factor resonance itself acts as a filter that ensures that this GB mode can be very pure. The cross-sectional shape and size of specimens for open-resonators is not important provided they are large enough to encompass the whole cross section of the GB at its waist, where it is narrowest. Specimens are flat laminas, their required minimum diameters depend on frequency. The following are typical: at 10 GHz typically 200 mm diameter, at 36 GHz 50 mm and at 72 GHz 35 mm diameter. The best practice is to compute the radius, w_0, of the GB at its waist and then use a specimen diameter of at least $5 \times w_0$. To achieve lower measurement uncertainties, specimens should be an integral number of half-wavelengths thick.

Though they are potentially very accurate, open-resonator measurements are prone to many errors that have to be checked for and so measurements can be quite time consuming. Significant sources of error include mode coincidence, warped specimens and gaps between specimens and mirrors. Errors in estimated loss can also occur if the specimen is anisotropic but is not known to be so. The errors and how to avoid them are described in detail in the Good Practice Guide [1].

The use of open-resonators for dielectric measurements has been described in great detail in a number of reviews, scientific papers and books [102,103,105,106], and the reader is referred to these sources for full details of methods and measurement theory. The microwave open-resonator technique came into its own for low-loss millimetre-wave measurements in the 1970s and it has been continuously developed since then. Work at NPL in the early 1990s addressed the measurement of medium-loss specimens and investigated new techniques for measuring specimen loss [107], while studies down to cryogenic temperatures in Germany [108] have much improved the resolution for loss by increasing resonator Q-factor through better coupling methods.

18.9.12 Time domain techniques

Time domain reflectometry (TDR) using pulse or step generators and sampling methods for detection, see Figure 18.23, was introduced in the second half of the twentieth Century [109] for dielectric measurements and is still widely used [110,111]. It is convenient to use for the purposes of scientific study, being a cost-effective broadband method which requires only small specimens, making temperature-control relatively easy. Solid specimens typically fit into a 7 mm coaxial line. The main limitation of the method is its absolute accuracy, which is not generally as good as that of reflectometry based on a calibrated ANA. The main reason for the continued popularity of TD methods using pulse/step generators is that they are cheaper than fully automated ANA measurements, and they are often more convenient to use.

These days it is possible to employ synthesised time domain techniques [38] because it is a feature available in ANAs. In both ANAs and conventional TDR systems time domain gating or de-embedding can be used to improve the accuracy of dielectric measurements: for example, by removing the effects of unwanted reflections in a cell by gating them out in the time domain.

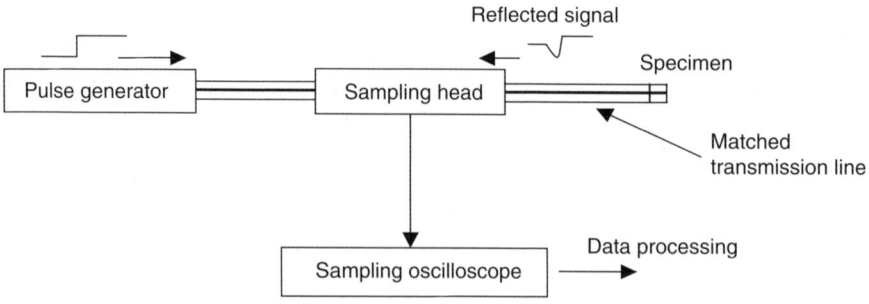

Figure 18.23   Block diagram of a typical time domain reflectometer

18.10 How should one choose the best measurement technique?

Answers to this question are covered in some detail in the Good Practice Guide [1] in the light of the policy (Section 18.6) that one should ideally match the method to the material. A checklist taken from the Guide demonstrates just how many parameters and issues one has to take into account:

- The frequency range of interest.
- The dielectric loss (high, medium or low) and expected permittivity range.
- The type of material, for example, hard, malleable or soft solids, volatile or viscous liquids.
- Specimen machining imperfections and tolerances and their influence on uncertainty.
- Specimen shape and size and their influence on measurement uncertainty.
- Specimen anisotropy and homogeneity and their influence on measurement uncertainty.
- Inhomogeneity and the presence of surface layers on specimens.
- The possibility that the specimen may be made from a magnetic and anisotropic material.
- The required uncertainties. What level of uncertainty can be achieved by available methods?
- The specimen composition (e.g. does the specimen have a laminated structure).
- The availability of suitable methods for machining and grinding specimens.
- The presence of surface inclusions and pores, surface conditions in solid specimens.
- Toxicity, contamination and evaporation of liquid specimens.
- The cost of machining specimens and performing measurements: cost-effectiveness.
- The time taken to perform the measurements – labour-intensiveness and the labour cost of measurements.

It is generally unlikely that one will have a completely free choice of method: more typically one will be asking questions like (1) how best can I press into service the facilities that I already have to perform the required measurement as accurately as possible? (2) Which techniques are best suited to checking consistency, rather than providing absolute accuracy. (3) Which techniques can be de-skilled? How proof are the various techniques against inexperience? (4) Which are most cost-effectively used in production control? All of these questions are addressed in the Guide [1].

18.11 Further information

The *Good Practice Guide* [1], scientific papers, textbooks and International Standards have all been discussed above but there are other sources of useful information such as manufacturers' literature, including catalogues, manuals and application notes, as in Reference 24. National Measurement Institutes (NMIs) other than NPL produce detailed technical notes, reports and guides on measurement techniques, notably NIST in the USA. NPL runs an Electromagnetics Measurements Club. Other societies, clubs and associations including the ARMMS RF & Microwave Society (general microwave topics) and the Ampere Association (microwave processing) can also provide support. A number of learned societies also run active dielectrics groups, notably the Institute of Physics, or groups that cover dielectric-related topics, notably the Institution of Engineering and Technology in the UK. Further details of all of these organisations can be found on the Internet.

References

1 Clarke, R. N. (ed.): *NPL good practice guide: a guide to the characterisation of dielectricmaterials at RF and microwave frequencies* (Institute of Measurement and Control, NPL, 2003)
2 Von Hippel, A. R.: *Dielectric Materials and Applications* (Technology Press of MIT, Wiley, New York, 1954; new edition Artech House, Dedham, MA, 1995)
3 Birch, J. R., and Clarke, R. N.: 'Dielectric and optical measurements from 30 to 1000 GHz', *The Radio and Electronic Engineer*, 1982;**52**:565–84
4 Afsar, M. N., Birch, J. R., Clarke, R. N., and Chantry, G W. (eds.): 'The measurement of the properties of materials', *Proceedings of the IEEE*, 1986; **74**:183–98
5 Anderson, J. C. (ed.): *Dielectrics* (Chapman and Hall, London, 1964)
6 Chamberlain, J., and Chantry, G. W. (eds.): 'High frequency dielectric measurement', *Proceedings of the NPL Conference*, Teddington, UK, 1972 (IPC Science and Technology Press, Guildford, 1973)
7 Grant, I. S., and Philips, W. R.: *Electromagnetism, Manchester Physics Series* (John Wiley, Chichester, 1996)
8 Ramo, S., Whinnery, J. R., and Van Duzer, T.: *Fields and Waves in Communication Electronics*, 3rd edn (John Wiley and Sons, New York, 1994)
9 Montgomery, C. G., Dicke, R. H., and Purcell, E. M.: *Principles of Microwave Circuits* (Peter Peregrinus, London, 1987)
10 Inan, U. S., and Inan, A. S.: *Electromagnetic Waves* (Prentice Hall, NJ, 2000)
11 Kittel, C.: *Introduction to Solid State Physics* (John Wiley and Sons, New York, 1996), Chapter 13
12 Neelakanta, P. S.: *Handbook of Electromagnetic Materials* (CRC Press, Boca Raton and New York, 1995)
13 Debye, P.: *Polar molecules* (Dover Publications Inc., New York, 1929)
14 Chamberlain, J.: 'Sub-millimetre wave techniques', *Proceedings of the NPL Conference*, Teddington, UK, 1972 (IPC Science and Technology Press, Guildford, 1973), pp. 104–116
15 Bailey, A. E. (ed.): *Microwave Measurement* (Peter Peregrinus Ltd (IEE), London, 1985 and later editions)
16 Somlo, P. I., and Hunter, J. D.: *Microwave Impedance Measurements*, IEE Electrical Measurement Series 2 (Peter Peregrinus Ltd (IEE), London, 1985)

17 Kibble, B., Williams, J., and Henderson, L., et al. (eds).: *A Guide to Measuring Resistance and Impedance Below 1 MHz* (NPL and the Institute of Measurement and Control, London, 1999)
18 Moreno, T.: *Microwave transmission design data* (Dover, NY, 1948 (original print); 1958 reprint)
19 Lorrain, P., Corson, D. R., and Lorrain, F.: *Electromagnetic Fields and Waves*, 3rd edn (W H Freeman and Co., New York, 1988)
20 Kajfez, D.: *Q Factor* (Vector Fields, Oxford, MS, 1994)
21 Birch, J. R., and Parker, T. J.: 'Dispersive fourier transform spectroscopy', in Button, K. J. (ed.), *Infrared and Millimetre Waves*, vol. 2 (Academic Press, New York, 1979), pp. 137–271
22 Daniel, V. V.: *Dielectric Relaxation* (Academic Press, London, 1967)
23 Jonscher, A. K.: *Dielectric Relaxation in Solids* (Chelsea Dielectrics Press, London, 1983)
24 Williams, G., and Thomas, D. K.: *Phenomenological and Molecular Theories of Dielectric and Electrical Relaxation of Materials* (Novocontrol Application Note Dielectrics 3, Novocontrol GmbH, 1998)
25 Pethig, R.: *Dielectric and Electronic Properties of Biological Tissues* (John Wiley & Sons, Chichester and New York, 1979)
26 Chikazumi, S.: 'Physics of ferromagnetism' in Birman, J. et al. (eds), *The International Series of Monographs on Physics*, 2nd edn (Clarendon Press, Oxford, 1997)
Crangle, J.: *Solid state magnetism* (Edward Arnold, London, 1991);
Jiles, D. *Introduction to Magnetism and Magnetic Materials* (Chapman and Hall, London, 1991)
27 Burfoot, J. C.: *Ferroelectrics: an introduction to the physical principles* (Van Nostrand, London, 1967)
28 Garg, S. K., and Smyth, C. P.: 'Microwave absorption and molecular structure in liquids LXII: the three dielectric dispersion regions of normal primary alcohols', *The Journal of Physical Chemistry*, 1965;**69**:1294–301
29 Gregory, A. P., and Clarke, R. N.: *Tables of complex permittivity of dielectric reference liquids at frequencies up to 5 GHz*, NPL Report CETM 33, NPL, 2001
30 Cole, K. S., and Cole, R. H.: 'Dispersion and absorption in dielectrics', *Journal of Chemical Physics*, 1941;**9**:341–51
31 Davidson, D.W., and Cole, R. H.: 'Dielectric relaxation in glycerol, propylene glycol, and n-propanol', *Journal of Chemical Physics*, 1951;**19**:1484
32 Havriliak, S., and Negami, S.: 'A complex plane analysis of α-dispersions in some polymer systems', *Journal of Polymer Science Part C*, 1966;**14**:99
33 James, J. R., and Andrasic, G.: 'Assessing the accuracy of wideband electrical data using Hilbert transforms', *IEE Proc. H, Microw. Antennas Propag.*, 1990;**137**:184–8
34 Lynch, A. C.: 'Relationship between permittivity and loss tangent', *Proc. Inst. Electr. Eng.*, 1971;**118**:244–6
35 Ide, J. P.: *Traceability for radio frequency coaxial line standards*, NPL Report DES 114, July 1992

36 Engen, G.: *Microwave Circuit Theory and Foundations of Microwave Metrology* (Peter Peregrinus Ltd (IEE), London, 1982)
Engen, G. F., and Hoer, C. A.: 'Thru-Reflect-Line: an improved technique for calibrating the dual six-port automatic network analyser', *IEEE Transactions on Microwave Theory and Techniques*, 1979;**MTT-27**:987–93
Hoer, C. A., and Engen, G. F.: 'On-line accuracy assessment for the dual six-port ANA: extension to non-mating connectors', *IEEE Transactions on Instrumentation and Measurement*, 1987; **IM-36**:524–9
37 Kerns, D. M. and Beatty, R. W.: *Basic Theory of Waveguide Junctions and Introductory Microwave Network Analysis* (Pergamon Press, London, 1967)
38 Ridler, N. M. (ed.): *Time domain analysis using network analysers: some good practice tips*, ANAMET Report 027, ANAMET Club, NPL, September 1999
39 Bartnikas, R.: 'Alternating-current loss and permittivity' in *Electrical Properties of Solid Insulating Materials: measurement techniques, Engineering Dielectrics* II B, ASTM STP926 (The American Society for Testing and Materials, Philadelphia, USA, 1987)
40 Schaumburg, G.: 'New broad-band dielectric spectrometers', *Dielectrics Newsletter*, (Novocontrol GmbH, Hundsangen, July 1994)
41 Waldron, R. A.: *The Theory of Waveguides and Cavities* (Maclaren & Sons Ltd, London, 1967)
42 Kajfez, D.: *Q Factor* (Vector Fields, Oxford, MS, 1994)
43 Jenkins, S., Hodgetts, T. E., Clarke, R. N., and Preece, A. W.: 'Dielectric measurements on reference liquids using automatic network analysers and calculable geometries', *Measurement Science and Technology*, 1990;**1**:691–702
44 *Guide to the Expression of Uncertainty in Measurement*, 1st edn (International Organisation for Standardisation (ISO), Geneva, Switzerland, ISBN 92-67-10188-9, 1993)
45 *The Expression of Uncertainty and Confidence in Measurement*, M3003 1st edn (UKAS, Feltham, Middx., 1997)
46 Control of Substances Hazardous to Health (COSHH), as specified by the UK government's Health and Safety Executive (HSE). An introduction is given in HSE leaflet INDG136rev2, COSHH – A brief guide to the regulations'. Available at: www.coshh-essentials.org.uk.
47 Hartshorn, L., and Ward, W. H.: 'The measurement of the permittivity and power factor of dielectrics at frequencies from 10^4 to 10^8 cycles per second', *Journal of the Inst. of Elec. Eng.*, 1936;**79**:597–609
48 Parry, J. V. L.: 'The measurement of permittivity and power factor of dielectrics at frequencies from 300 to 600 Mc/s', *Proceedings of the IRE*, 1951;**98** (Part III):303–31
49 Verweel, J.: 'On the determination of the microwave permeability and permittivity in cylindrical cavities', *Philips Research Reports*, 1965;**20**:404–14
50 Mensingh, A., McLay, D. B., and Lim, K. O.: 'A cavity perturbation technique for measuring complex permittivities of liquids at microwave frequencies', *Canadian Journal of Physics*, 1974;**52**:2365–9

51 Risman, P. O., and Ohlsson, T.: 'Theory for and experiments with a TM_{02n} applicator', *Journal of Microwave Power*, 1975;**10**:271–80
52 Collie, C. H., Ritson, D. M., and Hasted, J. B.: 'Dielectric properties of water', *Transactions of the Faraday Society*, 1946;**42A**:129–36
53 Horner, F., Taylor, T. A., Dunsmuir, R., Lamb, J., and Jackson, W.: 'Resonance methods of dielectric measurement at centimetre wavelengths', *J. Inst. Electr. Eng.*, 1946;**93** (Part III):53–68
54 Li, S., Akyel, C., and Bosisio, R. G.: 'Precise calculations and measurement on the complex dielectric constant of lossy materials using TM_{010} perturbation techniques', *IEEE Transactions on Microwave Theory Techniques*, 1981; **MTT-29**:1041–148
55 Cook, R. J.: 'Microwave cavity methods', in Anderson, J. C. (ed.), *Dielectrics* (Chapman and Hall, London, 1964), pp. 12–27
56 Stumper, U.: 'A TE_{01n} cavity resonator method to determine the complex permittivity of low loss liquids at millimetre wavelengths', *The Review of Scientific Instruments*, 1973;**44**:165
57 Cook, R. J., and Rosenberg, C. B., 'Dielectric loss measurements on low loss polymers as a function of temperature at 9 GHz', *Proceedings of the IEE Conference on Dielectric Materials, Measurements and Applications*, IEE Conference Publication No.129, Cambridge, 1975
58 Rosenberg, C. B., Hermiz, N. A., and Cook, R. J.: 'Cavity resonator measurements of the complex permittivity of low-loss liquids', *IEE Proc. H, Microw. Opt. Antennas*, 1982;**129**:71–6
59 Krupka, J., Geyer, R. G., Baker-Jarvis, J., and Ceremuga, J.: 'Measurements of the complex permittivity of microwave circuit board substrates using a split dielectric resonator and re-entrant cavity techniques', *Proceedings of the Conference on Dielectric Materials, Measurements and Applications*, Bath, UK, (IEE, London, 1996)
60 Krupka, J., Clarke, R. N., Rochard, O. C., and Gregory, A. P.: 'Split-Post Dielectric Resonator technique for precise measurements of laminar dielectric specimens – measurement uncertainties',*Proceedings of the XIII International Conference MIKON'2000*, Wroclaw, Poland, 2000, pp. 305–308
61 Bernard, P. A., and Gautray, J. M.: 'Measurement of dielectric constant using a microstrip ring resonator', *IEEE Transactions on Microwave Theory Techniques*, 1991; **MTT-39**:592–5
62 Tonkin, B. A., and Hosking,M. W.: 'Determination of material and circuit properties using superconducting and normal metal ring resonators', Institute of Physics conference series, *Proceedings of the 3rd European Conference on Applied Superconductivity*, vol. 1: *Small Scale and Electronic Applications*, 1997; **158**:291–294
63 Amey, D. I., and Horowitz, S. J.: 'Microwave material characterisation', *ISHM '96 Proceedings, SPIE* 1996, Part 2920, pp. 494–499
64 Tanaka, H., and Okada, F.: 'Precise measurements of dissipation factor in microwave printed circuit boards', *IEEE Transactions on Instrumentation and Measurement*, 1989;**IM-38**:509–14

65 Roberts, S., and Von Hippel, 'A new method for measuring dielectric constant and loss in the range of centimetre waves', *Journal of Applied Physics*, 1946; **17**:610–16

66 Baker-Jarvis, J.: *Transmission/Reflection and short-circuit line permittivity measurements* (NIST Technical Note 1341, The National Institute of Standards and Technology, Boulder, CO, 1990)

67 Baker-Jarvis, J., Vanzura, E. J., and Kissick, W. A.: 'Improved technique for determining complex permittivity with the Transmission/Reflection method', *IEEE Transactions on Microwave Theory Techniques*, 1990;**MTT-38**: 1096–103

68 Vanzura, E. J., Baker-Jarvis, J. R., Grosvenor, J. H., and Janezic, M. D.: 'Intercomparison of permittivity measurements using the Transmission/Reflection Method in 7 mm coaxial lines', *IEEE Transactions on Microwave Theory Techniques*, 1994;**MTT-42**:2063–70

69 UTE Standard 26-295, *Mesure de la permittivité et de la permeabilité de materiaux homogenes et isotropes a pertes dans le domaine des micro-ondes – Methode de mesure en guide coaxial circulaire* (UTE (Union Technique de l'Electricite et de la Communication), see C26-295, France, 1999)

70 Grant, J. P., Clarke, R. N., Symm, G. T., and Spyrou, N.: 'A critical study of the open-ended coaxial line sensor technique for RF and microwave complex permittivity measurements', *Journal of Physics E: Scientific Instruments*, 1989; **22**:757–70

71 Jenkins, S., Warham, A. G. P., and Clarke, R. N.: 'Use of an open-ended coaxial line sensor with a laminar or liquid dielectric backed by a conducting plane', *IEE Proc. H, Microw. Antennas Propag.*, 1992;**139**:1792

72 Jenkins, S., Hodgetts, T. E., Symm, G. T., Warham, A. G. P., Clarke, R. N., and Preece, A. W.: 'Comparison of three numerical treatments for the open-ended coaxial line sensor', *Electronics Letters*, 1990;**24**:234–5

73 Gregory, A. P., Clarke, R. N., Hodgetts, T. E., and Symm, G. T.: *RF and microwave dielectric measurements upon layered materials using a reflectometric coaxial sensor*, NPL Report DES 125, NPL, March 1993

74 Clarke, R. N., Gregory, A. P., Hodgetts, T. E., and Symm, G. T.: 'Improvements in coaxial sensor dielectric measurement: relevance to aqueous dielectrics and biological tissue', in Kraszewski, A. (ed.) *Microwave Aquametry* – papers from the IEEE 1993 MTTS Workshop on Microwave Moisture and Water Measurement (IEEE Press, Piscataway NJ, 1996)

75 Marsland, T. P., and Evans, S.: 'Dielectric measurements with an open-ended probe', *IEE Proc. H, Microw. Antennas Propag.*, 1987;**134**:341–9

76 Evans, S. 'The shielded open-circuit probe for dielectric material measurements', *Proceedings of the 8th International British Electromagnetic Measurements Conference*, NPL, Teddington, UK, 1997, Paper 5-2

Marsland, T. P., and Evans, S. 'Dielectric measurements with an open-ended coaxial probe', *IEE Proc. H, Microw. Antennas Propag.*, 1987;**134**:341–9

77 Gardiol, F. E.: 'Open-ended Waveguide: Principles and Applications', *Advances in Electronics and Electron Physics*, **63**:139–65

78 Sphicopoulos, T., Teodoridis, V., and Gardiol, F.: 'Simple nondestructive method for the measurement of material permittivity', *Journal of Microwave Power*, 1985;**20**:165–72
79 Sibbald, C. L., Stuchly, S. S., and Costache, G. I.: 'Numerical analysis of waveguide apertures radiating into lossy media', *International Journal of Numerical Modelling: Electronic Networks, Devices and Fields*, 1992;**5**:259–74
80 Clarke, R. N., Gregory, A. P., Hodgetts, T. E., Symm, G. T., and Brown, N.: 'Microwave measurements upon anisotropic dielectrics-theory and practice', *Proceedings of the 7th international British Electromagnetic Measurements Conference (BEMC)*, NPL, Teddington, 1995, Paper 57
81 Arai, M., Binner, J. G. P., and Cross, T. E.: 'Estimating errors due to sample surface roughness in microwave complex permittivity measurements obtained using a coaxial probe', *Electronics Letters*, 1995;**31**:115–17
82 Stuchly, S. S., Gajda, G., Anderson, L., and Kraszewski, A.: 'A new sensor for dielectric measurements', *IEEE Transactions on Instrumentation and Measurement*, 1986;**IM-35**:138–41
83 Preece, A. W., Johnson, R. H., and Murfin, J.: 'RF penetration from electrically small hyperthermia applicators', *Physics in Medicine and Biology*, 1987; **32**:1591–601
84 Johnson, R. H., Pothecary, N. M., Robinson, M. P., Preece, A. W., and Railton, C. J.: 'Simple non-invasive measurement of complex permittivity', *Electronics Letters*, 1993;**29**:1360–1
85 Kajfez, D., and Guillon, P. (eds), *Dielectric Resonators* (Vector Fields, Oxford MS, 1990)
86 Krupka, J., Derzakowski, K., Abramowicz, A., Tobar, M. E., and Geyer, R. G.: 'Whispering-gallery modes for complex permittivity measurements of ultra-low loss dielectric materials', *IEEE Transactions on Microwave Theory Techniques*, 1999;**MTT-47**:752–9
87 Krupka, J., Blondy, P., Cros, D., Guillon, P., and Geyer, R.: 'Whispering-gallery modes in magnetized disk samples, and their applications for permeability tensor measurements of microwave ferrites at frequencies above 20 GHz', *IEEE Transactions on Microwave Theory Techniques*, 1996;**MTT-44**: 1097–102
88 Krupka, J., Derzakowski, K., Riddle, B., and Baker-Jarvis, J.: 'A dielectric resonator for measurements of complex permittivity of low loss dielectric materials as a function of temperature', *Measurement Science and Technology*, **9**:1751–6
Krupka, J., Derzakowski, K., Abramowicz, A., *et al.*: 'Bounds on permittivity calculations using the $TE_{01\delta}$ dielectric resonator', *Proceedings of the XIV International Conference MIKON*, Gdansk, Poland, 2002
89 Clarricoats, P. J. B., and Olver, A. D.: *Corrugated horns for microwave antennas*, *IEE Electromagnetic Waves Series 18* (Peter Peregrinus on behalf of the IEE, London, 1984)
90 Wylde, R. J.: 'Millimetre-wave Gaussian beam modes optics and corrugated feed horns', *IEE Proc. H, Microw. opt. Antennas*, 1984;**131**:258–62

91 Kogelnik, H., and Li, T.: 'Laser beams and resonators', *Proceedings of the IEEE*, 1966;**54**:1312–29
92 LeSurf, J.: *Millimetre-wave Optics, Devices and Systems* (Adam Hilger, Bristol, 1990)
93 Lederer, P. G.: 'The fundamental principles of ram reflectivity measurement', *Symposium on the Measurement of Reflectivity of Microwave Absorbers* (DRA now QinetiQ), Malvern, February 1993)
94 Qureshi, W. M. A., Hill, L. D., Scott M., and Lewis, R. A.: 'Use of a Gaussian beam range and reflectivity arch for characterisation of radome panels for a naval application', *Proceedings of the International Conference on Antennas and Propagation (ICAP)*, University of Exeter (IEE, 2003)
95 Cook, R. J., and Rosenberg, C. B.: 'Measurements of the complex refractive index of isotropic and anisotropic materials at 35 GHz using a free-space microwave bridge', *Journal of Physics D: Applied Physics*, 1979;**12**:1643–52
96 Cook, R. J.: *The propagation of plane waves through a lamella*, NPL Report DES 52, August 1979
97 Campbell, C. K.: 'Free space permittivity measurements on dielectric materials at millimetre wavelengths', *IEEE Transactions on Instrumentation and Measurement*, 1978;**IM-27**:54–8
98 Shimabukuro, F. I., Lazar, S., Chernick, M. R., and Dyson, H. B.: 'A quasi-optical method for determining the complex permittivity of materials', *IEEE Transactions on Microwave Theory Techniques*, 1984;**MTT-32**:659–65, 1504
99 Gagnon, N., Shaker, J., Berini, P., Roy, L., and Petosa, A.: 'Material characterization using a quasi-optical measurement system', *IEEE Transactions on Instrumentation and Measurement*, 2003;**IM-52**:333–6
100 Li, S., Akyel, C., and Bosisio, R. G.: 'Precise calculations and measurement on the complex dielectric constant of lossy materials using TM_{010} perturbation techniques', *IEEE Transactions on Microwave Theory Techniques*, 1981;**MTT-29**:1041–148
101 Parkash, A., and Mansingh, A.: 'Measurement of dielectric parameters at microwave frequencies by cavity-perturbation technique', *IEEE Transactions on Microwave Theory Techniques*, 1979;**MTT-27**:791–5
102 Clarke, R. N., and Rosenberg, C. B.: 'Fabry-Perot and open-resonators at microwave and millimetre-wave frequencies, 2–300 GHz', *Journal of Physics E: Scientific Instruments*, 1982;**15**:9–24
103 Vaughn, J. M.: *The Fabry-Perot Interferometer – history, practice and applications*, Adam Hilger series on optics and optoelectronics (Adam Hilger, Bristol, 1989)
104 Jones, R. G.: 'The measurement of dielectric anisotropy using a microwave open-resonator', *Journal of Physics D: Applied Physics*, 1976;**9**:819–27
105 Cullen, A. L., and Yu, P. K.: 'The accurate measurement of permittivity by means of an open-resonator', *Proceedings of the Royal Society of London, Series A*, 1971;**325**:493–509
106 Jones, R. G.: 'Precise dielectric measurements at 35 GHz using a microwave open-resonator', *Proc. Inst. Electr. Eng.*, 1976;**123**:285–90

107 Lynch, A. C., and Clarke, R. N.: 'Open-resonators: improvement of confidence in measurement of loss', *IEE Proc. A Sci. Meas. Technol.*, 1992;**139**:221–5
108 Heidinger, R., Schwab, R., Königer, F., and Parker, T. J. (ed.): 'A fast sweepable broad-band system for dielectric measurements at 90-100 GHz', *Twenty-third International Conference on Infrared and Millimeter Waves*, Colchester, UK, 1998, pp. 353–4

Heidinger, R., Dammertz, G., Meiera, A., and Thumm, M. K.: 'CVD diamond windows studied with low- and high-power millimeter waves', *IEEE Transactions on Plasma Science*, 2002;**PS-30**:800–7

Danilov, I., and Heidinger, R.: 'New approach for open resonator analysis for dielectric measurements at mm-wavelengths', *Journal of the European Ceramic Society*, 2003;**23** (14):2623–6
109 van Germert, M. J. C.: 'Evaluation of dielectric permittivity and conductivity by time domain spectroscopy. Mathematical analysis of Felner-Feldegg's thin cell method', *Journal of Chemical Physics*, 1974;**60**:3963–74
110 van Germert, M. J. C.: 'Multiple reflection time domain spectroscopy ii. a lumped element approach leading to an analytical solution for the complex permittivity', *Journal of Chemical Physics*, 1975;**62**:2720–6
111 Feldman,Y., Andrianov, A., Polygalov, E. *et al*.: 'Time domain dielectric spectroscopy', *Review of Scientific Instruments*, 1996;**67**:3208–15
112 Berberian, J. G., and King, E. 'An overview of time domain spectroscopy', *Journal of Non-Crystalline Solids*, 2002;**305**:10–18
113 Baker-Jarvis, J.: *Transmission/Reflection and short-circuit line permittivity measurements* (NIST Technical Note 1341, National Institute of Standards and Technology, Boulder, CO, 1990)

Chapter 19

Calibration of ELF to UHF wire antennas, primarily for EMC testing

M. J. Alexander

19.1 Introduction

Wire antennas such as monopoles, biconical and log-periodic dipole array (LPDA) antennas are used for electromagnetic compatibility (EMC) testing, and typically, cover the frequency ranges 1 kHz to 30 MHz, 30–300 MHz and 200 MHz to 2 GHz, respectively. The primary parameter of interest is the maximum gain. EMC implies that the radiated emission from a product does not impair the performance of a 'victim' product, so that, for example, a radio and television set will operate satisfactorily when placed next to each other. It is useful to know the strength of the E-field that one product is 'bathing' the other product in. For this reason the antenna gain is given in terms of antenna factor (AF) which enables a direct conversion to E-field magnitude. Uncertainties for EMC-radiated emission measurements tend to be of several decibels; therefore, the AF data are generally not needed to uncertainties better than ±0.5 dB. The antenna return loss is usually measured during the calibration of AF so that the mismatch uncertainty of the receiver reading during an EMC test can be estimated.

Above 30 MHz, EMC measurements are made with the antenna both vertically and horizontally polarised so the cross-polar discrimination of the antenna is required as a component of the uncertainty budget. EMC measurements below 1 GHz are made over conducting ground planes or in free-space environments, either outdoors or in an anechoic chamber, whose imperfections will be 'seen' according to the directivity of the antenna, so knowledge of the radiation pattern is necessary. Monopole- and dipole-like antennas, including biconicals, have an omni-directional pattern in the H-plane which means that they are equally sensitive to E-fields arising from reflections in every direction about the H-plane. Below about 200 MHz it is difficult to achieve good free-space conditions in a fully lined anechoic chamber so measurements are generally made over a ground plane. The reflection from a good quality ground plane can be accurately calculated and taken into account when deriving AF from the measured coupling between two or more antennas. However, the presence of a metal plane in proximity to the antenna can also alter the AF from its free-space value because the antenna will couple with its own image in the ground plane. Knowledge of this alteration can be used to correct the antenna output voltage, or more commonly, to estimate its uncertainty. An EMC measurement is made at a specified distance from the product under test, so in order for an LPDA antenna to remain at a fixed location, it is necessary to correct the receiver reading for the variation of the LPDA phase centre with frequency.

This chapter covers the following topics: overview of traceability of E-field strength, measurement of free-space AFs, measurement of AFs over a ground plane, measurement of radiation pattern, cross-polar response, phase centre, balun imbalance and return loss.

19.2 Traceability of E-field strength

The units of E-field strength are volts per metre. The volt is traceable to the Josephson Junction and DC voltage can routinely be measured to an uncertainty of two parts in 10^7. The metre is traceable to the wavelength of a helium–neon laser and is routinely measurable to an uncertainty of one part in 10^7. To measure electric field, a sensor (e.g. an antenna) is placed in the field and a reading is obtained in volts, or more usually a power reading in watts, which can be converted into volts knowing the impedance characteristics of the transmission line. With the present state-of-the-art, field strength can be measured with wire antennas to an uncertainty of only about two parts in 10^2, which is equivalent to 0.17 dB.

Four methods, claimed to give traceability of field strength for the purpose of calibrating antennas or field probes, are all based on formulas that derive from Maxwell's equations. A simplified treatment will be given here to show the principles. This is followed by a typical uncertainty budget for field strength in an EMC-radiated emission measurement. A big advantage of the methods described in Sections 19.2.2 and 19.2.3 is that the quantity measured is attenuation, which can be measured to an uncertainty of better than ±0.01 dB.

19.2.1 High feed impedance half wave dipole

The first method is to place a half wave dipole in the field and measure the voltage at the feed point of the dipole. The open circuit voltage is related to the E-field and wavelength, λ, by the following formula:

$$E = V_{oc} \cdot \frac{\pi}{\lambda} \tag{19.1}$$

This can be realised in practice by placing a diode at the dipole feed point and measuring the rectified voltage [1]. It is necessary to calibrate the rectifying circuit by using an RF signal of known power level. The larger part of the uncertainty lies in this calibration. Standards Laboratories can measure power in the VHF and UHF ranges to an uncertainty of around ±1.5 per cent. Reference [1] has a refinement on the above formula, which takes into account the diameter of the dipole. Advantages of this method are its simplicity in concept and the fact that the high feed impedance of the dipole makes it insensitive to coupling with its image in a ground plane. Disadvantages are the loss of frequency selectivity and therefore reception of all fields present, and the need for high field levels to generate sufficient voltage, which are not permitted by the broadcasting regulators in some countries. A further disadvantage is that the combined uncertainty is higher than for the methods described in Sections 19.2.2 and 19.2.3.

19.2.2 Three-antenna method

The three-antenna method [2] uses the Friis formula to relate the insertion loss between two antennas to the product of their gains:

$$P_R = P_T G_T G_R \left(\frac{\lambda}{4\pi R}\right)^2 \tag{19.2}$$

where P_R and P_T are the powers received and transmitted, and G_R and G_T are the gains of the receive and transmit antennas. R is the separation distance between the antennas and λ is the wavelength, both in the same unit.

Uncertainties of gain as low as ±0.04 dB can be obtained [3] using the three-antenna method if the antennas have relatively high gain, for example, 20 dB standard gain horn antennas. This is possible because the high-directivity horns will radiate only a small amount of power away from the main beam, so reflections from the surroundings will be low, combined with affordable low-reflectivity absorber. This is a popular method for measuring the gain of EMC horn antennas to 40 GHz. If the extrapolation method is used, the strength of the launched E-field can be calculated at any given distance from the antenna for a known RF power into the antenna. The insertion loss (ohmic) of the antenna must be known, and this together with power is the main component of uncertainty.

By the principle of reciprocity an unknown incident E-field can be found by measuring the output power of the antenna. Realised gain, G, which includes mismatch into 50 Ω, can be converted to AF, using the following equation:

$$AF^2 = \frac{4\pi Z_F}{G\lambda^2 Z_0} \tag{19.3}$$

where λ is the wavelength, Z_F is free-space impedance (approximately 377 Ω) and Z_0 is the characteristic impedance of the antenna input transmission line, commonly 50 Ω.

The three-antenna method is also a very accurate method for measuring the AF of dipole antennas providing the ground plane is sufficiently large and flat. A straightforward formula [4] can be used for removing the ground reflected ray, provided the antennas are in each other's far-field – for dipoles a separation of three wavelengths is sufficient for an uncertainty of better than ±0.15 dB to be achieved [5].

It must be borne in mind that the measured AF applies for that height above a ground plane, that is, it includes the effects of coupling with its image; also the measurement uncertainties are generally greater for vertical polarisation because of reflections from structures such as the vertically dropping feed cable and the antenna mast. In addition to this, vertical polarisation is associated with greater sensitivity to site effects, such as edge diffraction (i.e. reflections from the edges of the ground plane).

19.2.3 Calculability of coupling between two resonant dipole antennas

The third method involves comparing the measured and calculated insertion loss between two antennas [6]. This method is useful for dipole antennas that have uniform H-plane patterns and for which it is difficult to avoid reflections from the surroundings. By configuring the antennas over a large flat ground plane outdoors this ensures one electromagnetically well-defined source of reflection and an upper hemisphere with no reflections. Using method of moments code [7], it is possible to predict the measured insertion loss between two wire antennas above a ground plane to an uncertainty of less than ±0.3 dB, and hence AF to less than ±0.15 dB [8]. This includes the effect of mutual coupling of the antenna to its image in the ground plane, which for normal usage of EMC antennas can alter the gain by as much as 100 per cent.

19.2.4 Calculable field in a transverse electromagnetic (TEM) cell

The fourth method of providing traceability for field strength is to deduce the field between the two conductors of a transmission line from the power input to the line. A TEM cell [9] is a coaxial line with an expanded cross section. Provided the characteristic impedance, Z_0, of the line is preserved with change in cross section, it is possible to achieve uncertainties better than 0.3 dB (3.5 per cent) in the E-field strength between the plates of the TEM cell [10]. A TEM cell is useful for the calibration of field probes at frequencies below 1 GHz. The field strength E (V m^{-1}) is given from the power input to the cell P (Watt) and the plate separation b (metre) by the following formula:

$$E = \frac{\sqrt{PZ_0}}{b} \qquad (19.4)$$

For best accuracy the probe being tested should be electrically small to reduce coupling with the sidewalls, and it should occupy a relatively small volume within the cell so that the TEM wave is perturbed as little as possible.

19.2.5 Uncertainty budget for EMC-radiated E-field emission

A statement of uncertainties implies traceability to national standards. In order for measurements to have a common meaning, and to support trade, within and across national boundaries, it is necessary that they are traceable to national standards. In turn, national standards laboratories take part in international intercomparisons to ensure that they are in step with the rest of the world [11]. Traceability of measurements in industry is assisted by the accreditation of test and calibration laboratories.

For this purpose Accreditation Bodies are set up whose job is twofold, first to ensure traceability of physical parameters and second to ensure that laboratories have a Quality System in place which ensures that the laboratory is able to deliver the traceability to the uncertainty which it claims it is capable of, and that this has been verified by appointed technical experts.

The magnitude of uncertainty commonly used is the Standard Uncertainty multiplied by a coverage factor of $k = 2$, providing a level of confidence of approximately 95 per cent. An EMC radiated emission test often involves a product with power and data cables that make it difficult to get a reproducible result, giving rise to large uncertainties, of the order of ± 10 dB. However the uncertainty contributions from the measuring site and equipment can be minimised, for example, uncertainties associated with the antenna can be quantified. Typical uncertainty components are listed in Table 19.1. Just because EMC uncertainties are very large compared with uncertainties for other physical parameters, such as voltage, does not mean that the evaluation of EMC uncertainties is pointless. Indeed the process of setting up an uncertainty budget highlights the largest components and directs the effort to reducing these.

Table 19.1 shows a budget comprising the main components that could affect the E-field magnitude measured during emission testing using either a biconical antenna or a LPDA antenna. For AF the uncertainty is likely to be given in a certificate. If it is given to a confidence level of 95 per cent the probability distribution will be Gaussian (or Normal) with a k value of 2. If the specification of the instrument does not refer to a standard uncertainty one has to assume that the measurement could lie anywhere within the specified accuracy limits, and without further information about the distribution of repeated results this is assumed to correspond to a rectangular distribution. The method by which uncertainty components with different probability distributions are summed is given in the ISO Guide [12]. However this Guide is comprehensive and can be a challenge on first reading. It has been presented in a more amenable form [13] for the subject of EMC.

The fourth component in Table 19.1 relates to antenna directivity, which is relative to a tuned dipole stipulated by CISPR 16-1-4:2004. Varying the height of the receive antenna over a ground plane means there are two ray paths between the equipment under test (EUT) and the antenna which diverge from the bore sight direction. For a biconical antenna the error is for vertical polarisation only, it being zero for horizontal polarisation because the H-plane directivity is uniform. The error is positive because it represents only loss of signal.

The 12th component, site imperfections, relates to normalised site attenuation (NSA) performance. Simply put, this term indicates how close the test site is to an ideal environment. Since the site is only required to meet a criterion of NSA within ± 4 dB of the theoretical value, strictly the value of this uncertainty component should be ± 4 dB. However, it has been set to ± 1 dB because this is the intention in Annex F of CISPR 16-1-4:2004 [14] and because good sites are likely to meet ± 1 dB.

The last components for ambient interference and cable layout have not been given values because these vary widely from site to site and operator to operator. These two components can increase by many decibels the expanded uncertainty of ± 3.8 dB in Table 19.1. Further explanation of this table can be found in Reference [13] and CISPR 16-4-2.

Table 19.1 Uncertainty budget for emission measurements on a 3 m open area test site

Component	Probability distribution	Uncertainty dB Biconical	Uncertainty dB LPDA
Antenna factor calibration	Normal ($k = 2$)	±1.0	±1.0
Cable loss calibration	Normal ($k = 2$)	±0.5	±0.5
Receiver specification	Rectangular	±1.5	±1.5
Antenna directivity	Rectangular	+0.5 / −0	+2 / −0
Antenna factor variation with height	Rectangular	±2	±0.5
Antenna phase centre variation	Rectangular	0	±2
Antenna factor frequency interpolation	Rectangular	±0.25	±0.25
Antenna balun imbalance	Rectangular	±1.0	±0
Measurement distance error ±2 cm	Rectangular	±0.1	±0.3
Height of antenna above ground plane (height error ±2 cm)	Rectangular	±0.1	+1.0 / −0
Height of EUT above ground plane (height error ±2 cm)	Rectangular	±0.05	±0.05
Site imperfections	Rectangular	±1.0	±1.0
Mismatch	U-shaped	±1.1	±0.5
System repeatability	Normal	±0.5	±0.5
Ambient interference	—	Large	Large
Reproducibility of EUT/cable layout	—	Poor	Poor
Combined standard uncertainty $u_c(y)$	Normal	+1.92 / −1.90	+2.1 / −1.8
Expanded uncertainty U	Normal ($k = 2$)	±3.8	+4.2 / −3.6

19.3 Antenna factors

Ideally AF should be independent of the antenna surroundings. However, the standard method for EMC testing is to measure emission over a ground plane at a distance of 10 m from the product under test. Signal nulls are caused by destructive interference of the direct and ground-reflected ray paths between the antenna and the product. To avoid measuring in nulls the antenna is height scanned between 1 and 4 m and the maximum signal is recorded. Height scanning brings into play the effects of the radiation pattern and mutual coupling of the antenna with its image, resulting in errors of around ±2 dB (refer to Table 19.1).

A widely used method for calibrating antennas is the American National Standard Institute (ANSI) procedure [15] which mimics the EMC radiated test method in that it uses the same geometry, including height scanning, to perform the three-antenna method. The product is replaced by a transmitting antenna at a fixed height of 1 or 2 m. This method will give low uncertainties for EMC testing in cases where the product

Figure 19.1 *The 60 × 30 m ground plane at the National Physical Laboratory, Teddington, UK, whose surface is flat to within ±15 mm over 95 per cent of its area*

behaves like that antenna, radiating at that fixed height, but more likely the product will radiate anywhere from the floor, via its cabling, to the top of the unit which may be more than 1 m from the ground. The height of the radiating source on the EUT varies with frequency and it is not practical to predict the height of maxima, as we do for a single antenna source during an ANSI calibration, and therefore the correct choice of receiving AF becomes an issue. A compromise would be to use an average of the receiving AF measured with the transmitting antenna at a range of heights. A study performed at NPL [17], in which the AF of a biconical antenna was both computed and measured (Figure 19.1) at a range of heights showed that the AF averaged over different heights was within ±0.5 dB of the free-space AF (AF_{FS}), and that the AF measured by the ANSI method on a 10 m site was also within ±0.5 dB of AF_{FS}.

There are techniques in which it is possible to measure AF_{FS} more efficiently and more accurately than by height scanning, and in view of the good comparisons cited, AF_{FS} is a sound basis from which to quantify antenna-related uncertainties of measurement. Furthermore AF_{FS} is a basic property of an antenna, with no built-in mutual coupling effects. Alternative methods of performing radiated emission tests in free-space conditions are being developed, such as the fully anechoic room, in which AF_{FS} will give the lowest uncertainties.

CISPR, a sub-committee of the IEC (International Electrotechnical Commission) is defining acceptable methods of calibration of antennas that are used for EMC testing, and the methods will be described in a future issue of CISPR 16-1-5.

19.3.1 Measurement of free-space AFs

Free-space conditions are defined as the illumination of an antenna in free-space by a plane wave, which implies that the antenna is in the far-field of the source. Antennas are often used in non-free-space conditions with the antenna above a ground plane or close to absorber lined walls and illuminated by a non-uniform field. It can be difficult to unravel these effects in order to quantify uncertainties but the best starting point is the free-space AF.

Absorbing material can be very effective above 200 MHz and a relatively small amount can be used to set up free-space conditions for the calibration of LPDA antennas. With ingenuity one can set up affordable methods for measuring AF_{FS} of dipole-like antennas below 200 MHz. These methods are outlined below. Uncertainties of ±0.5 dB can be routinely achieved for dipole, biconical and LPDA AFs [16,17].

19.3.2 The calculable dipole antenna

This method gives the lowest uncertainty obtainable for AF. The calculable dipole antenna is used by National Laboratories as a primary standard antenna. The antenna comprises two thin dipole elements fed in anti-phase by a 3 dB hybrid coupler. The verification of AF of the calculable dipole is given in References [6] and [8]. The AF is calculable either for free-space conditions or with the antenna mounted above a ground plane. The dipole antenna can be used as an accurate broadband dipole using numerical methods such as NEC [7]. As mentioned in Section 19.2.3 the uncertainty in the AF, above a ground plane or in free-space is better than ±0.15 dB. The calculable dipole is especially useful in providing traceability for tuned dipole and broadband antennas, and also evaluating the quality of EMC test sites.

19.3.3 Calibration of biconical antennas in the frequency range 20–300 MHz

In this frequency range, dipole-like antennas of length less than 1.5 m are fairly omni-directional and it is not so practicable to measure free-space AF by conventional methods. This is because the antennas have to be several wavelengths away from conducting surfaces, including the ground. This implies heights of greater than 10 m if the antennas are horizontally polarised – the problem with vertical polarisation is that it is difficult to reduce reflections from the input cable to an acceptable level, also the mast is a vertical structure and will be a source of reflection. At frequencies above 300 MHz the usual solution is to line a large room with pyramidal RF absorbing material (RAM). This is not practicable at 30 MHz. Ferrite tiles of about 6 mm thickness are used to line EMC chambers but their return loss is typically less than 15 dB, whereas reflections must be less than −25 dB to measure gain to uncertainties of less than ±0.5 dB.

It is possible to simulate free-space conditions without using absorber. One method is to mount the antenna vertically polarised at 2 m height above a ground plane. At this height mutual coupling to the ground image is negligible at the resonant frequency (where it is most sensitive) of the biconical antenna, around 70 MHz, and below resonance the high self-impedance makes the antenna insensitive to its ground image. A sufficiently plane wave to illuminate the antenna can be set up by placing a source antenna close to the ground at a distance of around 20 m. The standard antenna method is used with the broadband calculable dipole antenna as the standard. The cable must be extended horizontally behind the antenna several metres before dropping vertically in order to reduce the effect of reflections.

19.3.4 Calibration of LPDA antennas in the frequency range 200 MHz to 5 GHz

The traditional way to calibrate LPDA antennas is by the ANSI method at a distance of 10 m across a ground plane. A typical antenna for this frequency range is 0.65 m long. If the separation of the antennas is measured from their centres, the uncertainty in AF at the top and bottom frequencies is 0.3 dB due to the movement of the LPDA phase centre. Also in the UK, when limited to using the allowed transmit power, there is relatively high interference from TV transmissions. A more elegant method is to calibrate the antennas at fixed heights above RAM. If the phase centre is known at each frequency the separation can be reduced, overcoming ambient signals and reducing the amount of RAM required.

Phase centre can be found in a variety of ways. The methods used at NPL are (1) from the mechanical dimensions of LPDA elements, (2) from measurement of signal phase as the antenna is rotated in free-space and (3) using NEC modelling. The results agree well and phase centre is typically known to better than ± 1 cm. NPL uses a mid-antenna separation of 2.5 m which allows an uncertainty in AF of ± 0.5 dB to be achieved. The same method is used for conical log spiral antennas to an uncertainty of ± 1 dB.

19.3.5 Calibration of hybrid antennas

A conical-hybrid antenna is a physical combination of a biconical antenna and a log antenna into one antenna with a typical frequency range of 26 MHz to 2 GHz. At NPL these are calibrated to uncertainties of less than ± 0.7 dB using the two methods described in Sections 19.3.3 and 19.3.4. The AF data are 'sewn' together within the frequency overlap. One reason that the uncertainty is higher than ± 0.5 dB is that the phase centre at frequencies in the region between the 'biconical element' and the longest log-periodic element is only estimated by linear interpolation. Another reason is that hybrid antennas can be very large and day-to-day alignment on the mast is not so precise as for the smaller LPDA.

19.3.6 Calibration of rod antennas

Rod antennas are conventionally calibrated by replacing the monopole element with a capacitor of approximately 12 pF. This can give AF to within ± 1 dB below about 15 MHz but it does not work so well above this frequency. This method might be suited to antenna manufacturers because they can design a power splitter jig for their own model of antenna, but it is a big overhead for a calibration laboratory to develop the right jig for all types of rod antenna. Because this method is essentially a substitution of the real element there are some important aspects in the construction of the calibration jig, which may lead to incorrect AF values if wrongly done ([17], Section 9.10). NPL's principal method involves placing the rod antenna with its base on a large (60 × 30 m) ground plane and illuminating it with a source 20 m away. The standard antenna method is used. The standard is a calculable rod antenna whose AFs are calculated using NEC. Because of the difficulty with getting enough radiated signal below 10 MHz, calibrations below this frequency are done in a MEB1750 GTEM cell. The validity of using the GTEM cell was demonstrated by comparison with results obtained on the NPL ground plane. For this test a very large strip line, 2.5 m high, was built on the ground plane and the reception by a standard antenna was compared with that from the antenna under test (AUT).

19.3.7 Calibration of loop antennas

Loop antennas can be calibrated in a TEM cell to uncertainties of less than ±1 dB, typically over the frequency range 20 Hz to 30 MHz. The power output of the cell is measured and used to calculate the field strength between the plates, in which the loop is immersed. The validity of using the TEM cell was demonstrated by building a standard loop whose current was measured and the generated field could therefore be calculated; the AUT was placed on a common axis in a nearby parallel plane to the transmitting loop. The magnetic AF may then be calculated by the AUT response in the known field.

19.3.8 Other antenna characteristics

There are undesirable characteristics of antennas that can cause a great deal of trouble to practicing engineers. *A Measurement Good Practice Guide* [17] identifies the main problems and gives guidance on how to deal with them and more generally gives tips on calibrating antennas. This section deals with baluns, cross polar discrimination and breakdown of RF connection in antenna elements.

19.3.8.1 Balun imbalance

In the early stages of establishing a calibration service NPL discovered that some models of popularly used biconical antennas had severe balun imbalance. All one had to do was to invert the vertically polarised antenna and get a change in received signal of ±5 dB. The cause was imbalance of the balun which set up common mode currents on the cable, which radiated and interfered with the antenna. The effect is related to the size of current and the proximity of the vertically hanging cable to the vertically polarised antenna elements.

Since the 1970s engineers have noticed problems with the reproducibility of readings for certain models of antenna. A substantial literature including this topic has been spawned, giving advice on the orientation of the antenna, the layout of cables and the use of ferrites to suppress braid currents. Balun imbalance has been the cause of many man-days of wasted effort at many test sites, particularly with site validation. Putting ferrite toroids on the cable close to the antenna input can make some improvement, but alternative proprietary models of antenna that do not have this problem are readily available. The majority of dipole-like antennas pass the test in CISPR 16-1-4 with a balance of better than ±0.5 dB.

Text has been included in clause 4.4.2 of CISPR 16-1-4, which describes the measurement of balun imbalance and imposes a limit on the magnitude of the imbalance allowed.

19.3.8.2 Cross-polar performance

The following text has been included in clause 4.4.3 of CISPR 16-1-4:

> When an antenna is placed in a plane-polarised electromagnetic field, the terminal voltage when the antenna and field are cross-polarised shall be at least 20 dB below the terminal voltage when they are co-polarised. It is intended that this test apply to LPDA antennas for which the two halves of each dipole are in echelon. The majority of testing with such antennas is above 200 MHz, but the requirement applies below 200 MHz. This test is not intended for in-line dipole and biconical antennas because a cross-polar rejection greater than 20 dB is intrinsic to their symmetrical design. Such antennas, and horn antennas, must have a cross-polar rejection greater than 20 dB; a type test by the manufacturer should confirm this.

19.3.8.3 Mechanical construction of antennas

Some models of antenna have given poor repeatability of measured signal because of mechanical defects. The most common one is breakdown of RF contact between the elements on a log antenna and the transmission line they are screwed to. This can be caused by a build-up of metal oxide in the joint or simply a loose joint.

19.3.8.4 Return loss

It is assumed that the calibration of an antenna includes the measurement of return loss, with the antenna mounted in free-space conditions. This enables the operator to calculate the mismatch uncertainties of the emission result.

19.4 Electro-optic sensors and traceability of fields in TEM cells

There are two fundamental methods that are used to provide traceability for E-field strength. One is to use a calculable dipole to measure (or set up) the field, and the other is to generate the field in a TEM waveguide from a known input power. An electro-optic field sensor is an ideal device to make an intercomparison between the two methods because of its small size, high sensitivity. It is also non-intrusive to the field because of minimal use of metal parts and the use of an optical fibre to feed it.

The field in a TEM cell is calculable at frequencies below the resonant frequency of the first higher-order mode. The uncertainties of the field at the centre of the cell arise from measurements of the insertion loss from the input to the centre of the cell, the input power, the impedance of the (loaded) cell at the centre compared with the design impedance, and the effect on the field of placing the field sensor or other object at the centre of the cell. The onset of resonance dictates the maximum frequency of the cell, which is inversely proportional to the size of the cell. A cell that operates up to 1 GHz is small (of the order of 0.1 m). Such a cell has been developed at Physikalisch-Technischen Bundesanstalt (PTB) as a standard to calibrate small field probes [10] which in turn provide traceability for larger antennas.

NPL investigated the calibration of a TEM cell by using a transfer standard to trace the field strength to the calculable dipole [18]. The calculable dipole has a length of half a wavelength and by definition cannot fit between the plates of a cell. Also the current design is fed by a coaxial cable and this would cause uncertainties when inserting the dipole into the cell. The electro-optic transfer standard is a short dipole embedded in a lithium niobate crystal fed by optical fibres. An agreement of less than ±0.3 dB has been demonstrated between the field measured by a calculable dipole antenna and the field in a TEM cell. This should enable commercial field probes to be calibrated in TEM cells with uncertainties of less than ±0.5 dB.

Acknowledgements

Acknowledgements are due to the National Measurement System Policy Unit of the DTI for funding this work through several Electrical Programmes, and to staff of the RF & Microwave Group who contributed to the developments.

References

1. Camell, D. G., Larsen, E. B., and Ansen, W. J.: 'NBS calibration procedures for horizontal dipole antennas', *International Symposium on Electromagnetic Compatibility*, Seattle, 1988, pp. 390–394
2. IEEE Standard Test Procedures for Antennas, ANSI/IEEE Std 149-1979
3. Gentle, D. G., Beardmore, A., Achkar, J., Park, J., MacReynolds, K. and de Vreede, J.P.M.: *CCEM Key Comparison RF-K3.F, Measurement Techniques and Results of an Intercomparison of Horn Antenna Gain in IEC-R 320 at 26.5, 33.0 and 40.0 GHz*, NPL Report CETM 46, Sep 2003. Search for 'RF-K3.F' at http://kcdb.bipm.org/ appendixB/KCDB_ApB_search.asp [Accessed 2007]
4. Smith, A. A.: 'Standard-site method for determining antenna factors', *IEEE Transactions on Electromagnetic Compatibility*, 1982;**24**:316–22
5. Morioka, T., and Komiyama, K.: 'Measurement of antenna characteristics above different conducting planes', *IEEE Transactions on Instrumentation and Measurement*, 2001;**50** (2):393–6
6. Alexander, M. J., and Salter, M. J.: 'Low measurement uncertainties in the frequency range 30 MHz to 1 GHz using a calculable standard dipole antenna and national reference ground plane', *IEE Prococeedings-Science, Measurement and Technology*, 1996;**143** (4):221–8

7 Logan, J. C., and Burke, A. J.: *Numerical Electromagnetic Code* (Naval Oceans Systems Centre, CA, USA, 1981)
8 Alexander, M. J., Salter, M. J., Loader, B. G., and Knight, D. A.: 'Broadband calculable dipole reference antennas', *IEEE Transactions on Electromagnetic Compatibility*, 2002;**44** (1):45–58
9 Crawford, M. L.: 'Generation of standard EM fields using TEM transmission cells', *IEEE Transactions on Electromagnetic Compatibility*, 1974;**16** (4): 189–95
10 Münter, K., Pape, R., and Glimm, J.: 'Portable E-field strength meter system and its traceable calibration up to 1 GHz using a μGTEM cell', *Conference of Precision Electromagnetic Measurements*, Braunschweig, 1996, pp. 443–444
11 Alexander, M.: 'International comparison CCEM.RF-K7.b.F of antenna factors in the frequency range 30 MHz to 1 GHz', *Metrologia*, 2002;**39**:309–17
12 *Guide to the Expression of Uncertainty in Measurement*. International Organisation for Standardisation, Geneva, Switzerland, 1993
13 *The Treatment of Uncertainty in EMC Measurements*, LAB34, UKAS, April 2002 (update on NIS81 May 1994)
14 CISPR publication 16. *Specification for Radio Disturbance and Immunity Measuring Apparatus and Methods*, Part 1-4:2004 Apparatus, Part 160-2-3:2004 Methods, Central office of the IEC, 3 rue de Varembé, Geneva, Switzerland
15 ANSI C63.5:2004. Calibration of antennas used for radiated emissions measurements in electromagnetic interference (EMI) control
16 Alexander, M. J.: 'The measurement and use of free-space antenna factors in EMC applications', *Proceedings of 13th International Symposium on Electromagnetic Compatibility*, Zurich, 1999, paper F6
17 Alexander, M.J., Salter, M.J., Gentle, D.G., Knight, D.A., Loader, B.G., Holland, K.P.: *Measurement Good Practice Guide No. 73: Calibration and use of Antennas Focusing on EMC applications*, Dec 2004. www.npl.co.uk/publications
18 Loh, T.H., Loader, B., Alexander, M.: 'Comparison of electric field strentgh at VHF frequencies generated by dipoles and TEM cells', 18th International Zuric symposium on EMC, Sep 2007

Index

AFs (antenna factors) 464–9
  about AFs 459, 464–5
  ANSI procedure 464
  and Balun imbalance 468–9
  biconical antenna measurements 466–7
  calculable dipole antenna method 466
  CISPR acceptable calibration methods 465
  and cross-polar performance 469
  free-space measurement 466
  hybrid antenna calibration 467
  loop antennas calibration 568
  LPDA antenna calibration 467
  National Physical Laboratory (UK) ground plane 465
  and return loss 469
  rod antenna calibration 467
  see also E-field strength traceability; EMC (electromagnetic compatibility) measurement/testing
air lines (precision air-dielectric coaxial transmission lines) 188–93
  about air lines 188–9
  characteristic impedance 190–1
  conductor imperfections 192–3
  fully supported air lines 190
  partially supported air lines 189–90
  phase change 191–2
  RF impedance 194–8
    lossless lines 194–5
    lossy lines 195–8
    propagation constant 197
  skin depth issues 192–3
  standards 190–2
  unsupported air lines 189

amplitude modulation measurement, with spectrum analysers 383–4
antenna factors: see AFs (antenna factors)
attenuation measurement
  about attenuation measurement 91
  basic principles 91–3
  calibration standards 116
  definition of attenuation 91–2
  detector linearity 112–14
    measurement uncertainty budget 114
  insertion loss 91–3
  mismatch error/uncertainty 92–3, 110–12
    two-resistor power splitter 111–12
  repeatability 115–16
  RF leakage 112–13
  stability and drift 115
  system noise 115
  system resolution 115
attenuation measurement systems
  AF substitution method 104–5
  automatic network analyser 108–10
    vector network analyser 108–9
  IF substitution method 105–7
    piston attenuator 105–6
  inductive voltage divider 98–104
    attenuation 98
    automated system 99–100
    commercial attenuation calibrator 102–3
    construction 98
    dual channel system 101
    error 99
    gauge block system 100
    IFR 2309 FFT signal analyser 103–4

474   Index

attenuation measurement systems (*Cont.*)
   power ratio method 94–7
      power sensor linearity problem 94–5
      range switching/resolution problem
         95–7
      signal generator amplitude drift
         problem 94
      zero carry over problem 95
   RF substitution method 107–8
      rotary vane attenuator 107–8
   voltage ratio method 97–8
attenuation measurement worked example,
      30db attenuation 116–19
   contributions to uncertainty, Type A
      random uncertainty 117
   detector linearity 117
   leakage 117
   power meter resolution 117
   uncertainty spreadsheet 118
automatic network analysers (ANAs) 108–10,
      425–6
   see also calibration of automatic network
      analysers; network analysers;
      one-port devices/error models; scalar
      network analysers; TRL calibration;
      two-port error model for
      measurement; vector network
      analysers (VNAs); verification of
      automatic network analysers
avalanche diode noise sources 163–4

balanced device characterisation 305–28
   about balanced device characterisation
      305–7, 309–10
   about balanced and unbalanced systems
      305–6
   conversion parameters 310
   de-embedding 326–8
   differential structure issues 306–9
   differential through connection example
      321–6
   Far End Crosstalk (FEXT) measurements
      320
   ideal devices 307–8
   mixed-mode-S-parameter-matrix 318–19
   modal decomposition method 312–18
   Near End Crosstalk (NEXT) measurements
      320
   real devices 307–10
   SAW-filter measurement example 326–7
   self parameters 310
   single-ended to balanced device
      characterisation 319–20
   typical measurements 320–1
   using network analysis 310
   using physical transformers 310–11
   virtual ideal transformers 311
Balun imbalance 468–9
Bessel zero method/Bessel nulls 385–6
biconical wire atennas, with EMC testing
      459, 467
bolometers 333–4
   and microbolometers 249
Boltzmann's constant 159

calibration of automatic network analysers
      263–89
   about calibration 263
   see also network analysers; one-port
      devices/error models; scalar network
      analysers; TRL calibration; two-port
      error model for measurement; vector
      network analysers (VNAs)
calorimeters 334–6
cascaded receivers, noise 169
cascade matrix, S-parameter matrix 27–8
Cavity cell 443
Central Limit Theorem 46, 50–1
characterisation: *see* balanced device
      characterisation
characteristic impedance
   measurement 35–6
   non-TEM waveguides 33–5
   one-port devices 21–2
   real and imaginary parts 32–3
   and S-parameter measurements 30–6
   transmission lines, with losses 30–3
   transmission lines, no losses 4
coaxial air lines
   historical perspective 181
   see also air lines (precision air-dielectric
      coaxial transmission lines)
coaxial connectors 59–90
   about coaxial connectors 59–60, 66
   airline handling 61
   bead resonances 188
   buffer adapters 65
   cleaning 63–4
      normal procedure 64
      static sensitive devices 64
   connector recession 65–6
   connector savers 65

Index 475

dial gauges and test pieces 87–8
   calibrating 87
   gauge calibration blocks 88
   measurement resolution 87–8
   push on/screw on types 87
electrical characteristics 187–8
frequency ranges of common types 66–7, 188
future developments 200
gauging connectors 62–3
higher-order mode resonances 188
historical perspective 180–1
insertion loss repeatability for connector pairs 85
life expectancy 65
line sizes 60
repeatability issues 61–2
specifications 62
torque wrench setting values 86
'Traceability to National Standards' 59
coaxial connector types
   1.0 mm (Agilent W) connector 82–5
   1.85 mm (V^{TM}) connector 81–2
   2.4 mm (Type Q) connector 80–1
   2.92 mm K connector 79–81
   3.5 mm connector 77–9
   7/16 connector 73–6, 201
   7 mm precision connector 68–71
      connection procedure 68–70
      disconnection procedure 71
   14 mm precision connector 66–8
   compatibility possibilities 185–6
      electrical discontinuities from 186
   GPCs (general precision connectors) 60, 185
   line diameters summary 186
   LPCs (laboratory precision connectors) 60, 185
   mechanical characteristics 185–7
   N 7 mm connectors (rugged) 71–4
      dimensions chart 74
      gauging type N connectors 73
   precision/non precision 183–5
   sexed connectors 183
   sexless (hermaphroditic) connectors 60, 182–3
   SMA connectors 76–8
      dimensions chart 78
coaxial lines
   structures and properties 148–50
      applications 149–50

      dispersion characteristics 149
complex refractive index 412
complex relative magnetic permeability 412
connectors: *see* air lines (precision air-dielectric coaxial transmission lines); coaxial connectors
conversion parameters 310
coplanar waveguides (CPW)
   probes 231–2
   structures and properties 153–4
couplers, power measurement 343
coupling factor 209–10
Courtney Holder cell 443

Debye relaxation 419, 420, 421–2
decibels 330
delay line discriminators, for phase noise measurement 400–1
Dicke (switching) radiometer 164, 166
dielectric rod probe 249–50
dielectrics, basic concepts
   about dielectrics 409–10
   absolute permittivity 410
   basic parameters 410–13
   complex refractive index 412
   complex relative magnetic permeability 412
   equivalent circuits 411
   loss tangent 412
   magnetic hysteresis 412
   microradian 412
   permittivity of free space 410
   power absorption coefficient 413
   radiation absorbing materials (RAM) 412
   relative permittivity 410
dielectrics, basic measurement theory 413–17
   about dielectric measurement 413
   dielectric-test-set method 417
   frequency-change methods 417
   frequency coverage of the methods 417
   length-change methods 417
   lumped-impedance methods 414
      admittance cells 414
      cell admittance 414
      lumped equivalent circuits 414
   multi-pass techniques 417
   Q-factor (Quality Factor) 417
   resonance methods 416
      cavities 416
   wave methods 414–16
      attenuation constant 415

dielectrics, basic measurement theory (*Cont.*)
    Fresnel's equations 416
    guided wave media 414
    phase constant 415
    propagation/transmission parameters 415
    Scattering Parameters 414
    standing-wave methods 414
    travelling-wave methods 414, 416
dielectrics, loss processes 418–22
    dielectric relaxation 418–22
        Debye relaxation 419, 420, 421–2
        interfacial polarisation 418
        Kramers-Kronig relations 421–2
        Maxwell-Wagner effect 418
        polar/non-polar materials 418
        relaxation frequency 421
        rotational polarisation 418
    dielectric resonance 419–20
    electrical conduction/conductivity 418
    non-linear processes 420
dielectrics, measurement methods
    about choosing a method 449
    admittance methods 430–2
        Hartshorn and Ward (H & W) technique 432–3
        liquid dielectrics 430–1
        perturbation method 433–4
        resonant admittance cells 432–4
        three-terminal cells 431–2
        TM_{010}-mode cavity 434
        two-terminal cells 432
    Cavity cell 443
    coaxial probes 439–41
    coaxial transmission lines 437–9
    Courtney Holder cell 443
    dielectric probes 441
    dielectric resonators (DRs) 442–4
    free field methods 444–6
    Hakki-Colemen cell 443
    open-ended rectangular waveguide probes 442
    open resonators 446–8
    resonator perturbation technique 446
    ring resonators 437
    Roberts and Von Hippel method 438
    split-post dielectric resonators (SPDR) 436–7
    substrate methods 437
    TE_{01}-mode cavities 434–6
    time domain techniques 448
    waveguide probes 440–1
dielectrics, measurement practicalities 422–30
    about the need to measure 422–3
    anistropic materials 424–5
    automatic network analysers (ANAs) 425–6
    cleanliness aspects 424
    dimensions and preparation 424
    ferroelectrics 425
    frequency response analysers (FRAs) 426
    good practices 429–30
    high-permittivity dielectrics 425
    hygroscopic materials 424
    inhomogeneous materials 425
    international standard measurement methods 422
    low-loss materials 423
    magnetic materials 424–5
    matching method to material 423
    measurement cells 426–7
    medium/high-loss materials 423
    Q-factor (Quality Factor) measurement 427–9
dielectric waveguide 154–5
diode power sensors 333
dispersion, waveguides 148
dispersion effect, transmission lines with losses 9–10
DMMs (digital multimeters) 122, 124–5
    digitising DMMs 124–5
DVMs (digital voltmeters) 97–8

effective directivity 274, 277, 293–6
E-field strength traceability 460–4
    about E-field strength 460
    with electro-optic sensors 469–70
    high feed impedance half wave dipole 460–1
    TEM cell, calculable field 462
    three-antenna method 461–2
    two resonant dipole antennas, coupling calculability 462
    uncertainty budget 462–4
    see also AFs (antenna factors); EMC (electromagnetic compatibility) measurement/testing
electrical conduction/conductivity 418
electrical sampling scanning-force microscopy 254

Index 477

electric-field probing 251–4
electron beam probing 251
electro-optic sampling 252–4
electro-optic sensors, and E-field strength traceability 469–70
EMC (electromagnetic compatibility) measurement/testing 459–70
　about EMC testing 459–60
　with spectrum analysers 392, 392–3
　see also AFs (antenna factors); E-field strength traceability
ENR (excess noise ratio) 160, 163
equivalent circuit modelling (ECM) 226–8
error term verification for two port measurements 293–300
　effective directivity 293–6
　　effective source match 296–9
　　offset load/airline method 295–6
　　sliding load method 294–5
　　time domain gating 297–8
　effective isolation 299
　effective linearity 300
　effective load match 299
　time domain and de-embedding 300
　transmission and reflection tracking 299–300

Far End Crosstalk (FEXT) measurements 320
fast sampling DMMs 124–5
FFTs (fast Fourier transforms) 401
finline transmission line 154
flicker noise 159
FM discriminators 404–5
free space permeability 10–11
frequency modulation analysis, with spectrum analysers 384–5
frequency response analysers (FRAs) 426
frequency spectrum 122
frequency stability/phase noise: see phase noise/frequency stability measurement
Fresnel's equations 416

gas discharge tubes 163
Gaussian distributions 46–7
　probability density function 158
generalised scattering parameters 22–4
generator measurement tracking, with spectrum analysers 378–9
GPCs (general precision connectors) 160, 185

group velocity, waveguides 16
GSM pulse specification 346
GUM (*Guide to the Expression of Uncertainty in Measurement*) 43–52
　see also uncertainty and confidence in measurements
Gunn diodes 246

Hakki-Colemen cell 443
harmonic content, with voltage measurement 137–8
harmonic distortion measurement, with spectrum analysers 378
Hartshorn and Ward (H & W) technique 432–3
HEMT technology 248
hermaphroditic (non-sexed) connectors 60
　see also coaxial connectors
high feed impedance half wave dipole 460–1

impedance and admittance parameters 24–7
inductive voltage divider: see attenuation measurement systems
insertion loss 91–3
interfacial polarisation 418
intermodulation measurement/analysis, with spectrum analysers 380–2
intrinsic impedance 11
inverse Fourier transform (IFT) 221

Josephson Junction 460

Kramers-Kronig relations 421–2
Kuhn's rules, signal flow graphs 37

Lorentz reciprocity relation 37–8
losslessness, scattering parameters 39–40
loss tangent 412
LPCs (laboratory precision connectors) 160, 185
LPDA (log-period dipole array) antennas, with EMC testing 459–60, 467

magnetic-field probing 250–1
magnetic hysteresis 412
Maxwell's equations 11
Maxwell-Wagner effect 418

measurement verification 301–4
    about measurement verification 301
    customised verification example 301–2
    manufacturer supplied verification
        example 302–4
MEMS (Micro Electro-Mechanical Systems)
    336
microbolometers 249
microradian 412
microstrip transmission lines, structures and
        properties 151–2
    dispersion 152
microwave frequency spectrum 122
microwave network analysers: *see* network
        analysers
microwave voltage measurement: *see* voltage
        measurement
mismatched loads, one-port devices 20
mismatch error 92–3
mixed-mode-S-parameter-matrix 318–19
MMIC (monolithic microwave integrated
        circuit) (or RFIC) 217–55
    about S-parameter measurement 217–18,
        254–5
    bolometers/microbolometers 249
    cryogenic measurements 247–8
        HEMT technology 248
    dielectric rod probe 249–50
    electric-field probing 251–4
        electrical sampling scanning-force
            microscopy 254
        electron beam probing 251
        electro-optic sampling 252–4
        opto-electronic sampling 252
        photo-emissive sampling 252
    electromagnetic field probing 249–50
    magnetic-field probing 250–1
    thermal measurements 246–7
        Cascade Microtech Summit Evue
            system 247
        Cascade Microtech Summit
            S300-863 system 246
        Gunn diodes 246
MMIC/RFIC probe station measurements
        230–45
    about probe station measurements 230–1
    advantages of probe station measurements
        231
    DC biasing 240–1
    layout considerations 241–3

low-cost multiple DC biasing technique
    243
measurement errors 240
passive microwave probe design 231–6
    ACP probe (Cascade Microtech)
        233–5
    Picoprobe™ (GGB) 232
    tapered coplanar waveguide (CPW)
        probes 231–2
    waveguide input infinity probe
        231–5
probe calibration 236–40
    about probe calibration 236–8
    automated probes 239–40
    LRM technique 238–9
    Short Open Load Reflect routine 240
    SOLT technique 238
    stability checking 240
    TRL technique 238–9
upper-millimetre-wave measurements
    243–5
MMIC/RFIC test fixture measurements
    218–30
    about test fixture measurements 218–20
    calibration kits 219
    calibration methods summary 219–20
    one-tier calibration 229–30
    text fixture design guidelines 230
    two-tier calibration 220–8
        banded VNA 221
        broadband VNA 221
        equivalent circuit modelling (ECM)
            226–8
        in-fixture calibration 225–6
        synthetic-pulse TD reflectometry
            221
        T-D reflectometry (TDR) 221–5
        time domain gating 221–5
        VNA reference planes 220
modal decomposition method for
        characterisation 312–18
monopole wire antennas, with EMC testing
    459

Near End Crosstalk (NEXT) measurements
    320
network analysers 108–9, 207–16
    about network analysers 207–8
    block diagrams 208, 214–16
    built-in signal source 209
    coupling factor 209–10

Index 479

diode detectors 211–12
directional bridge 210–11
directional coupler 209–10
dynamic range 214
reference plane 208
sampler systems 213–14
signal separation hardware 209–11
tuned receivers 212–14
see also automatic network analysers (ANAs); scalar network analysers; vector network analysers
network analysis characterisation 310
noise 157–64
  about noise 157–8
  available noise power 160
  effective noise power 160
  ENR (excess noise ratio) 160, 163
  equivalent input noise temperature 161
  flicker noise 159
  Gaussian distribution probability density function 158
  noise factor/figure 161–2
  noise performance of receivers 161
  noise temperature 162
  quantum noise temperature 159
  and sensitivity 157
  shot noise 159
  thermal noise 158–9
noise measurement 164–76
  accuracy of measurement 166–71
  automated measurement 174–6
    noise figure meters/analysers 175
    on-wafer measurements 175–6
  cascaded receivers 169
  correlated noise 173
  Dicke (switching) radiometer 164, 166
  mismatch effects/factor 171–4
  noise resistance 173
  passive two-ports 169–71
  radiometer sensitivity 166
  receivers and amplifiers 172–4
  total power radiometer 164–6
  uncertainties (type A and B) 167–9
noise sources 162–4
  about thermal noise 162
  avalanche diodes 163–4
  gas discharge tubes 163
  temperature-limited diodes 163
Nomographs, with spectrum analysers 383
n-port devices, scattering parameters 24–7

Omni-Spectra SMA connector/wedge-shaped board socket 228
one-port devices/error models 19–22, 273–6
  characteristic impedance 21–2
  effective directivity 274, 277
  frequency response (tracking) error 275
  mismatched loads 20
  open circuit termination model 276, 278
  'perfect load' termination model 276
  phasor notation 21
  power 21–2
  reflection coefficient 20–1, 273–5
  short circuit termination model 276, 278
  see also transmission lines
opto-electronic sampling 252
oscillator phase noise performances 396–7
  see also phase noise/frequency stability measurement
oscilloscopes for voltage measurement 127–9
  analogue 128–9
  calibrator calibration 138–40
  digital 128–9
  sampling 129
  switched input impedance 129–30

permeability of a medium 10–11
permittivity 11, 410–11
  absolute permittivity 410
  permittivity of free space 11, 410
  relative permittivity 11, 410
phase constant/wave number 11
phase locked loops 402–3
phase noise/frequency stability measurement 395–407
  about phase noise 395–6, 406–7
  delay line discriminator technique 400–1
  FM discriminator method 404–5
  future possible methods 406
  measurement uncertainty issues 405–6
  oscillator phase noise performances 396–7
  quadrature technique 401–3
    fast Fourier transforms (FFTs) 401
    phase locked loops 402–3
  spectrum analyser techniques 397–9
    improvement with band-pass filters 399–400
    limitations 398
  summary of methods 406–7
phase velocity/phase constant, lossless transmission lines 5–6

480   Index

phase velocity, waveguides 15
phasor notation 21
photo-emissive sampling 252
Picoprobe™ (GGB) 232
Planck's constant 159
plane/transverse electromagnetic (TEM) waves 10–12
polar/non-polar materials 418
power flow, sinusoidal waves, lossless transmission lines 6–7
power measurement: *see* RF power measurement
power sensors 333–6
   acoustic meters 336
   calorimeters 334–6
   diode sensors 333
   flow calorimeters 334–6
   force and field based sensors 336
   MEMS (Micro Electro-Mechanical Systems) 336
   microcalorimeters 334
   thermistors/bolometers 333–4
   thermocouples/thermoelectric sensors 333
   twin load calorimeters 334
power splitters 339–43
   direct method for splitter output 341–3
   output match measurement 340–1
   splitter properties 340
   two resistor splitters 339–40
probe station measurements: *see* MMIC/RFIC probe station measurements
pulsed modulation display/analysis 389–91
pulsed power 344–6

Q-factor (Quality Factor) measurement 417, 427–9, 443
quantum noise temperature 159

radiation absorbing materials (RAM) 412
radio frequency integrated circuit (RFIC): *see* MMIC
radiometers: *see* noise measurement
reciprocity 37–8
rectangular metallic waveguides: *see* waveguides, rectangular metallic
reference plane 208
reflection coefficient
   lossless transmission lines 5
   one-port devices 20–1

reflectometers, power measurement 343–4
return loss, lossless transmission lines 7
RF frequency spectrum 122
RFIC (radio frequency integrated circuit): *see* MMIC
RF impedance
   air lines 195–8
      lossless lines 194–5
      lossy lines 195–8
   historical perspective 181–2
   terminations 198–200
      mismatched terminations 199–200
      near-matched terminations 199
      open-circuits 198–9
      short-circuits 198
RF millivoltmeters 125–6
RF power measurement 329–46
   basic theory 329–32
   calibration factor 332
   calibration/transfer standards 338–9
   couplers 343–4
   direct power measurement 337
   effective efficiency 332
   GSM pulse specification 346
   incident, reflected and delivered power 330–1
   mismatch uncertainty 332
   pulsed power 344–6
   ratio measurements 338–9
   reflectometers 343–4
   substitution techniques 330
   uncertainty budgets 337–8
   see also power sensors; power splitters
RF voltage measurement: *see* voltage measurement
ridged waveguides 150–1
ring resonators 437
Roberts and Von Hippel method 438
rotational polarisation 418

sampling RF voltmeters 126–7
scalar network analysers 263–6
   applied power level problem 269
   calibration
      reflection measurements 267–8
      transmission measurements 267
   fully integrated analysers 265–6
   limited dynamic range problem 269
   see also network analysers; vector network analysers (VNAs)

Index   481

scattering parameters: *see* S-parameters (scattering parameters)
self parameters 310
shot noise 159
signal flow graphs, scattering parameters 36–7
   Kuhn's rules 37
skin effect 2
slot guides, structure and properties 152–3
SMA printed circuit board socket 228
S-parameters (scattering parameters)
   about S-parameters 19
   generalised S-parameters 22–4
   impedance and admittance parameters 24–7
   losslessness 39–40
   measurements
      with MMIC/RFIC 217–18
      wave methods 414
   n-port devices 24–7
   reciprocity 37–8
   self parameters 310
   signal flow graphs 36–7
   two-port networks 22–4
   two-port transforms 40
   see also one-port devices
S-parameter matrix (scattering matrix)
   about S-parameters 23–4
   cascade matrix 27–8
   de-embedding 29–30
   mixed-mode-S-parameter-matrix 318
   network parameter examples 26–7
   renormalisation 28–9
   see also characteristic impedance
S-parameters equations, two-port error model 283–4
spectrum analysers, applications 376–93
   amplitude modulation measurement 383–4
   EMC measurements 392–3
   FM demodulation 386–7
   frequency modulation analysis 384–5
      with Bessel zero method 385–6
   generator measurement tracking 378–9
   harmonic distortion measurement 378
   intermodulation intercept point 382–3
   intermodulation measurement/analysis 380–2
   meter mode 379–80
   modulation AM/FM asymmetry 387–8
   nomograph usage 383
   overload dangers 392–3
   phase noise/frequency stability measurement 397–9
   pulsed modulation display/analysis 389–91
   square wave spectrum 388–9
   zero span mode 378–80
spectrum analysers, facilities and use 349–59
   amplitude modulation analysis 351–3
   basic usage 349–50
   block diagram/description 354, 356–8
   harmonic mixer concept 354–5
      multiple responses problem 355–6
      tracking generator 358–9
      tracking preselectors 356
   measurement domains 350
   and network analysers 207–8
   oscilloscope amplitude/time display 350–2
      for amplitude modulation 351–2
   pre-calibration (Auto Cal) 349–50
   spectrum analyser amplitude/frequency display 351
      for amplitude modulation 353
spectrum analysers, specification points 359–76
   about the main controls 359
   amplitude accuracy 372
   display detection mode 376–7
   dynamic range 366–72
      intermodulation and distortion 367–8
      internal distortion checking 371–2
      local oscillator phase noise 369–70
      noise 368–9
      sideband noise 370–1
   input attenuator/IF gain 360
   input VSWR effect 372–3
   noise/low-level signals 366
   residual FM 374–5
   residual responses 373–4
   resolution bandwidth 361–2
      shape factor 362–4
   sideband noise characteristics 373, 374
   sweep speed/span 360–1
   uncertainty contributions 375–7
   video averaging 365–6
   video bandwidths 365
split-post dielectric resonators (SPDR) 436–7
square wave spectrum analysis 388–9
standing waves from sinusoidal waves, lossless transmission lines 7–8
switching (Dicke) radiometer 164, 166
synthetic-pulse TD reflectometry 221

tapered coplanar waveguide (CPW) probes 231–2
TDD (time-domain duplex) techniques 396
TDMA (time-domain multiple access) techniques 396
T-D reflectometry (TDR) 221–5
   error sources 223–5
Telegraphist's equations
   transmission lines, lossless 3–4, 9
   transmission lines with losses 9
TEM: see transverse electromagnetic (TEM) cell/waves
temperature-limited diodes 163
test fixture measurements: see MMIC/RFIC test fixture measurements
thermal noise 158–9
thermal voltage converters (TVCs) 122–3
thermistors/bolometers 333–4
thermocouples/thermoelectric sensors 333
three-antenna method for E-field strength 461–2
time domain and de-embedding 300
time domain gating 221–5, 297–8
transmission lines, lossless waveguides: see one-port devices; waveguides, lossless
transmission lines, structures and properties 147–55
   about transmission lines 147–8
   coaxial lines 148–50
   coplanar waveguides 153–4
   dielectric waveguide 154–5
   finline 154
   higher mode operation 148
   microstrip 151–2
   rectangular waveguides 150
   ridged waveguides 150–1
   slot guides 152–3
   waveguide dispersion 148
transmission lines, two-conductor lossless
   basic principles 1–4
   characteristic impedance 4
   equivalent circuit 1–2
   phase velocity/phase constant 5–6
   power flow, sinusoidal waves 6–7
   reflection coefficient 5
   return loss 7
   skin effect 2
   standing waves from sinusoidal waves 7–8
   Telegraphist's equations 3–4

Voltage Standing Wave Ratio (VSWR) 7–8
wave equations 3–4
transmission lines, two-conductor with losses
   basic principles 8–9
   dispersion effect 9–10
   equivalent circuit 8
   pulses, effect on 9–10
   sinusoidal waves, general solution 10
   Telegraphist's equations 9
transverse electromagnetic (TEM) cell/waves 10–12, 19, 307
   calculable field 462
transverse electromagnetic (TEM) waveguides, and E-field traceability 469–70
TRL calibration 284–9
   basic principles 284–6
   calibration procedure 287–9
   four receiver operation 286–7
   isolation 286
   source match/load match 286
two-port error model for measurement 279–84
   error sources 279–80
   leakage/isolation 281–2
   S-parameters equations 283–4
   'through' measurement 283
   transmission coefficient 280
two-port networks, generalised scattering parameters 22–4
two-port transforms, scattering parameters 40

UHF measurement uncertainty 461
uncertainty analysis, voltage measurement 136–7
uncertainty budgets, RF power measurement 337–8
uncertainty and confidence in measurements 43–57
   Central Limit Theorem 46, 50–1
   combined standard uncertainty 50–1
   coverage factor 51–2
   expanded uncertainty 51
   expectation value 45
   flagpole example 48–51
   GUM Type A and Type B evaluations 44
   imperfect matching 45
   normal distributions 46, 49, 51–2
   normal/Gaussian probability distributions 47

Index 483

probability distributions 45–6
    and standard uncertainties 49
    purpose of measurement 44
    quantity (Q) in uncertainty evaluation 43
    sensitivity coefficients 48, 50
    standard uncertainty 44
    temperature uncertainty 48
    uncertainty budget 56
    U-shaped distributions 47
    voltage reflection coefficients (VRCs) 53
uncertainty sources in RF and microwaves 52–7
    calibration of coaxial power example 54–7
        uncertainty budget 56
    directivity 54
    RF connector repeatability 54
    RF mismatch errors 52–4
    test port match 54

vector network analysers (VNAs)
    about VNAs 108–9, 266
    calibration/accuracy enhancement 269–73
    correctable systematic errors 270
    directivity issues 270–1
    frequency response (tracking) 273
    isolation (crosstalk) 273
    load match 272–3
    non-repeatable random and drift errors 270
    source match 271–2
    VNA reference planes 220–1
    see also calibration of automatic network analysers; MMIC; network analysers; one-port devices/error models; two-port error model for measurement; verification of automatic network analysers
vector voltmeters 127–8
verification of automatic network analysers 291–304
    about verification 291–3
    calibration and verification 293
    definition 291
    see also error term verification; measurement verification; network analysers
VHF measurement uncertainty 461
virtual ideal transformers 311
voltage measurement 121–43
    about RF/microwave voltage measurement 121–2

capacitive loading 132–3
digital multimeters (DMMs) 122
    digitising DMMs 124–5
    fast sampling DMMs 124–5
    input impedance effects 130–2
    oscilloscopes, analogue/digital/sampling 127–9
        switched input impedance 129–30
    rectifier implementation 123–4
    RF millivoltmeters 125–6
    sampling RF voltmeters 126–7
    source loading and bandwidth 132–3
    thermal voltage converters (TVCs) 122–3
    traceability 133–5
        with micropotentiometers 134–5
        with thermal converters 133–4
    vector voltmeters 127–8
    voltage standing wave ratio (VSWR) 129–31
    wideband AC voltmeters 122–4
voltage measurement impedance matching/mismatch errors 135–43
    about impedance matching/mismatch 135–6
    harmonic content errors 137–8
    oscilloscope calibration example 138–41
    RF millivoltmeter calibration 140–3
    uncertainty analysis considerations 136–7
        oscilloscope bandwidth test example 137
    VSWR issues 136
VSWR (voltage standing wave ratio)
    impedance matching issues 136
    lossless transmission lines 7–8
    oscilloscope voltage measurement 129–31

wave equations 3–4
waveguide, dielectric 154–5
waveguide dispersion 148
waveguide input infinity probe 231–5
waveguides, coplanar, structures and properties 153–4
waveguides, lossless 10–12
    intrinsic impedance 11
    permeability 10–11
    phase constant/wave number 11
    plane/transverse electromagnetic (TEM) waves 10–12
waveguides, rectangular metallic 12–17, 150
    about rectangular waveguides 12–14

applications 150
cut-off frequency and wavelength 14–15, 17
general solution 16–17
group velocity 16
phase velocity 15
plane waves in 12–13

properties 150
reflection within 12–13
wave impedance 15–16
waveguides, ridged 150–1
wave impedance, waveguides 15–16
wave number 11
wideband AC voltmeters 122–4